D0216217

BACTERIAL ION CHANNELS

AND THEIR

EUKARYOTIC HOMOLOGS

BACTERIAL ION CHANNELS

AND THEIR

EUKARYOTIC HOMOLOGS

EDITED BY

Andrzej Kubalski
Department of Cell Biology
Nencki Institute of Experimental Biology
Warsaw, Poland

Boris Martinac
School of Biomedical Sciences
University of Queensland
Brisbane, Queensland, Australia

Washington, D.C.

Address editorial correspondence to ASM Press, 1752 N St. NW, Washington, DC 20036-2904, USA

Send orders to ASM Press, P.O. Box 605, Herndon, VA 20172, USA
Phone: (800) 546-2416 or (703) 661-1593
Fax: (703) 661-1501
E-mail: books@asmusa.org
Online: www.asmpress.org

Library of Congress Cataloging-in-Publication Data

Bacterial ion channels and their eukaryotic homologs / edited by
Andrzej Kubalski, Boris Martinac.
p. ; cm.
Includes bibliographical references and index.
ISBN 1-55581-328-3 (hardcover)
1. Ion channels. 2. Bacterial cell walls. I. Kubalski,
Andrzej.
II. Martinac, Boris.
[DNLM: 1. Ion Channels. 2. Bacterial Physiology. 3. Models,
Molecular. QH 603.I54 B131 2005]
QH603.I54B33 2005
571.6'4293—dc22
2004030702

10 9 8 7 6 5 4 3 2 1

Printed in the United States of America

Cover image: Top view of the structure of the mechanosensitive ion channel MscS from *Escherichia coli.*

To Professor Ching Kung for his pioneering work on structure and function in microbial ion channels

CONTENTS

CONTRIBUTORS

Aleksij Aksimentiev • Theoretical and Computational Biophysics Group, Beckman Institute for Advanced Science and Technology, University of Illinois at Urbana-Champaign, Urbana, IL 61801

Andriy Anishkin • Biology Department, University of Maryland, College Park, MD 20742

Christopher J. Bagley • Protein Laboratory, Hanson Institute, Institute of Medical and Veterinary Science, Frome Rd., Adelaide, South Australia 5000, and Department of Medicine, The University of Adelaide, Adelaide, South Australia 5005, Australia

Paul A. Bartley • Centre for Advanced Biomedical Studies, University of South Australia, North Terrace, Adelaide, South Australia 5000, Australia

Monica Betanzos • Biology Department, University of Maryland, College Park, MD 20742

Philip C. Biggin • Department of Biochemistry, University of Oxford, South Parks Road, Oxford OX1 3QU, United Kingdom

Christopher Bladen • Department of Physiology and Biophysics, University of Calgary, Calgary, Alberta T2N 4N1, Canada

Paul Blount • Department of Physiology, University of Texas Southwestern Medical Center, Dallas, TX 75390-9040

Ian R. Booth • School of Medical Sciences, Institute of Medical Sciences, University of Aberdeen, Foresterhill, Aberdeen AB25 2ZD, United Kingdom

Allan H. Bretag • Centre for Advanced Biomedical Studies, University of South Australia, North Terrace, Adelaide, South Australia 5000, Australia

Chien-Sung Chiang • Biology Department, University of Maryland, College Park, MD 20742

Jordi Cohen • Theoretical and Computational Biophysics Group, Beckman Institute for Advanced Science and Technology, University of Illinois at Urbana-Champaign, Urbana, IL 61801

Catherine Diao • Department of Physiology and Biophysics, University of Calgary, Calgary, Alberta T2N 4N1, Canada

Michelle D. Edwards • School of Medical Sciences, Institute of Medical Sciences, University of Aberdeen, Foresterhill, Aberdeen AB25 2ZD, United Kingdom

Robert J. French • Department of Physiology and Biophysics, University of Calgary, Calgary, Alberta T2N 4N1, Canada

Alessandro Grottesi • Department of Biochemistry, University of Oxford, South Parks Road, Oxford OX1 3QU, United Kingdom

Banuri Gunasekera • School of Medical Sciences, Institute of Medical Sciences, University of Aberdeen, Foresterhill, Aberdeen AB25 2ZD, United Kingdom

H. Robert Guy • Laboratory of Experimental and Computational Biology, Division of Basic Sciences, National Cancer Institute, National Institutes of Health, Bethesda, MD 20892-5567

Irene Iscla • Department of Physiology, University of Texas Southwestern Medical Center, Dallas, TX 75390-9040

Ching Kung • Laboratory of Molecular Biology and Department of Genetics, University of Wisconsin—Madison, Madison, WI 53706

Mario Meng-Chiang Kuo • Laboratory of Molecular Biology, University of Wisconsin—Madison, Madison, WI 53706

Chan Li • School of Medical Sciences, Institute of Medical Sciences, University of Aberdeen, Foresterhill, Aberdeen AB25 2ZD, United Kingdom

Yuezhou Li • Department of Physiology, University of Texas Southwestern Medical Center, Dallas, TX 75390-9040

Yi-Shiuan Liu • Department of Molecular Physiology and Biological Physics, Center for Structural Biology and Biophysics Program, University of Virginia Health Sciences Center, Charlottesville, VA 22908

Samantha Miller • School of Medical Sciences, Institute of Medical Sciences, University of Aberdeen, Foresterhill, Aberdeen AB25 2ZD, United Kingdom

Paul C. Moe • Department of Physiology, University of Texas Southwestern Medical Center, Dallas, TX 75390-9040

Ewan Murray • Bacterial Physiology Group, School of Medical Sciences, Institute of Medical Sciences, University of Aberdeen, Foresterhill, Aberdeen AB25 2ZD, United Kingdom

Evgeny Pavlov • Department of Physiology and Biophysics, University of Calgary, Calgary, Alberta T2N 4N1, Canada

Eduardo Perozo • Department of Molecular Physiology and Biological Physics, Center for Structural Biology and Biophysics Program, University of Virginia Health Sciences Center, Charlottesville, VA 22908

Christopher P. Ptak • Department of Molecular Physiology and Biological Physics, Center for Structural Biology and Biophysics Program, University of Virginia Health Sciences Center, Charlottesville, VA 22908

Heather R. Rickard • Centre for Advanced Biomedical Studies, University of South Australia, North Terrace, Adelaide, South Australia 5000, Australia

Yoshiro Saimi • Laboratory of Molecular Biology, University of Wisconsin—Madison, Madison, WI 53706

Mark S. P. Sansom • Department of Biochemistry, University of Oxford, South Parks Road, Oxford OX1 3QU, United Kingdom

Hildgund Schrempf • Fachbereich Biologie/Chemie, Universität Osnabrück, Barbarastr. 11, 49069 Osnabrück, Germany

Klaus Schulten • Theoretical and Computational Biophysics Group, Beckman Institute for Advanced Science and Technology, University of Illinois at Urbana-Champaign, Urbana, IL 61801

Indira H. Shrivastava • Laboratory of Experimental and Computational Biology, Division of Basic Sciences, National Cancer Institute, National Institutes of Health, Bethesda, MD 20892-5567

Marcos Sotomayor • Theoretical and Computational Biophysics Group, Beckman Institute for Advanced Science and Technology, University of Illinois at Urbana-Champaign, Urbana, IL 61801

Sergei Sukharev • Biology Department, University of Maryland, College Park, MD 20742

Emad Tajkhorshid • Theoretical and Computational Biophysics Group, Beckman Institute for Advanced Science and Technology, University of Illinois at Urbana-Champaign, Urbana, IL 61801

INTRODUCTION

Ion channels are molecular pores of excitable membranes. Since membranes of living cells are boundaries between the cell interior and "inanimate" extracellular space and form permeability barriers for hydrophilic ionized molecules, the role of ion channels is to facilitate passage of physiologically important ions such as sodium, potassium, chloride, or calcium ions across cellular membranes. Movement of these ions through ion channels (often > 10^6 ions/s) generates electrical signals, which can trigger a cascade of intracellular events (Hille, 2001). Ion channels thus function in numerous biological processes that range from sensory transduction, motility control, and synaptic connections to pacemaking and secretion. Although classification of ion channels has not been systematic and often depends on the channel properties that have been characterized, ion channels are usually classified according to (i) their gating mechanism (e.g., voltage gated, ligand gated, or mechano-gated [mechanosensitive]); (ii) their permeant ions (e.g., Na^+ channels, K^+ channels, Ca^{2+} channels, or Cl^- channels); or, most recently (iii) their molecular structure, which has helped to group them into families such as the superfamily of ligand-gated channels, the transient receptor potential channel family, or the voltage-gated family of ion channels.

HISTORICAL PERSPECTIVE: WHY BACTERIA CAME INTO PLAY

Microbes, including bacteria, have often been used to advance our knowledge of basic principles underlying diverse aspects of cellular biology. Examples include discoveries concerning genetic control of enzyme and virus synthesis (Luria and Delbrück, 1943; Jacob and Monod, 1961), the understanding of biological energy transfer through the formulation of the chemiosmotic theory (Mitchell, 1961, 1979), and elucidation of the enzymatic mechanism underlying the synthesis of ATP (Walker, 1994; Fillingame, 1990). As demonstrated in this book, bacteria have also served as an excellent model system for understanding the basic molecular design of ion channels.

Since the existence of ionic channels in neuronal membranes was postulated by A. Hodgkin and A. Huxley in the early 1950s (Hodgkin and Huxley, 1952a, 1952b), scientists have been looking for the means to document their molecular identity. Over the last 50 years, electrophysiology was the main technique used to characterize ion channels functionally and deduce information about their structural design from the electrophysiological functional analysis. Although most electrophysiologists worked on preparations from animal cells, some researchers turned their attention to microbial cells. As early as 1934, T. Kamada measured the membrane potential of a *Paramecium* cell intracellularly, which thus became the first documented intracellular recording from a living cell (Kamada, 1934). In 1972 Y. Naitoh and coworkers characterized a calcium action potential in paramecia

(Naitoh et al., 1972), which underlies the excitability of this ciliate, similar to the all-or-none action potential in neuronal cells. For excitability of *Paramecium* cells, calcium ions play a role similar to the role of sodium ions in neuronal excitability, whereas potassium ions have a membrane-repolarizing function as in neurons or muscles. For many years, C. Kung and coworkers have used the paramecium as a model cell with the aim of finding gene encoding ion channels found in the *Paramecium* membrane. Similar to the use of other model systems such as *Arabidopsis thaliana, Neurospora crassa*, or *Drosophila melanogaster* in mutagenesis studies, *Paramecium* was used for selecting behavioral mutants whose phenotype was dictated by defects in genes encoding ion channels (Saimi et al., 1983; Kung et al., 1992). To date, the gene and protein sequences of a number of calcium and potassium ion channels in paramecia have been identified (see Kung et al. [2000] and chapter 1 in this book).

In the early 1980s, studies of paramecia in Kung's laboratory were complemented by studies of other microorganisms, including *Saccharomyces cerevisiae* and *Escherichia coli*. A wealth of available information on the biochemistry, genetics, and molecular biology of these microorganisms promised to provide a fertile ground for functional and structural characterization of ion channels that might exist in their cellular membranes. This research direction was facilitated by the availability of the patch clamp method (Hamill et al., 1981), which allowed electrophysiologists to examine cells of almost any size (Fig. 1). The initial characterization of ion channels in cellular membranes of *S. cerevisiae* and bacteria by the patch clamp method (Gustin et al., 1986; Gustin et al., 1988; Martinac et al., 1987) has given researchers the opportunity to conduct structure and function studies of these membrane proteins. The advantage of using *S. cerevisiae* or bacteria for structural studies of ion channels is obvious. Both of these organisms offer distinct experimental advantages: the vast knowledge of their molecular biology and biochemistry, suitability for genetic manipulation, short doubling times, and large yields. Most importantly, these microorganisms can be grown in large quantities, yielding the milligram amounts of ion channel proteins required for structural studies by X-ray crystallography or magnetic resonance spectroscopy. Indeed, a few years after the isolation of the *kcsA* potassium channel gene from the bacterium *Streptomyces lividans* by H. Schrempf and coworkers (Schrempf et al., 1995) the X-ray crystal structure of this channel was solved in the laboratory of R. MacKinnon (Doyle et al., 1998); for his achievements in structural characterization and functional studies of bacterial ion channels, MacKinnon received the 2003 Nobel Prize in chemistry. Shortly after the KscA structure was published, structures of other bacterial channels, including three other potassium channels (MthK from *Methanobacterium thermautotrophicum* [Jiang et al., 2002], KirBac from *Burkholderia pseudomallei* [Kuo et al. 2003], and KvAP from *Aeropyrum pernix* [Jiang et al., 2003]), two ClC chloride channels from *Escherichia coli* and *Salmonella typhimurium* (Dutzler et al., 2002), the aquaporin AqpZ from *Escherichia coli* (Savage et al. 2003), and the mechanosensitive channels MscL and MscS from *Mycobacterium tuberculosis* and *Escherichia coli*, respectively (Chang et al., 1988; Bass et al., 2002), were solved.

PRESENT STATUS: THE SCOPE OF THIS BOOK

The aim of this book is to provide a summary of recent major achievements that have advanced our understanding of the structure and function of several types of ion channels

A

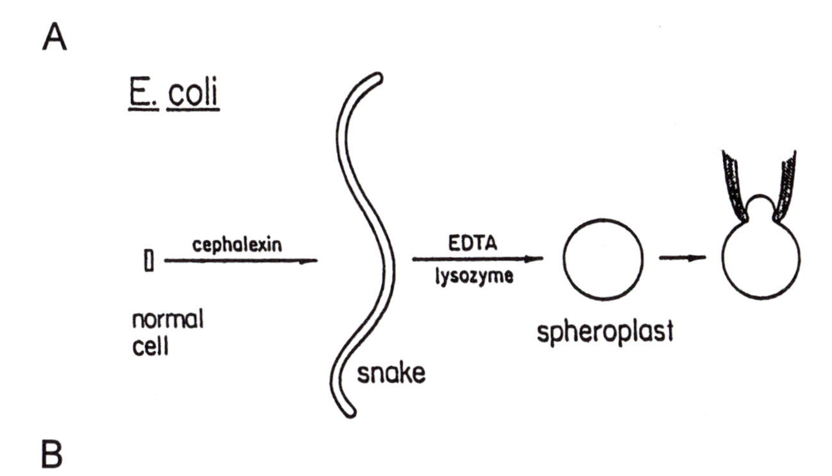

B

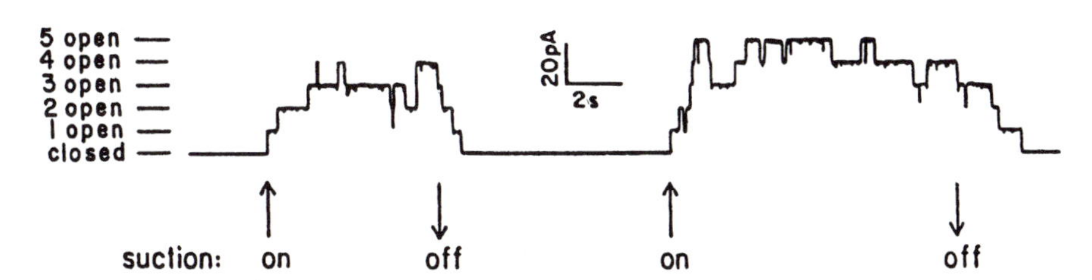

Figure 1. Studies of ion channels in *E. coli.* (A) A bacterial cell in the presence of the antibiotic cephalexin fails to septate and grows into a long filament (a "snake" up to 150 μm long). Treating the snake with EDTA and lysozyme transforms it into a giant spheroplast (5 to 10 μm in diameter) amenable to patch clamp experiments. (B) Activities of up to five mechanosensive channels of low conductance (~1 nS) recorded from a patch of a giant spheroplast. Channels opened when suction was applied to the patch-clamp pipette (on) and closed when suction was released (off). The zero-current level, when all channels are closed, and unitary current levels of channels 1 to 5 are indicated. Pipette voltage was +10 mV. Reprinted from Saimi et al. (1988) with permission.

including a variety of potassium channels, ClC chloride channels, and sodium ion channels. It also covers aquaporins facilitating membrane transport of water (Preston and Agre, 1991) and bacterial mechanosensitive channels serving as passages for various osmolytes including ions (Levina et al., 1999). This book can be seen as a compendium of the current knowledge of several types of channels, all of which were structurally characterized by using prokaryotic cells. Furthermore, the book emphasizes the importance of the use of a multidisciplinary approach employing molecular biology, genetics, protein chemistry, X-ray crystallography, magnetic resonance spectroscopy, electrophysiology, molecular dynamics (MD) simulations, and structural modeling, among other techniques, for the understanding of the molecular structure, biophysical properties, and physiological function of ion channels.

Chapter 1, by Kuo et al., describes the diversity of potassium channels in prokaryotic and eukaryotic cells. Particular attention is paid to the bacterial Kch potassium channel of six transmembrane domains, which has the same membrane topology as the eukaryotic K^+, Na^+, and Ca^{2+} channels. Kch is the first prokaryotic K^+ channel whose primary structure was identified in the genome of *Escherichia coli* (Milkman, 1994). Kch contains a large RCK (regulator of K^+ conductance) domain, which is very similar to the KTN domain (K^+ transport and NAD binding) in Kef- and Trk-type K^+ transport systems. The role of this domain as a channel and transport regulator is described in chapter 2 by Booth and colleagues.

The first bacterial potassium ion channel that was characterized at a molecular level is KcsA from *S. lividans*. It was cloned and sequenced by H. Schrempf and coworkers (see chapter 3). Shortly thereafter, in 1998, its crystal structure was solved in the laboratory of R. MacKinnon (Doyle et al., 1998). Today the structure of KcsA provides the best model of K^+ selectivity. The selectivity filter is determined by a structural motif made of five amino/acid residues, TVGYG. This motif is highly conserved among all K^+ channels known today. A crystal structure of KcsA in a (most likely) closed configuration provided a basis for an electron paramagnetic resonance spectroscopic study describing the molecular dynamics and open channel structure of the KscA channel (Perozo et al., 1999). The crystal structure of another bacterial K^+ channel, MthK from *M. thermautotrophicum*, which has a membrane topology very similar to that of KcsA, revealed the open state of a potassium channel. The molecular basis of gating in potassium channels is dealt with in chapter 4 by Ptak et al. Another bacterial K^+-selective channel—a glutamate receptor, GluR0, from the cyanobacterium *Synechocystis* strain PCC 6803—was investigated in the laboratory of E. Gouaux in 1999 (Chen et al., 1999). This channel is of particular interest because it combines the structural properties of potassium channels in its transmembrane domains with those of glutamate receptors in its extracellular domain. Both types of channels play essential roles in regulation of neuronal signal transduction (see chapter 5, by Shrivastava and Guy).

The voltage-gated channels, particularly Na^+ and K^+ channels, are essential for propagation of action potentials and neuronal activity. Although KcsA and MthK could serve as good models of structure and function of potassium channels, they did not provide any information about voltage sensing in ion channels. This situation changed when MacKinnon's laboratory published the crystal structure of KvAP, a voltage-dependent channel from *Aeropyrum pernix* (Jiang et al., 2003). Voltage sensing is associated with structural rearrangements of a voltage sensor, the amphipathic transmembrane helix S4, containing positive charges every three residues. Based on the KvAP structure, several models describing the movement of the voltage sensor upon activation were proposed. They are described in chapter 6 by Guy and Shrivastava.

Another important group of K^+ channels is the Kir channels. Their main function is to stabilize the membrane potential of cells, allowing inward K^+ currents upon membrane hyperpolarization. Sequence analysis of the bacterial genomes revealed a prokaryotic channel, KirBac, belonging to the Kir family of ion channels (Durell and Guy, 2001a). The crystal structure of this channel from *B. pseudomallei* was elucidated in 2003 (Kuo et al., 2003) (see chapter 7, by Shrivastava and Guy).

The availability of bacterial ion channel structures provided a fertile ground for theoretical biophysicists interested in molecular modeling of ion channels. Taking into account that crystal structures are snapshots of channel molecules providing no information on their dynamics, computational studies such as MD simulations play an important role in relating structure to function in ion channels. Two chapters in this book describe modeling and simulations of the conformational dynamics of several types of ion channels. Chapter 8, by Biggin and colleagues, deals with MD simulations of potassium channels and glutamate receptors. Chapter 9, by Tajkhorshid and colleagues, describes MD simulations of aquaporins, the alpha-hemolysin channel, ClC chloride ion channels, and MscS, a mechanosensitive channel of low conductance from *E. coli.*

The attention of ion channel biophysicists focused on potassium channels after several crystal structures of these channels became available. Because of the lack of crystal structures of Na^+ and Ca^{2+} channels, the studies of these channels have somewhat been neglected. This situation started changing after a putative calcium channel highly selective for Na^+ was described (Durell and Guy, 2001b; Ren et al., 2001). The channel was named NaChBac, for sodium channel from *Bacillus halodurans*. It consists of a single molecule having six transmembrane domains with an S4 domain containing several arginines placed every third residue apart, indicating a voltage-dependent channel (see chapter 10 by Pavlov et al.).

Much attention throughout the history of ion channel research has been paid to cation-selective channels. Anionic conductances have been studied to a much lesser extent, since they have been considered as passive leakage currents and therefore less important in cellular physiology. The first bacterial chloride channel, ClC0, a homolog of the members of the eukaryotic anion-selective ClC ion channel family, was found in the genome of *E. coli* (Blattner et al., 1997). Crystal structures of ClC0 from *Salmonella enterica* and *E. coli* provided new insights into the principles of anion selectivity (Dutzler et al., 2002) (see chapter 11, by Rickard et al.).

In addition to ion-specific channels there are channels in many cells whose primary function is not in ion conduction but in passing other solutes and water. Aquaporins (water channels), whose discovery was recognized by the awarding of the 2003 Nobel Prize in chemistry to P. Agre, serve as water permeation pathways in all types of organisms from bacteria to humans. The model of transport of water and glycerol through different types of aquaporins is described in chapter 9 by Tajkhorshid and colleagues. MD simulations of other nonspecific channels, including bacterial alpha-hemolysin (which forms nonspecific pores in the membranes of mammalian red blood cells) and mechanosensitive channels, are also described in chapter 9.

Chapters 12 and 13 describe bacterial mechanosensitive channels. MscS, a bacterial mechanosensitive channel of low conductance (see chapter 12, by Blount et al.), and MscL, a bacterial channel of high conductance (see chapter 13, by Sukharev et al.), are to date the best-characterized mechanosensitive channel molecules in terms of structure and function. Currently they provide an excellent model for the understanding of basic principles of mechanotransduction in living cells.

Chapter 14, by Booth et al., the last chapter in this book, describes some of the physiological roles that ion channels may play in the physiology of bacterial cells. Interestingly, the bacterial chloride channel functions as a Cl^--H^+ antiporter (Accardi and Miller, 2004), which indicates that the distinction between ion channels and transporters is not as obvious as previously believed. A specific physiological role for bacterial K^+ channels has not yet been established. They probably serve as K^+ efflux systems if fast K^+ efflux is required. In contrast, a physiological role of bacterial mechanosensitive channels has been well documented. They serve as emergency valves upon exposure of bacteria to hypoosmotic environments (Levina et al., 1999). In addition, recent evidence suggests that expression of bacterial mechanosensitive channels may be closely linked to cell wall remodeling during the stationary phase of cell growth (Stokes et al., 2004).

FUTURE PROSPECTS

Most of the information summarized in this book is a result of the last 10 years of research on ion channels, which brought novel insights into the structure and function as well as the evolutionary origins of these important molecules. The book reflects an interesting situation which emerged from this research. We know many details of the structures of prokaryotic channels but know little of their function in these microbes. In contrast, we have extensive knowledge of the function of ion channels in eukaryotes but little information about their structures. Most of the structural information on eukaryotic channels results from homology modeling based on the structure of their bacterial counterparts.

We may expect that in the future, structures of other prokaryotic channels will be solved, taking into account the availability of the large number of sequenced genomes from these microorganisms in databases. NaChBac and GluR0 are the most likely candidates for having their crystal structures resolved soon. One of the future challenges will be to determine structures of eukaryotic homologs of bacterial channels and establish how the complexity of these molecules contributes to the complexity of the physiology of eukaryotic cells. Also, for those interested in bacterial physiology, the challenge will be to establish further the function of bacterial channels whose structures are known. Other challenges will include elucidation of MD of ion channels by applying magnetic resonance or fluorescence spectroscopic techniques and computer modeling to the snapshots of the channel structures obtained by X-ray crystallography.

We hope that this book will serve as a reference for ion channel specialists and as a source of new knowledge for nonspecialists who want to learn about these interesting molecules.

Andrzej Kubalski
Boris Martinac

REFERENCES

Accardi, A., and C. Miller. 2004. Secondary active transport mediated by a prokaryotic homologue of ClC Cl– channels. *Nature* **427:**803–807.

Bass, R. B., P. Strop, M. Barclay, and D. C. Rees. 2002. Crystal structure of Escherichia coli MscS, a voltage-modulated and mechanosensitive channel. *Science* **298:**1582–1587.

Blattner, F. R., G. Plunkett III, C. A. Bloch, N. T. Perna, V. Burland, M. Riley, J. Collado-Vides, J. D. Glasner, C. K. Rode, G. F. Mayhew, J. Gregor, N. W. Davis, H. A. Kirkpatrick, M. A. Goeden, D. J. Rose, B. Mau, and Y. Shao. 1997. The complete genome sequence of Escherichia coli K-12. *Science* **277:**1453–1474.

Chang, G., R. H. Spencer, A. T. Lee, M. T. Barclay, and D. C. Rees. 1998. Structure of the MscL homolog from Mycobacterium tuberculosis: a gated mechanosensitive ion channel. *Science* **282:**2220–2226.

Chen, G. Q., C. Cui, M. L. Mayer, and E. Gouaux. 1999. Functional characterization of a potassium-selective prokaryotic glutamate receptor. *Nature* **402:**817–821.

Doyle, D. A., J. Morais Cabral, R. A. Pfuetzner, A. Kuo, J. M. Gulbis, S. L. Cohen, B. T. Chait, and R. MacKinnon. 1998. The structure of the potassium channel: molecular basis of K+ conduction and selectivity. *Science* **280:**69–77.

Durell, S. R., and H. R. Guy. 2001a. A putative prokaryotic member of the Kir family of potassium channels. *BMC Evol. Biol.* **1:**14.

Durell, S. R., and H. R. Guy. 2001b. A putative prokaryote voltage-gated Ca(2+) channel with only one 6TM motif per subunit. *Biochem. Biophys. Res. Commun.* **281:**741–746.

Dutzler, R., E. B. Campbell, and R. MacKinnon. 2002. Gating the selectivity filter in ClC chloride channels. *Science* **300:**108–112.

Fillingame, R. H. 1990. Molecular mechanics of ATP synthesis by F_1F_0-type H^+-transporting ATPases. *Bacteria* **12:**345–391.

Gustin, M. C., B. Martinac, Y. Saimi, M. R. Culbertson, and C. Kung. 1986. Ion channels in yeast. *Science* **233:**1195–1197.

Gustin, M. C., X. L. Zhou, B. Martinac, and C. Kung. 1988. A mechanosensitive ion channel in the yeast plasma membrane. *Science* **242:**762–765.

Hamill, O. P., A. Marty, E. Neher, B. Sakmann, and F. J. Sigworth. 1981. Improved patch-clamp techniques for high-resolution current recording from cells and cell-free membrane patches. *Pflügers Arch.* **391:** 85–100.

Hille, B. 2001. *Ion Channels of Excitable Membranes.* Sinauer Associates, Inc., Sunderland, Mass.

Hodgkin, A. L., and A. F. Huxley. 1952a. Currents carried by sodium and potassium ions through the membrane of the giant axon of Loligo. *J. Physiol.* (London) **116:**449–473.

Hodgkin, A. L., and A. F. Huxley. 1952b. The components of membrane conductance in the giant axon of Loligo. *J. Physiol.* (London) **116:**473–496.

Jacob, F., and J. Monod. 1961. Genetic regulatory mechanisms in the synthesis of proteins. *J. Mol. Biol.* **3:**318–356.

Jiang, Y., A. Lee, J. Chen, M. Cadene, B. T. Chait, and R. MacKinnon. 2002. Crystal structure and mechanism of a calcium-gated potassium channel. *Nature* **417:**515–522.

Jiang, Y., A. Lee, J. Chen, V. Ruta, M. Cadene, B. T. Chait, and R. MacKinnon. 2003. X-ray structure of a voltage-dependent K+ channel. *Nature* **423:**33–41.

Kamada, T. 1934. Some observations on potential differences across the ectoplasm membrane of Paramecium. *J. Exp. Biol.* **11:**94–102.

Kung, C., R. R. Preston, M. E. Maley, K. Y. Ling, J. A. Kanabrocki, B. R. Seavey, and Y. Saimi. 1992. In vivo Paramecium mutants show that calmodulin orchestrates membrane responses to stimuli. *Cell Calcium* **6–7:**413–425.

Kung, C., Y. Saimi, W. J. Haynes, K.-Y. Ling, and R. Kissmehl. 2000. Recent advances in the molecular genetics of Paramecium. *J. Eukaryot. Microbiol.* **47:**11–14.

Kuo, A., J. M. Gulbis, J. F. Antcliff, T. Rahman, E. D. Lowe, J. Zimmer, J. Cuthbertson, F. M. Ashcroft, T. Ezaki, and D. A. Doyle. 2003. Crystal structure of the potassium channel KirBac1.1 in the closed state. *Science* **300:**1922–1926.

Levina, N., S. Totemeyer, N. R. Stokes, P. Louis, M. A. Jones, and I. R. Booth. 1999. Protection of Escherichia coli cells against extreme turgor by activation of MscS and MscL mechanosensitive channels: identification of genes required for MscS activity. *EMBO J.* **18:**1730–1737.

Luria, S. E., and M. Delbrück. 1943. Mutations of bacteria from virus sensitivity to virus resistance. *Genetics* **28:**491–511.

Martinac, B., M. Buechner, A. H. Delcour, J. Adler, and C. Kung. 1987. Pressure-sensitive ion channel in Escherichia coli. *Proc. Natl. Acad. Sci. USA* **84:**2297–2301.

Milkman, R. 1994. An Escherichia coli homolog of eukaryotic potassium channel proteins. *Proc. Natl. Acad. Sci. USA* **91:**3510–3514.

Mitchell, P. 1961. Coupling of phosphorylation to electron and hydrogen transfer by a chemiosmotic type of mechanism. *Nature* **191:**144–148.

Mitchell, P. 1979. Keilin's respiratory chain concept and its chemiosmotic consequences. *Science* **206:**1148–1159.

Naitoh, Y., R. Eckert, and K. Friedman. 1972. A regenerative calcium response in Paramecium. *J. Exp. Biol.* **56:**667–681.

Perozo, E., D. M. Cortes, and L. G. Cuello. 1999. Structural rearrangements underlying K+-channel activation gating. *Science* **285:**73–78.

Preston, G. M., and P. Agre. 1991. Isolation of the cDNA for erythrocyte integral membrane protein of 28 kilodaltons: member of an ancient channel family. *Proc. Natl. Acad. Sci. USA* **88:**11110–11114.

Ren, D., B. Navarro, H. Xu, L. Yue, Q. Shi, and D. E. Clapham. 2001. A prokaryotic voltage-gated sodium channel. *Science* **294:**2372–2375.

Saimi, Y., R. D. Hinrichsen, M. Forte, and C. Kung. 1983. Mutant analysis shows that the Ca2+-induced K+ current shuts off one type of excitation in Paramecium. *Proc. Natl. Acad. Sci. USA* **80:**5112–5116.

Saimi, Y., B. Martinac, M. C. Gustin, M. R. Culbertson, J. Adler, and C. Kung. 1988. Ion channels in Paramecium, yeast and Escherichia coli. *Trends Biochem. Sci.* **13:**304–309.

Savage, D. F., P. F. Egea, Y. Robles-Colmenares, J. D. O'Connell III, and R. M. Stroud. 2003. Architecture and selectivity in aquaporins: 2.5 Å x-ray sructure of aquaporin z. *PLOS Biol.* **1:**334–340.

Stokes, N. R., H. D. Murray, C. Subramaniam, R. L. Gourse, P. Louis, W. Bartlett, S. Miller, and I. R. Booth. 2003. A role for mechanosensitive channels in survival of stationary phase: regulation of channel expression by RpoS. *Proc. Natl. Acad. Sci. USA* **100:**15959–15964.

Walker, J. E. 1994. The regulation of catalysis in ATP synthase. *Curr. Opin. Struct. Biol.* **4:**912–918.

Bacterial Ion Channels and Their Eukaryotic Homologs
Edited by A. Kubalski and B. Martinac

Chapter 1

K^+ Channels: a Survey and a Case Study of Kch of *Escherichia coli*

Mario Meng-Chiang Kuo, Ching Kung, and Yoshiro Saimi

THE INVISIBLE BIOLOGICAL WORLD

Seeing is believing. Out of sight, out of mind. The mind has evolved to deal with "sizable" parents, peers, predators, and preys, so it is wired to gauge objects by our own size, revering large things, downplaying small ones, and ignoring the invisibles. Except in science, we deal only with objects visible to the naked eye, counting them with our "digits" and sizing them by the length of our "feet." Whales and dinosaurs fascinate us, not because they are endangered (some are not) or ferocious (some are meek), but simply because they are much larger than we are. But there are a thousand ants for every human, and a thousand microbes for every ant. Our hubris and the anthropocentricity of our science notwithstanding, microbes in fact reign supreme on this planet (Gould, 1996) in diversity, in number, and in mass (Nee, 2004). They also reign in the past, present, and future. Their legacy dates back when the earth crust was stabilizing some 3.8×10^9 years ago (Woese, 1994), and we are only 10^5 years old. In all likelihood, microbes will remain and prosper after yet another "wave of massive extinction," natural or human-made. Today, they thrive deep underground, in arctic water, in hydrothermal vents, in the Dead Sea, as well as in or on plants, animals, and each other. Unfortunately, they are "invisible" even to professional microbiologists since most of them cannot be cultured as "colony-forming units" countable by the naked eye. Our inborn psychological bias to ignore the invisible, like many of our prejudices, will be difficult to overcome. The fact is clear enough. Recent large-scale DNA sequencing has verified beyond any doubt the overwhelming diversity of microbes from the Sargasso Sea to the space between our teeth (Pennisi, 2004).

When the mind tries to understand the brain, smaller animals such as the squid, the fly, and the mouse are only considered models. Microbes are judged irrelevant since they have no brains. Yet some of the nuts and bolts of the brain must have evolved early. While

---

Mario Meng-Chiang Kuo and Yoshiro Saimi • Laboratory of Molecular Biology, University of Wisconsin—Madison, Madison, WI 53706. ***Ching Kung*** • Laboratory of Molecular Biology and Department of Genetics, University of Wisconsin—Madison, Madison, WI 53706.

geneticists have long accepted that bacteria and viruses have genes, neurobiologists have long been in subconscious denial that ion channels exist in microbes. It is therefore ironic that the recent deep understanding of channel filtration and gating came from the atomic resolution of prokaryotic ion channels (e.g., Doyle et al., 1998) and that their physiological functions in the microbes themselves are unknown and do not seem to interest most ion channel biologists (Kung and Blount, 2004).

We believe that microbial channels are an interesting, important, and profitable topic for research. First, a small set of universal physicochemical principles underlies all life forms, despite their myriad dazzling diversities on the surface. In theory then, these principles may be discovered from any organism. Second, because microbes have diverged much earlier and much more than macrobes, commonalities among microbes likely represent basic biological mechanisms that have stood the test of evolutionary time. Third, microbes offer many experimental advantages, allowing us to discover these principles. Among them are short generation time, ease in mass culturing, clonal homogeneity, power of haploid genetics, relative biochemical simplicity, inexpensiveness, and absence of ethical concern in experimentation. It is no accident that DNA replication, transcription, translation, gene regulation, energy metabolism, cell cycling, etc., were first understood in microbes before the concepts were extended to animals and plants. Even within the realm of ion channels, two important basic principles have been revealed through the study of bacterial ion channels: the molecular mechanism of ion selectivity and the ability of lipids in force transmission, as reviewed by other chapters.

THE IMPORTANCE OF K^+

K^+ is the major cation in all cytoplasms, although it is often taken for granted and seems less important than the regulatory ions such as H^+, Ca^{2+}, and Zn^{2+} to students of modern biology. As the major solute, it draws in water and generates the turgor needed for growth. K^+ is considered a compatible solute, meaning that, even in high concentration, it does not interfere much with the structures and functions of macromolecules. How this compatibility comes about during macromolecular evolution is obscure. The early coevolution of the chemiosmotic lipid barrier, the negatively charged RNAs, K^+, and water, should make an interesting subject of investigation. The preeminence of water for life has been recently emphasized as we comb for evidence of its presence on Mars. K^+ is apparently crucial in dealing with biological water on Earth, even today. Dehydration, i.e., external osmotic upshift, demands an adjustment of cellular osmolarity, lest the pressure difference break the membrane. In cases more thoroughly studied, uptake of K^+ constitutes the first line of defense against dehydration. In *Escherichia coli*, for example, the Trk system, a constitutive uptake pump, increases its activity to pump in K^+ immediately when the turgor recedes, as when the cell is challenged with an osmotic upshift. Other mechanisms to increase internal osmolarity follow, including the production of the Kdp system, a high-affinity scavenger K^+ uptake pump, through transcriptional control (Bakker, 1993; Epstein et al., 1993). Trk and Kdp are membrane transport proteins that move K^+ against gradient by expending energy in the form of ATP or the proton motive force, not to be confused with K^+ channels, however. Whether K^+ channels first evolved to take up K^+ is debatable (see below).

Besides and beyond Excitation

Ion channels were first imagined in abstract to explain specific permeability changes during excitation. Being the basis of the action potential, the depolarization-activated Na^+ and Ca^{2+} channels are therefore prominent in the education and thinking of channel biologists. But even an animal is made mostly of nonexcitable cells. Patch-clamp and cultured cells now allow us to examine the unexcitable kidney cells, pancreatic cells, epidermal cells, mast cells, macrophages, osteoplasts, etc. As far as we know, cells from all animal tissues express K^+ channels but not necessarily the Na^+ and Ca^{2+} channels. Thus, K^+ channels seem to underlie some basic functions required by all cells, except a few extreme parasitic bacteria (below), and their variations far outnumber those of Na^+ and Ca^{2+} channels, which seem to have evolved by gene duplication and filter mutations of ancestral K^+ channels (Hille, 2001).

The dense and deep literature on animal K^+ channels classifies them by their electrophysiological properties into leak (resting), depolarizing activated (delayed rectifying), inward rectifying (anomalous), Ca^{2+}-activated, ATP-dependent, G-protein coupled, etc. However, the biological world outside the animal kingdom, especially that of the huge varieties of microbes, is largely inaccessible to the electrodes at present. Sorting microbial K^+ channels by their biophysical properties is therefore not practical at this point.

The Curious Numerology of K^+ Channel Genes

The ability to rapidly sequence large amounts of DNA presents a new vista. Those interested in K^+ channels can now survey different genomes by searching with the canonical K^+ filter sequence, followed by hydropathy plots and homology to known (mostly animal) K^+ channels. At first glance, the numbers of K^+ channel genes in fully sequenced genomes parallel our intuitive "evolutionary ladder" with humans being as complex as a community of 100 bacteria (Table 1). However, a worm has about the same number of K^+ channel genes as humans do. To rank the complexity of a nematode, an insect, and a flowering plant seems futile. The *Paramecium* genome-sequencing project (Dessen et al., 2001), recently completed, gave us a jolt. Though this genome (68 Mb) is much smaller than that of humans (2,650 Mb), it contains 298 K^+ channel genes, three times as many as in humans (W. J. Haynes, unpublished data). Most of these *Paramecium* K^+ channel (*PaK*) open

Table 1. Recognizable K^+ channel genes in various genomes

Organism	No. of K^+ channel genes[a]
Paramecium bursaria chlorella virus[b]	1
E. coli	1
Yeast *(Saccharomyces cerevisiae)*	1
Cress *(Arabidopsis thaliana)*	15
Fly *(Drosophila melanogaster)*	27
Worm *(C. elegans)*	76
Human *(Homo sapiens)*	91
Paramecium *(Paramecium tetraurelia)*	298

[a]There are different ways of counting K^+ channel genes. The data shown here are the numbers of ORFs that contain the K^+ filter sequence (TXXTXGYGD).
[b]K^+ channel genes are not common in viral genomes.

reading frames (ORFs) encode channels similar to the animal CNG type. They are not pseudogenes. Fifteen randomly chosen *PaK*s have been verified to be actually transcribed (Haynes et al., 2003). Though they have not been individually expressed and examined, a type of genetic interference experiment (Ling et al., 2001) showed that some are correlated with K^+ currents previously observed under voltage clamp. The plurality of K^+ channel genes is seen in the genome of another ciliate, *Tetrahymena pyriformis*, but not in other protists such as the free-living green alga *Chlamydomonas* (with 25 K^+ channel genes) or the parasitic malaria germ *Plasmodium* (two K^+ channel genes) or *Trypanosoma* (two K^+ channel genes). The *Paramecium* genome also has some 30 Na^+/Ca^{2+} channel genes, approximately the number found in the human genome (Haynes, unpublished data). We could argue that ciliates are more complex at the cellular level since these unicells have an animal life-style, having to move, seek, avoid, defend, eat, excrete, mate, etc. Nonetheless, it is humbling to find *Paramecium* species to be three times as "complex" as humans are, using the number of K^+ channel genes as the sole criterion.

Gene counting can be misleading. The one-gene–one-enzyme dogma notwithstanding, gene numbers do not necessarily reflect protein varieties. Differential splicing and heteromultimerization can greatly increase the possible number of different channel proteins. Much more research is needed to find out whether and how much different creatures practice these or other modes of multiplying channel-protein varieties.

A MENAGERIE OF K^+ CHANNELS

Sequencing tells more than gene counts. Conceptually translated ORFs provide some structural information. Although such characteristics as rectification and unitary conductance cannot be divined from the primary structure, an amino acid sequence tells us the length of the peptide, the number of possible transmembrane (TM) domains, the likely secondary structures (α, β, turns, etc.), and an increasing number of sites for possible glycosylation, phosphorylation, protein or DNA binding, etc.

The smallest K^+ channel ORF known is that of a virus (PBCV-1) that lives in a green alga (*Chlorella* sp.), which in turn lives in a paramecium (*Paramecium bursaria*). The viral gene, *kcv*, and many of its relatives each encode a 94-amino-acid peptide that forms a K^+ channel when expressed in *Xenopus* oocytes (Kang et al., 2004). The economy befits the viral lifestyle. However, genomes of most other viruses that invade animals, plants, and bacteria contain no recognizable K^+ channel genes, and the advantage *kcv* confers to PBCV-1 is obscure. The 94-amino-acid residues appear to span two TM domains with a pore helix and K^+ filter sequence in between. This S_1-P-S_2 (S, TM helical segment; P, pore-forming loop) structure appears to be the minimum needed for permeation, filtration, and gating (Fig. 1).

The S_1-P-S_2 core-structure motif is maintained from the mammalian inward rectifiers (Kir) to the KcsA of *Streptomyces lividans*. The S_1-S_2-S_3-S_4-S_5-P-S_6 of Shaker is another popular K^+ channel motif, where S_5-P-S_6 remains the core and S_1-S_2-S_3-S_4 apparently functions to regulate it. Covalently linking two cores forms S_1-P_1-S_2-S_3-P_2-S_4, the so-called two-pore-domain channel, which, among other functions, underlies the leak K^+ conductances of animal cells. This motif is particularly popular for the worm *Caenorhabditis elegans*. As described below, K^+ channels of S_1-S_2-S_3-S_4-S_5-P_1-S_6-S_7-P_2-S_8 motif are found in fungi and of S_1-S_2-S_3-S_4-S_5-P_1-S_6-S_7-S_8-S_9-S_{10}-S_{11}-P_2-S_{12} in ciliates. Besides enforcing a one-to-one subunit neighboring in the protein, the advantage of peptide bonding of subunits

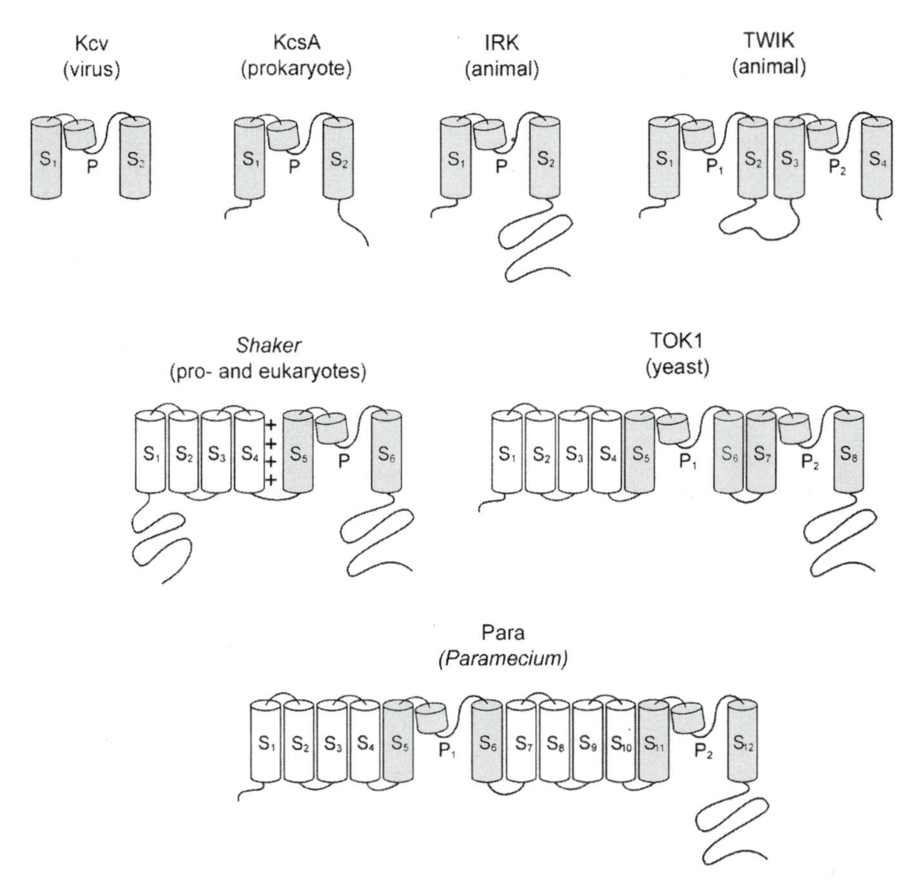

Figure 1. Topology of known K^+ channel pore-forming subunits. K^+ channels are multimers of such subunits, often with other auxiliary subunits. The minimal structure of the pore-forming subunit comprises an N-terminal TM α-helical segment (S_1), a short pore helix (P), the canonical K^+ filter sequence TXG(Y/F)GD, and the C-terminal TM segment (S_2) as seen in the viral form (upper left; see text). Such subunits form tetramers. The last TM segments from the four subunits converge to form the gate near the cytoplasmic side. Many prokaryotic or eukaryotic channels have this basic structure with variable lengths of peptide extending into the cytoplasm (top, center). Two such units are covalently joined to form the two-pore-domain-type K^+ channels found in animals (upper right) that assemble as dimers. The well-known Shaker-type subunit has four additional TM segments preceding the basic structure. In many cases, the fourth segment has positively charged amino acid residues spaced regularly at three-residue intervals and is considered the sensor for voltage-sensitive channels. Subunits with eight TM segments and two pore domains are found in fungi (middle row). All the above structures are known to actually support K^+ conductances in various experimental systems. There are genes in the *Paramecium* genome that are conceptually translated into subunits with 12 TM segments and two pore domains. Their corresponding conductances have not been examined (bottom row).

is not clear. In the opposite direction, even the smallest S_1-P-S_2 subunit could be assembled from noncovalently linked fragments. There are examples in prokaryotic genomes that adjacent ORFs within an operon encode peptides that can be conceptually assembled into K^+ channels (Derst and Karschin, 1998; Kumánovics et al., 2002). Though such examples are seen only by sequence gazing at the moment, their reality in the vastly diverse biological world seems inevitable.

In most instances, the assignment of the TM segments is not from direct study of the channel protein but is surmised from the hydropathy plot of the peptide conceptually translated from the gene. There are no rigid rules in assigning a segment or in specifying its length. The social tendency of our mind favors assignments conforming to the 6-TM topology of Shaker and the 2-TM Kir. Therefore, readers should keep in mind this caveat when writers assert channel peptide topologies, including those in this chapter.

Eukaryotic Microbes

Flowering plants and frolicking animals are the minority, even among eukaryotes. Much greater diversity exists among fungi, algae, flagellates, ciliates, amoebae, etc., in their lifestyle, metabolism, body plan, organelles, and molecules. Among them, some of the larger cells have been examined by electrophysiologists early on. The plant-like *Chara* sp. and *Nitella* sp., for example, exhibit a depolarization-activated K^+ current. The animal-like *Paramecium* sp., on the other hand, shows at least six types of K^+ conductances: K_{dep}, activated by depolarization; K_{hyp}, by hyperpolarization; $K_{[Ca]dep}$, by Ca^{2+} upon depolarization; $K_{[Ca]hyp}$, by Ca^{2+} upon hyperpolarization; K_{mec}, by posterior touch; and K_{leak}, the resting conductance (Preston et al., 1991). Genetics provided some additional information beyond the biophysical description, especially on the role of calmodulin as a channel subunit (Saimi and Kung, 2002). Of the 298 K^+ channel genes encountered in the *Paramecium* genome, 57 have recognizable similarity to the animal inward rectifier, de-polarization- or hyperpolarization-activated, as well as Ca^{2+}-activated BK or SK type. The remaining 239 are in the *PaK* family. Five predict the S_1-S_2-S_3-S_4-S_5-P_1-S_6-S_7-S_8-S_9-S_{10}-S_{11}-P_2-S_{12} topology.

Four laboratories independently recognized and named the same K^+ channel ORF in the genome of the budding yeast *Saccharomyces cerevisiae* (Ketchum et al., 1995; Zhou et al., 1995; Lesage et al., 1996; Reid et al., 1996). By priority, it is here called TOK1, for two-pore-domain outwardly rectifying K^+ channel. It corresponds to the depolarization-activated K^+ current previously registered with patch-clamp electrodes on the plasma membrane of yeast spheroplasts (Gustin et al., 1986). This ORF conceptually translates into a K^+ channel subunit with a S_1-S_2-S_3-S_4-S_5-P_1-S_6-S_7-P_2-S_8 topology. TOK1 homologs can be found in other fungi such as the infectious yeast *Candida albicans* and the filamentous bread mold *Neurospora crassa*, but this 8-TM topology has not been reported outside the fungal kingdom so far (Y. Saimi, unpublished data).

TOK1 is thoroughly analyzed, because it can be robustly heterologously expressed in *Xenopus* oocytes and subjected to rigorous mutagenic dissection by microbial genetics in yeast itself. The original notion that TOK1 acts as an animal inward rectifier inserted backward (Ketchum et al., 1995) seems no longer tenable since blockage by external Mg^{2+} could not be replicated (Zhou et al., 1995; Lesage et al., 1996). Analyses also showed that the outward rectification is not simply by voltage gating but is a function of $\Delta\mu_K$, the electrochemical motive force for K^+ ($\Delta\mu_K = V_m - E_K$). Q_{10} and other analyses indicate that this rectification likely reflects the instant collapse of the K^+ filter (the R state) when $\Delta\mu_K$ becomes inward and its instant restoration (the O state) when $\Delta\mu_K$ is outward (Loukin and Saimi, 1999). TOK1 also shuttles between R and a set of closed (C) states with more conventional Q_{10} and kinetics. Growth-stopping mutations (gain-of-function [GOF] mutations, see below), selected after random mutagenesis, remove the R state and were found to be located at the cytoplasmic ends of S_6 and S_8 (Loukin et al., 1997), now modeled to be at or near the point of inner gate closure by analogy to the KcsA structure (Doyle et al., 1998). Many mutations that subdue the above overly active channels were found to be deletions of the C-terminal cytoplasmic domain trailing S_8. Kinetic analyses best fit the notion that this cytoplasmic domain acts as a "foot in the door" that keeps the inner gate locked to stabilize the O state (Loukin et al., 2002). Interestingly, this domain, when separately expressed in addition to the channel body without its C-tail, still acts as the "foot" to the channel gate (Loukin et al., 2002). The current detailed model of TOK1 gating also describes the filter collapse as a prerequisite for the inner gate movement (Loukin et al., 2002).

Prokaryotic K^+ Channel Genes

The rapidly expanding sequence data from bacteria and archaea, including those that are currently not culturable, will soon make any gene survey obsolete. Nonetheless, we have scanned the 235 prokaryotic genomes whose entire predicted ORFs are available in the National Center for Biotechnology Information protein database as of July 2004, by BLAST search using the K^+ filter signature sequence as a query, followed by Kyte-Doolittle hydropathy analysis. Except for a few extreme parasitic bacteria, each genome contains at least one putative K^+ channel gene. The corresponding channels each have a TXGYGD or TXGFGD filter sequence as expected, since they are parts of the query sequence. There are exceptions, however. For example, in a *Prochlorococcus* species two putative K^+ channel proteins have nearly the same amino acid sequence. Yet, one has the usual filter sequence and the other one harbors a GHG tripeptide instead of the GYG or GFG. It is questionable whether the latter forms a functional K^+ channel.

Besides the filter and the putative transmembrane domains deduced from the hydropathy plots, clearly recognizable features include evenly spaced positive amino acid residues (arginine or lysine) in S_4 and a long (>100 residues) N-terminal tail before S_1, a long C-terminal tail after S_2/S_6, or both. Most but not all of these long tails contain regions homologous to the known RCK (below) or the cyclic-nucleotide-binding domain (CNBD). On the basis of these recognizable features, we sorted these K^+ channel gene products into 10 types. Although the majority of them have two or six TM helices (2- or 6-TM), we have encountered ORFs that encode the canonical K^+ filters attached to only one helix each ([1-TM] type) (Fig. 2). Whether they assemble with other subunits to form functional channels is not known. In the genomes of the two *Streptomyces* species, we also encountered ORFs that translate into K^+ channel subunits with four TM helices ([4-TM] type). It would be of interest to see whether each of these ORFs alone can form functional channels in *E. coli* or in oocytes. Among the 2-TM types are those with little cytoplasmic extensions, [2-TM]; and those with long tails at the N-ends terminal, +[2-TM]; at the C-terminal ends, [2-TM]+; or at both ends, +[2-TM]+. Among the 6-TM types are those that have little cytoplasmic extensions with or without the positively charged S4 ([6-TMv], [6-TM]), as well as those with large C-terminal extensions, again with or without the positively charged S4 ([6-TMv]+, [6-TM]+).

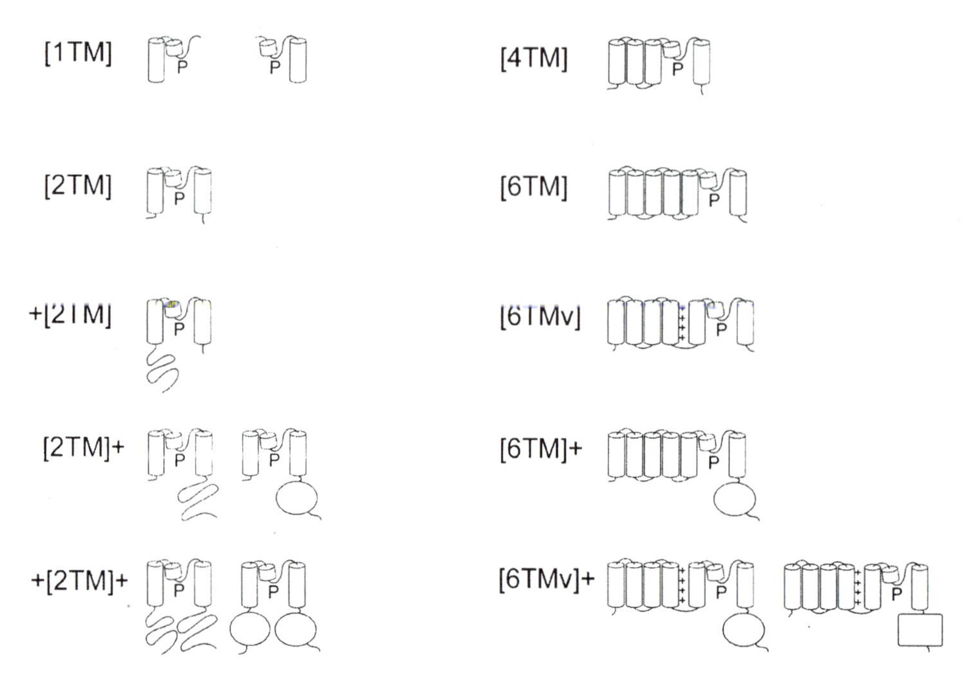

Figure 2. The diversity of prokaryotic K+ channel subunits. The putative prokaryotic K+ channel subunit proteins can be sorted into 10 types. All 10 types contain the K+ signature sequence with adjacent TM(s). These 10 types are different in the number of TM domains, the presence of a long tail at either or both the N and C terminus, and the presence of voltage-sensing charges in their S_4. Most of the long tails contain an RCK domain (oval), some of them contain CNBD (rectangle), and others do not match any known domain (line). See the text.

We usually group prokaryotes by their pathogenicity, convenient staining properties, or energy metabolism (chemosynthetic or photosynthetic, or respiratory; aerobic, anaerobic, etc.). Systematists classify them by their small-subunit rRNAs, lipid types, cell wall characteristics, etc. It would be informative if we could relate the above K+ channel types to these various classifications and gain insights into their functions. For example, if certain types of K+ channels were to parallel certain habits and habitats of the organisms, we would be able to speculate on their possible roles. Our survey, however, revealed no obvious recognizable patterns.

Our inability to discern a pattern between the recognizable structural types of K+ channels with any ordinary grouping of prokaryotes reflects our ignorance of the physiological roles of these channels, discussed below. Disparate organisms in disparate groups by our parochial or systematic criteria may be sensitive to the same set of biotic or abiotic environmental challenges such as a certain type of virus or a certain combination of conditions, say, osmolarity and pH. If different types of K+ channels are for different challenges, they will then reflect these contingencies, which require far more knowledge of the ecology and natural history of prokaryotes than we now have.

Prokaryotic versus Eukaryotic K^+ Channels

Prokaryotic and eukaryotic K^+ channels have a lot in common. RCK, common in prokaryote channels, is clearly a feature of the BK-type Ca^{2+}-activated K^+ channels in animals. (The RCK domain should not be confused with the "Ca^{2+} bowl" that binds Ca^{2+} to gate the BK channel, however.) We found six cases where the prokaryotic K^+ channels have the CNBD. CNBD also forms a regulatory domain of the CNG-gated animal channels, although the latter are not K^+-specific. On the other hand, some of the eukaryotic varieties have not been found in prokaryotes. So far, we have not encountered in the prokaryotic genomes any two-pore-domain-type K^+ channel subunits (4-TM, 8-TM, or 12-TM type) (Fig. 1) encoded in an ORF.

The overlap or the lack of it between the K^+ channel types from eukaryotes and those of prokaryotes should not be overinterpreted, since we can only use recognizable features to probe the unknown. In addition, we have seen only a small corner on a block at the tip of the iceberg of prokaryotic diversity. Nevertheless, one is left with the impression that most of the features that characterize the modern K^+ channel, such as K^+ filtration, ligand gating, voltage gating, and rectification, are very ancient, having arrived before bacteria and archaea diverged long before eukaryotes came into being 2 billion years ago. We will leave the experts to debate the importance of horizontal gene transfer in evolution and its implications in the evolution of K^+ channels.

Unknown Functions

In animals, the equilibrium potential of K^+ (E_K) is usually more negative than membrane potential (V_m), so K^+ channels open to let out K^+, making the cytoplasm more negative. Thus, we recognize the general functions of K^+ channels in stabilizing the resting V_m or in reestablishing the negativity after excitation to shorten the action potential and lower its frequency. In plants, where the resting potential is governed by the electrogenic H^+ pump, there are examples of K^+ channels being used to take up K^+. In plant leaves, an inwardly rectifying K^+ channel (KAT1) is apparently used to take up K^+ and thereby increases turgor of the guard cells to open the stomata (Kwak et al., 2001). In the roots, knocking out the gene for a K^+ inward rectifier channel (AKT1) reduces K^+ uptake and causes the mutant plant to grow poorly in a limiting supply of K^+ (Hirsch et al., 1998).

In contrast, the roles of K^+ channels in the myriad prokaryotes are entirely unknown at present. The disparity cannot be more extreme. On the one hand, we gain so much of our current understanding from the X-ray crystallographic structures from microbial K^+ channels. On the other hand, we cannot say anything about what these K^+ channels do for the microbes themselves. Because most channel specialists are animal or human physiologists, the reigning attitude is that their roles in the native scum and germs are of no concern. At the same time, most microbiologists do not understand ion channels and do not incorporate them into their thinking. Regardless of how we arrive at this peculiar stage of our intellectual history, the exploration into the virgin territory of microbial channel functions should be intellectually rewarding, but technically challenging.

TECHNICAL CHALLENGES

The challenges are largely technical. Examining the biophysical properties of prokaryotic channels is technically difficult. (Recall that electrophysiology starts with frog muscle and giant squid axons.) These properties can be examined if the channels can be reconstituted functionally into planar lipid bilayers. KcsA, a case where the K^+ current can be so registered, is reviewed in chapters by Schrempf and others. An additional case of functional reconstitution is the K^+ channel of the archaeon *Methanobacterium thermoautotrophicum* (Jiang et al., 2002). However, most ion channels have yet to be successfully reconstituted in this manner. The Kch of *E. coli*, for example, has not been successfully reconstituted to date, despite serious efforts. Direct patch-clamping of bacteria, made giant by genetic or pharmacological means, is another avenue to study channel biophysics. To date, the patch clamp has only been applied to a few bacteria, and routinely only to *E. coli*. Since other channels may be heterologously expressed in *E. coli*, one would hope to make giant *E. coli* the prokaryotic equivalent of *Xenopus* oocytes. Another technical limitation is that, in practice, we can only gate channels by voltage, stretch force, H^+, Ca^{2+}, or a few known ligands such as nucleotides. Microbial channels gated by unknown principles are therefore invisible to the experimenter. Even the electric activities of the K^+ channel of *E. coli*, Kch, have yet to be registered to date (see below).

The standard way of finding the function of a protein in vivo, especially with microbes, is to examine organisms whose protein has been mutated away or mutated to cause its activity to increase, decrease, or change in nature. Such "genetic dissections" have been enormously successful in solving biological mysteries, from embryonic development to biological clock. A large part of our knowledge of K^+ channels originated from Shaker, a mutant in an early genetic dissection of *Drosophila* behavior. It would seem that the application of genetics to microbial ion channels should be straightforward: knock out the channel genes and see their physiological changes in the mutants. In reality, the applications have so far been much less successful than one would have thought. Our experience points to two difficulties. First, phenotyping by microbiologists is largely limited to growth (colony formation), although colony morphology and tactic behavior are occasionally examined in some cases. Much of microbial genetics and molecular biology is built on the ability or inability to grow on various culture media, under various conditions, or in different genetic backgrounds. For ion channel mutant research, one can alter the concentrations of various ions in the medium, and this can be informative in some cases (see below). An important parameter for channel research is the membrane potential, but there is no reliable growth media one can use to control this parameter. Furthermore, if the gating principle is unknown, one cannot even rationally design the test media. Second, much of microbial physiology is quite robust. There are often redundant systems or backup alternative routes to achieve a metabolic end to ensure growth. Knocking out one pathway does not necessarily cripple the cell. This is one reason why only 5.7% of the *E. coli* genome is considered "essential" (from the Profiling of *E. coli* Chromosome Database [http://www.shigen.nig.ac.jp/ecoli/pec]). To date, there are only two successful cases, detailed elsewhere in this volume, where prokaryotic channel functions are clearly identified by mutations. In both cases, double mutations are needed to remove redundant pathways before the in vivo function can be revealed: *mscL*Δ *mscS*Δ shows the osmotic downshock protection provided by mechanosensitive channels, and *eriCΔ mriTΔ* shows the acid tolerance conferred by Cl^- channels (Iyer et al., 2002).

We are hopeful that the biological roles of ion channels in the common laboratory microbial species will be discovered in the near future. We can also hope that other channels can be heterologously expressed in laboratory creatures for biophysical or biochemical analyses.

Kch, THE K^+ CHANNEL OF *E. COLI*

Ion channels are not a topic of traditional microbial research. Emblematic of this peripheral status, the first prokaryotic K^+ channel gene, the *kch* of *E. coli*, was discovered not from a deliberate search but was encountered fortuitously by R. Milkman in 1994 (Milkman, 1994) when he, for comparative purposes, extended his DNA sequencing from the region of his central interest, the *trp* operon. The sequence conceptually translated into a protein that is unmistakably a pore-forming subunit of a Shaker-like channel, equipped with six TM α-helices and the K^+ "pore" sequence between S_5 and S_6.

Unlike the voltage-gated Shaker K^+ channel, Kch does not contain the periodic positive charged residues at S_4. On the other hand, Kch is equipped with a large domain extended from its S_6 with a 25-amino-acid-residue linkage in between. This domain, called RCK (regulator of K^+ conductance), is commonly found in the [2TM]+, +[2TM]+, and [6TM]+ types of prokaryotic K^+ channels in our BLAST search above (Fig. 2). The tertiary structure of the RCK domain in K^+ channels is similar to that of the KTN domain (K^+ transport and NAD-binding, Pfam02254) in Kef and Trk types of K^+ transport systems as revealed from the X-ray crystal structures (Jiang et al., 2001; Jiang et al., 2002; Roosild et al., 2002) (see also chapter 2). These domains, about 150 amino acid residues, consist of seven continuing (β-α) motifs with a Rossmann fold core (Fig. 3A). In the crystals, two domains dimerize though the last (α-β-α) motifs as a handshake (Fig. 3B). The Rossmann fold is recognized as an NAD(P)-binding domain (Rossmann et al., 1974), and several conserved amino acid residues form the dinucleotide pocket. These key residues are preserved in the KTN domain but are not in the RCK domain. This subtle difference seems to be the only distinction between RCK and KTN in domain terminology. In the RCK gating mechanisms proposed by either Jiang et al. (2002) or Roosild et al. (2002), a certain ligand binds to the RCK domain and causes conformational changes of the dimers, which in turn transduce the ligand-binding energy to a mechanical force to pull the four S_6s and open the gate. How the RCK dimers orient and how many dimers function to gate the channel in vivo are currently unclear (Kuo et al., 2003).

Deletion of *kch* Gives No Obvious Phenotype

E. coli is the established model organism. Its cell physiology is better documented than that of any other cells, and it is amenable to genetic or molecular biological manipulation. It is therefore reasonable to focus attention on Kch in *E. coli* in the hope of gaining insights into the functions of prokaryotic K^+ channels, even though the electric activities of Kch have not yet been reported.

In a genetically amenable system, the most direct way to define the role of a protein is to examine mutant phenotypes in vivo when the corresponding gene is deleted. For this

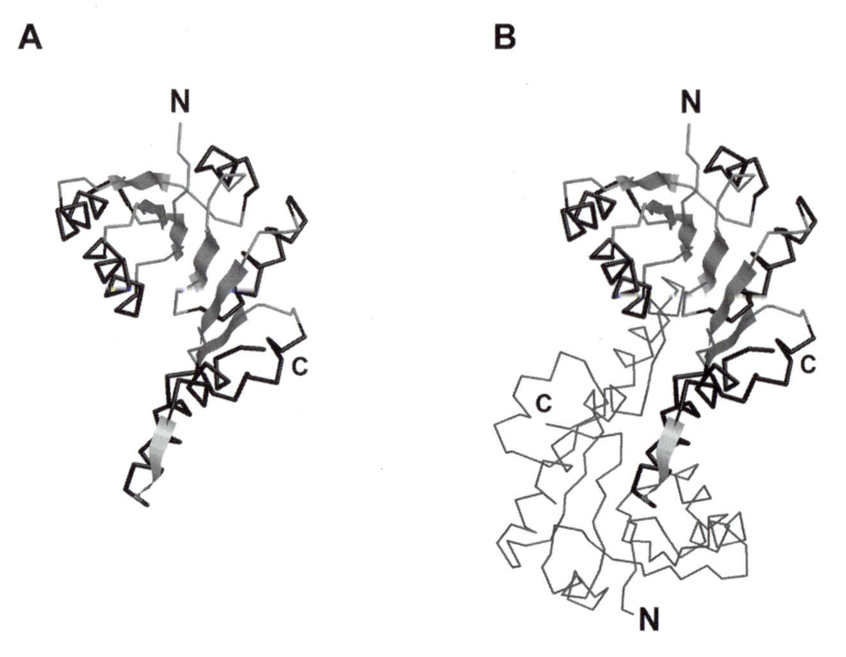

Figure 3. The structure of the RCK domain of Kch (PDB ID: 1ID1). (A) This domain consists of seven alternative β-sheets (ribbons) and α-helices (black backbones). The first five (β-α) motifs form the Rossmann fold core. (B) In the crystal, two RCK domains are dimerized through the last (α-β-α) motifs as a handshake (Jiang et al., 2001). The backbone of the second domain is shown in thin lines. The figure was produced with the RasMol program.

purpose, we generated a chromosomal deletion of the *kch* and examined the *kch*Δ strain (in comparison to its otherwise isogenic parent) in terms of cytosolic K^+ content, CFU on solid plates, or growth rate in liquid media under different conditions or treatments. The parameters that we mostly varied and tested are the pH, ionic strength, and osmolarity. No difference can be discerned in these three parameters in liquid or solid media. Besides continuous culturing in constant conditions, we also changed these parameters as short-term stresses on either growing or stationary-phase bacteria and continually monitored optical density or CFU to test their tolerance. With all these treatments, we were unable to detect any consistent difference between the *kch*Δ strain and its *kch*$^+$ parent.

Our, and apparently others', failure to find a *kch*Δ phenotype does not mean that Kch plays no role. On the contrary, the existence of *kch* in different strains of *E. coli* and the presence of K^+ channel genes in nearly all prokaryote genomes, which are far more streamlined than those of eukaryotes, strongly imply that K^+ channels confer certain selective advantages. Our inability to find the *kch*Δ phenotype likely indicates that there is one or more redundant "function" (in the genetics sense described below) in *E. coli*, even though the genome contains only one recognizable K^+ channel gene. Alternatively, Kch may be used in a condition or a combination of conditions in nature that we have not yet simulated in the laboratory.

Genetic Dissections of Microbial Ion Channels

The *kch*Δ mutant described above is what geneticists call a complete loss-of-function mutant. In genetics, the word "function" means what the gene product does that leads to the normal phenotype in vivo. Often the in vitro biochemical or biophysical activity of the protein or RNA in question is not known, even if the gene is cloned. The perversions of the phenotype by mutation allow geneticists to surmise whether that chemically undefined activity of the gene product has gone up, down, or changed in nature, and thereby to sort the mutations into the GOF, loss-of-function, or altered-function categories, respectively. Although these distinctions of mutations are most useful in the analyses of complex pathways such as animal development, they have been applied to microbial ion channel studies with some success.

In the context of microbial ion channels, the function refers to the presumed passage of ions in vivo, and the presumed normal activities of the channel correspond to the phenotype of normal growth. We reason that, if the ion passage through the mutant channel goes up uncontrollably, the mutation may slow or stop growth. Thus, by definition, such mutations are deemed GOF mutations. In biophysical reality, these GOF channels could have a larger unitary conductance or more likely a higher open probability, due to gating defects. GOF mutations in MscL identified the submolecular location of the gate constriction that closes the channel (Ou et al., 1998). This constriction was later confirmed by crystallography (Chang et al., 1998). Similarly, GOF mutations of TOK1 have led to the discovery of the foot-in-the-door phenomenon, described above (Loukin and Saimi, 2002). In these cases, the corresponding electrophysiological defects of the GOF mutant channels have been scrutinized to gain insights into the working of the channels.

The use of mutations to dissect biological pathways has an honorable history in genetics. This so-called forward genetics or genetic dissection entails random mutagenesis, selection of mutants by phenotypes, followed by scrutiny of these phenotypes toward understanding the pathway and its components. The famous Heidelberg screen, for example, breaks open the molecular analyses of the embryology of segmented animals. In much of the animal or plant research, genetics is used to identify the component gene products in the pathway. Though the philosophy is the same, our use of forward genetics is at a different level. We selected mutations to identify the parts of a single gene product crucial to its working. This type of "fine-structure genetic dissection" is used infrequently in animals and plants because of technical difficulties largely due to the relatively small populations that can be effectively handled in the laboratory. On the other hand, selecting mutants from 10^4 to 10^6 mutagen-treated microbes is routine.

The GOF Mutations of *kch*

To isolate the GOF Kch mutants (Kuo et al., 2003), we randomly mutated *kch* using an in vivo mutagenic system and screened these mutagenized *kch* genes in plasmids transformed into *E. coli* of the *kch*Δ background (chromosomal deletion) for the ability to stop bacterial growth only in the medium enriched with K^+ but not with Na^+ or sorbitol. We noticed that overproducing the wild-type Kch by isopropyl-β-D-thiogalactopyranoside induction causes the bacterium to become sensitive to high ionic strength, and this sensitivity cannot be suppressed even after the K^+ filter of the Kch is destroyed by mutations. To avoid the confusion of overproduction, the screen was carefully examined in the media without the inducer (only expressed at the leakage level of the *lacUV5* promoter). Seven Kch mutants conferring the K^+-only sensitivity were picked from ~8,000 colonies screened. They grow like their parent on medium enriched with Na^+ or sorbitol, but not one with K^+. Further examinations of these mutants with the promoter native to *kch* confirms this GOF phenotype (Fig. 4A). When these GOF mutants are tested on tryptone-based media, adding only a few millimoles of K^+ is enough to stop their growth (Fig. 4B). The kinetics of bacterial growth is better quantified by the increase of turbidity in liquid medium. We found that the GOF mutants stop growth immediately after the addition of KCl, within the resolution of this type of growth experiment (minutes) (Fig. 4C).

Recall that these mutations were not directed to any specific sites to test a preconceived hypothesis on the importance of any domains, even though the sequence of *kch* is at hand and we can delineate the various TM segments, the filter, and the RCK domain. The mutated *kch* of the seven mutants was sequenced. Each was found to have a single substitution within the RCK domain. This finding strongly supports the view that the RCK domain is part of the gating mechanism of *kch*, as deduced from comparative crystal-structure analyses. A likely scenario is that normally, in vivo, the binding of an unknown ligand to the cytoplasm of the RCK domains causes a conformational change that pulls open the constriction due to the convergence of the ends of the four S_6s. The GOF mutations in the RCK domain favor this conformation because either the mutant RCK domains bind the ligand more strongly or they assume the "open" conformation without binding. The abnormal opening of the channel leaks K^+, and this leakage stops growth, probably by perturbing the membrane potential (see below).

Kch Conducts K^+ In Vivo

If the above speculation is correct, a second mutation that stops the K^+ leak through the leaky gate should save the GOF mutants. Since we know that the K^+ filter and the gate are along the same conducting pathway, collapsing the filter should shut off the K^+ leakage even if the gate is loose. Indeed, an individual second mutation in either G187, Y188, or G189 of the TXGYGD canonical K^+ filter sequence of *kch* added to the GOF mutation in the RCK domain reverted the GOF phenotype and allowed the double mutant to grow in the presence of high [K^+]. Since *kchΔ* mutants can also grow in such media, the double mutant phenotype alone cannot be distinguished from bacteria that do not produce the Kch protein. To guard against the possibility that the doubly mutated gene simply no longer makes any proteins, we tested directly for the presence of the channel protein. An antibody produced by the Booth laboratory was used to assess the presence and the amount of the double mutant channel protein in an immunoblot. The result shows clearly that the double mutant channel protein is produced in the normal amount but is no longer toxic.

We focused our attention on the three key residues, GYG (187 to 189), and individually mutated each corresponding codon randomly. Seventeen different types of amino acid substitutions of GYG were found to suppress a GOF mutation at the RCK domain. This plurality indicates that the K^+ filter needs to be constructed precisely and is exquisitely sensitive to mutation, as is evident from the crystal structure. This precise architecture is also evident from the filter sequences of K^+ channels of different gating types and from different organisms. The "canonical" sequence is hardly variable. This sort of "zoo" comparison

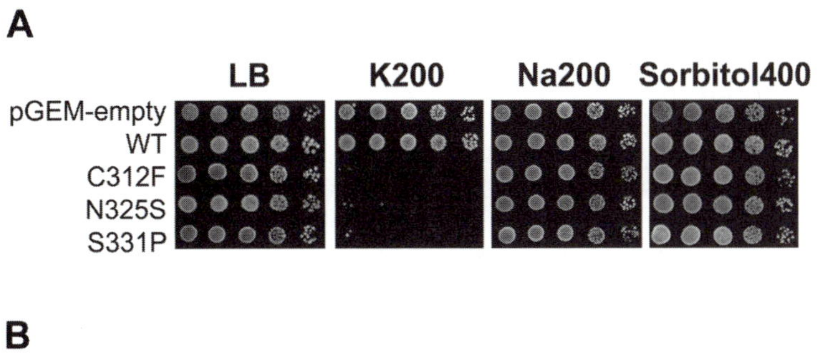

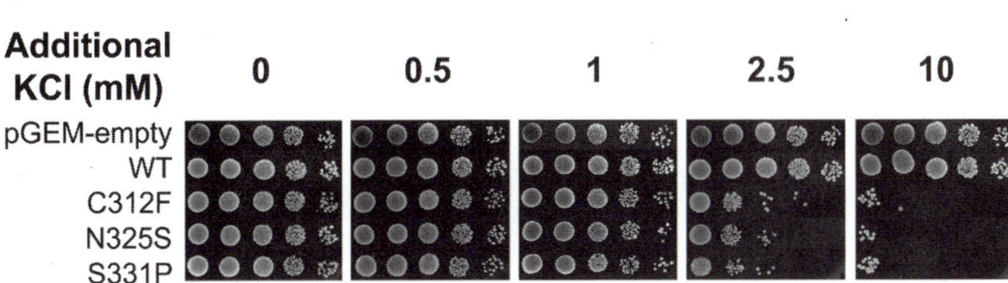

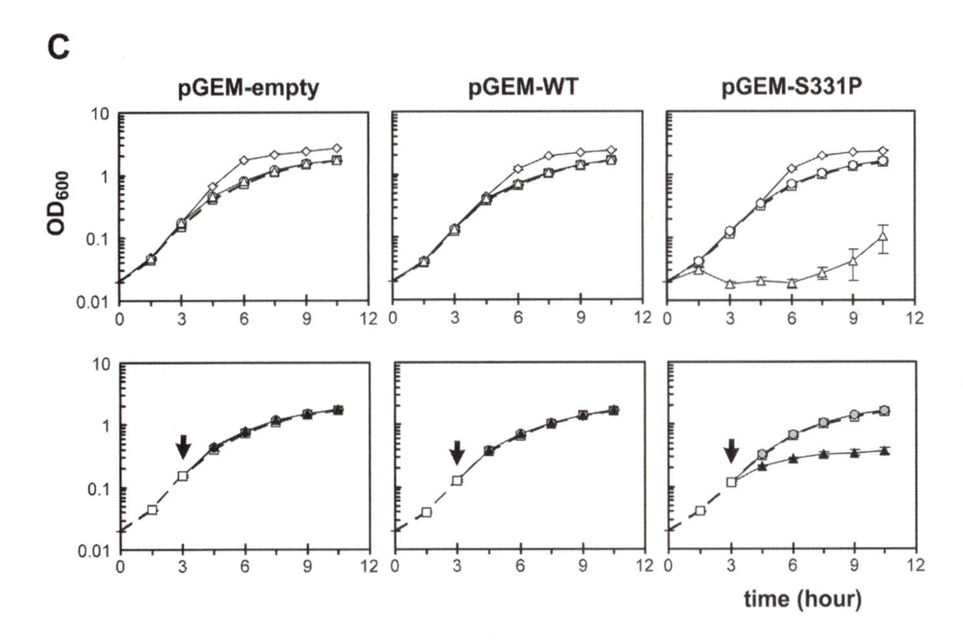

Figure 4. The K+-sensitive phenotype of the Kch GOF mutants. (A) The GOF Kch mutants cannot grow on LB agar plates enriched with 200 mM K^+ (K200). These were the conditions we used originally to screen for these GOF mutants. The *kch* genes are promoted by the native promoter of *kch* from pGEM vector. (B) The phenotyping was refined by using tryptone-agarose plates that have background [K^+] of ~0.5 mM (by flame photometry). A millimolar concentration of added K^+ stops GOF mutant growth. (C) Top row, when inoculated into fresh tryptone-based media, the stationary-phase GOF mutant starts to grow like the wild type in media enriched with 5 mM Na^+ (open circle) or 10 mM sorbitol (open diamond) but not in the one enriched with 5 mM K^+ (open triangle). Bottom, adding 10 mM K^+ (arrows) to the growing GOF mutant culture immediately suppresses growth (filled triangle). (Reprinted from Kuo et al. [2003] with permission.)

showed that the only variation of the GYG tripeptide allowed by nature is GFG, presumably because tyrosine and phenylalanine are both aromatic and function similarly. GFG is not among the 17 GYG mutations found to suppress the GOF phenotype. To rigorously test which variants in GYG can support the Kch filter, we selected from bacteria randomly mutagenized at Y188 those that do not suppress the GOF phenotype, i.e., those that continue to show K^+ stoppage of growth. Fifteen colonies whose replicas cannot grow on K^+-enriched plates were randomly picked and their filter sequences determined. Only two varieties were found: tyrosine (Y) and phenylalanine (F). Again, the experiment is free from preconceived hypothesis: any substitution at Y188 that preserves the filter conformation and suppresses growth in K^+-enriched plates was allowed to show. That only Y and F appeared in this selection after random mutagenesis indicates strongly that phenylalanine is the only other variant besides tyrosine in GYG that supports the integrity of the K^+ filter.

The main conclusion from this set of experiments is not about the structure and the importance of the filter or the RCK domain. That a K^+ specific growth phenotype can be generated in the Kch GOF mutants can only mean that a flux that passes through the K^+ filter takes place in these mutants. The geneticists' "phenotype" refers to characteristics of live organisms. The results therefore show that the GOF Kch channel indeed passes K^+ current in vivo, even though we have no direct electrophysiological evidence. This is the first demonstration that a K^+ channel actually passes K^+ current in a live prokaryote, although the evidence is indirect.

The Possible Use of K^+-Specific Flux of Kch

Our finding that Kch can in fact conduct K^+ current in vivo further begs the question of its biological role. We might speculate on how this current could affect the physiology, even though we have yet to find a relationship with any environmental conditions or stresses. A K^+ flux across the cell membrane either affects the K^+ concentration in the cytoplasm, $[K^+]_{in}$, or the membrane potential, V_m. We have cited above the former effects in plants and the latter effects in animals. Here we speculate that Kch functions to adjust the V_m instead of the $[K^+]_{in}$ in *E. coli*.

If Kch functions to take up K^+ in bulk or to guard against its loss, one would expect to find *kch* LOF mutants that require external K^+ supplement for growth. Mutants with this K^+-auxotrophy phenotype have been searched for and studied extensively by the laboratories of W. Epstein (Epstein and Davies, 1970; Epstein and Kim, 1971) and I. Booth (Bakker, 1993), among others. This bacterium uses a constitutive pump, Trk; an inducible pump, Kup; and an inducible scavenger, Kdp, to take up K^+. In triple mutant strains lacking all three uptake mechanisms, low residual uptake does not distinguish K^+ from Rb^+ and Cs^+. Two gene products, KefB and KefC, are found to be for K^+ efflux. Kch was not revealed in any of the searches for K^+ auxotrophy. We also directly measured the cytoplasmic content of K^+ in the wild-type and GOF mutants by flame photometry. The GOF mutant harboring the "loose-cannon" Kch channel appears to suffer no detectable loss in the $[K^+]_{in}$ (Kuo et al., unpublished result).

We found that increasing external K^+ by about 10 times from ~0.5 to ~5.5 mM (approximately from 10^{-4} to 10^{-3} M) had no effect on the wild-type mutant but stopped the growth of the GOF mutant (Fig. 4C). Interestingly, a concomitant increase of the external H^+ by about 10 times from ~pH 7.3 to ~5.8 (approximately 10^{-7} to 10^{-6} M) restored the mutant's growth to near normal (Kuo et al., 2003). This logarithmic relationship is best understood as the changes in equilibrium potential of K^+ (E_K), V_m, and proton motive force (Δp). Δp is apparently a key to growth since it is used for cell motility, ATP synthesis, nutrient or ion transport, and other works required for cell growth. Δp consists of two components, V_m and ΔpH ($pH_{in} - pH_{out}$, the pH difference across cell membrane), as described in the equation: $\Delta p \approx V_m - 60\Delta pH$. Changing either V_m or ΔpH alters Δp. Assume that the wild-type Kch channel is normally closed in growing cells, but the GOF mutant channels open uncontrollably and thereby create a resting K^+ permeability. This resting K^+ leakage through the GOF mutant channel would clamp the V_m at E_K. When external $[K^+]$ is increased (from ~0.5 to ~5.5 mM), the E_K and V_m of the GOF mutant become shallower (from ~ −150 to ~ −90 mV, assuming $[K^+]_{in}$ is 200 mM). The loss of internal negativity in turn reduces Δp to a degree that no longer supports growth. In the equation above, the shallowing of V_m can be compensated for by raising the other parameter, ΔpH, to keep the Δp up. We found that lowering pH_{out} from ~pH 7.3 to ~5.8 indeed restores the mutant growth. Our results and analyses therefore favor the view that the pathology of the GOF mutants originates from the perturbation of the V_m, and therefore the Δp, but not from any significant perturbations in internal K^+ concentration, $[K^+]_{in}$. By extension, it seems likely that wild-type Kch opens controllably to adjust the V_m and not $[K^+]_{in}$.

The influences of K^+ channels on the V_m and $[K^+]_{in}$ of bacteria can be calculated theoretically. Assuming a rod-shaped, growing *E. coli* as a cylinder 2 μm in length and 0.75 μm in diameter, this cell will have $2.5\pi \times 10^{-16}$ liter of volume and $1.6\pi \times 10^{-8}$ cm^2 of cell membrane. The specific capacitance of biological membranes is 1 $\mu F/cm^2$, and the total capacitance of our theoretical *E. coli* is $1.6\pi \times 10^{-14}$ F. K^+ channels only need to pass $\pi \times 10^4$ K^+ ions across the membrane to charge it by 100 mV. In the cell with a volume of $2.5\pi \times 10^{-16}$ liter, this amount of K^+ ions corresponds to 66 μM. For *E. coli*, whose $[K^+]_{in}$ is in the range of hundreds millimolar, this amount of change in $[K^+]_{in}$ is almost physiologically negligible. This calculation explains why the GOF Kch mutant has no measurable $[K^+]_{in}$ difference when compared with wild-type or *kch*Δ cells and supports the hypothesis that Kch functions to adjust V_m and not $[K^+]_{in}$.

PERSPECTIVE

Prokaryotic ion channels are important subjects of biological investigation. Crystal structures of prokaryotic K^+, Cl^-, glutamate-binding, and other channels have provided unprecedented quantum increases in our understanding of the workings of channels traditionally studied only in animals. MscL and MscS have taught us that physical forces can be delivered to membrane proteins from the lipid bilayers. We can look forward to gaining other insights on basic biological issues from further studies of prokaryotic channels.

The historical and ongoing separation between neurobiology and microbiology has made the exploration of microbial ion channels difficult, however (Kung and Blount, 2004). Innate human bias and pressure for medical relevance in research will continue to make neurobiologists shy away from microbial channels. The topic of prokaryotic channels will likely not attract many microbiologists' attention either, unless their biological roles are illustrated. That nearly every bacterial or archeal genome includes at least one K^+ channel gene, however, leaves little doubt that these channels provide selective advantages. The divergence of energy metabolism, cell structures, habits, and habitats of prokaryotes is much greater than that of animals and plants combined. It will be some time before we know what these advantages are, given our current ignorance.

It seems to us important to combine biophysical and genetic methods when dealing with prokaryotic channels. "Electrophysiological" recordings of the activities of these channels have been successful in only a few microbes. One also needs to be cautious in extending in vitro biophysical characteristics to understanding their role in "physiology." A major parameter missing in electrophysiological investigation of microbial channels is the native membrane potential. Optical methods of measuring membrane potential in vivo hold promise. Genetics can sometimes be used to advantage. As reviewed above, forward genetics has been useful in the study of MscL, TOK1, and Kch. On the other hand, the searches for the in vivo phenotype of mutants lacking the channels have so far been difficult. That double mutations are required to reveal the functions of the mechanosensitive channels and the Cl^- channels emphasizes the importance of redundant elements. In the case of Kch, for example, a "synthetic lethal" method may be needed to discover the redundant function (Bernhardt and de Boer, 2004).

While we encourage a multidisciplinary attack on prokaryotic channels, we also wish to emphasize the importance of respecting the hard-earned knowledge in traditional neurobiology and microbiology. Scientific traditions most often have rational bases. For example, unless rigid controls are used, bacteriologists do not try to fathom the physiological role of a gene by simply overexpressing it. This is because overexpression itself is harmful to the bacterium and causes pathologies by the congestion of misfolded or misplaced peptides and not by the activities of the increased number of normal protein cells as one might hope. Certain plasmids designed for massive production toward crystallization are unsuitable for physiological examination. Conclusions drawn from such overexpression experiments (Ungar et al., 2001; Munsey et al., 2002) need to be evaluated with care.

Acknowledgments. Work in our laboratory is supported by the Vilas Trust (C.K.) and NIH grant GM54867 (Y.S.).

Addendum in Proof. An update on the survey of prokaryotic K^+ channel genes has been submitted (M. M.-C. Kuo et al., submitted for publication).

REFERENCES

Bakker, E. P. 1993. *Alkali Cation Transport Systems in Prokaryotes*. CRC Press, Boca Raton, Fla.

Bernhardt, T. G., and P. A. J. de Boer. 2004. Screening for synthetic lethal mutants in *Escherichia coli* and identification of EnvC (YibP) as a periplasmic septal ring factor with murein hydrolase activity. *Mol. Microbiol.* **52:**1255–1269.

Chang, G., R. H. Spencer, A. T. Lee, M. T. Barclay, and D. C. Rees. 1998. Structure of the MscL homolog from *Mycobacterium tuberculosis:* a gated mechanosensitive ion channel. *Science* **282:**2220–2226.

Derst, C., and A. Karschin. 1998. Evolutionary link between prokaryotic and eukaryotic K^+ channels. *J. Exp. Biol.* **201:**2791–2799.

Dessen, P., M. Zagulski, R. Gromadka, H. Plattner, R. Kissmehl, E. Meyer, M. Betermier, J. E. Schultz, J. U. Linder, and R. E. Pearlman. 2001. *Paramecium* genome survey: a pilot project. *Trends Genet.* **17:** 306–308.

Doyle, D. A., J. M. Cabral, R. A. Pfuetzner, A. L. Kuo, J. M. Gulbis, S. L. Cohen, B. T. Chait, and R. MacKinnon. 1998. The structure of the potassium channel: molecular basis of K^+ conduction and selectivity. *Science* **280:**69–77.

Epstein, W., E. Buurman, D. McLaggan, and J. Naprstek. 1993. Multiple mechanisms, roles and controls of K^+ transport in *Escherichia coli*. *Biochem. Soc. Trans.* **21:**1006–1010.

Epstein, W., and M. Davies. 1970. Potassium-dependent mutants of *Escherichia coli* K-12. *J. Bacteriol.* **101:**836–843.

Epstein, W., and B. S. Kim. 1971. Potassium transport loci in *Escherichia coli* K-12. *J. Bacteriol.* **108:**639–644.

Gould, S. J. 1996. *Full House: the Spread of Excellence from Plato to Darwin.* Three Rivers Press, New York, N.Y.

Gustin, M. C., B. Martinac, Y. Saimi, M. R. Culbertson, and C. Kung. 1986. Ion channels in yeast. *Science* **233:**1195–1197.

Haynes, W. J., K.-Y. Ling, Y. Saimi, and C. Kung. 2003. PAK paradox: *Paramecium* appears to have more K^+-channel genes than humans. *Eukaryotic Cell* **2:**737–745.

Hille, B. 2001. *Ion Channels of Excitable Membranes.* Sinauer Associates, Inc., Sunderland, Mass.

Hirsch, R. E., B. D. Lewis, E. P. Spalding, and M. R. Sussman. 1998. A role for the AKT1 potassium channel in plant nutrition. *Science* **280:**918–921.

Iyer, R., T. Iverson, A. Accardi, and C. Miller. 2002. A biological role for prokaryotic ClC chloride channels. *Nature* **419:**715–718.

Jiang, Y. X., A. Lee, J. Y. Chen, M. Cadene, B. T. Chait, and R. MacKinnon. 2002. Crystal structure and mechanism of a calcium-gated potassium channel. *Nature* **417:**515–522.

Jiang, Y. X., A. Pico, M. Cadene, B. T. Chait, and R. MacKinnon. 2001. Structure of the RCK domain from the *E. coli* K^+ channel and demonstration of its presence in the human BK channel. *Neuron* **29:**593–601.

Kang, M., A. Moroni, S. Gazzarrini, D. DiFrancesco, G. Thiel, M. Severino, and J. L. Van Etten. 2004. Small potassium ion channel proteins encoded by chlorella viruses. *Proc. Natl. Acad. Sci. USA* **101:** 5318–5324.

Ketchum, K. A., W. J. Joiner, A. J. Sellers, L. K. Kaczmarek, and S. A. Goldstein. 1995. A new family of outwardly rectifying potassium channel proteins with two pore domains in tandem. *Nature* **376:**690–695.

Kumánovics, A., G. Levin, and P. Blount. 2002. Family ties of gated pores: evolution of the sensor module. *FASEB J.* **16:**1623–1629.

Kung, C., and P. Blount. 2004. Channels in microbes: so many holes to fill. *Mol. Microbiol.* **53:**373–380.

Kuo, M. M.-C., Y. Saimi, and C. Kung. 2003. Gain-of-function mutations indicate that *Escherichia coli* Kch forms a functional K^+ conduit in vivo. *EMBO J.* **22:**4049–4058.

Kwak, J. M., Y. Murata, V. M. Baizabal-Aguirre, J. Merrill, M. Wang, A. Kemper, S. D. Hawke, G. Tallman, and J. I. Schroeder. 2001. Dominant negative guard cell K^+ channel mutants reduce inward-rectifying K^+ currents and light-induced stomatal opening in Arabidopsis. *Plant Physiol.* **127:**473–485.

Lesage, F., E. Guillemare, M. Fink, F. Duprat, M. Lazdunski, G. Romey, and J. Barhanin. 1996. A pH-sensitive yeast outward rectifier K^+ channel with two pore domains and novel gating properties. *J. Biol. Chem.* **271:**4183–4187.

Ling, K.-Y., W. J. Haynes, L. Oesterle, C. Kung, R. R. Preston, and Y. Saimi. 2001. K^+-channel transgenes reduce K^+ currents in *Paramecium*, probably by a post-translational mechanism. *Genetics* **159:** 987–995.

Loukin, S. H., J. Lin, U. Athar, C. Palmer, and Y. Saimi. 2002. The carboxyl tail forms a discrete functional domain that blocks closure of the yeast K^+ channel. *Proc. Natl. Acad. Sci. USA* **99:**1926–1930.

Loukin, S. H., and Y. Saimi. 1999. K^+-dependent composite gating of the yeast K^+ channel, TOK1. *Biophys. J.* **77:**3060–3070.

Loukin, S. H., and Y. Saimi. 2002. Carboxyl tail prevents yeast K^+ channel closure: proposal of an integrated model of TOK1 gating. *Biophys. J.* **82:**781–792.

Loukin, S. H., B. Vaillant, X. L. Zhou, E. P. Spalding, C. Kung, and Y. Saimi. 1997. Random mutagenesis reveals a region important for gating of the yeast K^+ channel YKC1. *EMBO J.* **16:**4817–4825.

Milkman, R. 1994. An *Escherichia coli* homolog of eukaryotic potassium channel proteins. *Proc. Natl. Acad. Sci. USA* **91:**3510–3514.

Munsey, T. S., A. Mohindra, S. P. Yusaf, A. Grainge, M. H. Wang, D. Wray, and A. Sivaprasadarao. 2002. Functional properties of Kch, a prokaryotic homologue of eukaryotic potassium channels. *Biochem. Biophys. Res. Commun.* **297:**10–16.

Nee, S. 2004. More than meets the eye. *Nature* **429:**804–805.

Ou, X. R., P. Blount, R. J. Hoffman, and C. Kung. 1998. One face of a transmembrane helix is crucial in mechanosensitive channel gating. *Proc. Natl. Acad. Sci. USA* **95:**11471–11475.

Pennisi, E. 2004. The biology of genomes meeting: surveys reveal vast numbers of genes. *Science* **304:**1591.

Preston, R. R., J. A. Kink, R. D. Hinrichsen, Y. Saimi, and C. Kung. 1991. Calmodulin mutants and Ca^{2+}-dependent channels in *Paramecium*. *Annu. Rev. Physiol.* **53:**309–319.

Reid, J. D., W. Lukas, R. Shafaatian, A. Bertl, C. Scheurmann-Kettner, H. R. Guy, and R. A. North. 1996. The *S. cerevisiae* outwardly-rectifying potassium channel (DUK1) identifies a new family of channels with duplicated pore domains. *Receptors Channels* **4:**51–62.

Roosild, T. P., S. Miller, I. R. Booth, and S. Choe. 2002. A mechanism of regulating transmembrane potassium flux through a ligand-mediated conformational switch. *Cell* **109:**781–791.

Rossmann, M. G., D. Moras, and K. W. Olsen. 1974. Chemical and biological evolution of a nucleotide-binding protein. *Nature* **250:**194–199.

Saimi, Y., and C. Kung. 2002. Calmodulin as an ion channel subunit. *Annu. Rev. Physiol.* **64:**289–311.

Ungar, D., A. Barth, W. Haase, A. Kaunzinger, E. Lewitzki, T. Ruiz, H. Reilander, and H. Michel. 2001. Analysis of a putative voltage-gated prokaryotic potassium channel. *Eur. J. Biochem.* **268:**5386–5396.

Woese, C. R. 1994. There must be a prokaryote somewhere—microbiology's search for itself. *Microbiol. Rev.* **58:**1–9.

Zhou, X. L., B. Vaillant, S. H. Loukin, C. Kung, and Y. Saimi. 1995. YKC1 encodes the depolarization-activated K+ channel in the plasma membrane of yeast. *FEBS Lett.* **373:**170–176.

Bacterial Ion Channels and Their Eukaryotic Homologs
Edited by A. Kubalski and B. Martinac

Chapter 2

The Ktn Domain and Its Role as a Channel and Transporter Regulator

Ian R. Booth, Michelle D. Edwards, Banuri Gunasekera, Chan Li, and Samantha Miller

There is an increasing recognition that, in the bacteria and the archaea, K^+ movement across the membrane is controlled by systems that possess a semiconserved regulatory domain called Ktn or RCK (Biggin et al., 2000; Choe and Roosild, 2002). The proteins possessing this domain include well-defined "P-loop" channels, uptake systems, and specific efflux systems. The Ktn domain is also represented in K^+ transporters and channels in higher organisms (Jiang et al., 2001). In addition, this family has now been found to include a class of less well-defined transporters and potential signaling molecules that are structurally related, not just in their Ktn domains but also in their membrane domains, to one specific branch of this family, KefC (Booth, 2003; Fujisawa et al., 2004; Jiang et al., 2001; Wei et al., 2003). The KefC efflux system is one of the best understood Ktn-regulated systems, both in terms of its integration with cell physiology and the involvement of the Ktn domain with gating (Roosild et al., 2002). Other systems, such as the K^+ uptake systems KtrAB and TrkAEH, can also advance claims to be well characterized, but the understanding of the gating here is less well advanced (Durell et al., 1999; Holtmann et al., 2003; Nakamura et al., 1998; Schlosser et al., 1993; Tholema et al., 1999). In this chapter, we describe the Ktn domain, briefly consider the other bacterial systems that possess Ktn and Ktn-related domains, and, finally, present a more detailed analysis of our understanding of the structure and operation of the KefC system.

STRUCTURE OF THE Ktn DOMAIN

Ktn domains are found in a variety of channels, transporters, and efflux systems (Table 1) and are known to form ligand-binding domains. The domain can be found as an integral part of the main channel-forming protein (e.g., MthK, Kch, and KefC) (Jiang et al., 2001; Jiang et al., 2002a; Munro et al., 1991), as an ancillary extrinsic membrane protein (KtrA,

Ian R. Booth, Michelle D. Edwards, Banuri Gunasekera, Chan Li, and Samantha Miller • School of Medical Sciences, Institute of Medical Sciences, University of Aberdeen, Foresterhill, Aberdeen AB25 2ZD, United Kingdom.

Table 1. Channels and transporters utilizing Ktn domains

System	Role	Category	Ktn domain[a]		Ancillary subunit
			Fixed	Separate	
MthK	K^+ uptake[b]	Channel	Yes	Yes	–
Kch	Not known	Channel	Yes	Yes	–
KefC	K^+ efflux	Channel	Yes	No	KefF
TrkAEH	K^+ uptake	Transporter	No	Yes	TrkE
KtrAB	K^+ uptake	Transporter	No	Yes	–
AmhMT	K^+/NH_4^+ efflux	Regulated pore	No	Yes	–
YhaUST	K^+ efflux	Regulated pore	No	Yes	YhaS
RosB	K^+ efflux	Antiport[c]	Yes	No	–

[a]The Ktn domain is defined as "fixed" or "separate" to denote whether it is a part of the channel or transporter ion-translocating domain.
[b]Although not formally demonstrated, the observation that MthK is an inwardly rectified channel would be consistent with a role in K^+ uptake.
[c]Note that the role of RosB as an antiport has only been inferred from other data and not formally demonstrated by experiment (Bengoechea and Skurnik, 2000).

TrkA) (Nakamura et al., 1998; Schlosser et al., 1993), and as a protein attached to the membrane by a single transmembrane span (AmhM in *Bacillus pseudofirmus*) (Fujisawa et al., 2004) (Fig. 1). The protein is usually synthesized with a single Ktn domain, but covalent dimers, e.g., *Escherichia coli* TrkA and the mammalian BK channel, are also known and characterized (Jiang et al., 2001; Schlosser et al., 1993). In general, the Ktn domains form two-thirds of the full length of the domains or proteins that regulate K^+ movement. Ktn domains have at their core a Rossman fold (Moras et al., 1975), which exhibits variable conservation across the family (Jiang et al., 2002a; unpublished data). In the K^+ channels Kch and MthK, a degenerate Rossman fold is observed and this is shared with the recently described members of the RosB/AmhMT family of ion transporters (see below) (Fig. 1). However, there is a strongly conserved intradomain salt bridge between the end of the α4-helix and the start of the β6-sheet (Jiang et al., 2001). The Rossman fold has been characterized in many proteins and has been adapted to provide a binding site for ligands as diverse as NAD(H), leucine, Ca^{2+}, and Mg^{2+} (Jiang et al., 2001; Jiang et al., 2002a). The binding site is created by the carboxy-terminal ends of the β-sheets that lie at the core of the domain. In the case of the transporters TrkAEH and KtrAB, the fold provides a binding site for nucleotides (Roosild et al., 2002; Schlosser et al., 1993). Similarly, the Ktn domain of KefC can bind NADH (T. Roosild, personal communication), although cells can be extensively depleted of this nucleotide without altering K^+ efflux activity (McLaggan et al., 2000). In the MthK crystal structure, Ca^{2+} was observed bound to the Ktn domain, coordinated by three acidic groups (E212, D184, and E210) that lie below the Rossman fold (Jiang et al., 2002a). No ligand was detected attached to the RCK domain of Kch (Jiang et al., 2001).

Four crystal structures of Ktn domains have been published, with resolutions ranging from 2.3 to 3.3 Å (Jiang et al., 2001; Jiang et al., 2002a; Roosild et al., 2002). Several of the crystal structures are of truncated proteins, since sequences carboxy-terminal to the Ktn domain were often removed to aid the production of high-quality crystals. In particular, the last ~70 residues were removed from the carboxy-terminal ends of the *Methanococcus jannaschii* TrkA and *Bacillus subtilis* KtrA Ktn domains to obtain good crystallization

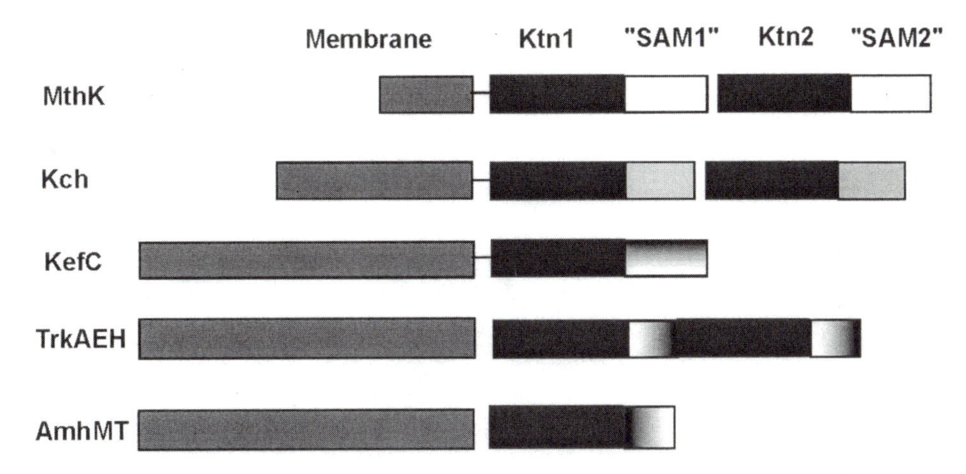

Figure 1. The organization of Ktn-bearing channels and transporters. The structures are shown schematically, with the sizes of the domains approximately to scale according to the numbers of amino acids. Key: dark gray, membrane domain; black, Ktn domain; white or shaded, SAM domains; black line, linker that covalently attaches Ktn to the membrane domain. Note that TrkA has two covalently linked Ktn and SAM domains. The numbers 1 and 2 refer to the first and second domains throughout (see text for details).

(Roosild et al., 2002). In addition, there are regions of the proteins that are disordered in the crystal structures. MthK gives the most complete structure, but the linker sequence is disordered. Both MthK and Kch are informative about the organization of sequences carboxy-terminal to Ktn (the so-called SAM domain) (Jiang et al., 2001; Jiang et al., 2002a). The SAM domain in KefC plays an essential role in transmitting the gating signal (S. Miller, C. Li, B. Gunasekera, and I. R. Booth, unpublished data) (see below).

Ktn domains form dimers, tetramers, and octamers utilizing two different dimerization interfaces. A helix-strand-helix motif (α5-β6-α6) enables dimerization of Ktn domains (Jiang et al., 2001; Jiang et al., 2002a; Roosild et al., 2002). The α6-helices are tightly associated with the opposite Ktn domain leading to ~1,500 to 1,800 Å^2 of buried surface area (Jiang et al., 2001; Roosild et al., 2002). However, the two Ktn domains are flexibly connected since there is only a single protein strand, the β6-α6 loop, joining each Ktn subunit to its α6-helix (Color Plate 1 [see color insert]). These short sequences form the hinge that allows the Ktn conformational flexibility that may arise from the occupation of the Rossman fold by different ligands (Roosild et al., 2002). In all four structures the Rossman folds are oriented toward each other across the hinge region and thus binding of a ligand in the Rossman fold has the potential to alter the hinge angle.

Higher-order structures (tetramer and octamer) can be constructed by the interaction at hydrophobic patches created by the α4- and α5-helices. Most Ktn proteins that are associated with transporters and channels have hydrophobic residues placed to create a patch that upon dimerization may bury up to ~800 Å^2 (Jiang et al., 2001; Roosild et al., 2002). Expression of a mutation M81R, at the center of the patch in the *M. jannaschii* Ktn domain, led to the production of insoluble aggregated protein (Roosild et al., 2002), and a similar observation was made with the RCK domain of the Kch channel (Jiang et al., 2001). Full-length KtrA and TrkA proteins are tetrameric in free solution at neutral pH in the presence of their ligands, NAD^+ and NADH, respectively (Roosild et al., 2002). Removal of the ligand and/or changes in pH of the buffer caused the proteins to precipitate, consistent with significant changes in conformation. These and similar observations led to the proposal of a conformational switch governed by ligand binding (see below).

Octameric assemblies, seen in MthK and Kch, use the same hydrophobic patches to create the higher-order oligomers but also utilize a novel expression system to generate separate as well as channel-attached Ktn domains. A soluble Ktn monomer is translated, via an internal ATG codon and ribosome-binding site, from the same mRNA from which it is also synthesized as an integral part of the membrane protein (Jiang et al., 2002a; Kuo et al., 2003b). In the MthK crystal, two tetrameric channel proteins have assembled via back-to-back dimerization of Ktn domains utilizing the two dimerization interfaces described above (Color Plate 2 [see color insert]). In cells it is envisaged that the extra Ktn domains, synthesized by intra-mRNA initiation of translation, would assemble in an analogous fashion to create an octamer of Ktn domains—four covalently attached to the channel-forming domains and four assembled onto this template by dimerization with individual channel-attached Ktn domains. This structure is in part stabilized by the dimerization of the SAM domain (Color Plate 2).

The separate Ktn domain was detected in cells overexpressing either the MthK protein (Jiang et al., 2002a) or the *E. coli* Kch channel (Kuo et al., 2003b). Recent work has demonstrated that the extra domain is not essential for activity. Mutant Kch channels were obtained that imposed a K^+ leak on *E. coli* cells due to a mutation in the Ktn domain (N325S, C312F, or S331P). The phenotype was not reversed when the secondary translation start site was removed by an M240L mutation. Western blotting with peptide-specific antisera, designed against the Ktn domain, established that the full-length Kch protein was still produced but that the independent Ktn domain was now absent (Kuo et al., 2003b) (see also chapter 1). Work in the author's laboratory has found that the extra Ktn domain appears to only be present when the *kch* gene is strongly overexpressed. When the separate Ktn domain is synthesized, it forms a stable association with the channel that is resistant to mild detergents, presumably by the formation of octameric structures as predicted by the MthK structure (R. Araujo, M. D. Edwards, P. Fraser, and I. R. Booth, unpublished data). No evidence for a separate expression of the Ktn domain has been observed with the KefC system (unpublished data).

Ktn domains may have different topologies with respect to their membrane partners. In MthK, and potentially Kch, the Ktn domains oligomerize such that their hinged regions are parallel to the axis of the channel pore and held at an angle of approximately 45° to the surface of the membrane (Jiang et al., 2002a). This assembly creates a hole that is coaxial with the channel pore, but it is not clear whether ions must traverse this extension or can pass sideways through portals created by the linker sequences that attach the Ktn domain to the channel domain (Color Plate 2). For other systems, such as KtrAB, YhaSTU, and TrkAEH, the separate expression of the Ktn domain that is intrinsic to these systems leaves possibilities of dimeric, tetrameric, and octameric assemblies (Choe and Roosild, 2002).

In KefC there is no evidence for a structure greater than the dimer, but there is an additional factor, the YabF (KefF) protein, that is required for full activity of the system (Miller et al., 2000). KefF is a 20-kDa homolog of human quinone oxidoreductase (QOR). Crystal structures of the human protein show QOR to be dimeric (Faig et al., 2000; Li et al., 1995), and the dimer interface appears to be conserved in KefF. Unpublished data suggest that KefF forms an intimate connection with the KefC Ktn domain in close proximity to the Rossman fold. KefC Ktn domains form inclusion bodies when expressed in the absence of KefF, and the KefC carboxy-terminal domain copurifies with KefF. Recent data suggest that hydrophobic regions may be exposed when KefC is expressed in the absence of KefF. The strongest candidates for these regions are the same α4- and α5-helices adjacent to the Rossman fold, despite these regions being less obviously hydrophobic than those in TrkA and KtrA. Thus, the homotetramer of Ktn domains that may be the central design of some K^+ channels, of KtrAB and TrkAEH, may be replaced by a heterotetramer of complexes of KefC dimers associated with dimers of KefF.

ROLE OF Ktn DOMAINS IN REGULATION OF TRANSPORT

Ktn domains have been implicated in regulation in a variety of systems. The earliest evidence came from the analysis of KefC (Miller et al., 1997; Ness and Booth, 1999) but has subsequently been supported by the analysis of Kch (Kuo et al., 2003b), YhaT (Fujisawa et al., 2004), and AmhM (Wei et al., 2003). KefC mutants were selected that increased the rate of K^+ leakage from cells and the sites of the mutations mapped (see below). The mutations were mapped to regions of the Ktn domain, some within the Rossman fold and others in the SAM domain (Miller et al., 1997). Mutations at similar positions were recently shown to alter the regulation of the Kch channel (Kuo et al., 2003b). During the analysis of the *B. subtilis* YhaUST system, which encodes a putative K^+ efflux system, and its homolog AmhM-AmhT in *B. pseudofirmus* strain OF4, it was demonstrated that deletion of YhaT or AmhM, which encode the Ktn domains, leads to enhanced K^+ loss from *E. coli* cells in which the partner membrane protein (YhaU and AmhT, respectively) was expressed (Fujisawa et al., 2004; Wei et al., 2003). In contrast, loss of the double Ktn domain, TrkA, which is an extrinsic component of the TrkAEH K^+ uptake system in *E. coli*, led to loss of ability to accumulate K^+ (Bossemeyer et al., 1989a; Epstein and Kim, 1971; Epstein, 2003). Clearly, therefore, Ktn domains regulate K^+ movement in both directions across the membrane, influx and efflux.

Potassium Transporters That Utilize Ktn Domains

Potassium uptake systems in bacteria essentially fall into three categories: K^+-transporting primary ATPases, such as Kdp; secondary transporters, such as Kup; and more complex systems, e.g., TrkAEH in *E. coli* and KtrAB in *B. subtilis*. Although the details of their mechanisms and their modes of regulation remain unclear, each shows some level of control over its activity.

1. Kdp is the most intensively studied and consequently one of the best understood K^+ transport systems (Bossemeyer et al., 1989b; Epstein, 2003; Rhoads et al., 1978). It is a P-type ATPase that drives the inward movement of K^+ at the expense of ATP hydrolysis that proceeds via a phosphorylated protein intermediate (Altendorf and Epstein, 1994; Altendorf et al., 1998). The system has been overexpressed and purified, and its biochemistry and genetics are well documented.

The KdpA subunit has been proposed to be folded into a channel-like organization, whereas KdpB is the ATPase and undergoes K^+-dependent reversible phosphorylation (Durell et al., 2000). The role of the KdpF and KdpC subunits is still not clear. In essence, this system serves to scavenge K^+ ions when they are only available at low concentrations. Kdp activity is primarily controlled at the level of expression, which is regulated in response to cell turgor and possibly also senses K^+ directly. There is also evidence that the activity of the Kdp system is inhibited by high external K^+, thus preventing flooding of the cell with K^+ in response to a sudden increase in the K^+ concentration in the environment (Rhoads and Epstein, 1978; Roe et al., 2000).

2. The Kup system is a 622-residue protein that consists of two domains, a 440-residue membrane protein and a carboxy-terminal soluble domain believed to reside in the cytoplasm (Bossemeyer et al., 1989b; Schleyer and Bakker, 1993). The system is responsible for Cs^+ sensitivity of *E. coli* cells, and this is the major route for uptake of this ion into *E. coli*. The carboxy-terminal domain is of unknown function, but it is implicated in the active conformation of the proteins since its removal leads to a significant loss of activity (Schleyer and Bakker, 1993). Homologs of Kup are found in a wide range of bacteria, in a fungus, and in plants. The length of the bacterial proteins is strongly conserved, and the only significant variations are at the amino-terminal end. From this it can be inferred that the carboxy-terminal domain is critical to the function of the transport system.
3. The *E. coli* Trk system has been subject to genetic analysis and is known to consist of at least three components encoded by the *trkA*, *trkE*, and either *trkG* or *trkH* genes (Epstein, 2003). TrkG and TrkH are membrane proteins that are conduits through which K^+ ions enter the cell (Schlosser et al., 1995). They are essentially duplicate copies of the same gene that has diverged. The *trkG* gene was acquired via prophage integration into the *E. coli* genome. The two systems (TrkAEH and TrkAG) display complex relationships to the energy status of the cell (Rhoads and Epstein, 1977; Stewart et al., 1985). They can be energized by the proton motive force but only in the presence of metabolism that generates ATP. The TrkE protein resembles an ATP-binding cassette ATPase domain and may be responsible for coupling the binding or hydrolysis of ATP to activation of Trk (Harms et al., 2001). However, TrkG does not require TrkE for activity, although absence of TrkE alters the kinetics of K^+ transport.

The specific interest in the Trk system derives from the fact that TrkG and TrkH can be folded in silico to create four K^+ selectivity loops, each separated by two transmembrane spans to create a structure equivalent to that of the K^+ channel KcsA (Durell et al., 1999). TrkA, which contains a dimer of Ktn domains (Bossemeyer et al., 1989a) (Fig. 1), docks onto the membrane, probably associating directly with TrkH or TrkG (Hamann et al., 1987). TrkA has been shown to bind NADH (Schlosser et al., 1993), and this provides a plausible mechanism for linking K^+ transport to metabolism (Roosild et al., 2002).

The Ktr systems are simpler but are considered to have the same potential for the membrane protein KtrB to fold into a channel-like conformation (Durell et al., 1999; Nakamura et al., 1998; Tholema et al., 1999). KtrB is proposed to interact with Ktn domains of the KtrA protein, but little is known about the interactions between these proteins. However, it is clear that these systems have in common with the *B. subtilis* YhaSTU system the interaction between a separately encoded Ktn-bearing protein with a membrane pore that may have channel-like properties, and thus similarities may exist in their mechanisms of gating.

CPA2: Cation-Proton Antiports, Pores, and Channels

The CPA2 family of proteins was originally defined around the monovalent cation-proton antiports but is now thought to contain channels as well as transporters (Saier et al., 1999). This diversity among a family of proteins with common organization at the level of hydrophobicity should not be surprising given the observation that the bacterial ClcA, a homolog of mammalian Clc-0 chloride channels, is a $2Cl^-/H^+$ antiport (Accardi et al., 2004). Sequence similarity is at best a crude guide to common evolutionary origin and a poor guide to mechanism as is exemplified by consideration of the breadth of function among CPA2 family members. The smallest members of this family are the NapA-related Na^+/H^+ antiports (Reizer et al., 1992). These proteins are ~380 amino acids long and contain 10 to 12 transmembrane spans. The sequence similarity between these proteins and the members of CPA2 that concern us here, namely the KefC and RosB families, is mainly in the amino-terminal 180 residues. It seems likely that the proteins evolved by gene duplication of this 180-residue element followed by evolution of the sequences toward the functions that the proteins display today (Saier et al., 1999). Alignment of the amino-terminal and carboxy-terminal halves of the membrane domain of these proteins for any member of this family shows significant sequence similarity consistent with an ancestral duplication event. This pattern is maintained in larger analyses of multiple sequences. The NapA branch is the simplest in this family, and those that have been studied biochemically appear to encode Na^+/H^+ and K^+/H^+ antiports (Ramirez et al., 1998; Reizer et al., 1992; Southworth et al., 2001).

The larger members of the CPA2 family (KefC, MagA, and RosB) are distinguished not only by the possession of an ~200-residue carboxy-terminal domain that contains the Ktn domain but also by the divergence of the amino acid sequence that corresponds to the second half of the membrane domain. In the KefC family of proteins this region is the most highly conserved and contains sequences that are important for the regulation of the activity. RosB (Bengoechea and Skurnik, 2000) and MagA (Nakamura et al., 1995) have been less intensively studied than KefC, and only limited information is available on sequences critical for activity and regulation. The linker sequences that attach the Ktn domains to the membrane protein are not strongly conserved in length or in sequence but are often strongly hydrophilic. The main factors used to separate the KefC and RosB branches of the family were based on sequence divergence, but increasingly they are seen to be mechanistically distinct.

The first member of the RosB branch to be described was the MagA transporter that enables Fe^{2+} uptake into membrane vesicles derived from a *Magnetospirillum* sp. (Nakamura et al., 1995). The protein is one of the shortest members of the family at 434 residues, truncation of the carboxy-terminal domain being responsible for the shorter length and leading to loss of the Ktn domain. In *E. coli* membrane vesicles it catalyzes the accumulation of Fe^{2+} in a proton motive force-dependent manner. Little is known about the mechanism of the transporter. Larger homologs, which retain a Ktn domain, have been identified in a number of bacteria and have been reported to have cation-transporting activity involving K^+ or NH_4^+ rather than Fe^{2+} (Fujisawa et al., 2004; Wei et al., 2003). Antiport activity has been reported to be associated with the *Yersinia* sp. homolog, RosB (Bengoechea and Skurnik, 2000). In *E. coli* the RosB homolog, YbaL, has been analyzed by both overexpression and deletion without an obvious phenotype being noted (C. Waugh and I. R. Booth, unpublished data). However, a clearer picture of the potential activity of these systems is emerging from their analysis in *Bacillus* species (Fujisawa et al., 2004; Wei et al., 2003). Here the analysis has been aided by the separation of the membrane and Ktn domains into individual proteins. The *yhaSTU* operon from *B. subtilis* encodes three proteins: YhaS, YhaT, and YhaU. YhaT is a degenerate Ktn domain, whereas YhaU is a homolog of the membrane domain. When the YhaU protein was cloned and expressed in *E. coli* strains lacking either the cation/proton antiports or the K^+ uptake systems, no antiport activity was detected, but expression introduced a K^+ leak into the strain that lacked K^+ uptake activities. The K^+ leak could be blocked by coexpression of YhaS and YhaT proteins with YhaU. Neither YhaS nor YhaT alone was sufficient to completely regulate the YhaU protein. In *B. subtilis* expression of the *yhaSTU* operon was induced at alkaline pH (Fujisawa et al., 2004). Previous analysis of a homologous system, AmhM-AmhT in *B. pseudofirmus* strain OF4, led to the proposal that the system forms a K^+/NH_4^+ pore that is regulated by the Ktn domain-bearing AmhM protein (Wei et al., 2003). The system is required for optimum growth at alkaline pH. Expression of the AmhT protein in *E. coli*, in the absence of AmhM, led to a K^+ leak that was blocked by coexpression of AmhM. A low level of K^+ uptake was observed via AmhT into K^+-depleted *E. coli* cells and could be blocked by NH_4^+ ions. In *B. pseudofirmus* the deletion of AmhM caused altered growth patterns and was associated with altered NH_4^+ pools (Wei et al., 2003). Clearly this is an interesting family of proteins. Not only do they create regulated pores, but, as with the TrkAEH and KtrAB K^+ uptake systems, they are capable of coupling a separate Ktn domain to the membrane protein to maintain the pore in a closed state. One of the key unresolved questions remaining is, What is their mechanism of activation?

KefC: K^+ Efflux Gated by Glutathione and Glutathione Adducts

KefC is the best documented and understood of the Ktn-bearing systems and was used as a test bed to examine some of the ideas arising from the structure of the Ktn domains. The system's discovery arose from three observations in *E. coli*. (i) *N*-ethylmaleimide (NEM) was observed to elicit rapid K^+ efflux from *E. coli* cells that could be reversed by reducing agents (Meury et al., 1980). This reversibility, which occurred in the absence of protein synthesis, led to the proposal that a thiol-containing peptide, or other nonprotein molecule, might regulate K^+ retention by *E. coli* cells. (ii) The analysis of *E. coli gshA* mutants, which are glutathione (GSH) deficient, demonstrated that these cells spontaneously leak K^+ and exhibit high rates of K^+ exchange (Meury and Kepes, 1982). The rate of K^+ leak was much slower than the efflux observed with NEM, but it was observed that this electrophile did not further stimulate K^+ efflux in *gshA* mutants (Elmore et al., 1990). GSH is a tripeptide, γ-glutamyl-cysteinyl-glycine, that is present in most gram-negative bacteria and accumulates to ~10 mM in the cytoplasm. Although a tripeptide, the presence of the γ-linkage inserts the side chain of glutamate into the length of the peptide, extending the distance between the Gly and Glu α-carboxyl groups to approximately 12 Å (see below). GSH was therefore a major candidate for the regulator of K^+ retention. (iii) Two mutant loci, *trkB* and *trkC*, that caused high rates of spontaneous K^+ loss were identified as encoding separate K^+ efflux systems responsible for both the GSH-gated K^+ efflux and the NEM-activated loss of K^+ (Bakker et al., 1987). The systems were subsequently renamed KefB and KefC (for $\underline{K}^+$ $\underline{ef}$flux), respectively. The structural genes were cloned, leading to the recognition of the presence of the Rossman fold domains at their carboxy terminus (Munro et al., 1991). Subsequent analysis has illuminated the regulation and physiological role of these systems (Ferguson et al., 1998).

KefC and KefB are complex multidomain proteins that are probably dimers. KefC is a 620-residue protein organized in three major domains: the membrane domain (residues 1 to 380), the Ktn domain (401 to 530), and the SAM domain (531 to 620). An ~19-residue linker connects the membrane domain to the Ktn domain (Fig. 2). The hydrophobicity plot reveals 10 to 12 regions that could form transmembrane (TM) segments. The analysis is complicated by the observation that three TM segments are not consistently hydrophobic: TM7 (residues 191 to 226) resembles a voltage sensor from a conventional K^+ channel, having the core structure $(Rxxx)_{4-5}$ where x is a hydrophobic residue, and TM11 and TM12 exhibit a mixed hydrophobic-hydrophilic character. KefB is shorter (601 residues) but is similar in sequence (42% identity and 64% similarity; e^{-136}). The major differences between the two proteins are, first, that KefB is more strongly activated by methylglyoxal (MG) than is KefC (Ferguson et al., 1993; MacLean et al., 1998). Second, a strong K^+ leak results from overexpression of KefB, whereas KefC can be expressed at high level without significantly influencing the retention of K^+ (Ness and Booth, 1999). At the present time there are no clear insights into the location of the pore-lining helices, and even the mechanism of ion translocation remains in doubt. The earliest measurements in whole cells demonstrated that K^+ efflux was accompanied by Na^+ and H^+ influx, such that each K^+ lost was balanced by a net of 0.7 Na^+ ions and 0.3 H^+ ions (Bakker and Mangerich, 1982). It was assumed that the system encoded a K^+/H^+ antiport with Na^+ entry taking place via reversal of the Na^+/H^+ antiport (NhaA or NhaB) (Padan and Schuldiner, 1994). However, these ion movements are measured indirectly on whole cells and thus represent the summation of ion fluxes rather than mechanistic ion movements, and it would have been equally legitimate to propose a K^+/Na^+ or H^+ antiport or indeed a channel. Subsequent more detailed analysis suggested that the rate of K^+ movement was close to that of an ion channel: some channel mutants display very fast K^+ efflux that would be difficult to reconcile with antiport function (Miller et al., 1997; Ness and Booth, 1999). To date it has not proved simple to provide experimental support for either mechanism of ion transport. We refer to the system as a channel because it displays many of the characteristics of these systems.

Analysis of mutations in KefC and KefB identified four regions that are critical for the gating of K^+ efflux (Miller et al., 1997; Ness and Booth, 1999) (Fig. 2). Mutations were

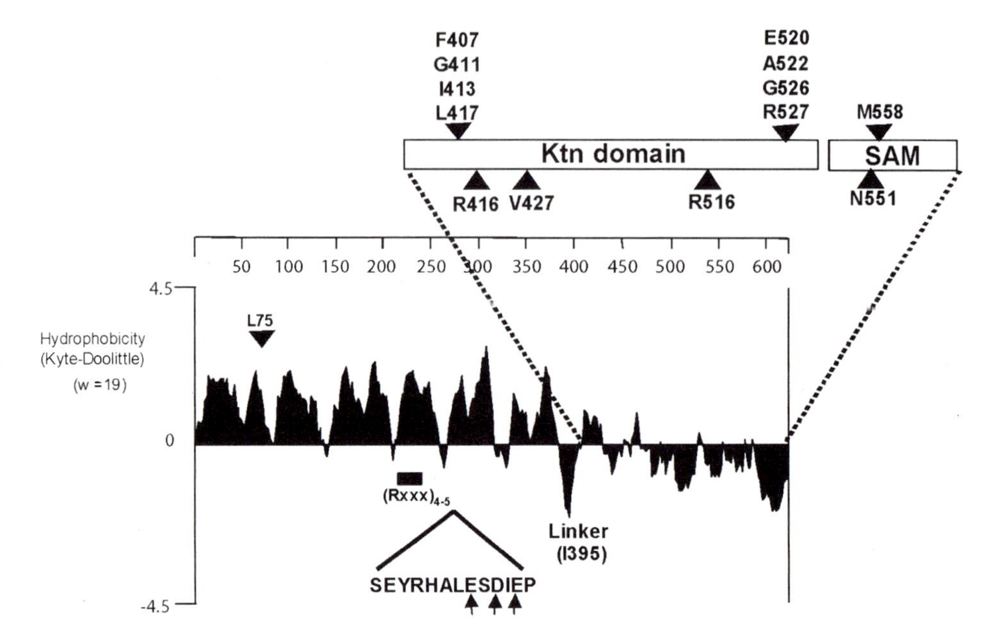

Figure 2. Structure and mutations in KefC. The figure depicts the hydrophobicity of KefC (Kyte and Doolittle, 1982) and the locations of the different structural features. The positions of mutations that alter the gating of the KefC channel are indicated (Miller et al., 1997; Ness and Booth, 1999; Roosild et al., 2002; unpublished data). Residues indicated above the Ktn and SAM domains inhibit KefC function when mutated and those below the domain increase spontaneous activity. For the linker the single diminished function mutation is shown in brackets. The conserved sequence between helices 8 and 9 is SEYRHALESDIEP, which extends beyond the core HALESDIEP sequence that has been analyzed previously (Miller et al., 1997; Ness and Booth, 1999; Roosild et al., 2002). Arrows indicate the three critical acidic residues that when mutated alter KefC and KefB activity. The position of the putative voltage sensor is indicated $(Rxxx)_{4-5}$.

originally identified after UV mutagenesis, enrichment for inability to grow on low K^+ (100 μM) agar, and DNA sequencing of the chromosomal genes in the mutants. This analysis identified mutations that lay in an acidic loop (HALESDIEP; residues 259 to 267) between helices 8 and 9 of KefC (D264A, E262K) and in the Ktn domain (V427A, R416S, and R516C). The E262K mutant was subsequently found to be a double mutant, E262K/N551D (C. Li and I. R. Booth, unpublished data). In KefB only a single mutation, L75S, was identified, but subsequent site-directed mutagenesis showed that mutations introduced into the loop between helices 8 and 9 (HELETAIDP) and in the Ktn domain altered the regulation of KefB in a similar way to those in KefC (Ness and Booth, 1999).

Suppression analysis using the *E. coli* KefC mutations V427A, D264A, and R416S led to the identification of mutations in the carboxy-terminal domain that inactivate KefC (see below) but in addition identified the *yabF* gene, described above (Miller et al., 2000). A similar ancillary subunit, YheR, is associated with full activity of KefB. These proteins have been renamed KefF and KefG for YabF and YheR, respectively (Miller et al., 2000). The *yabF* gene encodes a 20-kDa flavoprotein that is required for maximum activity of

KefC. In *E. coli* the *yabF* and *kefC* genes overlap (as do the equivalent *yheR* and *kefB* genes), with the ribosome-binding site for *kefC* being within the *yabF* coding sequence in a region of potential secondary structure (Miller et al., 2000). Coexpression of the two genes is almost certain since some of the mutations that suppress KefC V427A and R416S mutations lie in the promoter region 5′ to *yabF* or insert transcriptional terminators within the *yabF* gene. Most of the mutants in *yabF* inactivate expression of the protein due to insertion of an insertion sequence element in the gene. However, two missense mutations in this gene have also been obtained that suppress the R416S Ktn domain mutation. These mutations specifically block the spontaneous K^+ leak, but not the activation by electrophiles (C. Li, R. Lamb, S. Miller, and I. R. Booth, unpublished data). The KefF and KefG proteins belong to the QOR family of proteins, and KefF has been shown to associate with the carboxy-terminal domain of KefC (T. Roosild, C. Li, S. Choe, and I. R. Booth, unpublished data). The role of the KefF protein appears to be predominantly to fully activate the KefC system, possibly by altering its stability or assembly. We have not been able to detect a specific role in the gating process for this protein since regulation by GSH and electrophiles is normal when expression of the KefC protein takes place in the absence of KefF (B. Gunasekera and I. R. Booth, unpublished data).

Gating of KefC by Glutathione and Glutathione Adducts

In *E. coli gshA* mutants the rate of K^+ loss is not dramatic but can lead to growth inhibition under appropriate circumstances (Elmore et al., 1990). Mutations that eliminate KefB and KefC activity block the leak, and overexpression of either KefB or KefC greatly enhances the rate of K^+ loss. Recent work has shown that all of the regulation by GSH and its adducts resides in the KefC protein, i.e., that it does not require KefF (Gunasekera and Booth, unpublished data). No redox chemistry is directly involved in the gating by GSH since a cysteine-free KefF-KefC complex has been found to retain full activity. Further, ophthalmic acid (γ-glutamylaminobutyrylglycine), which has the central cysteine of GSH replaced by aminobutyrate, blocks the K^+ leak in a GSH mutant (Meury and Kepes, 1982), but the system can no longer be activated by electrophiles, such as NEM. We infer from these observations that GSH binds to KefC in a ligand-specific way to create an inactive conformation of the KefC system.

Full activation of the KefC system is achieved only by the formation of specific GSH adducts. Some electrophiles react spontaneously with GSH (e.g., MG, NEM, and iodoacetate), whereas others require the action of a glutathione *S*-transferase (e.g., 1-chloro-2,4-dinitrobenzene). The best activators of both KefC and KefB are those that modify the cysteine of GSH with a large hydrophobic group (Elmore et al., 1990), although KefB is also strongly activated by the MG adduct, *S*-lactoylglutathione (SLG) (MacLean et al., 1998). Both KefC and KefB have a number of cysteine residues, but their replacement with serine does not significantly alter the regulation of either system, suggesting that either they cannot be modified directly or that the modification of the protein cysteine residues does not induce a change in activity. Similarly, cysteine residues can be introduced at a number of positions without changing KefC activity in the presence of NEM (J. Whyte, S. Miller, and I. R. Booth, unpublished data). Recently, we determined that KefC could be made more readily activated by SLG through the introduction of the gating mutations described above (S. Kinghorn, C. Li, and I. R. Booth, unpublished data). Wild-type

KefC is only mildly activated by SLG, and the effect of this compound is only readily appreciated when KefC is expressed from a multicopy plasmid (MacLean et al., 1998). In contrast, SLG formation elicits rapid K^+ loss from the Ktn-located gating mutants, V427A and R416S. Similar observations have been made with D264A and E262K, which lie in the helix 8-9 loop. Mutations that inactivate the Ktn domain suppress the activation of KefC by SLG (Kinghorn et al., unpublished data). We infer from these observations that, in wild-type KefC, the binding of SLG does not perturb the Ktn structure sufficiently to gate the efflux system. However, in mutants with a weakened interaction between the Ktn domain and the channel forming domain, further perturbation by binding SLG is sufficient to "trip" the channel into the open configuration.

The Structure of the KefC System

All channels and regulated transporters exhibit two critical properties: selectivity and gating. For each system the objective is to determine the structural basis of selectivity and the molecular mechanism of gating. Our understanding of the basis of selectivity for classical K^+ channels and for chloride has been greatly enhanced by the crystal structures for these channels (Doyle et al., 1998; Dutzler et al., 2002; MacKinnon, 2000; Morais-Cabral et al., 2001). However, KefC does not have regions that are immediately obvious equivalents for the K^+ selectivity filter. The generality of gating mechanisms, expressed in terms of an allosteric transition in the Ktn (or equivalent) domains that triggers helix movement in the membrane, has also been established for the classical K^+ channels (Jiang et al., 2002a; Kuo et al., 2003a; Roosild et al., 2002). Much less certainty surrounds the triggering of the allosteric transition and of the precise mechanism of transmission of the signal to the channel domain. As has been demonstrated for the K^+ channels, an understanding of structure amply informs models for gating signal transmission (Jiang et al., 2002a; Kuo et al., 2003a; Roosild et al., 2002). For KefC genetic rather than structural analysis has been most illuminating to date. Mutations at four regions of the protein have been found to influence gating: the linker, the Ktn domain, HALESDIEP, and the SAM domain (Fig. 2).

Linking the Ktn Domain to the KefC Membrane Domain

The membrane domain of KefC is joined to Ktn by a hydrophilic linker that is similar in length to that linking the Ktn domain to the pore-lining helix of MthK. It is generally the case that linkers are designed to facilitate the interactions of domains, and this is the case for the KefC linker. Its length can be increased without impairing function, but shortening the linker even by as little as three residues can lower the overall activity of the system (Gunasekera and Booth, unpublished data). Deletion of either five or seven residues almost completely blocks the system without significantly affecting protein expression. Shortening the linker by five amino acids also impairs the activity of gating mutants, such as D264A and R416E. The linker also appears to contain positional information. The 19-residue linker is composed principally of acidic residues, particularly in the central region (EQSSTEEAREADEIDEEQP), and is not strongly conserved in either length or with regard to specific residues. The effect of deletion of three amino acids depends on which three are removed. Removal of residues 390 to 392 (ΔREA) has the greatest effect, removal of residues 387 to 389 (ΔEEA) has the least effect, and removal of residues 384 to 386 (ΔSST) has an intermediate effect. There is a single basic residue (R390) in the linker that is removed in the ΔREA deletion. It seemed that this residue might play a critical role. However, in a full-length linker an R390E mutant exhibits full activity, suggesting that the basic residue is not essential. In contrast, insertion of an Arg residue in a shortened linker (ΔREA deletion), such that it retains an equivalent position relative to the membrane domain but now 3 Å closer to Ktn, causes a sixfold loss of activity. This argues that charge and context are critical. The data can be rationalized by considering the central part of the linker to form an α-helix, for which there is some support from combined structure prediction algorithms (unpublished data). However, in MthK the linker is disordered in the crystal structure, which implies a lack of strong structure, whereas in KirBac1.1, where the sequence is an α-helix, it is clearly resolved (Jiang et al., 2002a; Kuo et al., 2003a).

If MthK and KefC gating mechanisms were similar, one might expect the linkers to have similar properties, and the observation that the length of the linker in KefC can be increased without impairing function implies that structure might not be important. Other data, however, argue for significant structure close to the boundaries with the membrane and Ktn domains. These two regions exhibit greater sequence conservation than the central region. A pair of mutations in the linker, which were isolated as suppressers of the Ktn V427A mutation in KefC, lie close to Ktn. These mutations resulted in substitution of the sole hydrophobic residue in the linker (I395) with a Ser or Asn, which are helix-breaking residues. The mutations reduced both the spontaneous leak from the V427A mutant and lowered the NEM activation of the channels, both when combined with V427A and when I395N was combined with a wild-type Ktn domain. These data support the importance of the regions proximal to the pore and regulatory domains.

The SAM Domain

The region immediately carboxy-terminal to the exchanged α6-helix is variable both in length and in sequence. The sequence conservation in this region is limited between different groups of Ktn-bearing systems. We have chosen to call this domain SAM, and the sequence in KefC bears no strong similarity to the equivalent regions of TrkA, KtrA, and Kch/BK but is similar to the SAM domain of MthK (Jiang et al., 2002a). The only crystal structure in which this sequence has been resolved is MthK, but the partial structure seen in Kch suggests a similar organization (Jiang et al., 2001; Jiang et al., 2002a). In both structures the peptide backbone undergoes a sharp turn at the end of the exchanged α6-helix, with the α7-helix doubling back on α6 at an acute angle so that it packs against both the Ktn domain and the adjacent SAM domain to create a dimer of SAM domains (Color Plate 1b and c).

The SAM domain plays a critical role in KefC in facilitating the gating of the channel. Deletion of the SAM domain and the α6-helix (ΔE517–S620) does not prevent assembly of the channel but prevents the system from responding to the formation of GSH adducts. Moreover, a point mutation, M558R, which lies within the SAM domain, has been isolated as a suppresser of V427A. In addition, a point mutation in this region, N551D, exhibits a small GSH-insensitive spontaneous K^+ leak but normal activation by NEM. Substitution of Cys, Arg, or Lys at N551 led to complete inactivation of KefC. However, a double mutant that combines N551D with a HALESDIEP mutation, E262K, caused a rapid K^+ leak that is further stimulated by removal of GSH (unpublished data). As will be discussed below, the SAM domain is an important component of the structure of KefC, despite its relative lack of conservation.

Ktn Mutations

A great many mutations have now been identified in the Ktn domain. These mutations fall into two categories: those that increase the frequency of the open state leading to a K^+ leak even in the absence of an electrophile and those that decrease the activation of KefC (Fig. 2). The former include V427A, R416S, and R516C, whereas mutations that inactivate are F407L, G411S, I413A, L417A, E520G, A522V, R527E, and G526V (unpublished data). The positions of these residues have been mapped onto Ktn crystal structures, and they cluster around the Rossman fold or at the interface between the Ktn and SAM domains. Some suppresser mutations have quite specific effects. The F407L suppresser lies within the $\beta1$-$\alpha1$-$\beta2$-$\alpha2$ pocket, which lies at the core of the Rossman fold that is the binding site for NADH (Color Plate 3 [see color insert]). The substitution is likely to restore the correct fold in the presence of the V427A mutation (Roosild et al., 2002). Thus, a V427A / F407L mutant has an almost wild-type phenotype, whereas F407L in a wild-type background has reduced KefC activity; in the double mutant the two changes almost completely compensate for each other. Leu and Ala residues are found naturally in the *B. subtilis* Ktn domain (L14/A34) at the equivalent positions of the KefC mutant V427A and its suppresser, F407L, which emphasizes the compatibility of these two residues at this position in the structure.

Three Arg residues in the Ktn domain have very specific roles in determining the level of KefC activity (Roosild et al., 2002). There are seven surface-exposed Lys or Arg residues in the Ktn domain of KefC (note that none of these residues is involved in the salt bridge that is required for stability of the Rossman fold [Jiang et al., 2001]). Mutation of four of them to Ser was without effect on KefC activity, but three of them had specific consequences—R416, R516, and R527 (Color Plate 1b). The R416S mutation causes increased K^+ leak through KefC. A variety of amino acid substitutions can be made at this position with similar effects on the channel phenotype; even the conservative R416K mutation is altered in its activity. The critical quality of the R416 mutants is that, unlike the wild-type, HALESDIEP, and other Ktn mutants, removal of GSH diminishes the leak of K^+ from the cells expressing R416 mutations. These data raised the possibility that R416 interacted directly with GSH. Similarly, R516 mutations cause a K^+ leak, but activation by NEM is impaired. Charge-switch mutants at R416 and R516 (R416E and R516E) have the most severe leak and, in the case of R516E, the most limited activation by NEM.

The third important basic residue is R527, which is believed to form a salt bridge to the HALESDIEP sequence in the membrane domain. An R527E mutation exhibits no K^+ leak but is severely impaired in NEM-activated K^+ efflux (Roosild et al., 2002). Creating the double mutants R527E/D264R or R527E/E262R led to the observation of restoration of full KefC activity and established the likelihood of a salt bridge between the HALESDIEP sequence and the Ktn domain that is essential for activity. Refinement of the analysis by inserting the shorter basic residue, Lys, at positions 262 and 264 has shown that the most probable partnership is with E262 since R527E/E262K is active but R527E/D264K is not (unpublished data). The charge-swap mutants in the HALESDIEP sequence are in themselves rather unusual. In the presence of GSH, they behave essentially like the wild type: a very slow leak and strong activation by NEM. In the absence of GSH, they exhibit a rapid K^+ leak, suggesting that KefC E262K is very dependent on GSH for the closed state. The E262K mutant can also be activated by SLG, which again is in marked contrast to the wild-type protein. The implication of the salt bridge is that R527 and E262 are within ~3 Å of each other in KefC; how this might be achieved is easily seen from a model of KefC based on the structure of MthK.

The three critical Arg residues create two clusters of basic residues located in the interface between the Ktn and SAM domains (Color Plate 1b). In MthK the residues filling the cleft between the two domains essentially create a flat surface. A similar organization in KefC, which is consistent with the residues filling this region, would create a shallow depression that is flanked by the three Arg residues separated by ~19 Å, which has the potential to be the GSH-binding site. R416 is contributed to the site by the opposite subunit; R516 and R527 are located in the hinge and on the exchanged α6-helix, respectively. Glutathione is ~12 Å in length in the most frequent conformation found in crystal structures of glutathione-*S*-transferases (Color Plate 4a [see color insert]) and would neatly salt-bridge to the most extreme Arg residues in the KefC Ktn-SAM pocket. This creates a significant dilemma for the model for KefC. Two models can be suggested. Previously it was suggested that the Ktn hinge would face the membrane with the two Arg clusters displayed below the hinge and making contact with the HALESDIEP sequence (Roosild et al., 2002). The problem with this model is where to place the SAM domains. KefC has significant similarity to MthK across the Ktn-SAM interface, and this suggested an alternative model in which the SAM domains lie immediately below the hinge. In this model the Ktn-SAM interface is laid parallel to the HALESDIEP sequence at the membrane surface (Color Plate 4b). The effect of this model would be that one Arg cluster faces the membrane but the other is oriented to the cytoplasm. If the Arg cluster really is the GSH-binding site, then one molecule of GSH (and by inference GSX, where X is derived from the electrophile) is trapped between the Ktn-SAM domain and the membrane and the other binds at a site freely accessible from the cytoplasm. Such a model places the KefF (YabF) dimer close to the membrane surface in a position that would allow electron flow from NADH via the flavoprotein to the electron transport chain. This raises the possibility of an as yet undiscovered dual function for the channel.

The location of the GSH-binding site is not clear from the current studies. One possible location for the GSH-binding site is the Rossman fold, which can be occupied by NADH but which has been found previously to be occupied in enzymes by GSH adducts (Puder and Soberman, 1997). However, genetic evidence favors the Ktn-SAM interface, since many of the mutations that influence gating and alter the sensitivity to GSH and GSX lie in this region. Such a model has significant consequences in that the two GSH-binding sites in the dimer would not necessarily be equivalent. One binding site at the lower surface has the potential to be involved in direct interaction with the HALESDIEP loop in either one or both membrane domains (Color Plate 4b). Diffusion into this site might be constrained by the protein packing around the GSH molecule. The other site, as suggested above, would be exposed to the cytoplasm and thus the GSH should be freely exchangeable with GSX molecules that are synthesized in the cytoplasm. Asymmetric complexes, one site occupied by GSH and the other by GSX, might then achieve gating.

Models for Ktn Gating

Recent work on K^+ channel structure has generated insights into the helix movements that are required to gate channels (Cortes et al., 2001; Jiang et al., 2002b; Jiang et al.,

2003; Liu et al., 2001). In most models the selectivity filter is a relatively fixed structure with dehydrated K^+ ions bound by rings of carbonyl groups and separated from each other by water molecules (Doyle et al., 1998). When the channel is open, the K^+ ions move simultaneously with water molecules essentially at the rate of diffusion in free solution (Morais-Cabral et al., 2001; Roux and MacKinnon, 1999; Zhou et al., 2001). The block on ion movement is the seal to the vestibule located immediately below the selectivity filter (Doyle et al., 1998; Jiang et al., 2002b). The four vestibule-lining helices meet to form the seal, each usually contributing one hydrophobic residue to the seal. Similar principles of seal formation apply in other channels (Bass et al., 2002; Chang et al., 1998). Gating then involves helix bending at critical Gly residues within the vestibule-lining helices such that the seal is broken, allowing ions to flow (Jiang et al., 2002b). In the voltage-gated bacterial channel, KvAp, and in KirBac, the inward rectifier channel, gating transitions probably move a second helix (TM1 or TM5 for channels with two or six TM spans, respectively) that constrains the vestibule-lining helix from bending (Jiang et al., 2003; Kuo et al., 2003a). A change in the conformation of the sensor domain (voltage sensor or ligand-binding domain, such as Ktn) pulls the second helix away from the axis of the pore and allows the vestibule-lining helix to bend, leading to channel opening.

How can such a model be applied to KefC? The presence of GSH is critical to the inactive state of KefC, and the formation of GSH adducts is required for activation. All of the studies with KefC mutants suggest that GSH plays a critical function in holding the channel in a closed state. Mutants that block KefC gating lie both in the HALESDIEP sequence and in the Ktn domain. E266 is a critical residue: creation of E266R led to almost complete block of KefC activity, but the mechanism of this mutation is not clear. In contrast, it is clear that mutations that interfere with the hinging motion of the Ktn domain strongly disrupt KefC activity. It seems reasonable, therefore, to place conformational changes in Ktn at the heart of the activation mechanism. A gating transition caused by a change of the hinge angle, arising from binding of GSH adducts, would be transmitted to the membrane domain. Shortening the linker always impairs activity, demonstrating the critical role of this sequence in positioning the Ktn domains so that they can control gating. Our studies have shown that the juxtaposition of the Ktn and SAM domains proximal to the HALESDIEP loop is also critical for maintaining the closed state of the channel. Any loosening of this contact, as for example in the charge reversal mutants in HALESDIEP, makes the channel easier to gate. Such mutants look normal until GSH is removed, suggesting that the tripeptide is critical for maintaining the interaction of Ktn and HALESDIEP. It is relatively easy to envisage a model in which the HALESDIEP loop maintains the channel closed by its interaction with Ktn, possibly via GSH, and that changing the hinge angle in the Ktn dimer simultaneously allows motion in helices 8 and 9 that flank the HALESDIEP loop, generating an open pore. We do not know whether the channel pore is formed by the interaction of the two membrane domains that are brought into proximity by dimerization of Ktn domains. An alternative model would envisage two separate pores regulated in a coordinated fashion.

CONCLUSIONS

The analysis of the crystal structures of K^+ channels has generated great insights into the mechanism of ion selectivity. It seems almost certain that the analysis of the biochemistry of channels such as MthK and KirBac will provide new insights into the gating mechanism. Given the importance of channels in mammalian physiology it seems appropriate that most efforts will be concentrated on systems that are homologs of the channels from higher organisms. However, it seems equally clear that the gating mechanisms evolved first in bacteria and have been adapted to a multitude of purposes. Indeed, it is frequently overlooked that bacteria have a host of regulated transporters that utilize novel domains with the potential of new gating mechanisms (Wood, 1999). The Ktn domain is found in four different classes of ion-translocating proteins (MthK, KefC, AmhMT, and TrkAEH) and is therefore a major determinant of bacterial cell behavior. Since related domains are sometimes found in solute-binding proteins, the influence of Ktn is even more widely spread. It is a domain that demonstrates the sophistication of the Rossman fold as an adaptable structure able to bind different ligands coupled with a potential array of conformations and configurations that make it the ultimate gating module.

Acknowledgments. We thank The Wellcome Trust (grant 040174) and the BBSRC, which have funded research on the KefC system.

We extend special thanks to Tarmo Roosild, Senyon Choe, Jim Naismith, Changjiang Dong, Marcel Jaspars, and Wenjian Wang for their insights on protein structure. Two generations of research students and postdoctoral workers have fueled the insights into the KefC mechanism, and it is with pleasure that we thank them for their contributions.

REFERENCES

Accardi, A., L. Kolmakova-Partensky, C. Williams, and C. Miller. 2004. Ionic currents mediated by a prokaryotic homologue of CLC Cl- channels. *J. Gen. Physiol.* **123:**109–119.

Altendorf, K., and W. Epstein. 1994. Kdp-Atpase of *Escherichia coli. Cell. Physiol. Biochem.* **4:**160–168.

Altendorf, K., M. Gassel, W. Puppe, T. Mollenkamp, A. Zeeck, C. Boddien, K. Fendler, E. Bamberg, and S. Drose. 1998. Structure and function of the Kdp-ATPase of *Escherichia coli. Acta Physiol. Scand.* **163:** 137–146.

Bakker, E. P., I. R. Booth, U. Dinnbier, W. Epstein, and A. Gajewska. 1987. Evidence for multiple K^+ export systems in *Escherichia coli. J. Bacteriol.* **169:**3743–3749.

Bakker, E. P., and W. E. Mangerich. 1982. *N*-ethylmaleimide induces K^+-H^+ antiport activity in *Escherichia coli* K-12. *FEBS Lett.* **140:**177–180.

Bass, R. B., P. Strop, M. Barclay, and D. C. Rees. 2002. Crystal structure of *Escherichia coli* MscS, a voltage-modulated and mechanosensitive channel. *Science* **298:**1582–1587.

Bengoechea, J. A., and M. Skurnik. 2000. Temperature-regulated efflux pump/potassium antiporter system mediates resistance to cationic antimicrobial peptides in *Yersinia. Mol. Microbiol.* **37:**67–80.

Biggin, P. C., T. Roosild, and S. Choe. 2000. Potassium channel structure: domain by domain. *Curr. Opin. Struct. Biol.* **10:**456–461.

Booth, I. R. 2003. Bacterial ion channels, p. 91–112. *In* J. K. Setlow (ed.), *Genetic Engineering: Principles and Methods*, vol. 25. Kluwer Academic/Plenum Publishers, New York, N.Y.

Bossemeyer, D., A. Borchard, D. C. Dosch, G. C. Helmer, W. Epstein, I. R. Booth, and E. P. Bakker. 1989a. K^+ transport protein TrkA of *Escherichia coli* is a peripheral membrane protein that requires other *trk* gene products for attachment to the cytoplasmic membrane. *J. Biol. Chem.* **264:**16403–16410.

Bossemeyer, D., A. Schlosser, and E. P. Bakker. 1989b. Specific cesium transport via the *Escherichia coli* Kup (TrkD) K^+ uptake system. *J. Bacteriol.* **171:**2219–2221.

Chang, G., R. H. Spencer, A. T. Lee, M. T. Barclay, and D. C. Rees. 1998. Structure of the MscL homolog from *Mycobacterium tuberculosis*: a gated mechanosensitive ion channel. *Science* **282:**2220–2226.

Choe, S., and T. Roosild. 2002. Regulation of the K channels by cytoplasmic domains. *Biopolymers* **66:**294–299.

Cortes, D. M., L. G. Cuello, and E. Perozo. 2001. Molecular architecture of full-length KcsA: role of cytoplasmic domains in ion permeation and activation gating. *J. Gen. Physiol.* **117:**165–180.

Doyle, D. A., J. M. Cabral, R. A. Pfuetzner, A. L. Kuo, J. M. Gulbis, S. L. Cohen, B. T. Chait, and R. MacKinnon. 1998. The structure of the potassium channel: molecular basis of K^+ conduction and selectivity. *Science* **280:**69–77.

Durell, S. R., E. P. Bakker, and H. R. Guy. 2000. Does the KdpA subunit from the high affinity K^+-translocating P-type KDP-ATPase have a structure similar to that of K^+ channels? *Biophys. J.* **78:**188–199.

Durell, S. R., Y. L. Hao, T. Nakamura, E. P. Bakker, and H. R. Guy. 1999. Evolutionary relationship between K^+ channels and symporters. *Biophys. J.* **77:**775–788.

Dutzler, R., E. B. Campbell, M. Cadene, B. T. Chait, and R. MacKinnon. 2002. X-ray structure of a CIC chloride channel at 3.0 angstrom reveals the molecular basis of anion selectivity. *Nature* **415:**287–294.

Elmore, M. J., A. J. Lamb, G. Y. Ritchie, R. M. Douglas, A. Munro, A. Gajewska, and I. R. Booth. 1990. Activation of potassium efflux from *Escherichia coli* by glutathione metabolites. *Mol. Microbiol.* **4:** 405–412.

Epstein, W. 2003. The roles and regulation of potassium in bacteria. *Prog. Nucleic Acid Res. Mol. Biol.* **75:** 293–320.

Epstein, W., and B. S. Kim. 1971. Potassium transport loci in *Escherichia coli* K-12. *J. Bacteriol.* **108:**639–644.

Faig, M., M. A. Bianchet, P. Talalay, S. Chen, S. Winski, D. Ross, and L. M. Amzel. 2000. Structures of recombinant human and mouse NAD(P)H:quinone oxidoreductases: species comparison and structural changes with substrate binding and release. *Proc. Natl. Acad. Sci. USA* **97:**3177–3182.

Ferguson, G. P., A. W. Munro, R. M. Douglas, D. Mclaggan, and I. R. Booth. 1993. Activation of potassium channels during metabolite detoxification in *Escherichia coli. Mol. Microbiol.* **9:**1297–1303.

Ferguson, G. P., S. Totemeyer, M. J. MacLean, and I. R. Booth. 1998. Methylglyoxal production in bacteria: suicide or survival? *Arch. Microbiol.* **170:**209–218.

Fujisawa, M., Y. Wada, and M. Ito. 2004. Modulation of the K^+ efflux activity of *Bacillus subtilis* YhaU by YhaT and the C-terminal region of YhaS. *FEMS Microbiol. Lett.* **231:**211–217.

Hamann, A., D. Bossemeyer, and E. P. Bakker. 1987. Physical mapping of the K^+ transport *trkA* gene of *Escherichia coli* and overproduction of the TrkA protein. *J. Bacteriol.* **169:**3138–3145.

Harms, C., Y. Domoto, C. Celik, E. Rahe, S. Stumpe, R. Schmid, T. Nakamura, and E. P. Bakker. 2001. Identification of the ABC protein SapD as the subunit that confers ATP dependence to the K^+-uptake systems Trk(H) and Trk(G) from *Escherichia coli* K-12. *Microbiology* **147:**2991–3003.

Holtmann, G., E. P. Bakker, N. Uozumi, and E. Bremer. 2003. KtrAB and KtrCD: two K^+ uptake systems in *Bacillus subtilis* and their role in adaptation to hypertonicity. *J. Bacteriol.* **185:**1289–1298.

Jiang, Y. X., A. Lee, J. Y. Chen, M. Cadene, B. T. Chait, and R. MacKinnon. 2002a. Crystal structure and mechanism of a calcium-gated potassium channel. *Nature* **417:**515–522.

Jiang, Y. X., A. Lee, J. Y. Chen, M. Cadene, B. T. Chait, and R. MacKinnon. 2002b. The open pore conformation of potassium channels. *Nature* **417:**523–526.

Jiang, Y. X., A. Lee, J. Y. Chen, V. Ruta, M. Cadene, B. T. Chait, and R. MacKinnon. 2003. X-ray structure of a voltage-dependent K^+ channel. *Nature* **423:**33–41.

Jiang, Y. X., A. Pico, M. Cadene, B. T. Chait, and R. MacKinnon. 2001. Structure of the RCK domain from the *E. coli* K^+ channel and demonstration of its presence in the human BK channel. *Neuron* **29:**593–601.

Kuo, A., J. M. Gulbis, J. F. Antcliff, T. Rahman, E. D. Lowe, J. Zimmer, J. Cuthbertson, F. M. Ashcroft, T. Ezaki, and D. A. Doyle. 2003a. Crystal structure of the potassium channel KirBac1.1 in the closed state. *Science* **300:**1922–1926.

Kuo, M. M. C., Y. Saimi, and C. Kung. 2003b. Gain-of-function mutations indicate that *Escherichia coli* Kch forms a functional K^+ conduit in vivo. *EMBO J.* **22:**4049–4058.

Kyte, J., and R. F. Doolittle. 1982. A simple method for displaying the hydropathic character of a protein. *J. Mol. Biol.* **157:**105–132.

Li, R., M. A. Bianchet, P. Talalay, and L. M. Amzel. 1995. The 3-dimensional structure of Nad(P)H:quinone reductase, a flavoprotein involved in cancer chemoprotection and chemotherapy: mechanism of the 2-electron reduction. *Proc. Natl. Acad. Sci. USA* **92:**8846–8850.

Liu, Y. S., P. Sompornpisut, and E. Perozo. 2001. Structure of the KcsA channel intracellular gate in the open state. *Nat. Struct. Biol.* **8:**883–887.

Mackinnon, R. 2000. Mechanism of ion conduction and selectivity in K channels. *J. Gen. Physiol.* **116:**17.

MacLean, M. J., L. S. Ness, G. P. Ferguson, and I. R. Booth. 1998. The role of glyoxalase I in the detoxification of methylglyoxal and in the activation of the KefB K^+ efflux system in *Escherichia coli. Mol. Microbiol.* **27:**563–571.

Martz, E. 2002. Protein explorer: easy yet powerful macromolecular visualization. *Trends Biochem. Sci.* **27:**107–109.

McLaggan, D., H. Rufino, M. Jaspars, and I. R. Booth. 2000. Glutathione-dependent conversion of *N*-ethylmaleimide to the maleamic acid by *Escherichia coli*: an intracellular detoxification process. *Appl. Environ. Microbiol.* **66:**1393–1399.

Meury, J., and A. Kepes. 1982. Glutathione and the gated potassium channels of *Escherichia coli*. *EMBO J.* **1:**339–343.

Meury, J., S. Lebail, and A. Kepes. 1980. Opening of potassium channels in *Escherichia coli* membranes by thiol reagents and recovery of potassium tightness. *Eur. J. Biochem.* **113:**33–38.

Miller, S., R. M. Douglas, P. Carter, and I. R. Booth. 1997. Mutations in the glutathione-gated KefC K^+ efflux system of *Escherichia coli* that cause constitutive activation. *J. Biol. Chem.* **272:**24942–24947.

Miller, S., L. S. Ness, C. M. Wood, B. C. Fox, and I. R. Booth. 2000. Identification of an ancillary protein, YabF, required for activity of the KefC glutathione-gated potassium efflux system in *Escherichia coli*. *J. Bacteriol.* **182:**6536–6540.

Morais-Cabral, J. H., Y. F. Zhou, and R. MacKinnon. 2001. Energetic optimization of ion conduction rate by the K^+ selectivity filter. *Nature* **414:**37–42.

Moras, D., K. W. Olsen, M. N. Sabesan, M. Buehner, G. C. Ford, and M. G. Rossmann. 1975. Studies of asymmetry in the three-dimensional structure of lobster D-glyceraldehyde-3-phosphate dehydrogenase. *J. Biol. Chem.* **250:**9137–9162.

Munro, A. W., G. Y. Ritchie, A. J. Lamb, R. M. Douglas, and I. R. Booth. 1991. The cloning and DNA sequence of the gene for the glutathione-regulated potassium-efflux system KefC of *Escherichia coli*. *Mol. Microbiol.* **5:**607–616.

Nakamura, C., J. G. Burgess, K. Sode, and T. Matsunaga. 1995. An iron-regulated gene, MagA, encoding an iron transport protein of *Magnetospirillum* sp. strain Amb-1. *J. Biol. Chem.* **270:**28392–28396.

Nakamura, T., R. Yuda, T. Unemoto, and E. P. Bakker. 1998. KtrAB, a new type of bacterial K^+-uptake system from *Vibrio alginolyticus*. *J. Bacteriol.* **180:**3491–3494.

Ness, L. S., and I. R. Booth. 1999. Different foci for the regulation of the activity of the KefB and KefC glutathione-gated K^+ efflux systems. *J. Biol. Chem.* **274:**9524–9530.

Padan, E., and S. Schuldiner. 1994. Molecular biology of Na^+/H^+ antiporters: molecular devices that couple the Na^+ and H^+ circulation in cells. *Biochim. Biophys. Acta* **1187:**206–210.

Puder, M., and S. Soberman. 1997. Glutathione conjugates recognize the Rossmann fold of glyceraldehyde-3-phosphate dehydrogenase. *J. Biol. Chem.* **272:**10936–10940.

Ramirez, J., O. Ramirez, C. Saldana, R. Coria, and A. Pena. 1998. A *Saccharomyces cerevisiae* mutant lacking a K^+/H^+ exchanger. *J. Bacteriol.* **180:**5860–5865.

Reizer, J., A. Reizer, and M. H. Saier. 1992. The putative Na^+/H^+ antiporter (NapA) of *Enterococcus hirae* is homologous to the putative K^+/H^+ antiporter (KefC) of *Escherichia coli*. *FEMS Microbiol. Lett.* **94:**161–164.

Rhoads, D. B., and W. Epstein. 1977. Energy coupling to net K^+ transport in *Escherichia coli* K-12. *J. Biol. Chem.* **253:**1394–1401.

Rhoads, D. B., and W. Epstein. 1978. Cation transport in *Escherichia coli*. IX. Regulation of K^+ transport. *J. Gen. Physiol.* **72:**283–295.

Rhoads, D. B., L. Laimins, and W. Epstein. 1978. Functional organization of the *kdp* genes of *Escherichia coli* K-12. *J. Bacteriol.* **135:**445–452.

Roe, A. J., D. McLaggan, C. P. O'Byrne, and I. R. Booth. 2000. Rapid inactivation of the *Escherichia coli* Kdp K^+ uptake system by high potassium concentrations. *Mol. Microbiol.* **35:**1235–1243.

Roosild, T. P., S. Miller, I. R. Booth, and S. Choe. 2002. A mechanism of regulating transmembrane potassium flux through a ligand-mediated conformational switch. *Cell* **109:**781–791.

Roux, B., and R. MacKinnon. 1999. The cavity and pore helices the KcsA K^+ channel: electrostatic stabilization of monovalent cations. *Science* 285:100–102.

Saier, M. H., B. H. Eng, S. Fard, J. Garg, D. A. Haggerty, W. J. Hutchinson, D. L. Jack, E. C. Lai, H. J. Liu, D. P. Nusinew, A. M. Omar, S. S. Pao, I. T. Paulsen, J. A. Quan, M. Sliwinski, T. T. Tseng, S. Wachi, and G. B. Young. 1999. Phylogenetic characterization of novel transport protein families revealed by genome analyses. *Biochim. Biophys. Acta* **1422:**1–56.

Schleyer, M., and E. P. Bakker. 1993. Nucleotide sequence and 3'-end deletion studies indicate that the K^+-uptake protein Kup from *Escherichia coli* is composed of a hydrophobic core linked to a large and partially essential hydrophilic C terminus. *J. Bacteriol.* **175:**6925–6931.

Schlosser, A., A. Hamann, D. Bossemeyer, E. Schneider, and E. P. Bakker. 1993. NAD^+ binding to the *Escherichia coli* K^+-uptake protein TrkA and sequence similarity between TrkA and domains of a family of dehydrogenases suggest a role for NAD^+ in bacterial transport. *Mol. Microbiol.* **9:**533–543.

Schlosser, A., M. Meldorf, S. Stumpe, E. P. Bakker, and W. Epstein. 1995. TrkH and its homolog, TrkG, determine the specificity and kinetics of cation-transport by the Trk system of *Escherichia coli*. *J. Bacteriol.* **177:**1908–1910.

Southworth, T. W., A. A. Guffanti, A. Moir, and T. A. Krulwich. 2001. GerN, an endospore germination protein of *Bacillus cereus*, is an Na^+/H^+-K^+ antiporter. *J. Bacteriol.* **183:**5896–5903.

Stewart, L. M. D., E. P. Bakker, and I. R. Booth. 1985. Energy coupling to K^+ uptake via the Trk system in *Escherichia coli*: the role of ATP. *J. Gen. Microbiol.* **131:**77–85.

Tholema, N., E. P. Bakker, A. Suzuki, and T. Nakamura. 1999. Change to alanine of one out of four selectivity filter glycines in KtrB causes a two orders of magnitude decrease in the affinities for both K^+ and Na^+ of the Na^+ dependent K^+ uptake system KtrAB from *Vibrio alginolyticus*. *FEBS Lett.* **450:**217–220.

Wei, Y., T. W. Southworth, H. Kloster, M. Ito, A. A. Guffanti, A. Moir, and T. A. Krulwich. 2003. Mutational loss of a K^+ and transporter affects the growth and endospore formation of alkaliphilic *Bacillus pseudofirmus* OF4. *J. Bacteriol.* **185:**5133–5147.

Wood J. M. 1999. Osmosensing by bacteria: signals and membrane-based sensors. *Microbiol. Mol. Biol. Rev.* **63:**230–262.

Zhou, Y. F., J. H. Morais-Cabral, A. Kaufman, and R. MacKinnon. 2001. Chemistry of ion coordination and hydration revealed by a K^+ channel-Fab complex at 2.0 angstrom resolution. *Nature* **414:**43–48.

Bacterial Ion Channels and Their Eukaryotic Homologs
Edited by A. Kubalski and B. Martinac

Chapter 3

Deciphering *Streptomyces lividans* KcsA as a K Channel Model

Hildgund Schrempf

In the course of studying variable regions of the genome of the bacterium *Streptomyces lividans*, a small gene encoding an unknown protein was discovered. Through extended studies, the small gene product was revealed as the first bacterial potassium channel, named KcsA (K channel of streptomycetes A). After optimization, channel activity was identified in a newly established protoplast-vesicle system and, after reconstitution of the purified KcsA, in a bilayer system. In addition, KcsA proved to be the first K channel protein that could be gained after expression in a bacterial host in large quantities. (Schrempf et al., 1995). These findings created the most essential basis for crystallization studies. Meanwhile, KcsA has been developed to be the most important channel model to analyze structure-function relationships. The discovery of KcsA convincingly proved that channels are not a "privilege" of eukaryotic organisms and thus induced a search for other types of channel proteins from bacteria. At the same time, the principal concept for analyzing the KcsA channel protein has been successfully adopted for a range of prokaryotic and eukaryotic proteins.

GENERAL FEATURES OF STREPTOMYCETES

As a member of the family *Streptomycetaceae, S. lividans* is a gram-positive soil bacterium with a complex growth cycle (Kutzner, 1981). In the presence of nutrients this growth is initiated by the germination of spores. Replication of the large linear chromosomal DNA does not correlate (as in most bacteria) with cell division. As a consequence, long hyphae with numerous copies of the chromosomal DNA develop. Extension of the hyphae and branch formation lead subsequently to an extended filamentous network, the substrate mycelium (Fig. 1A). Upon depletion of nutrients, aerial hyphae arise and within these hyphae septations occur, and subsequently compartments containing genomic DNA develop into spores. These spores withstand dryness as well as cold conditions and are in a dormant stage as long as nutrients are not available (Kutzner, 1981).

Hildgund Schrempf • Fachbereich Biologie/Chemie, Universität Osnabrück, Barbarastr. 11, 49069 Osnabrück, Germany.

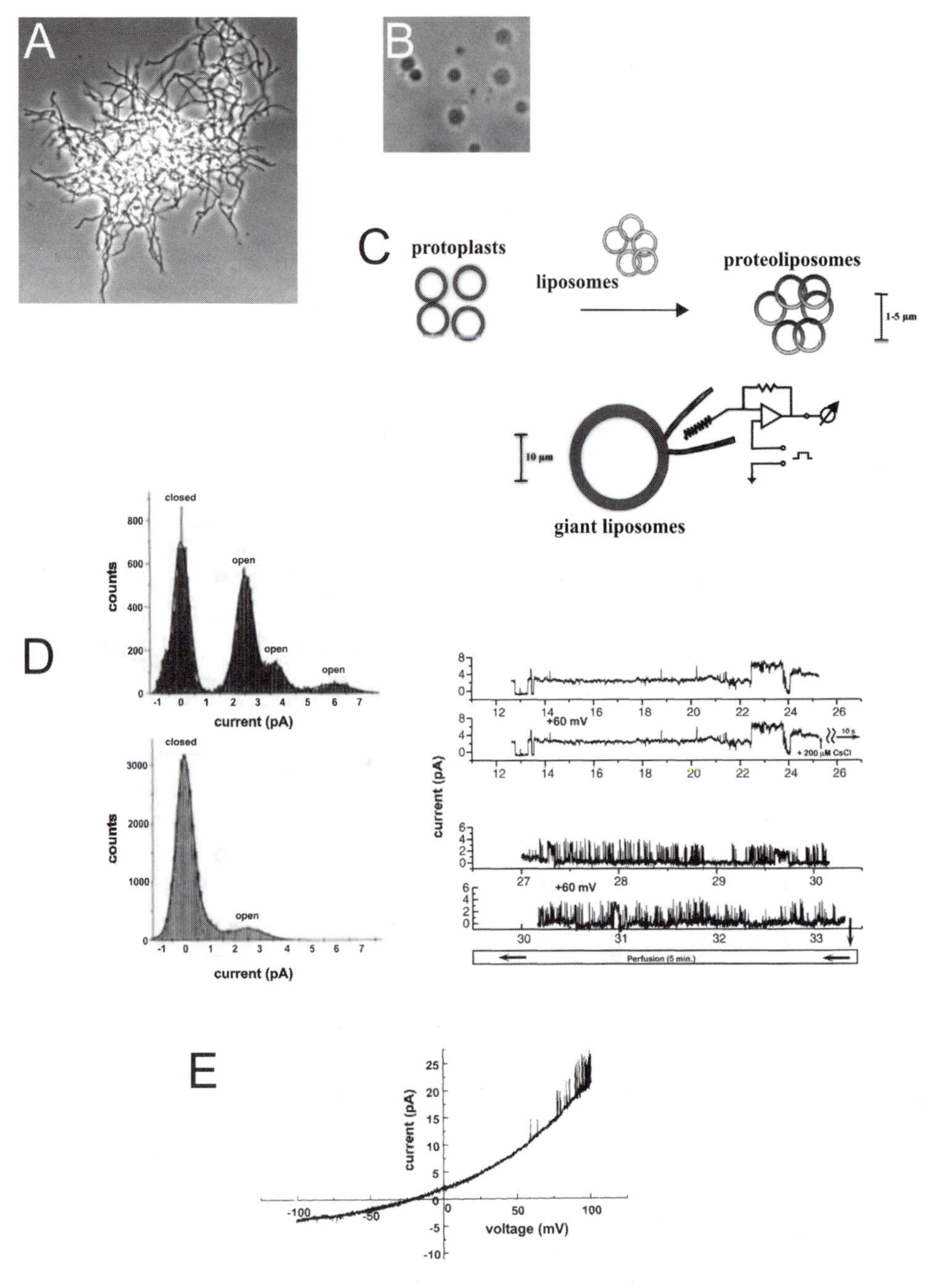

Figure 1. Features of *S. lividans* and the generation of giant protoplasts. (A) Vegetative network of hyphae. (B) Protoplasts of *S. lividans* were obtained by digesting (with lysozyme) the cell wall of the hyphae. (C) General scheme to fuse protoplasts with liposomes to generate giant protoplast vesicles. (D) Current recording at pH 7.2 under asymmetric conditions and the deduced histogram to present the proportion of closed and open channels (top). Current recordings and all point histograms after the addition of CsCl (bottom). (E) Current recordings (excised patch) under asymmetric conditions within ranging potentials.

Within the natural habitat streptomycetes (of which close to 1,000 species have been classified) play a major role in the turnover of organic material as they have a large repertoire of enzymes leading to the degradation of macromolecules, including cellulose, lignocellulose, chitin, starch, xylan, proteins, and lipids. Hence these enzymes are of great ecological and biotechnological value. Streptomycetes have a huge capacity to synthesize chemically different secondary metabolites, many of which have antibacterial, antifungal, or cytostatic activities and are being used in medicine (Kutzner, 1981; Schrempf, 1999).

IDENTIFICATION OF KcsA

Streptomyces species have a large linear chromosomal DNA with a high G+C content. The chromosomal DNA of *Streptomyces* species has been found to undergo frequent changes leading to chromosomal deletion and/or amplification of local region(s); hence spontaneous variants arise frequently (Kessler et al., 1989). In the course of an analysis of various genes from variable regions of the *S. lividans* genome, an open reading frame of 483 bp with a high G+C content of 70.6% (typical of *Streptomyces* DNA) and a coding capacity for a 17.6-kDa protein was discovered (Schrempf et al., 1995). Comparative studies revealed that the deduced protein (Fig. 2A) is the first of its type as it did not share a significant number of amino acids with those known at the time. The deduced protein was found to possess a short N-terminal region, followed by two potential transmembrane segments flanking a short central region and a C terminus (Fig. 2A). The N-terminal part comprises seven leucine and two arginine residues and a short C terminus, which is rich in arginine as well as glutamic and aspartic acid residues (Schrempf et al., 1995).

At the time of discovery, within a stretch of 116 amino acids of KcsA, 32.8 to 31% were found to be identical with those from deduced K^+ channels, such as ShC1 (Shaker) from *Drosophila melanogaster* (Schwarz et al., 1988), Kv3.2c from rats (Luneau et al., 1991), HBK2 from humans (Grupe et al., 1990), and others. Of the identical amino acids, 66.7% are present in a stretch of 24 amino acids (Fig. 2A), which corresponds to the H5 domain of 25 amino acids (later named P region) of the K^+ channel proteins described above and others. Interestingly, the two transmembrane helices of KcsA share the relative positioning of some identical amino acids within the S5 segment (i.e., ShC1 > KAT1 > Slo) and S6 segment (ShC1 > Slo) of voltage-gated channel families (note the nomenclature refers to those given in the original papers) and are most distantly related to the M1 and M2 segments from inwardly rectifying channels (i.e., ROMK1). The S4/5 linkers of deduced Shaker proteins and their N termini (ShA > ShC1) share a few identical amino acids with KcsA. The KcsA protein lacks a stretch of amino acids corresponding to the complete S4 motif.

After a collection of *S. lividans* mutants was screened, one type (N1), which had a larger chromosomal deletion including a gene encoding a protein of unknown function, was identified. The novel gene was cloned together with its upstream region in a multicopy *Streptomyces* vector. The resulting hybrid plasmid (pKCS1) was transformed into the spontaneous *S. lividans* mutant strain N1. Comparative growth studies suggested that the novel protein might present the first identified K channel of a bacterium (Schrempf et al., 1995). It is interesting that the deduced KcsA protein differs considerably from bacterial proteins, which function as potassium transporters.

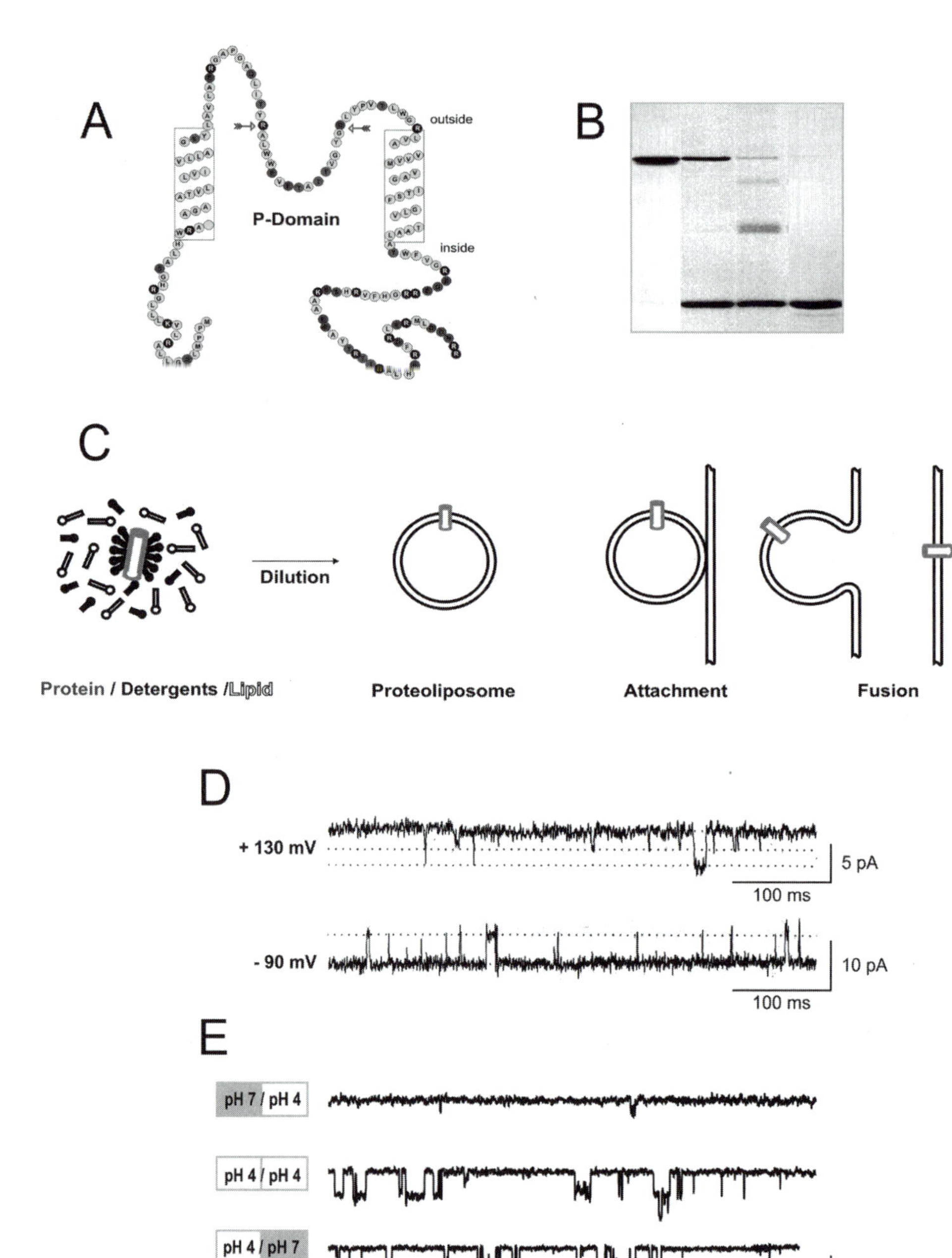

Figure 2. Characteristics of the deduced and the isolated KcsA protein as well as electrophysiological analysis of using a bilayer. (A) The characteristic regions of the KcsA protein deduced from the corresponding gene. (B) The KcsA protein assembles to a stable tetramer (left), which upon increase of temperature, can be resolved via intermediates (middle; above 70°C) to the monomeric form (right; melting point close to 85°C). (C) Generation of proteoliposomes and subsequent fusion with a bilayer. (D) Current recordings of a bilayer containing KcsA at pH 7.2 under asymmetric conditions. (E) Current recordings at different combinations of pH at *cis* and *trans* sites under symmetric conditions.

Subsequently prerequisites (see below) were established to test whether the novel protein could have characteristics of eukaryotic channels. Based on the results derived from further studies (see below), the novel protein was indeed identified as the first bacterial K channel and was named KcsA.

CHANNEL ACTIVITY WITHIN GIANT PROTOPLAST VESICLES

Various K^+ ion channels from different eukaryotes have been analyzed after the translation of the corresponding mRNA within oocytes from *Xenopus laevis* (Hille, 1992) by patch-clamp analysis. Thus, it was obvious to test a similar strategy for KcsA. After a polyadenylation tail (which is absent within the original *Streptomyces* host) was attributed through an in vitro transcription system, the resulting mRNA was injected into oocytes. However, no extra channel activity was detected (Schrempf et al., 1995).

Due to their small diameter as well as the rigid cell wall (consisting of multilayered peptidoglycan), the hyphae (Fig. 1A) of *S. lividans* were deemed to be unsuitable for direct patch clamping. Thus the cell wall was enzymatically (with lysozyme) digested. The resulting protoplasts (stabilized in sucrose-containing buffer) had diameters that usually ranged from 1 to 3 μm, and rarely 5 to 10 μm (Fig. 1B). However, because of these small dimensions, all attempts to obtain high-resistance seals between the outer protoplast membrane and the patch pipette (GΩ-seals) failed. Therefore, protoplasts were fused (Fig. 1C) with small unilamellar liposomes, which were subsequently enlarged to giant vesicles (Criado and Keller, 1987; Enz et al., 1993). After preparations were optimized, seals with resistances between 10 and 50 GΩ were reproducibly obtained in the course of patch-clamp measurements (Schrempf et al., 1995).

Giant liposome-protoplast vesicles derived from the *S. lividans* mutant strain carrying the plasmid pKCS1 (with the *kcsA* gene, see above) showed ion channel activity under neutral pH (7.2) and asymmetric conditions (Fig. 1D, E). Current recordings at a holding potential (V_h) of +40 mV yielded channel activities with different amplitudes. Similar data were obtained at all positive pipette potentials, whereas at negative values almost no channel activity could be observed. The current-voltage relationship revealed that the ion channel currents could be grouped into three different classes of open channel amplitudes with slope conductance values congruent to 20 ± 7 pS ($n = 5$), 40 ± 8 pS ($n = 11$), and 90 ± 18 pS ($n = 6$). The channel chord conductance values for the different peaks in the amplitude histogram were congruent to 16, 48, and 72 pS, whereas the reversal potential (E_{rev}) was ≅–30 mV in this particular experiment. In addition, transitions indicating subconductance states were noticed. The sequence shown in Fig. 1D of channel currents could only be explained by the presence of four simultaneously active channels, two of them approaching a conductance congruent to 20 pS. The other conductances were ≅100 and ≅60 pS, respectively. However, if more than one active channel protein was present in the patch, these channels would have to open simultaneously within <1 ms, which is statistically extremely unlikely (Patlak, 1988). Different current amplitudes, sequential openings, and closures of two or more active channels were not observed. Therefore, the temporal sequence of the current data indicated that the different amplitudes were due to subconductance states rather than to the simultaneous activities of two or more channels. The E_{rev} for the different channel transitions was identical with a value of –33 mV, near to an E_{K^+} of –40.6 mV. As expected, channel activity was absent in the membrane patches from giant protoplast vesicles prepared from a *S. lividans* mutant control strain lacking the *kcsA* gene (Schrempf et al., 1995).

In summary, the data demonstrated that the *kcsA* gene encodes a K^+-selective ion channel. Its maximal conductance of ≅90 pS is considerably less often observed than its smaller subconductance states of 20 and 40 pS. All conductance states had a high selectivity for potassium ions. When K^+ was partially substituted by Na^+ (bath, 200 mM K^+ + 100 mM Na^+; pipette, 40 mM K^+), the E_{rev} changed from ≅ −33 to −37 mV. Assuming that the K^+ permeability did not alter after the addition of Na^+, the permeability ratio P_{K^+}/P_{Na^+} can be calculated as $P_{K^+}/P_{Na^+} \cong 2.87$. In independent experiments, Mg^{2+} ions (1 mM, absolutely necessary only in the bath, not within the pipette) were required for channel activity. Channel gating was only ascertained at mainly positive pipette potentials, suggesting that the KcsA channels were unidirectionally incorporated into the membranes of the protoplast-liposome vesicles. Since the probability of channel openings was close to zero at negative pipette potentials, currents must have rectified. Assuming that the orientation of the protoplast membrane had not changed during the fusion process, the bacterial membranes will be oriented right side out in the liposome-protoplast vesicles. Consequently, the observed potentials of the pipette and the bath would refer to the outside and the cytoplasmic site, respectively. In this case, the suspected Mg^{2+}-binding site was predicted to be located in the cytosol. In the presence of CsCl, the overall open probability ($P_{open} \cong 0.52$) was drastically reduced ($P_{open} \cong 0.1$), and only the 40 pS open amplitude and the flickering of currents were observed (Fig. 1D).

Based on the results with this newly established protoplast-liposome vesicle system, it had to be concluded that KcsA acts as a K^+ ion channel, exhibiting a rectifying voltage relation as well as several subconductance states, with the main open state of ≅40 pS. Channel gating is dependent on Mg^{2+} ions, and the channel is blocked by Cs^+ ions (Schrempf et al., 1995).

HIGH SYNTHESIS LEVELS AND PURIFICATION OF KcsA

To obtain large quantities of KcsA, the *kcsA* gene was expressed in the heterologous *Escherichia coli* host. At the time, such a strategy was novel for a channel protein. The DNA fragment containing the *kcsA* gene (without upstream region and start codon) was cloned into the polylinker of the *E. coli* pQE-32 vector. This vector includes a ribosomal binding site, a start codon, and six codons encoding histidine. A transformant of the selected *E. coli* strain that carried the correct construct was further tested for synthesis of the K^+ channel protein containing the His-affinity tag at its N terminus. The expected fusion protein could be identified after cultivation time, temperature, and induction conditions for the strain were optimized. The dominant portion of the protein was found neither within inclusion bodies nor in the cytoplasm, but in the membrane fraction. The protein was extractable with various detergents (i.e., Mega9, dodecylmaltoside, and others) and could be purified by subsequent affinity chromatography (Fig. 2B). From one liter of culture several milligrams of the protein could be obtained routinely. The data revealed, for the first time, that a K channel protein could be obtained in large quantities. Hence, the major basis (Schrempf et al., 1995) for subsequent biochemical and electrophysiological studies as well as crystallization had been established.

FUNCTIONAL RECONSTITUTION IN A BILAYER

KcsA (Fig. 2B and C) was reconstituted by a dilution method into liposomes from L-α-azolectin. Planar lipid bilayers were produced with the painting technique (Mueller et al., 1962). Liposomes containing reconstituted KcsA were added to the *cis* compartment, and the solution was stirred to promote fusion; then perfusion was performed to change the electrolytes to the desired composition.

At neutral pH a low level of voltage-dependent single channel activity was observed (Fig. 2D). The current-voltage relationship of the reconstituted channel was linear, and its E_{rev} congruent to +59 mV was close to the Nernst potential for potassium ($E_{K^+} \cong$ +64 mV). In contrast to the giant protoplast vesicle system (see previous section), the activity of the reconstituted channel was affected neither by Mg^{2+} nor by Cs^+. These data suggested that the conformation of the reconstituted channel differs from that expressed within the liposome-protoplast vesicles (Fig. 1D and E). This result was not surprising, as the artificial composition of the bilayer is different from that of the protoplasts (generated from the mycelia of the *Streptomyces* strain), and the bilayer also lacks other components present in natural membrane of the protoplasts. However, these initial data demonstrated unambiguously the first and successful reconstitution of a K^+ ion channel protein and paved the way for determining parameters affecting in vitro channel activity.

Subsequent results indeed showed (see following sections) that the composition of the lipids and the presence of additional components and parameters considerably affect channel activity within bilayers containing KcsA.

OLIGOMERIZATION OF KcsA

KcsA (mostly histidine-tagged KcsA was used) assembled spontaneously to tetramers, which remained stable at room temperature for extended periods (Fig. 2B). The tetramer was stably maintained in a range of pH values or in the presence of various detergents. In the range of 20 to 70°C only ~5% of the wild-type KcsA protein was found as a monomer and ~95% were assembled as tetramers (Cuello et al., 1998; Meuser et al., 1999). Additional studies have implicated the C terminus as a contributor to the stability of the tetramer (Perozo et al., 1999). Furthermore, amino acids, which support tetramerization (Splitt et al., 2000), have been identified experimentially.

Solubilization of KscA (containing a His tag) from an *E. coli* strain overexpressing the channel protein was achieved in the presence of detergents (Schrempf et al., 1995) with the most nonionic ones being very effective in retaining the degree of the KcsA oligomerization (Cuello et al., 1998). As a prerequisite for the determination of ion-flux activity, liposomes containing different lipids and KcsA were analyzed. Like recording of the channel activity (Schrempf et al., 1995), flux measurements could be carried out if asolectin (crude soy lipids) was used to generate the protein-containing liposomes (Cuello et al., 1998). In addition, native lipids or phospholipids from *E. coli* were also found to be effective. Phosphatidylethanolamine (PE) and a negatively charged phospholipid were found to be most important (Heginbotham et al., 1998). On the basis of these findings, generally tetramers were used to analyze features of the KcsA channel. More recent studies (Demmers et al., 2003) led to precise insights about the interaction between KcsA tetramers and defined lipids as well as with the native *E. coli* membrane.

GENERAL FEATURES OF THE RECONSTITUTED FUNCTIONAL KcsA CHANNEL

After fusion of proteoliposomes containing KcsA (His tag) tetramers (subsequently named KcsA) into planar bilayers (Fig. 2C), voltage-dependent currents were observed with larger mean overall currents in one direction. Owing to the different gating characteristics in either direction, the relative orientation of KcsA within the bilayer could be determined during each experiment. In most cases, the orientation of the individual channels was unidirectional; however, the orientation could vary from fusion to fusion. This finding was attributed to the sonication procedure prior to the addition to the *cis* chamber, which may have led, at least in part, to an inversion of the proteoliposomes (Meuser et al., 1999). Based on hydrophobicity predictions (Schrempf et al., 1995) and the crystal structure of KcsA (Doyle et al., 1998), it was concluded that the tyrosine residue at position 82 resides in the outer region of KcsA (Fig. 2A and 3A).

An important feature of eukaryotic channels is the block of ion currents by external agents. Tetraethylammonium (TEA) has been shown to bind strongly to K channels and hence has been a probe to study the inhibition of ion currents (Coronado et al., 1980). The effective inhibition of the wild-type KcsA by TEA prompted the exchange of the codon determining the tyrosine residue in position 82 (corresponding to position 449 in the Shaker K channel) by one for valine (Meuser et al., 1999; Heginbotham et al., 1999). The characteristics of the Y82V mutant protein as to tetrameric assembly, conductance, and internal block by tetrabutylammonium (TBA) corresponded to those ascertained for the wild-type KcsA. At the side of the wild-type and the mutant protein, which was more conducting, the 50% inhibitory concentration (IC_{50}) value for TEA was almost identical. In contrast, at the opposite side (later allocated as external) the blocking effect by TEA for the mutant channel decreased 24-fold as compared to the wild-type channel. These results allowed the unequivocal allocation of the two electrophysiologically determined distinct sides to the external or the internal face within KcsA. Moreover, gating was faster in the wild-type than in the mutant KcsA. The wild-type and the mutant channel had about identical mean open durations (τ_1 = 1 ms and τ_2 = 8.8 ms), but the amplitude ratios of the fast and the slow gating components for KcsA ($A_1:A_2 = 0.79:0.21$) and for the mutant protein ($A_1:A_2 = 0.39:0.61$) significantly differed.

Electrophysiological characteristics of the wild-type and the Y82V mutant protein disclosed that KcsA is an intrinsically outward-rectifying channel. Thus, by convention the *trans* chamber was set as an internal compartment, with outward-directed potassium currents at positive voltages (Fig. 2E).

Analysis of single channel currents revealed at least three conductance states at positive and negative potentials (Meuser et al., 1999). All point histograms of traces at ±100 mV showed that rectification is a consequence of different occupancies of the individual open states at opposite voltages with almost the same overall open probabilities ($NP_O^{out} = 0.45$ and $NP_O^{in} = 0.43$). Outward K^+ currents were mainly carried by the open 3 state, inward currents primarily due to the open 1 state. As the conductance of open 3 is higher ($\Lambda_3 = 97$ pS at 100 mV) than that of open 1 ($\Lambda_1 = 41$ pS), the mean overall currents are rectified outward (Fig. 3E). The observed subconductance states are likely the result of conformational changes (Fox, 1987). The peaks from the three open states within the all-point histograms showed rather broad current distributions, indicating that these open states correspond

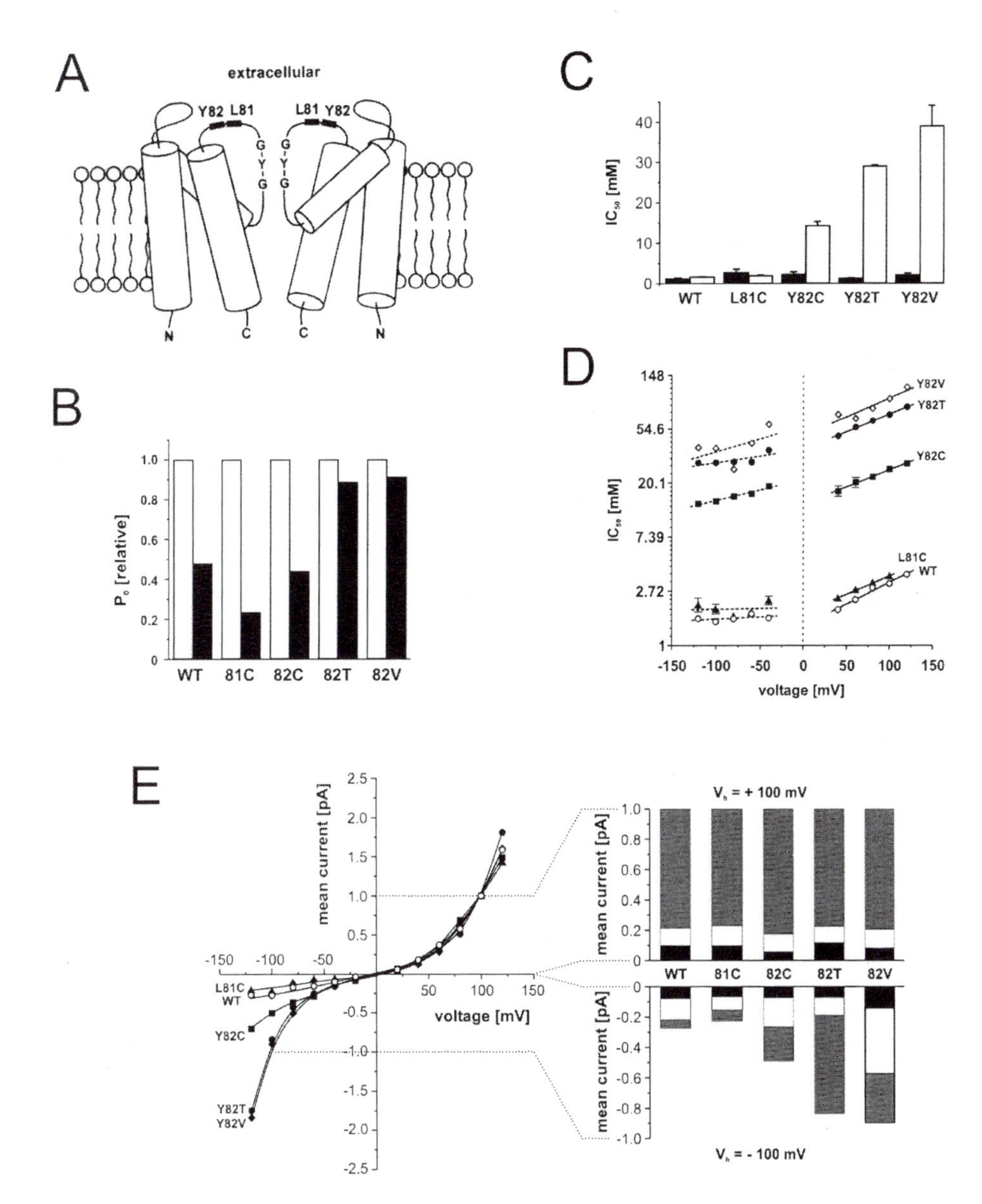

Figure 3. Comparative electrophysiological characteristics of KcsA wild-type (WT) and mutant proteins generated after exchange of residues L81 and Y82. (A) The relative positioning of the relevant amino acids within KcsA. (B) Relative open probabilities of the channels (white bars, −100 mV; black bars, +100 mV). (C) Comparative sensitivity to TEA of WT and mutant channels (white bars, −100 mV; black bars, +100 mV). (D) Voltage dependence of WT and mutant channels of external TEA block. (E) Rectification properties of different proteins.

to rather flexible conformations (Lauger, 1985). Moreover, key elements of the gating machinery in KcsA are sensitive to the direction of the electric field. Subconductance states were also characteristic of single KcsA channels within the giant protoplast-vesicle system (see earlier section of this chapter).

One property of the channel pore can be deduced from the observation that Ba^{2+} ions blocked K^+ currents from the internal side by entering but not passing the selectivity filter. With an IC_{50} of 0.52 ± 0.03 mM at +100 mV, this side was about 120 times more sensitive than the opposite one (IC_{50} = 60 ± 20 mM at −100 mV). The internal block was highly voltage dependent; it was nearly absent at negative values and considerably increased at positive potentials, when the Ba^{2+} ions were forced into the channel pore. The voltage dependence of the channel block can be described by the classical model (Woodhull, 1973) with a single binding site at 94% of the electrical distance. Analysis of the K^+ conduction using the constant rate approach revealed the necessity of at least two binding sites. Possibly Ba^{2+} ions crossing most of the electrical distance from the internal side interact with the external binding site.

To explore further characteristics of the open channel, the block of KcsA by quaternary ammonium ions (QAs) with different chain lengths (Le Dain et al., 1994; Pascual et al., 1995; Hille, 1992) was systematically studied at both sides of the KcsA-containing bilayer (Meuser et al., 1999). Significant differences could be observed at the internal side of KcsA, where the block increased (up to 10^7-fold) in correlation with the chain length (up to C4) of the QAs, with TBA being the strongest inhibitor (IC_{50} = 31 ± 2 nM). Compared to TBA, the block efficiency decreased for QAs with chain lengths of 5 to 7 atoms. The block at the internal side by tetramethylammonium (TMA) and at the external side by TMA, TEA, and tetrapropylammonium, TPrA, respectively, was fast ($\tau_{on} < 100\ \mu s$). At the external side, TEA showed the strongest inhibition (IC_{50} = 1.7 ± 0.1 mM).

To account for the movement of QAs into the cavity from the internal entrance of the pore, this has to widen to a diameter larger than that found in the crystal structure. This could be achieved before or during opening by rotations of TM1 and TM2 identified by electron paramagnetic resonance (EPR) studies (Perozo et al., 1999). Ammonium (d = 3.1 Å), methylammonium (d = 3.8 Å), dimethylammonium (d = 4.6 Å), and trimethylammonium (d = 5.2 Å) ions were found to pass the pore of the reconstituted KcsA channel, but not TMA (d = 5.5 Å) ions (Meuser et al., 1999). When the dimensions of these compounds were taken into account, it had to be deduced that the crystal structure of KcsA pictures the pore (Doyle et al., 1998) not in the open but in a closed state (Meuser et al., 1999).

CHANNEL OPENING AND CATION SELECTIVITY AT LOW pH

During initial experiments (Fig. 2D) KcsA activity was determined at pH 7.2 (Schrempf et al., 1995). Subsequently fluxes within liposomes have been recorded under neutral pH (Heginbotham et al., 1998), whereas other authors found fluxes preferentially under acidic conditions (Cuello et al., 1998). On the basis of additional studies of bilayers containing reconstituted KcsA, proton activation has been implicated and subsequently considered by several investigators.

Within the KcsA-containing planar bilayer system, currents were recorded at either side at a pH of 4 or in a configuration where one side was kept at pH 7 and the other at pH 4. Perfusion of the internal side with a buffer of pH 7 led to a rapid inactivation of KcsA;

however, gating could be reactivated by replacing the buffer again with one with a pH of 4. KcsA, in contrast, remained active after perfusion of the external side with a buffer of pH 7. The data clearly demonstrate that only the internal side of reconstituted KcsA is pH sensitive (Fig. 2E). Due to this effect most of the subsequent experiments considered these conditions.

Further basic electrophysiological properties of the channel such as ion permeation and selectivity were studied. The relative permeabilities (P_X/P_K) of several monovalent and divalent cations to those of potassium ions were calculated from the bi-ionic reversal potentials in 250 mM K^+/250 mM X^+ (or 125 mM X^{2+}) using the constant field approach (Goldman-Hodgkin-Katz voltage equation) (Hille, 1992). The ratios of anion and cation selectivity were determined from the reversal potentials in asymmetric potassium solution (245 mM KCl/20 mM KCl). KcsA has no detectable permeability for chloride ions. It is strongly selective for cations; among those tested, it showed highest preference for Rb^+ (1.15 ± 0.03), K^+ (set as reference), and Cs^+ (1.0 ± 0.2), but was less permeable for NH_4^+ (0.19 ± 0.02), Na^+ (0.09 ± 0.01), and Li^+ (0.03 ± 0.003). Thus selectivity follows the Eisenman III sequence, implying a weaker field interaction at the selectivity filter and a larger radius of the cation-binding site. The reversal potentials and conductances determined in solutions with different concentration ratios of potassium and sodium ions (internal 250 mM K^+/external x mM K^+, y mM Na^+; $x + y = 250$) displayed independence of ionic movement through the channel. Notably, the relative permeabilities for Mg^{2+} and Ca^{2+} were comparatively low (0.022 ± 0.002 and 0.015 ± 0.003).

The K^+ currents carried by KcsA in the open 3 state were measured in symmetrical solutions ranging from 250 mM to 1.5 M KCl. In contrast to a linear current-voltage relationship, which would be expected for ionic movement governed solely by electrodiffusion, currents saturated in dependence on the applied voltage as well as on the potassium concentration (Meuser et al., 1999).

Under asymmetric conditions the selectivity of K^+:Na^+ was determined to be 11:1 (Meuser et al., 1999). In constrast, under symmetrical conditions at low pH the selectivity of K^+:Na^+ was documented as 150:1 (LeMasurier et al., 2001); however, subsequent analyses of these data revealed the results were inconsistent with the proposed kinetic model (Nelson, 2003). Transport properties, which were determined with the aid of flux measurements under neutral conditions, revealed the following permeability: $K^+ \approx Rb^+ \approx Cs^+ > Na^+ \gg Li^+$. Na^+ was found to be at most 10% as permeant as K^+. Cs^+ appeared more permeable when present externally (Heginbotham et al., 1998). In contrast, other authors could not record fluxes at neutral pH, but at low pH (less than 5) at one side the selectivity was $K^+ > Rb^+ > NH_4^+ \gg Na^{2+} > Li^+$ (Cuello et al., 1998). It can be suspected that the stability of the phospholipids may be altered at low pH, and in addition, the charges of amino acids will differ from those at neutral pH. This accounts especially for the histidines at the C terminus, which will be postively charged at low pH. As a consequence, repulsive forces may be expected to affect the bundle formation of the C termini within the KcsA tetramer.

Together the data showed that channel characteristics are considerably influenced by symmetric or asymmetric conditions or different pH values. Based on subsequent EPR studies, pH-induced opening of the intracellular part of the pore by conformational changes in the C-terminal part of TM2 was deduced. In contrast, Y82 (located at the external side of the selectivity filter) (see Fig. 3A) showed no spin-label mobility shifting from pH 7 to pH 4 (Perozo et al., 1999). Since KcsA is mainly present in a nonconducting state, even at pH 4

(P_{closed} > 90%), it was assumed that complete opening of the channel cannot solely be related to the conformational changes deduced from those studies (Meuser et al., 2001).

EFFECTS OF ELECTROCHEMICAL GRADIENT, pH, AND TEMPERATURE ON OPENING AND SELECTIVITY

A reflection of the divergent results for channel activity (see above) induced the investigation of the effect of the electrochemical gradient (Zakharian and Reusch, 2004b). For this KcsA was reconsituted and analyzed at neutral pH as initially described (Schrempf et al., 1995). Internal concentrations were kept at 200 mM and varying concentrations (200 to 10 mM) were applied to the external side. High concentrations of K^+ (200 to 120 mM) did not lead to an opening of the channel. The potential required for channel gating was found to be decreased in correlation with an increasing K^+ gradient. The open probability rose from 0.4 to 0.9 while the K^+ gradient rose from 2.5:1 to 20:1 (Zakharian and Reusch, 2004b).

Channel activity had been recorded under different pH conditions, and depending on the laboratory, asymmetric and/or symmetric conditions had been used. As a consequence, the channel features varied considerably (see previous section). Selectivity was shown to be affected by the application of symmetric or asymmetric conditions (Schrempf et al., 1995; Meuser et al., 1999; Heginbotham et al., 1999). Based on different results obtained in the bilayer system compared to the giant protoplast system, which is expected to mimic the physiological conditions more closely (Schrempf et al., 1995) (see above), careful comparative reinvestigations of the bilayer system were pursued at about neutral pH conditions (Zakharian and Reusch, 2004a). Under asymmetric conditions (200 mM KCl intracellular, 20 mM KCl extracellular at pH 7.2), the KcsA channel within a bilayer was confirmed to be cation-selective. Under bio-ionic conditions divalent ions, Mg^{2+} or Ca^{2+}, were found to be impermeable from the extracellular side but permeable from the opposite one. Though permeable from the intracellular side, the presence of high concentrations of Ca^{2+} or Mg^{2+} blocked the currents rapidly under bi-ionic conditions. Experiments in the presence of EDTA showed that chelating Mg^{2+} at the intracellular side (but not on the other side) leads to channel closure. Additional experiments revealed that the reversal potential abruptly changes from −27 to −54 mM at pH 7.2, but not at slightly higher (up to pH 7.5) or lower (pH 6.8–6.5) values. At 22°C conductances in the range of 12 pS (pH 7.2), 38 pS (pH 7.0), and 44 pS (pH 6.8) were observed. During testing of a range of temperature conditions (22 to 30°C), an abrupt increase in selectivity for K^+ ions at the internal side was noticed to occur at 26°C. This effect is likely due to a conformational change. At 30°C at pH 7.2, high K^+ selectivity occurs at both sides and chord conductances for outward current corresponded to 100 pS (open probability [*Po*], 0.35 ± 0.05) and with a subconducting state of 20 pS (*Po*, 0.9 ± 0.05). Under these conditions the Mg^{2+} and Ca^{2+} were impermeable. Under bi-ionic conditions (200 mM KCl and 200 mM KCl or NaCl) selectivity ratios with respect to Rb^+ or Na^{2+} were 1.2:1 or 20:1.

Considering all optimized parameters, it can be concluded that even in a bilayer system the highest activity of KcsA can be achieved if a combination of several parameters is applied: pH 7.2, ≥26°C, intracellular Mg^{2+}, and an outward current. The measured chord conductances are in a similar range compared to those observed by the patch-clamp analysis of the giant protoplast vesicles (Schrempf et al., 1995) (see previous section) at pH 7.2 under asymmetric conditions, which match the optimal growth conditions of the vegetative mycelium of *S. lividans* hyphae. The sharp maximal activity at ≥26°C suggests a conformational transition. Channel activity was sensitive to both the membrane potential and the size of the gradient, indicating that gating depends on both components of the electrochemical gradient (Zakharian and Reusch, 2004a; Zakharian and Reusch, 2004b).

AMINO ACIDS AFFECTING ASSEMBLY AND ACTIVITY

To identify amino acids that may affect the tetrameric assembly (Fig. 2A and B, Fig. 3A), several designed mutations within the *S. lividans kcsA* gene resulted in a set of mutant proteins, which were characterized with respect to their assembly and channel activities (Splitt et al., 2000; Meuser et al., 2001). (i) The amino acid residue leucine 81 (L81) located at the external side of KcsA was found to be exchangeable by a cysteine residue without affecting the assembly and the channel characteristics. (ii) The substitution of the tyrosine residue 82 by cysteine, valine, or threonine, but not by glycine, led to functional channel types. Like the wild-type and an L81C channel, the mutant channels exhibit an internal pH-sensitive side and are cation-selective. (iii) Substitution of the first glycine (G77) residue within the GYG motif by an alanine (A) or substitution of the tyrosine (Y) residue 78 by a phenylalanine (F) led to mutant proteins that form tetramers of reduced stability. In contrast to the mutant protein, containing the AVG motif, GFG functions as an active K^+ channel whose characteristics correspond to those of the wild-type KcsA channel. (iv) The investigated mutant proteins, which carry different mutations (T72A, T72C, V76A, V76E, G77E, Y78A, G79A, G79D, G79E) within the signature sequence of the pore region, do not or only to a very small degree assemble as tetramers and lack channel activity (Splitt et al., 2000).

Comparisons of these experimental results with those deduced from earlier studies of eukaryotic channels and the crystal structure as well as EPR of KcsA led to the following insights. An exchange of threonine T72 located within the signature motif of KcsA by alanine induced the formation of the mutant protein T72A, which did not assemble as a stable tetramer in vitro and did not function as an active channel in a bilayer system. This is in agreement with the lack of detectable currents within *Xenopus* oocytes after injection of mRNA derived from each of the mutant genes predicted to encode the corresponding mutant Shaker ShB protein T439A or T439S (Yool and Schwarz, 1991) or each of the Shaker H4 mutant proteins T439A or T439V (Heginbotham et al., 1994). Only a very small portion of the KcsA mutant protein T72C assembled as tetramer; its lack of functionality was thus not surprising. Also, no currents were detected within *Xenopus* oocytes after injection of mRNA derived from a mutant gene predicted to encode the mutant T370C Kv 2.1 channel (Pascual et al., 1995) (in a position corresponding to T72C in the mutant KcsA channel). As shown by EPR studies, the T72C-substituted KcsA displayed increased steric contacts while moving away from the symmetry axis (Perozo et al., 1999). Inspection of the eight-residue sequence of the pore signature of the known deduced potassium channel proteins revealed that most of them comprise a threonine residue in the first position of the signature motif. A substitution by cysteine is encountered within the deduced *Drosophila* EAG channel protein (Warmke et al., 1991), the signature motif of which, however, considerably differs from that of KcsA and all members of the Kv channels. Replacement of the valine residue in position 76 of KcsA by either alanine or glutamic acid induces the formation of proteins that lack tetrameric assembly and functional channel properties in a bilayer. It had previously been shown that an exchange of the corresponding valine residue (V443) by alanine in Shaker H4 leads to a nonselective channel, whereas a substitution by glutamic acid did not entail functional expression (Heginbotham et al., 1994). Whether this may also be due to the lack of assembly or a low stability of the mutant tetramer has, however, not been investigated. Within the crystal structure of KcsA (Doyle et al., 1998), V76 is oriented toward the pore helix. Its exchange by glutamic acid may lead to a repulsion of the negatively charged glutamic acid in position 71, which has not yet been positioned in the crystal structure.

Most K^+ channels contain the conserved GYG motif, which is involved in ion selectivity (Heginbotham et al., 1994; Slesinger et al., 1996; Navarro et al., 1996). KcsA mutant proteins G77E, G79A, G79D, and G79E neither assemble in vitro nor form a functional channel. Replacement of the aromatic residue Y78 by alanine stops tetrameric assembly; substitution by another aromatic residue (F), on the other hand, leads to the formation of tetramers with reduced stability. Contrary to the Y78A protein, Y78F still functions as a K^+ channel whose electrophysiological properties correspond to those of the wild-type KcsA. On the basis of the crystal structure (Doyle et al., 1998), the aromatic tyrosine 78 residue was shown to interact with the two tryptophane residues in positions 67 and 68. It is thus evident that a similar interaction is caused by a phenylalanine residue, but not by an alanine residue in position 78. In the position corresponding to Y78 in KcsA, a phenylalanine residue is present in a few deduced channel proteins, including EAG (Warmke et al., 1991), HERG (Trudeau et al., 1995), Kir6.2 (Inagaki et al., 1995), and the deduced TREK-1 monomeric protein (Fink et al., 1996) with two predicted pore regions. The side chains of the mutated G79A subunits are assumed to be oriented toward the pore in the KcsA crystal structure. Considering this structure, it is not obvious why G79A does not form a stable tetramer. Since the protein G77A still assembles as a nonfunctional tetramer, this exchange may lead to a smaller diameter of the pore and thus hinder the permeation of K^+ ions. The exchanges of G77E, G79E, and G79D might be unstable, due to repulsion effects with other amino acids within the pore region. Mutations in corresponding positions (CYG and GYC) within the Kv 2.1 channel exhibited no currents, either (Pascual et al., 1995). In view of these findings, it is surprising that functionality was previously reported for the mutants GYA and AYG from the Shaker channel (Heginbotham et al., 1994).

Substitution of the amino acid residue L81 of KcsA by cysteine leads to the formation of a protein that still assembles as tetramer but has a much lower melting point than the wild-type protein. In a previous attempt to define pH-dependent molecular movements near the selectivity filter (Perozo et al., 1999), many amino acids had been replaced by cysteine residues. No significant change could be ascertained for exchanges of amino acid residues (including L81C) at the external side of the selectivity filter in probe mobility or spin-spin interaction. The corresponding mutant V378C from Kv 2.1 also formed an active channel (Pascual et al., 1995).

In the crystal form that has been used to elucidate the structure of the KcsA channel, the fourfold symmetry cannot be seen within the X-ray diffraction patterns. Therefore it cannot be concluded that the four subunits supply identical sets of atoms in a symmetrical manner to provide the pore lining (Choe et al., 1999). However, it could be deduced that the TVGYG motif (amino acids 75 to 79) forms a stretch of 12 Å in length with the narrowest opening of the pore (Doyle et al., 1998). As the dimensions of the selectivity filter are determined by lining the groups, hydrogen bonds, and van der Waals interactions with the remaining protein, it is conceivable that already subtle alterations of the amino acid composition in this region affect the stability of the tetrameric arrangement.

In this context it is interesting to note that on the basis of simulations of molecular dynamics a salt bridge between D80 and R89 of neighboring subunits has been deduced that possibly plays an important part in the stabilization of the KcsA tetrameric structure (Guidoni et al., 1999).

DYNAMIC CHANGES WITHIN THE PERMEATION PATHWAY

K^+ currents mediated by the KcsA wild-type channel are composed of distinguishable subconductance levels, which can be allocated to at least three classes. Outward K^+ currents were mainly carried by the open 3 state. Inward currents also show these three subconductances; however, the open states 1 and 2 are dominant (Meuser et al., 1999). The relatively broad current distributions of the three open states picture the dynamic changes during the opening process. The occupancy of the subconductance levels depends on the direction of the transmembrane electric field resulting in intrinsic rectification of the currents. The characteristic gating properties, in accordance with previous EPR studies (Perozo et al., 1999), point to a high plasticity of the pore.

Detailed analyses showed that the currents of the analyzed mutant proteins (Fig. 3) had wild-type-like subconductance states, and outward they were mainly carried by the open 3 state ($\Lambda_3 \cong 97$ pS at $V_h = +100$ mV in 250 mM K^+). The inward current of the mutant L81C channel corresponded to that of the wild-type channel, and here the open 1 state ($\Lambda_1 \cong 41$ pS at $V_h = -100$ mV in 250 mM K^+) dominated. The inward currents of Y82C, Y82T, and Y82V, in contrast, considerably differed in the relative abundance of each of the three open states. Compared to the wild-type channel, the current of the mutant channels is most frequently carried by open 3 (Y82T), open 2 (Y82V), or equally by open 1, 2, and 3 (Y82C). Furthermore, the ratio of open and closed areas within the histograms for the mutant channels at a V_h of −100 mV was two times lower (L81C), nearly identical (Y82C), or about two times higher (Y82V, Y82T) as compared to that of the wild-type KcsA. The rectification properties diverge scarcely (L81C), little (Y82C), or considerably (Y82T, Y82V) from those of the wild-type KcsA (Fig. 3E). The mean inward and outward currents are identical for Y82T and Y82V, respectively; thus they have lost the characteristic pronounced outward rectification property of the wild-type KcsA. The importance of position 82 is underlined by the fact that a substitution to glycine leads to loss of channel activity. It is evident that the amino acid type in position 82 plays a crucial part (directly or due to its interaction with other amino acids) in stabilizing an open conformation(s) at negative potentials. This may also explain the varying degrees of rectification (Meuser er al., 2001).

At the internal side of the channel a slow block is caused by TEA ($IC_{50}^{int} = 1.2 \pm 0.3$ mM in 250 mM K^+, $V_h = +100$ mV), reducing the open probability of the channel and leading to a reduced mean current with an unaltered single channel current. The internal block of all mutant proteins was nearly identical to that of the wild-type channel. The external block is extremely quick; consequently, due to instrumental limitations, fast current fluctuations (< 50 μs) could not be resolved and resulted in apparently reduced single channel current amplitudes. The IC_{50} of the external block of the wild-type ($IC_{50}^{ex} = 1.7 \pm 0.1$ mM in 250 mM K^+, $V_h = -100$ mV) and of the L81C channel were almost identical; the TEA sensitivities of the Y82 mutant channels, in contrast, were reduced on this side, in the following order: Y82 > Y82C > Y82T > Y82V. Thus the IC_{50} for TEA was the highest for the Y82V channel; recently calculated binding energies are in agreement with this finding (Luzhkov et al., 2003).

The external TEA block for each channel type was dependent on the voltage. When outward currents of the dominating open 3 state at positive voltages were evaluated (Woodhull, 1973), the relative electrical distances (δ) for the external block were 20% ± 2% (wild-type), 18% ± 4% (Y82V), 17% ± 1% (Y82T), 17% ± 1% (L81C), and 16% ± 1% (Y82C). The TEA block for the inward currents, as deduced from the maximum of the averaged open states 1 to 3, suggests an apparent shallow site. However, the apparent relative electrical distance for the mutant channels Y82C, Y82V, and Y82T was larger than that of the wild-type and the L81C channel (Fig. 3B and C). These effects may be due to a voltage-dependent conformation of the outer vestibule to modify the relative electrical distance of the TEA-binding side.

After *Shaker* mutant channels were analyzed, the diameter of the cage formed by the tyrosine residues (T449Y) trapping one TEA ion has been proposed to amount to about 8 Å (Heginbotham et al., 1994). Within the crystal structure of KcsA, the Y82 residues (corresponding to T449Y in the Shaker channel) are located at the external vestibule but are about 10.8 to 15 Å apart. However, due to spatial inhomogeneity, the resolution in this crystal region was relatively poor and thus dimensions could not be unambiguously determined (Doyle et al., 1998). As the crystal structure has been deduced to represent the closed form (see above and Meuser et al., 1999), the precise position of Y82 is not yet known for the open channel. However, studies with KcsA mutant channels show clearly that the extent to which an open conformation(s) is stabilized at negative potentials varies, depending on the amino acid type in position 82 (Meuser et al., 2001).

The subsequent in silico calculations show that both TMA, TEA, and the larger TPrA can be accommodated at the external binding site. The docked position of the small TMA lies near one of the four Y82 residues, in which case the nitrogen carbon is displaced away from symmetry axis of the channel. The docked TEA and TPrA ions find positions in the central part of the outer activity between the Y82 phenyl rings. In molecular dynamics (MD) trajectories of the channel-TMA complex the ligand fills only a part of the tyrosine cage space, and its position is subject to substantial variation. Thus the ligand drifts inside the outer vestibule and interacts with different pairs of Y82 residues. In MD trajectories the ligand positions for the channel complexes with TEA remain stable. Further investigations suggest that the ligand does not have an entirely rigid position in the outer ring. This can be seen from partial rotations of the Y82 phenyl rings to maximize the contact surface with the ligand. The absolute solvation-free energies of the blocker increase when the size of the ammonium ion becomes smaller. The results support the above-outlined findings (and are in contrast to the common view that the outer cavity precisely matches to bind TEA) about the binding of larger ions and thus reflect that the available volume is larger. Based on further calculations for the wild-type and the Y82V channel, it has been revealed that for absolute binding affinities, the combination of polar and hydrophobic (nonpolar) interactions is important for stabilizing tetraalkylammonium ions at the external binding site (Luzhkov et al., 2003).

INTERACTION OF KcsA WITH PHOSPHOLIPIDS

The stability of KcsA is not only caused by protein interaction between KcsA subunits but also by the lipids. The activity of KcsA within a bilayer has been found to depend on the type of lipid being used; the presence of negatively charged lipids is especially important. Some recent studies have elaborated more details regarding the interaction of KcsA with lipids of a designed composition.

The negatively charged lipid phosphatidylglycerol (PG) and the nonbilayer lipid PE support tetramerization and membrane association of KcsA better than the zwitterionic bilayer lipid phosphatidylcholine (PC). PG was found to be important for membrane assembly and thermostability of the KcsA tetramer, and this effect likely occurs via electrostatic interactions. Systematic comparative studies suggested that the stabilization of the tetramer within a reconstituted bilayer occurs dominantly by the surrounding lipids. The effect of PE might be explained by its capacity for hydrogen bonding to the protein. Alternatively, PE could possibly surround KcsA with its relatively small headgroup and may lead to altered and perhaps more favorable packing properties at the protein-lipid interphase as compared to PC (Demmers et al., 2003).

Tetramerization and membrane association of newly synthesized KcsA were found to be most efficient in vesicles containing small amounts of the negatively charged PG together with a large amount of zwitterionic PE. In contrast, in vesicles prepared from pure dioleoylphosphatidylcholine (DOPC), no tetramerization of KcsA was detected and membrane association was slow. Possibly PG (up to 20 to 30%) could assist targeting to the membrane and proper folding. The inhibition of tetramerization by higher concentrations of PG was explained by a possible too-high negative charge density. PE assists tetramerization very efficiently; this effect could be due to the small size of the PE headgroup, which may facilitate the initial interfacial insertion. Subsequently PE could possibly promote oligomerization as a consequence of high packing density in the hydrophobic part of the membrane as compared to that of a bilayer containing only PC. A unifying principle for protein-lipid interactions is the lateral pressure, which varies with the depth within a bilayer and which is localized at the interphase between the headgroups and the acyl chain region, due to interfacial tension. The positive lateral pressure in the headgroup region is the result of repulsions among the headgroups and a positive lateral pressure due to acyl chain repulsions. The pressure profile is affected by the lipid type as well as small solutes.

The addition of 2.2.2-trifluoroethanol (TFE) leads to dissociation of tetrameric KcsA to the monomeric form. This process was found to be independent of the presence of the C terminus of the KcsA protein, suggesting that stabilizing interactions occur mainly within the transmembrane domain. This dissociation was shown to be caused in part by interactions between TFE and KcsA and dominantly by an indirect effect via changes of the stabilizing interactions of the protein with lipids. Based on these observations, it has been proposed that TFE inserts in the headgroup region of the lipid bilayer. This would result in an increased lateral pressure in the headgroup region, which is compensated by a lower lateral pressure in the acyl chain region. Small alcohols (i.e., methanol, ethanol, and propanol) were observed to induce also dissociation of the KcsA tetramer. This effect correlated with the octanal-water partitioning coefficient and hence the ability of the alcohols to partition into the bilayer. Based on the fact that ethanol localizes preferentially in the lipid headgroup region, it has been concluded that other small alcohols also have this feature. As a consequence, a redistribution of the lateral pressure appears likely and will subsequently lead to dissociation of the tetramers of KcsA (van den Brink-van der Laan et al., 2004).

ASSEMBLY OF KcsA WITHIN THE INNER MEMBRANE OF AN *E. COLI* HOST

Very little is known about how oligomeric proteins are integrated and correctly assembled into a membrane in vivo. To study this process, *E. coli* carrying a plasmid with the *kcsA* gene (under the control of an inducible promoter) was analyzed at different time periods after expression was induced with in vivo labeling conditions (pulse-chase experiments with labeled methionine). KcsA monomers were found to assemble within the *E. coli* host directly into tetramers without the detectable levels of dimers and trimers. Further studies indicated that the monomer directly interacts with the inner membrane in which the assembly to the tetramer occurs. This process was found not to be dependent on ATP hydrolysis. The proton motive force has been proved to be essential for the efficiency of oligomerization (van Dalen et al., 2000). In vitro studies using a transcription-translation system revealed that no tetramer is formed in a membrane-free reaction but only after the addition of *E. coli* inner membrane vesicles. Dissipation of both ΔpH and ΔΨ on the CCCP (an uncoupler of the proton motive force) leads to inhibition of tetramerization. The addition of nigericin (which destroys only ΔpH but not ΔΨ) does not affect the tetramerization. Taken together, the data suggest that ΔΨ is the essential component of proton motive force during the oligomerization process. Possibly, ΔΨ could have an effect on stabilizing the highly charged C terminus but might also affect charges around the P region, which has been shown to be important for tetramerization (Splitt et al., 2000). The results also indicate a novel role of the proton motive force in assembly of a membrane protein (van Dalen et al., 2000).

With *E. coli* mutants it could be shown that the single recognition particle pathway is important for the efficient assembly of the KcsA, most likely for correct insertion of the monomer allowing tetramerization. At reduced SecEYG levels, assembly of KscA occurs with reduced efficiency. With a coupled in vitro transcription-translation system, highest tetramerization was recorded in the presence of pure lipid vesicles, demonstrating that a phospholipid bilayer is the minimal requirement to form the KcsA tetramer (van Dalen et al., 2002a; van Dalen et al., 2002b).

INTERACTION OF KcsA WITH NONPROTEIN MOLECULES FROM AN *E. COLI* HOST

Polyhydroxybutyrate (PHB) and inorganic phosphate (polyP) are widely distributed among prokaryotic and eukaryotic organisms. PHB is a linear polymer of (R)-3-hydroxybutyrate and is an amphiphilic polyester that forms ion-conducting complexes with salts. PolyP is a linear polymer of phosphoryl units joined by high-energy phosphoanhydride bonds and has the capacity for ion exchange and the ability to discriminate among cations by charge (Reusch, 1999). After extraction (from the heterologous *E. coli* host) PHB was found to be bound to each KcsA monomer; the amount corresponded to about 28 units per tetramer. PHB is highly insoluble in water and thus it has been suspected to be located in the hydrophobic region of the channel. PolyP (about 15 units) was suggested to be held by ionic forces within the KcsA tetramer and to be bound to positive residues on the outside of the assumed intracellular portion of the tetramer. Within the truncated protein used for crystal structure the theoretically calculated pI value of KcsA is

10.3. The considerably lower value (6.5 to 7.5), which was determined experimentally, would be consistent with the inclusion of a polyanion. Such a compound was, however, not identified within the crystal. The partial X-ray structure (Doyle et al., 1998), however, depicts only 9 of the 32 charged residues of each subunit, since it lacks the C terminus. In the structure, the acidic residues face inward and the basic ones outward, indicating that the channel has a positive surface charge. However, the pI of the tetramer indicates that the surface charge of the integral channel is nearly neutral. More experimental data are required to test the proposition that the supramolecular channel contains, in addition to the protein, PHB and polyP. The crystal structure has been generated from a protein lacking the C-terminal 40 amino acids. Within bilayers a protein lacking these amino acids has been shown to lack functionality (Perozo et al., 1998; Perozo et al., 1999). The data reveal an important role of the C terminus. The increasing numbers of modeling studies on ion transport in KcsA are all limited to data deduced from the crystal structure. As a consequence, the experimentally identified important contribution that the C terminus is supposed to have an intracellular position has not yet been considered. It has recently been suggested (Zakharian and Reusch, 2004b) that KcsA may recruit PHB and polyP to form a conductive core that selects and transports K^+ to the inner face of the selective filter. Incorporation of the polyanionic polyP within the intracellular bundle of polypeptides would reverse the net charge from positive to negative, resulting in attraction of cations. In particular, the eight arginine residues at the intracellular mouth may create an environment for the selection of monovalent cations. The cation-binding cavities formed by phosphoryl oxygens of polyP and ester oxygens of PHB are suspected to displace water of hydration and select for K^+ over Na^+ in a similar fashion as the amide carbonyl oxygens of the selectivity filter.

Fluorescent quenching studies suggest that during reconstitution into anionic phospholipid no large conformational change of KcsA is occurring (Alvis et al., 2003). Two classes of binding site for lipid on KcsA have been deduced from the quenching data: one set corresponds to annular binding sites around KcsA with an equal high affinity for DOPC and two chain anionic phospholipids. The nonannular sites are those at which anionic phospholipids bind. Additional data suggest that electrostatic interactions between the anionic phospholipid headgroup and positively charged residues on KcsA are important for binding at the nonannular site. This site is assumed to correspond to the binding site of the lipid molecule found in a deep cleft at each monomer interface in the tetrameric KcsA crystal structure. The headgroup of the lipid is, however, not resolved in the crystal structure. Initially it has been modeled as a diacyl-glycerol but later it was presumed to be phosphatidylglycerol (Valiyaveetil et al., 2002).

MODULE SHUFFLING AMONG KcsA AND EUKARYOTIC CHANNEL PROTEINS

Various K^+ ion channels from different eukaryotes have been analyzed after the translation of the corresponding mRNA within oocytes from *Xenopus laevis* (Hille, 1992) using patch-clamp analysis. Thus it was obvious to test a similar strategy for KcsA. After a polyadenylation tail (which is absent within the original *Streptomyces* sp. host) was attributed using an in vitro transcription system, the resulting mRNA was injected into oocytes. However, no extra channel activity could be detected (Schrempf et al., 1995).

In the course of more recent studies, the pore region of other channels has been replaced by that of KcsA. To achieve this, chimeras had been designed, in which the region encoding the S4-pore-S5 region of two eukaryotic genes for channels with six transmembrane regions (Kv1.1 and Shaker) had been replaced with the corresponding region of the *kcsA* gene. After injection of the mRNA into oocytes of *X. laevis*, translation was recorded, but channel activities were not detectable. The results of additional immunological experiments suggested that though the chimeras assemble into stable complexes that are delivered to the cell surface, they do not exhibit channel activity or gating currents (Caprini et al., 2001). Taking into account the importance of the C terminus of the Shaker channel for its activity, a range of additional chimeras was studied (Lu et al., 2002). Increasing the length of the replacing KcsA pore correlated with the reduction of the voltage dependence of the chimeras. It could be concluded that the voltage dependence reflects a property of the voltage-sensing modules in the Shaker protein but not in the substituting KcsA part. This result is in agreement with the finding that the pore region of an inward rectifier can be replaced by that of KcsA and it still acts as an inward rectifier.

The DNA region of the *kcsA* gene being situated among the M1 and M2 encoding regions could be exchanged by the S5-S6 linker region of the eukaryotic Kv-channel Kv1.3. The chimeric genes were expressed in *E. coli.* A sensitive binding assay revealed that the tetrameric form of the KcsA-Kv1.3 chimeric protein interacts with the scorpion toxins kalliotoxin and homotoxin (Legros et al., 2000).

CLOSE AND DISTANT RELATIVES OF KcsA

Multiple aligments had suggested that KcsA has closest kinship to the P-region and its neighboring transmembrane helices (S5 and S6) of eukaroytic voltage-gated K channel families (Schrempf et al., 1995). The sequencing of an increasing number of bacterial genomes revealed an increasing number of prokaryotic genes encoding putative channel proteins with two or more transmembrane regions and sometimes accessory domain(s). As outlined in the following, the functions of some of these proteins have been investigated.

A search for segments of a bacterial homolog corresponding to one eukaryotic type of glutamate receptor (GluR) resulted in the identification of a deduced protein (named GluR0) from the cyanobacterium *Synechocystis* strain PCC6803 (Chen et al., 1999). Subsequent investigations revealed that the protein had in addition two binding sites for glutamate two-transmembrane segments with a connecting region including a TVGVG motif. The corresponding gene was cloned in *E. coli* and the protein was obtained in a soluble form. The protein had strongest binding capacity to glutamate. Competition experiments revealed the most important determinants for ligand binding. Using the oocytes of *X. laevis* or HEK cells, no channel activity was identified after injection of the mRNA corresponding to the gene for GluR0. After the region corresponding to the signal sequence was substituted by a 5′ untranslated region of GluR6 (a kainate-preferring eukaryote GluR), expression was achieved. Subsequent studies revealed that cyanobacterial GluR0 encodes a glutamate-gated channel that exhibits the ion selectivity properties of a potassium channel. The findings led to the following conclusions. (i) The M1, P, and M2 pore-forming regions in GluRs and potassium channels have a similar architecture, and the difference lies primarily in the selectivity filter. (ii) The relationship of the bacterial GluR0 and the eukaryotic GluRs suggests also tetrameric assembly of GluRs. (iii) The membrane-spanning

M3 region within GluRs and lacking in GluR0 may play a specific role for GluRs. (iv) GluR0 or a homolog is suspected to have been an evolutionary precursor of eukaryotic ionotropic glutamate receptors (Chen et al., 1999).

Compared to bacteria, which encode none or two putative channels, the genome of the hyperthermophilic archaeon *Methanococcus jannaschii* was found to determine a significant number of diverse predicted channel types (Derst and Karschin, 1998). Two of their deduced proteins, MjK_1 and MjK_2, have two transmembrane helices and a region predicted to present a pore region TXGYGD with similarity to KcsA. Both proteins contain a cytoplasmatic RCK domain, which consists of regions that were initially identified within the *E. coli* potassium transport system Trk. One of these regions (TrkA-N) includes a Rossmann motif known to bind NAD or NADH. After the corresponding gene was cloned into an *E. coli* vector, the MjK_1 protein was identified as a tetramer (166 kDa) within the membrane of the *E. coli* transformant. The expression of the MjK_1 encoding gene in an *E. coli* host (which is defective in its three major potassium uptake systems) led to growth of the cells in a range of 0.5 to 5 mM KCl. The amount of protein produced after induction was very low. Hence, it has been concluded that the protein is in the open configuration to allow K^+ uptake. The gene encoding MjK_2 was also cloned into an *E. coli* strain lacking its free major potassium uptake systems. The expression rate was high, but the protein did not complement the defect in the *E. coli* triple mutant. In contrast to KcsA, the isolated MjK_2 protein was found not to form stable tetramers in vitro (Hellmer and Zeilinger, 2003).

The genome of the archaebacterium *Aeropyrum pernix* has been found to encode a homolog (KvAP) of voltage-gated Kv channels comprising the helices S5 and S6 connected by the P-region and additionally the helices S1 to S4. The corresponding gene has been overexpressed in *E. coli* (Jiang et al., 2003) and subsequently crystallized (see next section). MthK is determined by a gene from the archaeon *Methanobacterium thermautotrophicum* that has homology to mammalian BK voltage Ca^{2+}-gated channels. After the gene was cloned in *E. coli*, the MthK protein was accessible in larger quantities and subsequently crystallized (see next section).

The gene *lctB* from *Bacillus stearothermophilus* encodes a protein of 134 amino acids, which has a predicted M1-P-M2 topology. Except for the signature sequence, the protein does not share a relevant number of identical amino acids with other channel proteins including KcsA. After the gene was cloned in *E. coli*, the corresponding His tag protein was found to be produced only in low amounts. Channel activity was recorded after reconstitution in a bilayer and after expression in oocytes. Due to the low level of production within *E. coli* and the instability of the tetramers, further biochemical studies have not been done (Wolters et al., 1999).

The open reading frame (*kch*) was found within the sequence information of the bacterium *E. coli* and its predicted protein included a general topology with similarity to channel proteins with six putative transmembrane segments (Milkman, 1994). Further alignments of Kch revealed that its deduced P segment is related to those within IRKs from eukaryotes, but that its transmembrane segments share only up to a few amino acids with those from various eukaryotic channel proteins and KcsA (Schrempf et al., 1995). In line with these observations is the recent finding that Kch does not function as a channel (as previously implied by several researchers). Results (see elsewhere in this book) of gain-of-function mutations indicate that Kch is not required for in vivo channel activity in *E. coli* but forms a functional K conduit (Kuo et al., 2003).

The evaluation of more recently sequenced bacterial genes revealed several deduced bacterial proteins (KirBac), which have the overall characteristics of the eukaryotic inward rectifier family (IRKs or Kir channel proteins) but contain additional C-terminal extensions. These proteins were found within several species of *Burkholderia* (see below), *Magnetospirillum magnetotacticum*, and *Nostoc punctiforme* (Durell and Guy, 2001). One of these proteins (KirBac 1.1. from *Burkholderia pseudomallei*) has been crystallized (see below). Recently, the KirBac 1.1. protein has been reconstituted (as previously established for KcsA) in liposomes. Subsequently K^+-selective permeation, which is inhibited by Ba^{2+} and Ca^{2+} ions, has been recorded (Enkvetchakul et al., 2004).

The increasing numbers of bacterial proteins (deduced from genomes) that could be related to different classes of eukaryotic channel proteins are being classified according to the pore regions, the number and composition of transmembrane helices, and accessory domains (for details, see other chapters). Through three-dimensional computer modeling conserved sequences within transmembrane regions as well as the putative pore elements within K^+ transporters and channels have been quantified and classified. From the analyses of results from mutational studies, it has been concluded that K^+ symporter families from prokaryotes and eukaryotes have evolved from the prokaryotic K channel proteins (Durell et al., 1999; Durell and Guy, 2001).

THE CRYSTAL STRUCTURE OF KcsA: AN OPEN OR CLOSED CHANNEL?

As outlined in a previous section, optimizing the expression levels of the *kcsA* gene in the heterologous *E. coli* host led to the synthesis of large quantities of the KcsA protein, which could easily be purified (Schrempf et al., 1995). From these findings, the crystallization of the protein was pursued. Crystals were initially obtained from a C-terminally and N-terminally truncated KcsA protein in the presence of 150 mM K^+ ions at pH 7.5. The X-ray structure was determined from an anisotropic crystal. The highest resolution reached within some parts was 3.2 Å, whereas in other regions the resolution was considerably lower (Doyle et al., 1998). Within the resolved tetrameric complex, the relative orientation of the key channel elements can be allocated. Importantly, the four P-segments (determining ion selectivity) are oriented within the fourfold symmetry around the axis of the transmembrane pore. The first portion of each P-element is tilted and points directly toward a water-filled cavity in the center of the transmembrane region of the protein. As outlined in detail in chapters 4 and 8, the overall configuration of the KcsA channel allowed, for the first time, the ability to deduce the structure of the key elements within an ion channel. Despite this limited resolution, it was concluded that the crystal structure corresponds to the open state of the KcsA channel. Being the first X-ray structure of a K^+ channel, it has been used to deduce the functioning of KcsA as a general model. As summarized in chapter 4, results of EPR studies with the full-length KcsA protein led to the suggestion that the C terminus of KcsA forms a four-helix bundle with likely fourfold symmetry (Cortes et al., 2001).

Numerous molecular dynamic simulation studies have used the coordinates of the X-ray structure as the open state of the channel. As outlined in this chapter, questions about permeation, selectivity, and gating of the KcsA channel have been mostly investigated by use of the crystal structure described above. A second, more refined X-ray structure of tetrameric KcsA has been obtained from a crystal grown in the presence of a high K^+ concentration (Zhou et al., 2001). Surprisingly, the deduced coordinates only recently have been initiated for use (Compoint et al., 2004) in molecular dynamics. As outlined in some previous publications, several channel characteristics, which had been determined experimentally, led us to the conclusion that the X-ray structure pictures the dimensions of a closed rather than an open configuration (Meuser et al., 1999; Meuser et al., 2001). This view was subsequently supported by results from EPR studies (Cortes et al., 2001). The findings also implied that details about ion permeation cannot be directly deduced from a closed channel.

The recently solved crystal structure of the MthK protein presents the open channel configuration. Within the protein crystal contacts between intracellular (RCK) domains are formed. However, it has been noted that this open configuration is inconsistent with a high-energy state (Gulbis and Doyle, 2004). The electron density maps for the pore region are, however, poorly defined. Another complication is that the asymmetric unit contains a complex of two K channels; hence it is difficult to conclude if this affects the relative arrangement of the RCK rings to the pore. Homology modeling of MthK and KcsA led subsequently to the elucidation of the open state of KcsA (Holyoake et al., 2003).

The voltage-dependent KvAP channel has six transmembrane spanning regions (Jiang et al., 2003). The crystal structure of the integral membrane part has been resolved to moderate resolution and that of the isolated voltage sensor portion by a high resolution. Though some insights for a voltage-gated channel can be made, controversies exist with previous biophysical data (Gulbis and Doyle, 2004).

The crystal structure of the KirBac1.1 protein includes its predicted intracellular domains. The detailed analysis clearly indicates that this structure corresponds to the closed state of the channel. Furthermore, it can be shown which intramolecular contacts are responsible for complete occlusion of the pore at the intracellular side of the membrane (Gulbis and Doyle, 2004).

The results of experimental and comparative structural data described above support the conclusion that the crystal structure of the tetrameric KcsA does not present the open state. Hence several implications about the details of ion permeation (including the hydration status of the permeating K^+ ion) need to be reinvestigated. Thus it will be an important future task to generate the crystal structure of the full-length protein trapped within the open state.

VIRAL Kcv CHANNELS ARE SMALLER THAN KcsA

The *Chlorella* viruses belong to the genus *Chlorovirus*, which comprises, to date, three different species. They infect eukaryotic chlorella-like algae. The Pbi or the NC64A virus type infects *Chlorella* strain Pbi or, respectively, *Chlorella* strain NC64A, which are hereditary endosymbionts within green isolates of the protozoan *Paramecium bursaria*. In the endosymbiotic stage the algae are surrounded by a host-derived membrane and are resistant to virus infection. Algae can be virus-infected during the nonendosymbiotic stage. Both *Chlorella* isolates can also grow under laboratory conditions free of the virus and free of the protozoan. Members of the third Chlorovirus type infect symbiotic chlorellas within the coelenterate *Hydra viridis*. So far this host has not been cultivated free of the virus.

The model virus PBCV-1 attaches to the outer surface of the algal cell wall, which is subsequently locally degraded. The viral DNA including probably associated proteins is taken up into the nucleus of the infected host cell. After transciption of early viral genes, the viral DNA is replicated and subsequently transcription of its late genes follows. The majority of the virus progeny is released about 8 h after infection. The double-stranded viral DNA of about 330 kb has been predicted to have 375 protein-encoding genes. Such a large number has never been encountered in any virus type to date. One of the genes has been found to encode a small protein comprising 94 amino acids. Two putative transmembrane domains are separated by a stretch of 44 amino acids including a THSTVGFG motif. The N-terminal regions are predicted to be cytoplasmatic and comprise a putative phosphorylation site. In contrast to KcsA, the protein lacks a predicted cytoplasmic terminus. In vitro transcribed Kcv mRNA has been injected into ooytes of *X. laevis*. Subsequently K channel activity with a moderate voltage dependence has been recorded. Electrophysiological studies revealed that about 70% of the channels are always open: at hyperpolarizing voltages about 30% of the channels open in a voltage-dependent fashion and reach half-maximal activation at about –70 mV. The Kcv channel is blocked by Ba^{2+} but is not sensitive to Cs^{+}. The effect of inhibitors during infection of *Chorella* strain NC64A cells by PBCV-1 supports the view that the viral Kcv channel has an essential role during the infection process. The current model assumes that Kcv is located in the internal membrane of the virus and will be inserted into the plasma membrane of the host cell. The subsequent changes in K^{+} conductance and membrane depolarization might lead to a reduction of further pressure of the host cell and allow viral DNA to be released (Mehmel et al., 2003).

CONCLUSIONS

The discovery of the bacterial KcsA channel has opened new avenues to explore channel proteins. The novel strategies have proved to be powerful and applicable for other proteins. The results of the studies have led to important novel insights. (i) The cloning and expression procedure within *E. coli*, which has been introduced for the *kcsA* gene, can also be used for other channel genes. (ii) Within the inner membrane of the *E. coli* host KcsA assembles as tetramer and the process is dependent on the proton motive force. (iii) Within the membrane KcsA interacts preferentially with phospholipids. (iv) Large quantities of KscA tetramers are extractable from the *E. coli* membrane in the presence of a range of detergents. (v) KcsA can be fused into liposomes and reconstituted functionally into a bilayer. (vi) Based on comparisons of the channel features within a bilayer and those within a prototoplast-vesicle system, it has to be concluded that opening of KcsA is governed by the electrochemical gradient. The marked effect of temperature and pH implies the relevance of membrane fluidity and charge distribution within the protein, respectively. (vii) The establishment of conditions to obtain large quantities of pure protein was the basis to crystallize KcsA and subsequently other channel proteins. (viii) The crystal structure of KcsA pictures a tetrameric assembly and allowed for the first time the allocation of key elements of the channel. (ix) Comparative functional studies of KcsA and its mutants led to the conclusion that the dimensions within the crystal structure do not picture the open form of the channel. Subsequent EPR studies as well as comparisons with other channel structures support this conclusion. (x) To generate a more complete view of the KcsA channel, it will be necessary to generate crystals of the channel in the open state. Subsequently the hydration state of K^+ ions and their permeation need to be reinvestigated. (xi) The comparative studies of KcsA and its closely and distantly related proteins allowed significant conclusions about the evolution of channel proteins, certain receptors, and symporters. (xii) Considering that it took considerable time for a number of channel biologists to appreciate functional studies of bacterial channel proteins, it is rewarding that KcsA has evolved to be an outstanding channel model. In the future it will be of great value to elucidate even more "hidden secrets" of KcsA and its other bacterial or viral counterparts.

Acknowledgment. I am grateful to K. Siemienewicz for preparing files of the figures.

REFERENCES

Alvis, S. J., I. M. Williamson, J. M. East, and A. G. Lee. 2003. Interactions of anionic phospholipids and phosphatidylethanolamine with the potassium channel KcsA. *Biophys. J.* **85:**3828–3838.

Caprini, M., S. Ferroni, R. Planells-Cases, J. Rueda, C. Rapisarda, A. Ferrer-Montiel, and M. Montal. 2001. Structural compatibility between the putative voltage sensor of voltage-gated K+ channels and the prokaryotic KcsA channel. *J. Biol. Chem.* **276:**21070–21076.

Chen, G. Q., C. Cui, M. L. Mayer, and E. Gouaux. 1999. Functional characterization of a potassium-selective prokaryotic glutamate receptor. *Nature* **402:**817–821.

Choe, S., A. Kreusch, and P. J. Pfaffinger. 1999. Towards the three-dimensional structure of voltage-gated potassium channels. *Trends Biochem. Sci.* **24:**345–349.

Compoint, M., P. Carloni, C. Ramseyer, and C. Girardet. 2004. Molecular dynamics study of the KcsA channel at 2.0-A resolution: stability and concerted motions within the pore. *Biochim. Biophys. Acta* **1661:**26–39.

Coronado, R., R. L. Rosenberg, and C. Miller. 1980. Ionic selectivity, saturation, and block in a K+-selective channel from sarcoplasmic reticulum. *J. Gen. Physiol.* **76:**425–446.

Cortes, D. M., L. G. Cuello, and E. Perozo. 2001. Molecular architecture of full-length KcsA. Role of cytoplasmic domains in ion permeation and activation gating. *J. Gen. Physiol.* **117:**165–180.

Criado, M., and B. U. Keller. 1987. A membrane fusion strategy for single-channel recordings of membranes usually non-accessible to patch-clamp pipette electrodes. *FEBS Lett.* **224:**172–176.

Cuello, L. G., J. G. Romero, D. M. Cortes, and E. Perozo. 1998. pH-dependent gating in the *Streptomyces lividans* K^+ channel. *Biochemistry* **37:**3229–3236.

Demmers, J. A., A. van Dalen, B. de Kruijff, A. J. Heck, and J. A. Killian. 2003. Interaction of the K+ channel KcsA with membrane phospholipids as studied by ESI mass spectrometry. *FEBS Lett.* **541:**28–32.

Derst, C., and A. Karschin. 1998. Evolutionary link between prokaryotic and eukaryotic K+ channels. *J. Exp. Biol.* **201:**2791–2799.

Doyle, D. A., J. M. Cabral, R. A. Pfuetzner, A. Kuo, J. M. Gulbis, S. L. Cohen, B. T. Chait, and R. MacKinnon. 1998. The structure of the potassium channel: molecular basis of K^+ conduction and selectivity. *Science* **280:**69–77.

Durell, S. R., and H. R. Guy. 2001. A family of putative Kir potassium channels in prokaryotes. *BMC Evol. Biol.* **1:**14.

Durell, S. R., Y. Hao, T. Nakamura, E. P. Bakker, and H. R. Guy. 1999. Evolutionary relationship between K(+) channels and symporters. *Biophys. J.* **77:**775–788.

Enkvetchakul, D., J. Bhattacharyya, I. Jeliazkova, D. K. Groesbeck, C. A. Cukras, and C. G. Nichols. 2004. Functional characterization of a prokaryotic Kir channel. *J. Biol. Chem.* **279:**47076–47080.

Enz, C., T. Steinkamp, and R. Wagner. 1993. Ion channels in the thylakoid membrane (a patch clamp study). *Biochim. Biophys. Acta* **1143:**76.

Fink, M., F. Duprat, F. Lesage, R. Reyes, G. Romey, C. Heurteaux, and M. Lazdunski. 1996. Cloning, functional expression and brain localization of a novel unconventional outward rectifier K^+ channel. *EMBO J.* **15:**6854–6862.

Fox, J. A. 1987. Ion channel subconductance states. *J. Membr. Biol.* **97:**1–8.

Grupe, A., K. H. Schroeter, J. P. Ruppersberg, M. Stocker, T. Drewes, S. Beckh, and O. Pongs. 1990. Cloning and expression of a human voltage-gated potassium channel. A novel member of the RCK potassium channel family. *EMBO J.* **9:**1749–1756.

Guidoni, L., V. Torre, and P. Carloni. 1999. Potassium and sodium binding to the outer mouth of the K^+ channel. *Biochemistry* **38:**8599–8604.

Gulbis, J. M., and D. A. Doyle. 2004. Potassium channel structures: do they conform? *Curr. Opin. Struct. Biol.* **14:**440–446.

Heginbotham, L., L. Kolmakova-Partensky, and C. Miller. 1998. Functional reconstitution of a prokaryotic K+ channel. *J. Gen. Physiol* **111:**741–749.

Heginbotham, L., M. LeMasurier, L. Kolmakova-Partensky, and C. Miller. 1999. Single *Streptomyces lividans* K^+ channels. Functional asymmetries and sidedness of proton activation. *J. Gen. Physiol.* **114:**551–559.

Heginbotham, L., Z. Lu, T. Abramson, and R. MacKinnon. 1994. Mutations in the K^+ channel signature sequence. *Biophys. J.* **66:**1061–1067.

Hellmer, J., and C. Zeilinger. 2003. MjK1, a K+ channel from *M. jannaschii*, mediates K+ uptake and K+ sensitivity in *E. coli. FEBS Lett.* **547:**165–169.

Hille, B. 1992. *Ionic Channels of Excitable Membranes*. Sinauer Associates Inc., Sunderland, Mass.

Holyoake, J., C. Domene, J. N. Bright, and M. S. Sansom. 2003. KcsA closed and open: modelling and simulation studies. *Eur. Biophys. J.* **33:**238–246.

Inagaki, N., T. Gonoi, J. P. I. Clement, N. Namba, J. Inazawa, G. Gonzalez, L. Aguilar-Bryan, S. Seino, and J. Bryan. 1995. Reconstitution of IKATP: an inward rectifier subunit plus the sulfonylurea receptor. *Science* **270:**1166–1170.

Jiang, Y., A. Lee, J. Chen, V. Ruta, M. Cadene, B. T. Chait, and R. MacKinnon. 2003. X-ray structure of a voltage-dependent K+ channel. *Nature* **423:**33–41.

Kessler, A., W. Dittrich, M. Betzler, and H. Schrempf. 1989. Cloning and analysis of a deletable tetracycline-resistance determinant of *Streptomyces lividans* 1326. *Mol. Microbiol.* **3:**1103–1109.

Kuo, M. M., Y. Saimi, and C. Kung. 2003. Gain-of-function mutations indicate that *Escherichia coli* Kch forms a functional K conduit in vivo. *EMBO J.* **15:**4049–4058.

Kutzner, H. J. 1981. The family *Streptomycetaceae*, pp. 2028–2090. *In* M. P. Starr, H. Stolp, H. G. Trüper, A. Balows, and H. Schlegel (ed.), *The Prokaryotes: a Handbook on Habitats, Isolation and Identification of Bacteria*. Springer-Verlag, Berlin, Germany.

Lauger, P. 1985. Structural fluctuations and current noise of ionic channels. *Biophys. J.* **48:**369–373.

Le Dain, A. C., P. J. Anderton, D. K. Martin, and T. J. Millar. 1994. A tetraethylammonium-insensitive inward rectifier K+ channel in Muller cells of the turtle (*Pseudemys scripta elegans*) retina. *J. Membr. Biol.* **141:**239–245.

Legros, C., V. Pollmann, H. G. Knaus, A. M. Farrell, H. Darbon, P. E. Bougis, M. F. Martin-Eauclaire, and O. Pongs. 2000. Generating a high affinity scorpion toxin receptor in KcsA-Kv1.3 chimeric potassium channels. *J. Biol. Chem.* **275:**16918–16924.

LeMasurier, M., L. Heginbotham, and C. Miller. 2001. KcsA: it's a potassium channel. *J. Gen. Physiol.* **118:** 303–314.

Lu, Z., A. M. Klem, and Y. Ramu. 2002. Coupling between voltage sensors and activation gate in voltage-gated K+ channels. *J. Gen. Physiol.* **120:**663–676.

Luneau, C., R. Wiedmann, J. S. Smith, and J. B. Williams. 1991. *Shaw*-like rat brain potassium channel cDNA's with divergent 3′ ends. *FEBS Lett.* **288:**163–167.

Luzhkov, V. B., F. Osterberg, and J. Aqvist. 2003. Structure-activity relationship for extracellular block of K+ channels by tetraalkylammonium ions. *FEBS Lett.* **554:**159–164.

Mehmel, M., M. Rothermel, T. Meckel, J. L. Van Etten, A. Moroni, and G. Thiel. 2003. Possible function for virus encoded K+ channel Kcv in the replication of chlorella virus PBCV-1. *FEBS Lett.* **552:**7–11.

Meuser, D., H. Splitt, R. Wagner, and H. Schrempf. 1999. Exploring the open pore of the K^+ channel KcsA from *Streptomyces lividans. FEBS Lett.* **462:**447–452.

Meuser, D., H. Splitt, R. Wagner, and H. Schrempf. 2001. Mutations stabilizing an open conformation within the external region of the permeation pathway of the potassium channel KcsA. *Eur. Biophys. J.* **30:**385–391.

Milkman, R. 1994. An *Escherichia coli* homologue of eucaryotic potassium channel proteins. *Proc. Natl. Acad. Sci. USA* **91:**3510–3514.

Mueller, P., D. O. Rudin, H. Tien, and W. C. Wescott. 1962. Reconstitution of cell membrane structure in vitro and its transformation into an excitable system. *Nature* (London) **194:**979–980.

Navarro, B., M. E. Kennedy, B. Velimirovic, D. Bhat, A. S. Peterson, and D. E. Clapham. 1996. Nonselective and $G_{\beta\gamma}$-insensitive *weaver* K^+ channels. *Science* **272:**1950–1953.

Nelson, P. H. 2003. Modeling the concentration-dependent permeation modes of the KcsA potassium ion channel. *Phys. Rev. E.* **68:**061908.

Pascual, J. M., C.-C. Shieh, G. E. Kirsch, and A. M. Brown. 1995. Multiple residues specify external tetraethylammonium blockade in voltage-gated potassium channels. *Biophys. J.* **69:**428–434.

Patlak, B. J. 1988. Sodium channel subconductance levels measured with a new variance-mean analysis. *J. Gen. Physiol.* **98:**413–430.

Perozo, E., D. M. Cortes, and L. G. Cuello. 1998. Three-dimensional architecture and gating mechanism of a K^+ channel studied by EPR spectroscopy. *Nature Struct. Biol.* **5:**459–469.

Perozo, E., D. M. Cortes, and L. G. Cuello. 1999. Structural rearrangements underlying K^+-channel activation gating. *Science* **285:**73–78.

Reusch, R. N. 1999. *Streptomyces lividans* potassium channel contains poly-(R)-3-hydroxybutyrate and inorganic polyphosphate. *Biochemistry* **38:**15666–15672.

Schrempf, H. 1999. Investigations of streptomycetes using tools of recombinant DNA technology, p. 501–510. *In* A. L. Demain and J. E. Davies (ed.), *Manual of Industrial Microbiology and Biotechnology*, 2nd ed. ASM Press, Washington, D.C.

Schrempf, H., O. Schmidt, R. Kümmerlen, S. Hinnah, D. Müller, M. Betzler, T. Steinkamp, and R. Wagner. 1995. A prokaryotic potassium ion channel with two predicted transmembrane segments from *Streptomyces lividans*. *EMBO J.* **14:**5170–5178.

Schwarz, T. L., B. L. Tempel, D. M. Papazian, Y. N. Jan, and L. Y. Jan. 1988. Multiple potassium-channel components are produced by alternative splicing at the *Shaker* locus in *Drosophila*. *Nature* **331:**137–142.

Slesinger, P. A., N. Patil, Y. J. Liao, Y. N. Jan, L. Y. Jan, and D. R. Cox. 1996. Functional effects of the mouse weaver mutation on G protein-gated inwardly rectifying K^+ channels. *Neuron* **16:**321–331.

Splitt, H., D. Meuser, I. Borovok, M. Betzler, and H. Schrempf. 2000. Pore mutations affecting tetrameric assembly and functioning of the potassium channel KcsA from *Streptomyces lividans*. *FEBS Lett.* **472:**83–87.

Trudeau, M. C., J. W. Warmke, B. Ganetzky, and G. A. Robertson. 1995. HERG, a human inward rectifier in the voltage-gated potassium channel family. *Science* **269:**92–95.

Valiyaveetil, F. I., Y. Zhou, and R. MacKinnon. 2002. Lipids in the structure, folding, and function of the KcsA K+ channel. *Biochemistry* **41:**10771–10777.

van Dalen, A., S. Hegger, J. A. Killian, and B. de Kruijff. 2002a. Influence of lipids on membrane assembly and stability of the potassium channel KcsA. *FEBS Lett.* **525:**33–38.

van Dalen, A., H. Schrempf, J. A. Killian, and B. de Kruijff. 2000. Efficient membrane assembly of the KcsA potassium channel in *Escherichia coli* requires the protonmotive force. *EMBO Rep.* **1:**340–346.

van Dalen, A., L. M. van der Laan, A. J. Driessen, J. A. Killian, and B. de Kruijff. 2002b. Components required for membrane assembly of newly synthesized K+ channel KcsA. *FEBS Lett.* **511:**51–58.

van den Brink-van der Laan, E., V. Chupin, J. A. Killian, and B. de Kruijff. 2004. Small alcohols destabilize the KcsA tetramer via their effect on the membrane lateral pressure. *Biochemistry* **43:**5937–5942.

Warmke, J., R. Drysdale, and B. Ganetzky. 1991. A distinct potassium channel polypeptide encoded by the *Drosophila eag* locus. *Science* **252:**1560–1562.

Wolters, M., M. Madeja, A. M. Farrell, and O. Pongs. 1999. *Bacillus stearothermophilus lctB* gene gives rise to functional K+ channels in *Escherichia coli* and in *Xenopus* oocytes. *Receptors Channels* **6:**477–491.

Woodhull, A. M. 1973. Ionic blockage of sodium channels in nerve. *J. Gen. Physiol.* **61:**687–708.

Yool, A. J., and T. L. Schwarz. 1991. Alteration of ionic selectivity of a K+ channel by mutation of the H5 region. *Nature* **349:**700–704.

Zakharian, E., and R. N. Reusch. 2004a. Functional evidence for a supramolecular structure for the *Streptomyces lividans* potassium channel KcsA. *Biochem. Biophys. Res. Commun.* **322:**1059–1065.

Zakharian, E., and R. N. Reusch. 2004b. *Streptomyces lividans* potassium channel KcsA is regulated by the potassium electrochemical gradient. *Biochem. Biophys. Res. Commun.* **316:**429–436.

Zhou, Y., J. H. Morais-Cabral, A. Kaufman, and R. MacKinnon. 2001. Chemistry of ion coordination and hydration revealed by a K+ channel-Fab complex at 2.0 A resolution. *Nature* **414:**43–48.

Bacterial Ion Channels and Their Eukaryotic Homologs
Edited by A. Kubalski and B. Martinac

Chapter 4

The Molecular Basis of K^+ Channel Gating

Christopher P. Ptak, Yi-Shiuan Liu, and Eduardo Perozo

Potassium (K^+) channels utilize the electrochemical gradient across a cell membrane to participate in the cell's signaling processes. On a functional level, control of channel opening and closure in response to external stimuli is defined as gating while K^+ flows through the pore by passive diffusion of the open channel. The basic sequence of gating involves two steps: initial stimulation, typically controlled by a regulatory domain, coupled to opening of the pore domain, which acts as a pathway for ions to flow in and out of the cell. The actual kinetics of gating become much more complex as multiple stimulatory and inhibitory signals effect opening and often an inactivation event can lead to additional closed states. Defects in the ability of a channel to sense a signal or to mechanically open affect a channel's functional properties and are the molecular basis for a number of diseases. Therefore, understanding the molecular process by which channels control the flow of ions across the membrane is a critical question in membrane physiology.

Until recently, electrophysiology served as an indirect window into our understanding of channel gating and structure. Functional data provided the basis for conceptual models, while molecular details were left waiting for advances in high-resolution structural studies to open a new window into membrane proteins. This progress has been aided significantly by the study of prokaryotic channels. So far, eukaryotic channels have offered little hope of providing the amount of protein and stability needed for high-level purification and structural studies. Through the course of evolution, the constraints controlling potassium selectivity evolved early and efficiently; thus, the general blueprint that underlies K^+ selectivity is extremely well conserved from bacteria to humans. Consequently, the stable primary sequence of the selectivity filter and secondary structure of the channel have made for extremely efficient genome mining of K^+ channels and also have provided a wealth of structural data relevant for channels in all domains of life.

Structural data have provided critical information about the intrinsic conformational changes occurring during channel gating, and from this, a general picture has emerged. All ion-selective cation channels are thought to pinch off the conduction pathway through a

Christopher P. Ptak, Yi-Shiuan Liu, and Eduardo Perozo • Department of Molecular Physiology and Biological Physics, Center for Structural Biology and Biophysics Program, University of Virginia Health Sciences Center, Charlottesville, VA 22908.

similar underlying process. Recently, a clearer picture of protein movements involved in gating has emerged from the merger of crystallographic and spectroscopic studies with functional analysis. We discuss these and other emerging results from the perspective that understanding the molecular details of gating helps explain how a wide variety of effectors can function to open or close a target channel, allowing for the large diversity of channels.

CORE CHANNEL DOMAIN STRUCTURE

The core domain of a K^+ channel is formed by the selectivity filter and the surrounding transmembrane helices, which together create the conduction pathway. To understand gating, the molecular architecture of the K^+ channel's core domain needs to be known in three dimensions. The basic domain was first revealed at high resolution by the structure of the first K^+ channel, KcsA (Doyle et al., 1998). As predicted in previous work (Cortes and Perozo, 1997), the KcsA crystal structure forms a tetramer, which shields an aqueous tunnel connecting the cytoplasm to the extracellular space through the membrane. A short (12-Å) region dedicated to the selectivity filter only occupies the upper half of the membrane-spanning channel followed by a large, water-filled cavity holding a single ion. Eight membrane-spanning helices maintain and surround the position of the filter but narrow around the lower half of the conduction pathway. The inner helices reach their closest point near the membrane-water interface, and the crossover point between the four inner helices is an obvious candidate to physically close off the K^+ channel. These gating movements can only be inferred from the crystal structure of KcsA, which occupies a nonconducting conformation. Although the KcsA crystal structure lacks the nontransmembrane regions preventing interpretation of the role in the gating process of elements beyond the ion-conducting domain, some inferences can be drawn from electron paramagnetic resonance (EPR)-derived structural information and associated three-dimensional (3-D) models (Cortes et al., 2001). Although the initial structure of KcsA left a great deal of questions to be answered, it provided many long-sought answers to the structural questions plaguing channel physiologists.

The possibility of dissecting the structural movements involved in gating was strengthened when it was discovered that decreasing the internal pH increases the open probability of KcsA and promotes ion conduction through the channel (Cuello et al., 1998; Heginbotham et al., 1999). Attempts to obtain crystals of an open KcsA channel have been limited by the inability to obtain crystals at low pH. Although closed KcsA crystals require the chymoptrypsin removal of the cytoplasmic C terminus, the C-terminal helix is thought to play a role in stabilizing the channel conformation, which may be required to trap a stable open state (Cortes et al., 2001; Molina, 2004).

KcsA CHANNEL GATING

Two potential locations for the gate were suggested by the KcsA structure. Both the selectivity filter and the inner helix crossing are points of narrowing of the conduction pathway. Large conformational changes relating to channel gating were first detected at medium resolution by using EPR (Perozo et al., 1998, 1999), although significant structural changes had been suspected in the K^+ channel inner helix from cysteine reactivity studies. Studies on eukaryotic channels show that cysteines at sites along the inner helices are more reactive with reagents in the open state detected as a greater effect on subsequent function (Liu et al., 1997; Webster et al., 2004).

Site-directed spin labeling in combination with EPR allows the monitoring of the local and global environment at specific positions along a protein sequence using a nitroxide probe (Hubbell et al., 1998). Single cysteine point mutations are generated to provide a fixed location for a spin label probe to attach and report on spectral parameters. The spectral lineshape provides a direct measure of the nitroxide side chain's relative mobility (McHaourab et al., 1996) and can be used to calculate spin-spin distances between 5 and 20 Å (Hustedt et al., 1997; McHaourab et al., 1997; Rabenstein and Shin, 1995). Additional experiments provide information on hydrophobic (oxygen) and hydrophilic [nickel(II) ethylenediaminediacetate (Ni-Edda)] accessibility (Farahbakhsh et al., 1992), which can be used as a measure of solvent exposure. A specific advantage of EPR is that it can be performed on a membrane protein in its native lipid environment and does not depend on the size of the molecule. Structural elements of the closed KcsA were characterized using EPR as a template for identifying conformational changes (Perozo et al., 1998) and were in agreement with the structure determined by X-ray diffraction (Gross et al., 1999; Perozo et al., 1998). The helical nature of the transmembrane segments of KcsA was detected from periodicity analysis of EPR parameters (Cornette et al., 1987). Helix packing was then aided by spin-spin interactions for the C-terminal end of the inner helix (TM2).

EPR was used to locate major conformational changes attributed to KcsA gating movements induced by increasing the proton concentration. A detailed look at the point of inner helix crossing revealed major spectral changes (Perozo et al., 1999). Dipolar coupling increases with a dependence of $1/r^3$ (where r is the distance between two electrons) so distance changes are reflected in broadening changes in the EPR spectra. Movements within the inner helix are easily detected by the broadening of the spectra at increased dipolar coupling, which can be quantitated with low temperature spectra using the Fourier deconvolusion method (Liu et al., 2001; Rabenstein and Shin, 1995) or with pulsed EPR approaches (Borbat et al., 2002). In addition, changes in probe mobility, which relates to the environment as a function of the attached spin label's movement, increased the molecular resolution of the specific movements of the helices already observed by the interaction parameter. For KcsA, the lower portion of the inner helices separates significantly. Little change was observed for the upper half of the helices above the conserved glycine (G99) (Perozo et al., 1999).

A computational model of the movement of the inner helices was developed using experimental distance constraints from the EPR interaction parameter taken at low temperature conditions (Liu et al., 2001; Sompornpisut et al., 2001). The restraint-driven Cartesian transformation computational method, which samples rotational and translational movements of the TM2 helix, was developed under a number of simplifying assumptions. First, the helix was thought to be a rigid to semirigid body, which moves together in three dimensions and the inner helices would move with fourfold symmetry. In addition, it was assumed that there is no reorientation of the spin-label side chain from the closed to open state. Finally, the crystal structure represents the channel in the closed state and was used as the starting point for searching the best-fit open structures. The assumptions were necessary to simplify the computation but may be somewhat inaccurate. Spin-label side chain reorientation may be inevitable in EPR even in a frozen environment, and

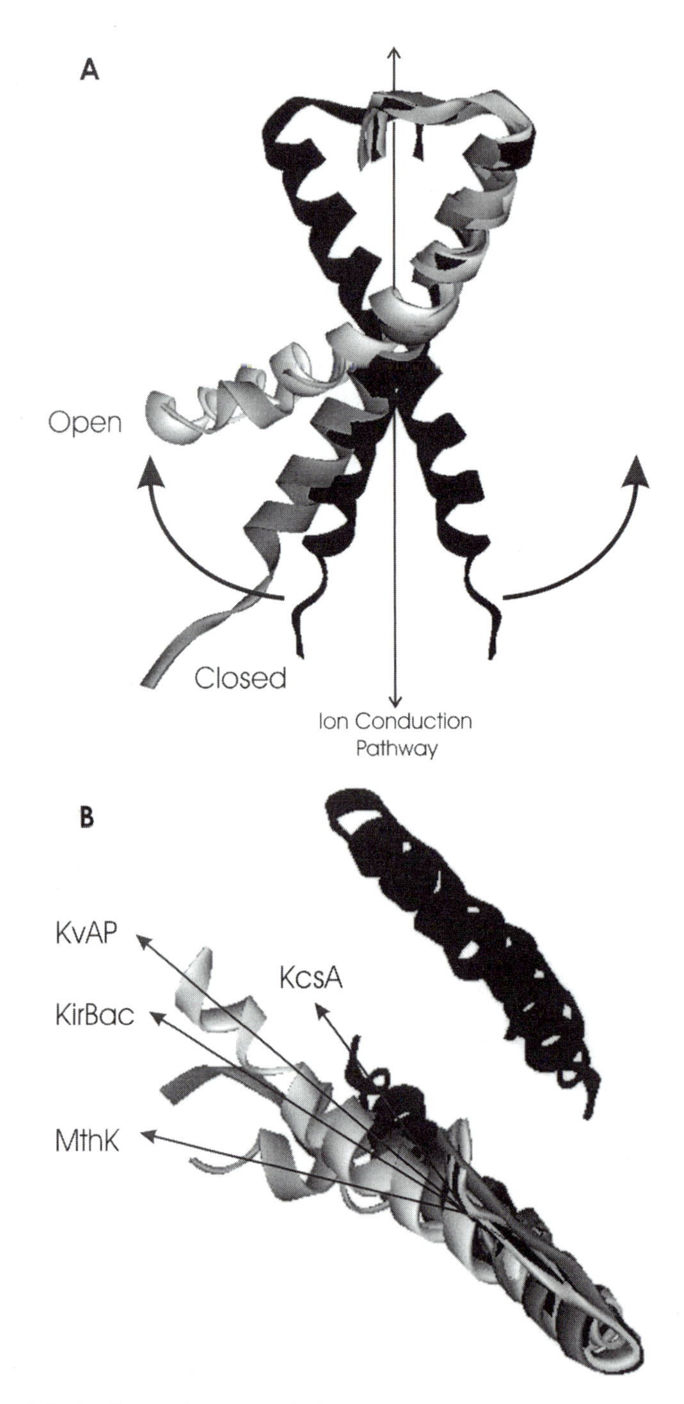

Figure 1. Inner helix bending motion. (A) A single inner helix from the four known K^+ channel structures (closed: KcsA, black; KirBac, dark gray; open: MthK, gray; KvAP, white) was aligned from the selectivity filter to the glycine hinge. The opposite KcsA inner helix is shown to provide the relative position of the bending helix within the tetramer. (B) A top view of the separation of inner transmembrane helices. (C) Cα-Cα distances for residues along the inner helices of K^+ channel structures provide an idea of the size of the ion conduction pathway. Major differences between open and closed channels occur after the glycine hinge.

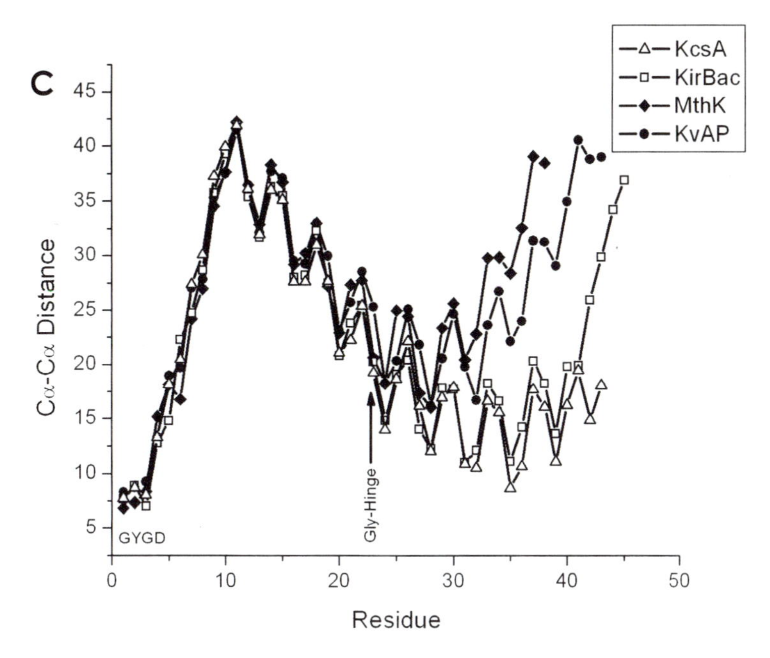

Figure 1. *Continued.*

the rigidness of the whole helix has been shown by recent crystal structures to be broken at a glycine hinge point (equivalent to G99 in KcsA). Nevertheless, the specific details of the separation of the inner helices measured by EPR are a good representation of the movements involved in KcsA gating. In a complementary experiment using site-directed mass tagging of pore-lining residues in KcsA, an increase in cysteine residue accessibility to methanethiosulfonate reagents was observed at lower pH values (Kelly and Gross, 2003).

K^+ CHANNEL OPEN STRUCTURES

Regulatory elements affect conformational movements of the inner helices. Our understanding of gating has been limited, as each of the channel structures has been determined in a single conformation. Several groups are currently concentrating their efforts to tackle this issue, but as of now spectroscopic approaches offer the most reliable experimentally based structural model of a channel in both an open and closed conformation.

Crystal structures of additional channels have been solved and provide insight into changes in helical arrangements upon gating (Jiang et al., 2002; Jiang et al., 2003a; Kuo et al., 2003a). An inner helix from the four crystal structures was overlapped in Fig. 1A and B. An unmistakably closed channel, KirBac, shows a similar distance between the inner transmembrane helices as KcsA (Kuo et al., 2003a). Both KcsA and KirBac have straight unbroken helices lining the ion conduction pathway and are packed tightly enough to pinch off the pore. In KirBac, the inner helices are brought together by multiple constraints from the cytoplasmic domain. The MthK crystal structure was first to offer a glance at an open ion channel and its associated regulatory domain (Jiang et al.,

2002). The inner helices align with the closed channels until a conserved glycine induces a kink, which promotes a very large structural change. The kink in the inner helix allows the lower half of the helix to separate farther from the central axis. The upper portion of the helix and the selectivity filter remain relatively fixed in a stable position. The lower portion of the inner helices is propped open by the cytoplasmic domain. A second open channel, KvAP, was solved having a voltage-sensitive regulatory domain (Jiang et al., 2003a). The KvAP structure is also open with a kink at the conserved glycine and the inner helices are nearly as separated as MthK. The alpha-carbon distances for the K^+ channel crystal structures reveal how far the central cavity opens to allow K^+ to flow through (Fig. 1C). The contributions of channel structures from different proteins have indicated that the glycine at the kink is in a conserved position and appears to represent a common feature of gating.

For a range of channels including Shaker, BK, NachBac, and Kir, the conserved glycine has been implicated in opening (bending) as most mutations at this location fail to open (Jin et al., 2002; Magidovich and Yifrach, 2004; Yi et al., 2001; Zhao et al., 2004). Proline residues provide a fixed kink, which forces the channel to be open. A simple model, where helix bending occurs only at the conserved glycine, is still under debate, as most channels contain one or two glycines beyond this point, which could easily function as helix breakers and decrease the tension beyond the first glycine (Gulbis and Doyle, 2004). A glycine farther down the helix is thought to play a major role in gating of Kir channels, and it is located near a proline-valine-proline (PVP) helix breaker in Kv channels (del Camino et al., 2000; Labro et al., 2003). Bilayer thickness, lipid composition, and relative location of the regulatory domain(s) may have an effect on the most energy-efficient position for a TM2 helix-breaking hinge. Regardless of the position along the helix, a TM2 hinge appears to be necessary for gating to occur. A comparison between the EPR-based KcsA open model and a KcsA model based on MthK suggests a similar general model for gating; however, because the original calculations of an open state from EPR did not consider the TM2 glycine hinge, a number of details are different (Color Plate 5 [see color insert]). EPR data mapped onto the structure reveal that little change is detected by EPR above the TM2 glycine hinge with large structural changes beyond the hinge, and KcsA opening can largely be explained by a model based on the MthK open structure. However, the potential for a second glycine to slightly unravel the inner helix may explain an additional helix rotation suggested by EPR.

Another controversial point is the extent of helix bending and whether major helical bending is necessary. In comparison to the EPR distances observed for KcsA, MthK and KvAP open much larger. A large helix opening may not be a requirement for channel gating as a small helix bend can allow K^+ ions an adequate path for flow. For KcsA, the extent of the bend is thought to be smaller on average. Part of this may be due to the low open probability seen electrophysiologically at steady state (Po, ~0.15) (Cuello et al., 1998; Heginbotham et al., 1998). However, both MthK and KvAP have only two glycines whereas KcsA has a third glycine, which may function to delocalize the extent of the bend along the helix.

Gating is also thought to occur at the selectivity filter. High-resolution structures stabilized with antibodies under high and low K^+ concentrations show slight differences in the ion coordination sites (Zhou et al., 2001). At very low K^+ concentrations, the selectivity filter begins to undergo a conformational movement. A collapse of the normally stable filter structure causes difficulty for ions to enter. Tiny mobility and interaction changes within the residues below the selectivity filter on the pore helix have been detected by EPR between an open and closed channel (Perozo et al., 1999). KirBac has an altered selectivity filter, which may pinch the entryway for ions (Kuo et al., 2003a). The selectivity filter's possible function as a second gate would maintain a higher degree of regulation on the channel and has been suggested to be directly coupled to the large movement of the pore-lining transmembrane segment. This will also complicate interpretations correlating the size of TM2 conformational changes and the apparent low open probability in KcsA.

RCK DOMAIN-CONTAINING CHANNELS

The MthK structure presents the possibility of a new set of regulatory tools to be employed by prokaryotic channels (Jiang et al., 2002). The cytoplasmic gating ring of the channel is formed by an octamer of RCK domains. A second methionine start site is located between the channel and the RCK domain, generating a C-terminal domain only product from the same gene. The cytoplasmic octamer is formed by alternating RCK domains from the channel and from a second start site after the channel and is actually a tetramer of heterodimers. The crystal structure was determined in a channel lacking the second start site and formed from back-to-back channels. The methionine start site is consistently present in a large number of RCK domain channels somewhere along the end of the channel or within the linker but before the RCK domain. A fixed location of the second methionine does not appear to be necessary as the linker itself appears to vary in length. The MthK homolog genes from *Escherichia coli*, Kch, and *Methanococcus jannaschii*, MjK2, both express two gene products in *E. coli*, which correspond to the full channel and C-terminal domain proteins (Hellmer and Zeilinger, 2003; Kuo et al., 2003b; Ptak et al., 2005). Little is known about the function of the second domain. Mutating the methionine can abolish the expression of the second start site for all of the channels. Yet, almost no functional difference was observed for the expression of these methionine mutants with in vivo assays compared with wild-type channels. The possibility exists that the second gene product is not necessary for proper functioning of the channel but is merely a regulatory partner or that a second gene at another region of the genome may be conferring the same function as part of the octameric channel. In some cases, channels lacking the second methionine have a second RCK domain after the first (Jiang et al., 2002). The crystal structure of MthK appears to require an octameric cytoplasmic domain to stabilize the open channel. It is possible that another RCK domain-containing protein could rescue the function of a channel. An alternative possibility suggests that the channel can function as a dimer of dimers and not require the second open reading frame.

The RCK domain provides a ligand-binding pocket, which can accommodate a large variety of activators. Both Ca^{2+} and NAD^+ have been found to populate the open state of RCK domain channels. However, the residues, which are coordinating Ca^{2+} in MthK, are not conserved among RCK domain-containing channels with two transmembrane segments. In fact, the position of the Ca^{2+}-coordinating negative charges is not found in any other channel protein. Although Ca^{2+} induces channel opening in bilayer recordings and the removal of a Ca^{2+}-coordinating aspartate residue decreases gating by Ca^{2+}, the high level (200 mM) of Ca^{2+} used in crystallization conditions may be assigned at additional weak binding sites (Jiang et al., 2002). The pocket has been shown to accept NADH and NAD^+ as ligands in RCK domains that possess the classical Rossmann fold signature (GXGXXG and a trailing negatively charged residue 8 or 9 amino acids away) (Roosild et al., 2002).

The RCK domain is connected to a peripheral domain of unknown function through a linker region. The peripheral domain is diverse and is highly variable from channel to channel; however, the domain presented in the MthK structure is the most common structural motif. Little is known about the function of the peripheral domain although it has been suggested as a small molecule-binding domain (Anantharaman et al., 2001). A VAST (3-D-structure) search (Gibrat et al., 1996) of the peripheral domain from MthK revealed fold similarities with transcriptional regulators (the non-DNA-binding domain), PDZ domains (Hung and Sheng, 2002; Im et al., 2003), and interestingly, a phosphocarrier histidine-containing protein (Jones et al., 1997; Waygood, 1998). Although the phosphorylated histidine does not align with any serine/threonine groups in the primary sequence of MthK, a structural alignment reveals the similar 3-D positioning of a serine/threonine group conserved among related channels. Thus, phosphorylation may be involved in the regulation of the peripheral domain of some RCK domain channels.

CONFORMATIONAL MOVEMENT OF RCK DOMAINS AND GATING

The RCK domain dimer appears to undergo a conformational change at the linker between the RCK domain and the peripheral domain (Fig. 2). The region interlinks the two monomers in the dimer. An analysis of the related crystal structures reveals a bending that

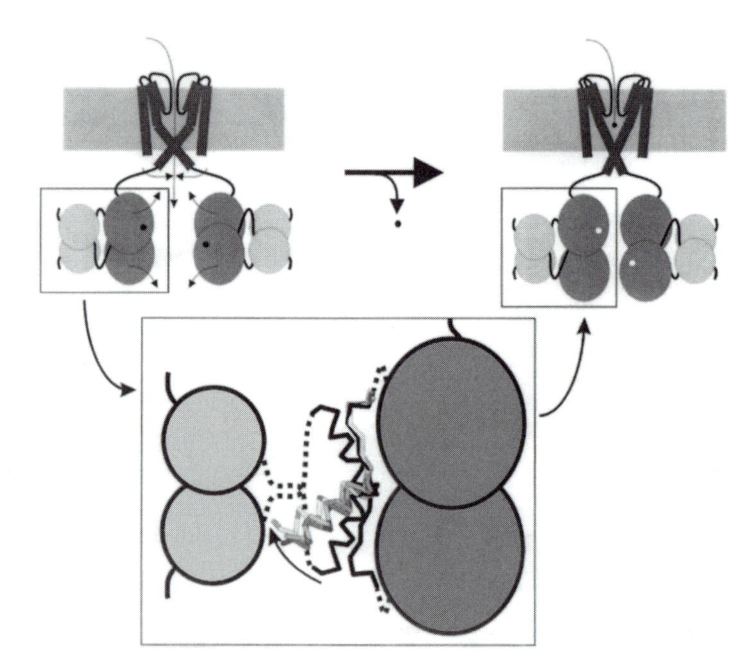

Figure 2. MthK gating movements. Ligand-induced conformational changes in the cytoplasmic gating ring control the open state of MthK. A structural alignment (MthK-Ca^{2+}, black; KtnBsu-nad^{+}, dark gray; *E. coli* Kch-empty, gray; KtnBsu-nadh, white) of the hinge between the RCK domain and the peripheral domain provides insight into the physical movements within the cytoplasmic domain dimer that occur in response to the release of ligand.

takes place in the region at a conserved proline residue at the RCK hinge joint connection between an α-helix and a β-sheet. The RCK hinge may open in response to ligand binding or may be affected by the peripheral domain. The ligand-binding pocket of the *E. coli* RCK domain is empty (Jiang et al., 2001), and the RCK hinge is bent further than the hinge from MthK, which has a ligand-filled pocket (Jiang et al., 2002). The conserved proline hinge decreases its bending between two KTN domains, a structurally related domain, when the domain has NADH bound instead of NAD^+ (Roosild et al., 2002). In all four structures besides MthK, the peripheral domain is absent due to either poor resolution or a removal for crystallization, and in each case the bend is much greater than in MthK. Whatever causes the bending of the RCK hinge, it appears to be related to the gating movement of MthK. The RCK hinge movement alters the position of the RCK domains relative to each other and between the internal monomer interface of the dimer domain. In the octomeric structure, the interdimer contacts appear to remain fixed while the intradimer contacts are flexible. A relative separation of the dimer interface is induced in part by ligand binding as suggested by the representative set of structures.

The electrostatic nature of MjK2, a relative of MthK, may influence the position of the RCK domain. The MjK2 RCK domain alone has been shown to bind electrostatically to negatively charged lipids (Ptak et al., 2005). Adding a membrane-bound tether should only increase the attachment of the domain to negatively charged lipids. A model of the MjK2 RCK domain as an octameric or a tetrameric arrangement can be docked to the membrane around the pore domain with little movement, which would restrict the range of movements for the cytoplasmic multimer. However, binding to the membrane would increase the contacts for a particular gating state and may increase the energetic barrier to transition between the closed and open state and may increase accessibility to membrane-bound stimulus. An increasing number of channels and transporters have cytoplasmic domain interactions with anionic lipids (Hilgemann et al., 2001); for instance, in Kir channels, the C-terminal domain interacts directly and specifically with phosphatidylinositol-bisphosphate to stabilize the open conformation (Hilgemann and Ball, 1996; Huang et al., 1998).

Like MthK, MjK2 has sequence similarity to a eukaryotic family of Ca^{2+}-activated K^+ channels, BK_{Ca}. The BK_{Ca} family describes channels with subunits having six transmembrane segments and a large cytoplasmic domain containing two RCK domains immediately following the final transmembrane segment. The complex gating mechanism of BK_{Ca} channels, involving a transmembrane voltage-sensing domain controlling voltage-dependent gating and the cytoplasmic domain controlling divalent cation-dependent gating, is not well understood (Magleby, 2003). Voltage and Ca^{2+} gating are synergistic and relatively independent with the gate at the base of the inner helices as a point of indirect interaction between the RCK domain of BK_{Ca} channels and the voltage sensor (Cox and Aldrich, 2000; Cui and Aldrich, 2000; Horrigan and Aldrich, 2002; Niu et al., 2004). However, Mg^{2+} activates BK_{Ca} channels through a molecular location on the RCK domain, which is different from the Ca^{2+} activation site (Shi et al., 2002; Xia et al., 2002). Adding to this complexity, it was recently shown that Mg^{2+}, and not Ca^{2+}, activation in BK_{Ca} channels is influenced by mutations in the voltage sensor, suggesting a direct interaction between the RCK domain and the voltage sensor (Hu et al., 2003). Analogously, the RCK domain of MjK2 directly interacts with a patch of anionic lipids adjacent to the pore in the same relative location as the voltage sensor (Jiang et al., 2003a; Jiang et al., 2003b). In this way, additional gating states may exist beyond the two suggested by the MthK structure for Ca^{2+}-activated channels, thereby opening the door for many complex interactions governing BK_{Ca} channel gating.

Some questions remain as to how energy input is translated to gating and whether the inner helices are pulled apart or pushed shut. A recent article on the Ca^{2+}-activated K^+ channel tried to address this issue by monitoring the effect of the linker length on Ca^{2+} activation (Niu et al., 2004). Shortening the linker shifted the curve of activation to lower Ca^{2+} levels, while lengthening the linker had the opposite effect. The results suggest that mechanical tension imposed on the inner helix by the gating ring is necessary to force the channel open. The requirement for energy input to achieve an open state may explain that all open channel structures have been obtained in either the presence of a tension-inducing antibody or, in the case of MthK, the tension from the opposite RCK linked channel. However, if tension is necessary to open the MthK channel, the linker should be resolved in the crystal structure as it is expected to be rigid. Usually, only disordered regions of a protein have unresolved structure from the X-ray data. Still, it is possible that energy is required to transition from a low-energy stabilized open state to a low-energy stabilized closed state. A high-energy transition state could explain the poor resolution for the MthK linker as the cytoplasmic contacts may be required to force the inner helices past the transition state, but the low-energy open state with a wider helix separation may be the most favorable to occupy, therefore alleviating the tension on the linker. The glycine TM2 hinge along the inner helix seems to be a key switch in the transition between closed and open states. From a stabilized closed conformation, bending at the TM2 hinge may induce alternate contacts to form in which the open state is stabilized. In the case of anionic lipid interactions, the electrostatic binding contacts may stabilize a particular gating conformation.

CONCLUSION

The mechanism of channel gating lies at the core of our understanding of how channels respond to specific stimuli. Information extracted from functional, crystallographic, and spectroscopic studies of prokaryotic channels has revealed molecular details of how the inner helices and the selectivity filter are involved in channel gating. Catalyzed by an activating signal, a large separation of the inner helices is associated with the channel's open state. For MthK, these gating movements within the inner transmembrane helix occur as bending at a glycine hinge and are directly linked to constraints imposed on the inner helix by the regulatory domain. The cytoplasmic RCK domain undergoes local conformational changes in response to effects of ligand-binding energies. Stimulus-induced rearrangements within the regulatory domain affect the gating state of the channel through the transduction of physical forces exerted on the inner pore helices. Focusing the gating forces at a consistent position along the ion conduction pathway allows channels to exist with a large diversity of regulatory domains but maintain a conserved core architecture necessary for efficient function.

REFERENCES

Anantharaman, V., E. V. Koonin, and L. Aravind. 2001. Regulatory potential, phyletic distribution and evolution of ancient, intracellular small-molecule-binding domains. *J. Mol. Biol.* **307:**1271–1292.

Borbat, P. P., H. S. McHaourab, and J. H. Freed. 2002. Protein structure determination using long-distance constraints from double-quantum coherence ESR: study of T4 lysozyme. *J. Am. Chem. Soc.* **124:**5304–5314.

Cornette, J. L., K. B. Cease, H. Margalit, J. L. Spouge, J. A. Berzofsky, and C. DeLisi. 1987. Hydrophobicity scales and computational techniques for detecting amphipathic structures in proteins. *J. Mol. Biol.* **195:** 659–685.

Cortes, D. M., L. G. Cuello, and E. Perozo. 2001. Molecular architecture of full-length KcsA: role of cytoplasmic domains in ion permeation and activation gating. *J. Gen. Physiol.* **117:**165–180.

Cortes, D. M., and E. Perozo. 1997. Structural dynamics of the *Streptomyces lividans* K^+ channel (SKC1): oligomeric stoichiometry and stability. *Biochemistry* **36:**10343–10352.

Cox, D. H., and R. W. Aldrich. 2000. Role of the beta1 subunit in large-conductance Ca^{2+}-activated K^+ channel gating energetics. Mechanisms of enhanced Ca^{2+} sensitivity. *J. Gen. Physiol.* **116:**411–432.

Cuello, L. G., J. G. Romero, D. M. Cortes, and E. Perozo. 1998. pH-dependent gating in the *Streptomyces lividans* K^+ channel. *Biochemistry* **37:**3229–3236.

Cui, J., and R. W. Aldrich. 2000. Allosteric linkage between voltage and Ca^{2+}-dependent activation of BK-type mslo1 K^+ channels. *Biochemistry* **39:**15612–15619.

del Camino, D., M. Holmgren, Y. Liu, and G. Yellen. 2000. Blocker protection in the pore of a voltage-gated K^+ channel and its structural implications. *Nature* **403:**321–325.

Doyle, D. A., J. Morais Cabral, R. A. Pfuetzner, A. Kuo, J. M. Gulbis, S. L. Cohen, B. T. Chait, and R. MacKinnon. 1998. The structure of the potassium channel: molecular basis of K^+ conduction and selectivity. *Science* **280:**69–77.

Farahbakhsh, Z. T., C. Altenbach, and W. L. Hubbell. 1992. Spin labeled cysteines as sensors for protein-lipid interaction and conformation in rhodopsin. *Photochem. Photobiol.* **56:**1019–1033.

Gibrat, J. F., T. Madej, and S. H. Bryant. 1996. Surprising similarities in structure comparison. *Curr. Opin. Struct. Biol.* **6:**377–385.

Gross, A., L. Columbus, K. Hideg, C. Altenbach, and W. L. Hubbell. 1999. Structure of the KcsA potassium channel from *Streptomyces lividans*: a site-directed spin labeling study of the second transmembrane segment. *Biochemistry* **38:**10324–10335.

Gulbis, J. M., and D. A. Doyle. 2004. Potassium channel structures: do they conform? *Curr. Opin. Struct. Biol.* **14:**440–446.

Heginbotham, L., L. Kolmakova-Partensky, and C. Miller. 1998. Functional reconstitution of a prokaryotic K^+ channel. *J. Gen. Physiol.* **111:**741–749.

Heginbotham, L., M. LeMasurier, L. Kolmakova-Partensky, and C. Miller. 1999. Single *Streptomyces lividans* K^+ channels: functional asymmetries and sidedness of proton activation. *J. Gen. Physiol.* **114:** 551–560.

Hellmer, J., and C. Zeilinger. 2003. MjK1, a K^+ channel from *M. jannaschii*, mediates K^+ uptake and K^+ sensitivity in *E. coli. FEBS Lett.* **547:**165–169.

Hilgemann, D. W., and R. Ball. 1996. Regulation of cardiac Na^+,Ca^{2+} exchange and KATP potassium channels by PIP_2. *Science* **273:**956–959.

Hilgemann, D. W., S. Feng, and C. Nasuhoglu. 2001. The complex and intriguing lives of PIP_2 with ion channels and transporters. *Sci. STKE* **2001:**RE19.

Horrigan, F. T., and R. W. Aldrich. 2002. Coupling between voltage sensor activation, Ca^{2+} binding and channel opening in large conductance (BK) potassium channels. *J. Gen. Physiol.* **120:**267–305.

Hu, L., J. Shi, Z. Ma, G. Krishnamoorthy, F. Sieling, G. Zhang, F. T. Horrigan, and J. Cui. 2003. Participation of the S4 voltage sensor in the Mg^{2+}-dependent activation of large conductance (BK) K^+ channels. *Proc. Natl. Acad. Sci. USA* **100:**10488–10493.

Huang, C. L., S. Feng, and D. W. Hilgemann. 1998. Direct activation of inward rectifier potassium channels by PIP_2 and its stabilization by $G\beta\gamma$. *Nature* **391:**803–806.

Hubbell, W. L., A. Gross, R. Langen, and M. A. Lietzow. 1998. Recent advances in site-directed spin labeling of proteins. *Curr. Opin. Struct. Biol.* **8:**649–656.

Hung, A. Y., and M. Sheng. 2002. PDZ domains: structural modules for protein complex assembly. *J. Biol. Chem.* **277:**5699–5702.

Hustedt, E. J., A. I. Smirnov, C. F. Laub, C. E. Cobb, and A. H. Beth. 1997. Molecular distances from dipolar coupled spin-labels: the global analysis of multifrequency continuous wave electron paramagnetic resonance data. *Biophys. J.* **72:**1861–1877.

Im, Y. J., J. H. Lee, S. H. Park, S. J. Park, S. H. Rho, G. B. Kang, E. Kim, and S. H. Eom. 2003. Crystal structure of the Shank PDZ-ligand complex reveals a class I PDZ interaction and a novel PDZ-PDZ dimerization. *J. Biol. Chem.* **278:**48099–48104.

Jiang, Y., A. Lee, J. Chen, M. Cadene, B. T. Chait, and R. MacKinnon. 2002. Crystal structure and mechanism of a calcium-gated potassium channel. *Nature* **417:**515–522.

Jiang, Y., A. Lee, J. Chen, V. Ruta, M. Cadene, B. T. Chait, and R. MacKinnon. 2003a. X-ray structure of a voltage-dependent K^+ channel. *Nature* **423:**33–41.

Jiang, Y., A. Pico, M. Cadene, B. T. Chait, and R. MacKinnon. 2001. Structure of the RCK domain from the *E. coli* K^+ channel and demonstration of its presence in the human BK channel. *Neuron* **29:**593–601.

Jiang, Y., V. Ruta, J. Chen, A. Lee, and R. MacKinnon. 2003b. The principle of gating charge movement in a voltage-dependent K^+ channel. *Nature* **423:**42–48.

Jin, T., L. Peng, T. Mirshahi, T. Rohacs, K. W. Chan, R. Sanchez, and D. E. Logothetis. 2002. The βγ subunits of G proteins gate a K^+ channel by pivoted bending of a transmembrane segment. *Mol. Cell* **10:**469–481.

Jones, B. E., P. Rajagopal, and R. E. Klevit. 1997. Phosphorylation on histidine is accompanied by localized structural changes in the phosphocarrier protein, HPr from *Bacillus subtilis. Protein. Sci.* **6:**2107–2119.

Kelly, B. L., and A. Gross. 2003. Potassium channel gating observed with site-directed mass tagging. *Nat. Struct. Biol.* **10:**280–284.

Kuo, A., J. M. Gulbis, J. F. Antcliff, T. Rahman, E. D. Lowe, J. Zimmer, J. Cuthbertson, F. M. Ashcroft, T. Ezaki, and D. A. Doyle. 2003a. Crystal structure of the potassium channel KirBac1.1 in the closed state. *Science* **300:**1922–1926.

Kuo, M. M., Y. Saimi, and C. Kung. 2003b. Gain-of-function mutations indicate that *Escherichia coli* Kch forms a functional K^+ conduit in vivo. *EMBO J.* **22:**4049–4058.

Labro, A. J., A. L. Raes, I. Bellens, N. Ottschytsch, and D. J. Snyders. 2003. Gating of *shaker*-type channels requires the flexibility of S6 caused by prolines. *J. Biol. Chem.* **278:**50724–50731.

Liu, Y., M. Holmgren, M. E. Jurman, and G. Yellen. 1997. Gated access to the pore of a voltage-dependent K^+ channel. *Neuron* **19:**175–184.

Liu, Y. S., P. Sompornpisut, and E. Perozo. 2001. Structure of the KcsA channel intracellular gate in the open state. *Nat. Struct. Biol.* **8:**883–887.

Magidovich, E., and O. Yifrach. 2004. Conserved gating hinge in ligand- and voltage-dependent K^+ channels. *Biochemistry* **43:**13242–13247.

Magleby, K. L. 2003. Gating mechanism of BK (Slo1) channels: so near, yet so far. *J. Gen. Physiol.* **121:**81–96.

McHaourab, H. S., M. A. Lietzow, K. Hideg, and W. L. Hubbell. 1996. Motion of spin-labeled side chains in T4 lysozyme. Correlation with protein structure and dynamics. *Biochemistry* **35:**7692–7704.

McHaourab, H. S., K. J. Oh, C. J. Fang, and W. L. Hubbell. 1997. Conformation of T4 lysozyme in solution. Hinge-bending motion and the substrate-induced conformational transition studied by site-directed spin labeling. *Biochemistry* **36:**307–316.

Molina, M. L., J. A. Encinar, F. N. Barrera, G. Fernandez-Ballester, G. Riquelme, and J. M. Gonzalez-Ros. 2004. Influence of C-terminal protein domains and protein-lipid interactions on tetramerization and stability of the potassium channel KcsA. *Biochemistry* **43:**14924–14931.

Niu, X., X. Qian, and K. L. Magleby. 2004. Linker-gating ring complex as passive spring and Ca^{2+}-dependent machine for a voltage- and Ca^{2+}-activated potassium channel. *Neuron* **42:**745–756.

Perozo, E., D. M. Cortes, and L. G. Cuello. 1998. Three-dimensional architecture and gating mechanism of a K^+ channel studied by EPR spectroscopy. *Nat. Struct. Biol.* **5:**459–469.

Perozo, E., D. M. Cortes, and L. G. Cuello. 1999. Structural rearrangements underlying K^+-channel activation gating. *Science* **285:**73–78.

Ptak, C. P., L. G. Cuello, and E. Perozo. 2005. Electrostatic interaction of a K^+ channel RCK domain with charged membrane surfaces. *Biochemistry* **44:**62–71.

Rabenstein, M. D., and Y. K. Shin. 1995. Determination of the distance between two spin labels attached to a macromolecule. *Proc. Natl. Acad. Sci. USA* **92:**8239–8243.

Roosild, T. P., S. Miller, I. R. Booth, and S. Choe. 2002. A mechanism of regulating transmembrane potassium flux through a ligand-mediated conformational switch. *Cell* **109:**781–791.

Shi, J., G. Krishnamoorthy, Y. Yang, L. Hu, N. Chaturvedi, D. Harilal, J. Qin, and J. Cui. 2002. Mechanism of magnesium activation of calcium-activated potassium channels. *Nature* **418:**876–880.

Sompornpisut, P., Y.-S. Liu, and E. Perozo. 2001. Calculation of rigid-body conformational changes using restraint-driven cartesian transformations. *Biophys. J.* **81:**2530–2546.

Waygood, E. B. 1998. The structure and function of HPr. *Biochem. Cell Biol.* **76:**359–367.

Webster, S. M., D. Del Camino, J. P. Dekker, and G. Yellen. 2004. Intracellular gate opening in Shaker K^+ channels defined by high-affinity metal bridges. *Nature* **428:**864–868.

Xia, X. M., X. Zeng, and C. J. Lingle. 2002. Multiple regulatory sites in large-conductance calcium-activated potassium channels. *Nature* **418:**880–884.

Yi, B. A., Y. F. Lin, Y. N. Jan, and L. Y. Jan. 2001. Yeast screen for constitutively active mutant G protein-activated potassium channels. *Neuron* **29:**657–667.

Zhao, Y., V. Yarov-Yarovoy, T. Scheuer, and W. A. Catterall. 2004. A gating hinge in Na^+ channels; a molecular switch for electrical signaling. *Neuron* **41:**859–865.

Zhou, Y., J. H. Morais-Cabral, A. Kaufman, and R. MacKinnon. 2001. Chemistry of ion coordination and hydration revealed by a K^+ channel-Fab complex at 2.0 A resolution. *Nature* **414:**43–48.

Bacterial Ion Channels and Their Eukaryotic Homologs
Edited by A. Kubalski and B. Martinac

Chapter 5

Glutamate-Activated Channels

Indira H. Shrivastava and H. Robert Guy

Ligand-gated ion channels (LGICs) are a major class of ion channels. Postsynaptic LGICs generate electrical signals in response to specific chemical neurotransmitters such as acetylcholine, glutamate, glycine, or γ-aminobutyric acid (Hille, 2001). Due to the ubiquitous nature of these receptors in phylogenetic terms (they occur in nematodes, insects, and vertebrates), they serve as a good candidate to study from both the structural and evolutionary points of view.

Glutamate receptors (GluRs) are LGIC proteins that are nonselectively permeable to Na^+, K^+, and Ca^{2+} cations. In most animals, the neuron is a signal-transducing device that connects stimuli to response. Glutamate is a major excitatory neurotransmitter in neurons of vertebrate brains. Mutations of genes encoding these proteins cause malfunctioning of these proteins, leading to debilitating diseases of the central nervous system (CNS). Malfunctioning of GluR channels is implicated in diseases associated with CNS disorders such as epilepsy; Alzheimer's, Parkinson's, and Huntington's diseases; schizophrenia; Rasmussen's encephalitis; anxiety; pain perception; learning; cognition; and even stroke (Armstrong et al., 1998). Thus, knowledge of the structural and functional relationships of these ion channels is expected to lead to a better understanding of the pathogenesis of the diseases in which they are implicated. Sequence homology screening has of late become an increasingly popular method of cloning genes of interest, given the availability of large amounts of sequence data of various prokaryotic and eukaryotic organisms. The sequencing of the human, mouse, and rat genomes now allows a comprehensive bioinformatics approach to study gene families and their variation and conservation patterns. This information is useful in identifying new members of gene families, functionally related sequence conservation, and evolutionary effects of selection. Also, understanding the polymorphism of the genes encoding the GluRs in particular should increase our understanding of the role of these receptors in neurogenetic variations (Lipsky and Goldman, 2003). In addition, the wide diversity and heterogeneous distribution of GluR subtypes could provide opportunities for development of pharmacological agents specific to the CNS for therapeutic effects (Osborne et al., 2000). Thus, a genetic approach combined with pharmacological tools

Indira H. Shrivastava and H. Robert Guy • Laboratory of Experimental and Computational Biology, Division of Basic Sciences, National Cancer Institute, National Institutes of Health, Bethesda, MD 20892-5567.

may lead to major advances in understanding the roles of GluRs in regulating the CNS systems and animal behavioral patterns (Conn, 2003).

GluRs occur not only in animals but also in plants (Lam et al., 1998) and prokaryotes (Chen et al., 1999), both of which lack a CNS. The functional role of GluRs in plants and unicellular organisms remains unknown; however, they are important in the evolutionary scheme of things, and several theories linking the prokaryotic GluRs to potassium channels have been proposed, based on structural and sequence homologies. Eukaryotic GluRs belong to two distinct superfamilies: the ionotropic glutamate receptors (iGluRs) form ion channels, while the metabotropic glutamate receptors (mGluRs) are G-protein-coupled receptors that act by binding and releasing G proteins that affect a variety of pathways and targets, including ion channels. The mGluRs respond to glutamate by activating proteins in the cells that affect neuronal excitability, synaptic transmission, and cell metabolism functions (Conn, 2003). The iGluRs and mGluRs are further classified into subgroups based on their sequence homology, agonist pharmacology, and intracellular transduction mechanisms. The membrane topology of GluRs typically consists of three domains—the extracellular domain, the intracellular domain, and the transmembrane domain. The structural and sequence similarity of the different domains of GluRs to proteins from different types of organisms gives rise to some interesting implications in the evolutionary relationship between prokaryotes and eukaryotes (discussed below).

ANIMAL GLUTAMATE RECEPTORS

Animals possess both mGluRs and iGluRs.

mGluRs

The first evidence of existence of mGluRs was produced in the 1980s (Sugiyama et al., 1987), and the first mGluR receptor DNA was cloned several years later (Conn, 2003). Further search for mGluR-related cDNA resulted in the isolation of seven other genes and several splice variants encoding mGluRs (Conn and Pinn, 1997), though there is evidence that further subtypes of mGluRs may yet be identified (Osborne et al., 2000).

The eight mGluRs are classified into three major groups on the basis of sequence homologies, coupling to second messenger systems, and selectivities of various agonists (Fig. 1). The group I mGluRs are associated with postsynaptic inositol phosphate metabolism and intracellular calcium signaling. This group includes two subtypes, mGluR1 and mGluR5. The group II and group III mGluRs are both linked to inhibition of adenylyl cyclase activity and have been suggested to mediate presynaptic actions of glutamate in some brain areas. The subtypes in group II are mGluR2 and mGluR3, while those in group III are mGluR4, mGluR6, mGluR7, and mGluR8. The topology of mGluRs (Fig. 2A) consists of seven transmembrane (TM) helices spanning the bilayer and an unusually long (~500-amino-acid) extracellular amino-terminal domain (ATD), connected to the TM domain by a hinge region and an intracellular C-terminal domain. The intracellular C-terminal domain generally undergoes splice variations, giving rise to alternative splice variants, and also serves as a binding site for regulatory proteins belonging to the Homer family (Brakemann et al., 1997). Agonists bind to the open form of ATD, inducing a closure of the domain. This structural change in the ATD is "relayed" to the cysteine-rich hinge and thereby to the TM

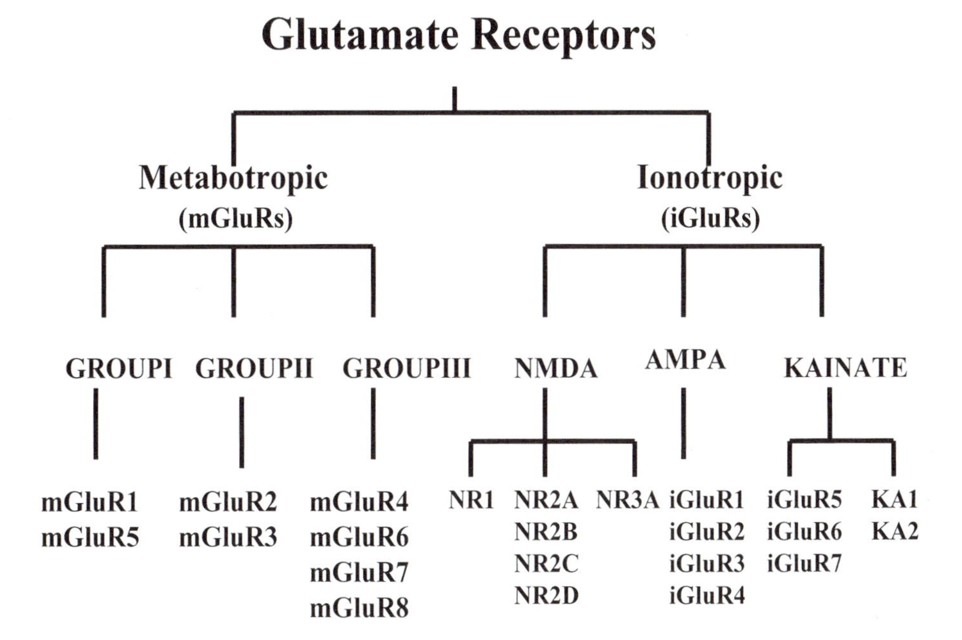

Figure 1. Classification of mGluRs and iGluRs based on their pharmacological activities and structural similarities.

region, which gets activated. The ATD domain of mGluRs from mammalian brain has sequence homology with bacterial periplasmic binding protein (PBP) (O'Hara et al., 1993). Within a group, there is an ~60% sequence identity in the proteins while in between two groups, the sequence identity falls to about 40 to 50% (Osborne et al., 2000). Current understanding of the roles and involvement of mGluR subtypes in physiological and pathophysiological functions remains fairly limited, though research is being stimulated by the discovery of potent pharmacological agents selective for mGluRs (Schoepp et al., 1999).

iGluRs

Experiments on cloning and functional characterization of animal GluRs have revealed a complex and diverse superfamily of iGluRs. They are encoded by at least six gene families (Lipsky and Goldman, 2003). The iGluRs are classified into subtypes based on the ligands by which they are activated (Fig. 1). The ligands activating iGluRs include *N*-methyl-D-aspartic acid (NMDA), kainate (KAI), and α-amino-3-hydroxy-5-methyl-4-isoxazolepropionic acid (AMPA). The iGluRs are thus classified as NMDA receptors or non-NMDA receptors, which include KAI and AMPA receptors. The AMPA receptors consist of GluR1 and GluR4 subunits, while the KAI receptors consist of GluR5, GluR7, KA1, and KA2 subunits. The NMDA receptors are formed by NR1 and NR2A-NR2D subunits, which encode glutamate-binding sites. They also coassemble less commonly with NR3A and NR3B gene families, which encode glycine-binding sites (Mayer and Armstrong, 2004). In addition, there exists a seventh group of iGluRs, the δGluRs, consisting of

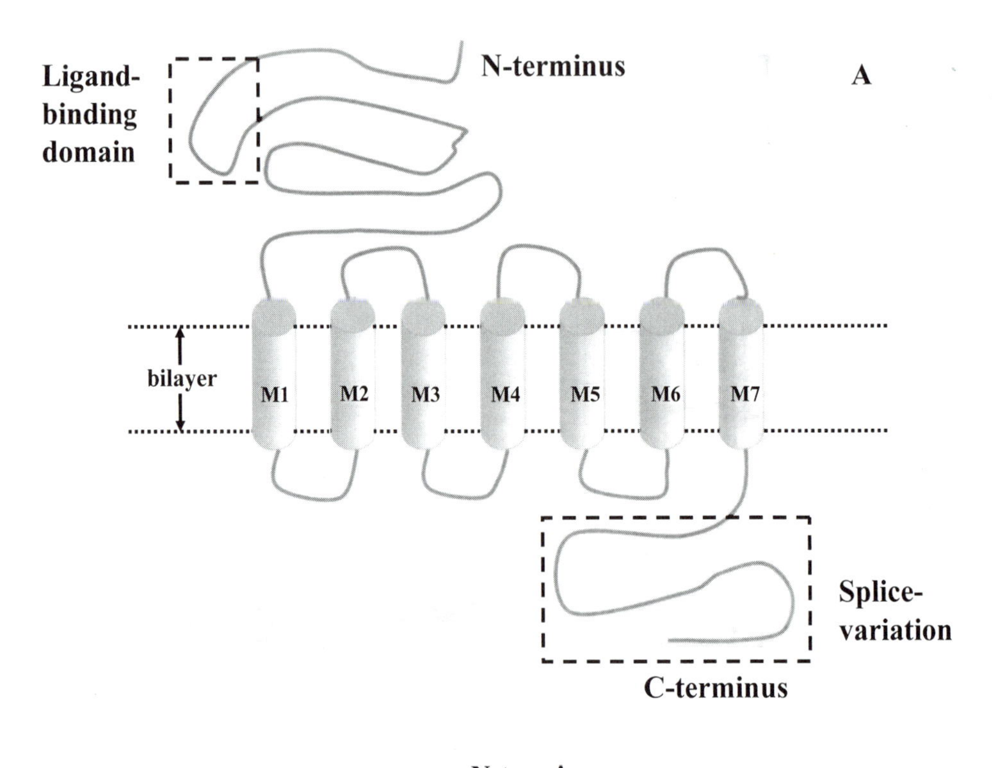

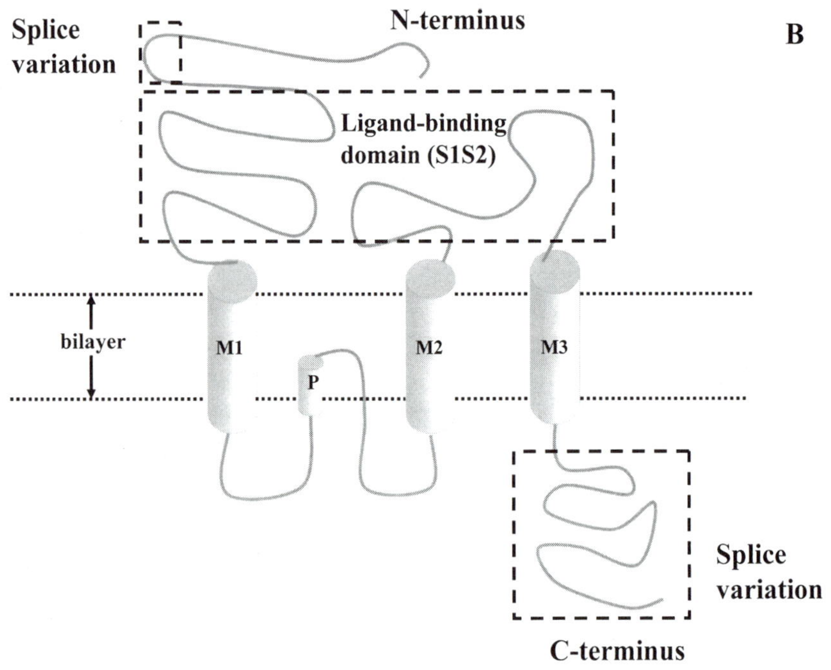

Figure 2. General structure of the membrane topology of a single subunit of mGluRs (A) and iGluRs (B).

two subunits, δ1 and δ2. However, these proteins neither form channels nor associate with other GluRs (Lomeli et al., 1994), though some experimental observations suggest that the δ2 subunit may form functional receptors (Zuo et al., 1997).

At present, six NMDA receptor subunits and four AMPA receptor subunits have been cloned and characterized, and five subunit building blocks for KAI receptors have been identified (Osborne et al., 2000; Lipsky and Goldman, 2003). This, along with alternative splicing of the messenger tRNA, editing of mRNA, and variable mixing of the different subunits in the receptor complexes, gives rise to a complex and diverse variety of NMDA and non-NMDA receptors (Dingledine et al., 1999).

The iGluRs, like the mGluRs, also have a modular topology and are composed of four major domains, an amino-terminal domain (NTD), the extracellular ligand-binding domain (S1S2), the TM domain, and an intracellular C-terminal domain (Fig. 2B). The NTD, made up of ~400 amino acids, is a major determinant of the subtype-specific assembly within the iGluR family (Ayalon and Stern-Bach, 2001). The NTD shows sequence similarity to the leucine/isoleucine/valine-binding protein, a bacterial PBP (Madden, 2002). This domain contributes to the binding site of many NMDA receptor modulators and receptor desensitization (Choi and Lipton, 1999; Krupp et al., 1998). For the non-NMDA receptors, the NTD contributes to receptor assembly (Wo and Oswald, 1994). The two-domain fold of the NTD suggests that it might also encode a binding site, though no ligand binding to the NTD has been identified (Mayer and Armstrong, 2004).

The S1S2 domain consists of the N terminus of the first TM helix, M1, and the extracellular loop connecting M2 (*M3*) and M3 (*M4*). (The letters in italics are the conventional nomenclature of GluRs, which is discussed below.) The crystal structure of the ligand-binding domain of eukaryotic and prokaryotic GluRs has been determined, both in the ligand-bound and ligand-free (apo) states (Armstrong et al., 1998; Armstrong and Gouaux, 2000; Mayer et al., 2001). The ligand-binding domain forms a bilobed binding pocket within which the ligand-binding site is embedded. This domain is proposed to be in a dynamic equilibrium between the closed and open forms. Models for activation and desensitization have been proposed based on experimental data (Armstrong and Gouaux, 2000; Mayer et al., 2001). The ligand-binding domain has sequence and structural similarities to the glutamine-binding proteins (Wo and Oswald, 1995; Nakanishi et al., 1990) and to lysine/arginine/ornithine-binding protein (Paas, 1998). PBPs are a family of soluble receptors for amino acids and sugars in the periplasmic space of bacteria that interact with membrane signaling proteins. The PBP also consists of two distinct globular domains, interconnected by a hinge region and separated by a cleft (Quiocho and Ledvina, 1996). The structural similarities between the S1S2 domain and the PBPs imply an evolutionary relationship between these two proteins.

Editing of subunit-specific pre-mRNA within AMPA and KAI GluRs results in the generation of two isoforms that have either an arginine (R) or a glutamine (Q) at the position where the helical part of the P segment (*M2*) terminates (Sommer et al., 1991). This site is called the Q/R site and is considered to be an important determinant in the channel permeation and blockade (Dingledine et al., 1999). The hydrogen-bonding patterns at the Q/R site have been used to explain the differences in the pore diameters and channel sensitivity to blockers of NMDA and non-NMDA channels (Villarroel et al., 1995; Burnashev et al., 1996). Experimental data on this Q/R site have been helpful in building homology models of GluRs (Tikhonov et al., 2002). In addition, there is a splice variation region in the M2-M3

(*M3-M4*) extracellular loop region that affects the receptor desensitization, thereby modulating the processing of synaptic information and plasticity of synapses. These splice variants are known as flip-flop splice forms, and they impart different gating kinetics to the respective receptor forms. The flip variant occurs more frequently in the early development, while the flop variant is found in greater abundance in mature animals (Bigge, 1999). The C-terminal domain of vertebrate iGluRs interacts with cytoskeletal protein and is considered important for protein trafficking, though the amino acid sequence alignments do not reveal any significant homology with other proteins (Mayer and Armstrong, 2004).

The topology of glutamate receptors has been the topic of many controversies due to the absence of a crystal structure of its transmembrane domain, analogous to the animal iGluRs. Initially, GluRs were placed in the group of postsynaptic receptors such as gamma-aminobutyric acid and the nicotinic acetylcholine receptor, the subunits of which have a 4-TM segment topology. Phosphorylation and glycolysation experiments later revealed that the putative *M2*-helix does not span the bilayer entirely but in fact forms a reentrant loop, similar to the P-loop of 2-TM K^+ channels (Hollman et al., 1994). Protease sensitivity studies (Bennet and Dingledine, 1995) and cysteine substitution studies using sulfhydryl reagents in NMDA receptors (Kuner et al., 1998) support models in which the TM domain is made up of three TM helices (M1, M2, and M3) and a reentrant membrane loop, or P segment, on the cytoplasmic side. On the basis of sequence alignment studies of K^+ channels and iGluRs, the P-loop and M2-helix of K^+ channels were considered to be similar, albeit with a low identity score (20 to 25%) (Wo and Oswald, 1995; Wood et al., 1995). Thus, *M2* in iGluRs is equivalent to the P-helix of 2-TM K^+ channels and *M3* to M2. (Here, italics indicate the standard GluR nomenclature for *M1-M4* segments, and these segments will henceforth be denoted as M1, P-loop, M2, and M3, respectively, when the GluRs are compared to potassium channels.) The most compelling evidence for a common structural ancestor for iGluRs and K^+ channels was provided by the identification of a prokaryotic K^+-selective ion channel, GluR0 (Chen et al., 1999). In bacterial iGluRs, the S2 segment of ligand-binding domain is followed by a short, extracellular C terminus, while in vertebrates, the S2 segment is linked to a third TM helix (M3), making the C terminus intracellular (Mayer et al., 2001). Homology models have been developed of the TM region, based on the homology between the sequences of the TM domain of GluRs and K^+ and on the crystal structure of KcsA (Arinaminpathy et al., 2002; Tikhonov et al., 2002). K^+ channels and iGluRs share other important structural motifs, including tetrameric subunit stoichiometry (MacKinnon, 1991; Rosenmund et al., 1998). Thus, the different domains of iGluRs are seen to have structural similarities with proteins having different functions and origin. This modular homology of iGluRs indicates that iGluRs may have been assembled from components, in accord with the "genes-in-pieces" hypothesis (Gilbert, 1978).

PROKARYOTIC GLUTAMATE RECEPTORS

The first prokaryotic GluR to be discovered was GluR0 from *Synechocystis* sp. strain PCC 6803 (Chen et al., 1999). The membrane topology of GluR0 consists of an extracellular ligand-binding domain, S1S2, and a TM domain made up of 2-TM (M1 and M2) and a reentrant P-loop, similar to that observed in the K^+ channel, KcsA (Doyle et al., 1998). GluR0 does not have a domain homologous to the N-terminal domain nor a third *M4*, TM helix, and its pore is selective for K^+ ions. (The fact that NTD is absent in GluR0 has led to suggestions that NTD has only a modulatory role in eukaryotic GluRs and does not play a role in channel gating.) When the amino acid sequence of the extracellular S1S2 domain of GluR0 was used as a probe to search the sequence database, significant similarity to the corresponding S1S2 domain of eukaryotic GluRs was observed. However, when the TM region of the protein was used as a probe, the sequence was found to be homologous to sequences from K^+ channels (Chen et al., 1999). Thus, it is a sort of "hybrid" of eukaryotic GluRs and K^+ channels. Based on sequence homologies and functional relationships between prokaryotic GluR0 and eukaryotic GluRs, this protein has been proposed to be a precursor to eukaryotic GluRs (Chen et al., 1999). Since GluR0 shares some common structural features with both eukaryotic iGluRs and bacterial potassium channels (Fig. 3), it has been considered to be an evolutionary link between the two classes of proteins (Chen et al., 1999). This protein thus likely provides an example of evolution of a protein from prokaryotes to eukaryotes.

A search of the microbial sequence database by Kuner et al. (2003) revealed eight more putative bacterial GluRs. Four of these sequences (from *Prochlorococcus marinus*, *Synechococcus* sp. strain WH8102, *Synechococcus* sp. strain PCC 7002, and *Trichodesmium erythraeum* IMS101) have signature sequences prevalent in K^+ ion channels. The other four (from *Nostoc punctiforme* [two sequences], *Silicibacter pomeroyi*, and *Magnetospirillum magnetotacticum*) have sequences in the P segment, similar to those of eukaryotic GluRs. An interesting observation was made based on scores of the similarity of profiles of the P-loop and M2 helix. The scores were higher for the bacterial GluR profiles than for the eukaryotic GluRs, suggesting a lesser divergence from the ancestral protein of bacterial

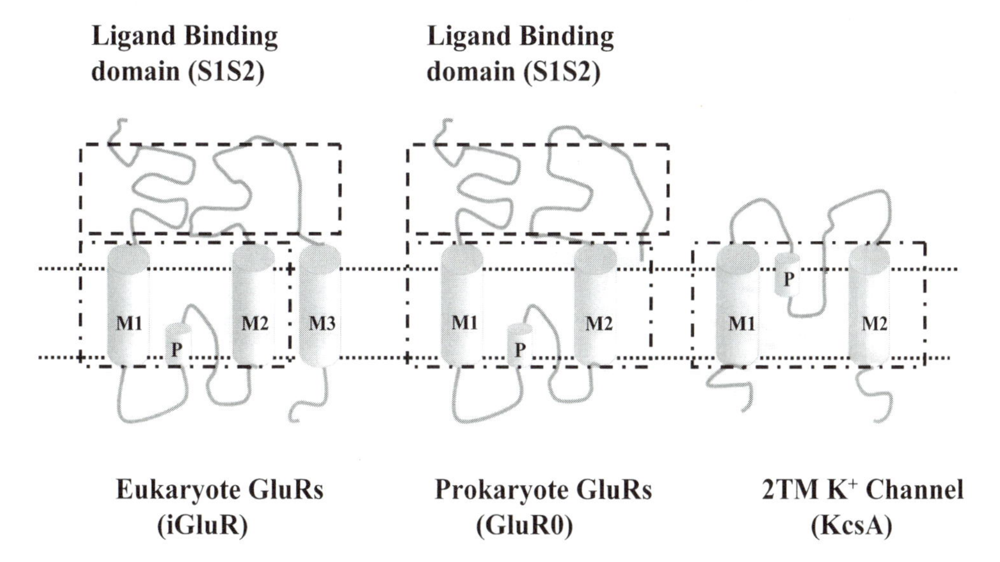

Figure 3. The predicted membrane topologies of eukaryotic GluRs, prokaryotic GluRs, and KcsA. The boxed regions indicate the ligand-binding domain (dashes) and M1-P-M2 motif (dash-dot-dash).

GluRs than from eukaryotic GluRs. Thus, the P-loop and M2 are structurally related in GluRs and K^+ channels, representing a functionally important domain required for ion selectivity and channel gating (Kuner et al., 2003). The sequences of these bacterial GluR0s are thus intermediate between bacterial K^+ channels and eukaryotic GluRs and therefore support the hypothesis that eukaryotic GluRs and K^+ channels indeed have evolved from a common ancestor of prokaryotic origin.

PLANT GLUTAMATE RECEPTORS

Plant glutamate receptors (GLRs) were first identified in *Arabidopsis thaliana* (Lam et al., 1998). The sequencing of the complete genome of *A. thaliana* revealed the existence of not 1 but 20 genes that encoded putative GLRs (Lam et al., 1998). Two full-length DNAs, GLR1 and GLR2, were isolated from this plant, each of which encoded a putative homolog for an animal iGluR. Initially, different prediction programs, including hydropathy plots, transmembrane prediction, and protein sorting, predicted that *Arabidopsis* GLRs encode a plasma membrane signal peptide and four TM domains (*M1-M4*); however, it is likely that they have the M1-P-M2-M3 topology of iGluRs. Plants produce neurotoxins such as kainate (from *Digenia simplex*), quisqualic acid (from *Quisqualis indica*), nicotine, cocaine, and caffeine, which are known to activate specific iGluRs. Experiments carried out to study the effect of 6,7-dinitroquinoxaline-2,3-dione on *A. thaliana* indicated that putative plant iGluRs might play a role in light signal transduction (Lam et al., 1998) and in calcium homeostasis (Kim et al., 2001). It has been hypothesized that plants may synthesize iGluR antagonists to regulate their endogenous iGluRs. GLR-related genes also exist in higher plants such as dicotyledons (tobacco and pea) and monocotyledons (rice and maize) (Lam et al., 1998).

EVOLUTIONARY RELATIONSHIP

A famous quote of Theodosius Dobzhansky, "Nothing in biology makes sense, except in the light of evolution," emphasizes the importance of evolutionary studies in biology. Evolutionary inferences essentially rely on diversity among organisms, where the differences are accumulated randomly from some common ancestor. For a highly adaptive yet complex system, the evolutionary pattern can be followed from either fossilized transitional forms or early-divergent extant organisms (Knight and Landweber, 2000). Paleontologists often search for evidence of links between distantly related species; e.g., the discovery of dinosaurs with feathers established a link between dinosaurs and modern-day birds (Ji et al., 1998). The discovery of a prokaryotic glutamate-activated K^+ channel was the molecular equivalent of discovering a feathered dinosaur.

Whereas ion channels are known to constitute a large class of proteins, phylogenetic analyses of channels in completed genome sequences have only just begun. A phylogenetic analysis of the gene families yields a wealth of information, including conservation of functionally important residues, functional specificity, and hierarchical position of the genes in the evolutionary history (Harte and Ouzounis, 2002). The availability of genomes from the three domains of life, *Archaea*, *Bacteria*, and *Eukarya*, provides an excellent opportunity for such an analysis. The iGluRs have been identified in prokaryotes (Chen et al., 1999), plants such as *A. thaliana* (Lam et al., 1998), invertebrates such as *Caenorhabditis elegans* (Marciq et al., 1995) and *Drosophila melanogaster* (Schuster et al., 1991), and vertebrates (Sprengel and Seeburg, 1995).

Chiu et al. (2002) performed a phylogenetic analysis of prokaryote GluRs, plant GLRs, and animal iGluRs. As expected, the prokaryotic and eukaryotic GluRs are distant from each other. The phylogenetic tree (Fig. 4), generated from the parsimony analysis, resulted in two equally parsimonious trees, one for plant GLRs and one for animal GluRs. They also performed a modular amino acid percent identity (AAPI) using different regions as a "probe." When the entire protein was used as a probe, there was a clear separation between plant GLRs and animal iGluR classes. The AAPI was higher (29 to 56%) within the same animal iGluR class (except NMDA) than between genes from two different classes (as low

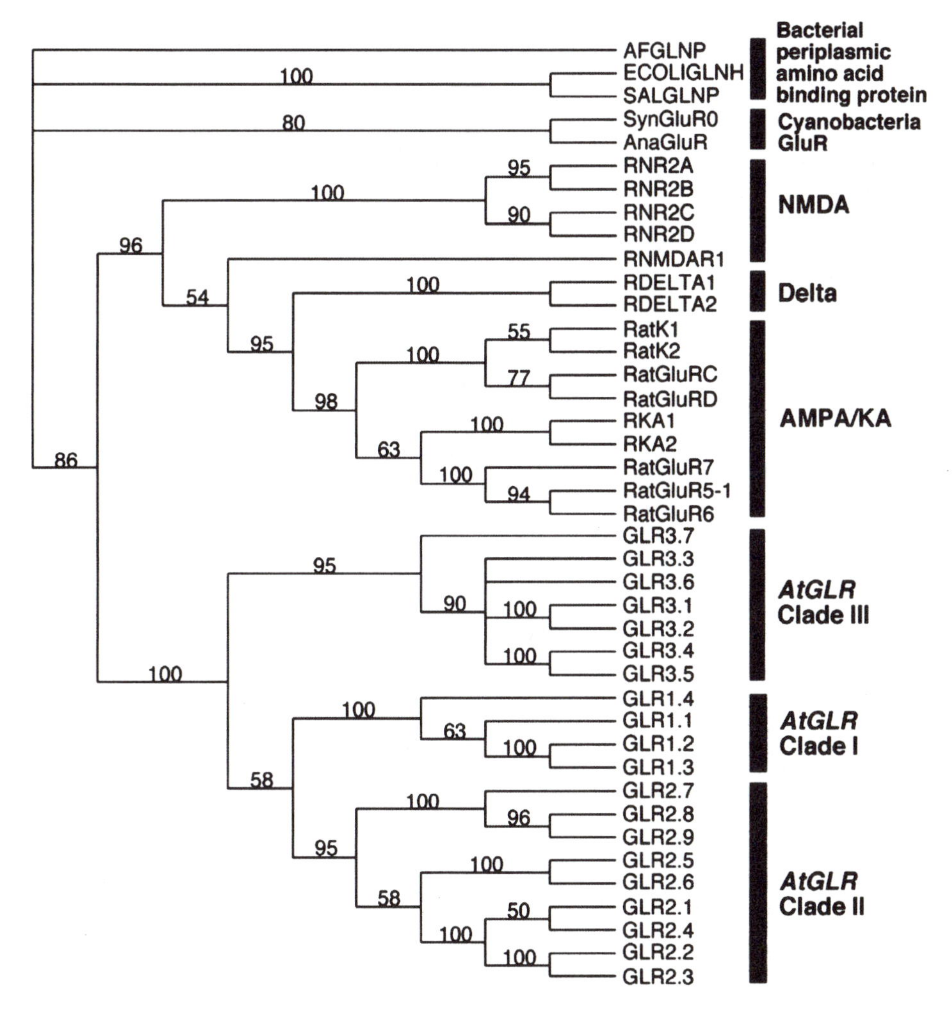

Figure 4. Phylogenetic tree generated from parsimony analysis of amino acid sequences of rat iGluRs, *Arabidopsis* GLRs, and two prokaryotic cyanobacterial iGluRs. (Reprinted from Chiu et al. [2002] with permission.)

as 10 to 50%). When only the ligand-binding domain was used to generate the AAPIs, the same trend as for the whole protein was observed. But when only the pore region consisting of the M1-P-M2 motif was used, the clear boundary between the different clades in plant GLRs disappeared, suggesting an overlap of functionality (Chiu et al., 1999). For animal iGluRs, however, the boundary between the classes remained distinct (excluding NMDA) (Chiu et al., 2002). This modular similarity of GluRs with proteins from different classes and different functions led to the speculation that the functional diversity of modern ion channels may be based on the combinatorial assembly of individual motifs, each resulting from distinct evolutionary events (Ranganathan, 1994).

Functional diversity of ion channels is also known to be achieved by alternative splicing in different types of channels, including potassium channels (Toro et al., 1998; Nichols and Lopatin, 1997), sodium channels (Plummer and Meisler, 1999), cyclic nucleotide-gated channels (Bönigk et al., 1996), and iGluRs (Wenthold et al., 1992). Functional and physiological variance can also be achieved by formation of either homomeric or heteromeric assemblies of subunits. The functional variants of iGluRs have been linked to both behavioral differences and to intermediate phenotypes in the brain. Identification of such intermediate phenotypes in the CNS can help in identifying the effect of the candidate gene on phenotype (Lipsky and Goldman, 2003). The evolution of eukarya into complex organisms seems to have increased the diversity of ion channels in eukaryotes. It is interesting that the more complex the organism, the greater the number of ion channels, especially if the nervous system of the organism is complex. For example, *C. elegans* has a much larger number of neurotransmitter-gated channels than *D. melanogaster*, while *Streptomyces lividans* and *Haemophilus influenzae* have only three ion channels and one ion channel, respectively (Harte and Ouzounis, 2002). The pathogenic species have no or very few ion channels, as compared to their nonpathogenic, free-living counterparts. Though bacteria and archaea do not have a wide range of ion channel proteins, they also have other transport proteins that perform similar functions. The habitat and physiology of the organisms should be taken into consideration to explain the presence or absence of certain families of ion channels.

One of the most important roles of ion channels is to establish a network of signaling pathways in the nervous system of higher organisms. K^+ channels and GluRs are two of the most important channels that play a vital role in regulating neuronal signal transductions. Therefore, understanding the structural and functional characteristics of these proteins will help in understanding the molecular and genetic determinants of neurological disorders. A combined pharmacological and genetic approach to understanding the regulatory mechanisms of ion channels may lead to major developments in the treatment of neurological disorders. Moreover, animal iGluRs can be used as a representative case in linking phylogenetic analysis to physiological function, since there is no dearth of experimental data available on these genes. The technological advances in genome sequences (genomics) and protein identification (proteomics) have led to a proliferation of databases with billions of data entries. With the exponential increase of computer power and disk storage space, bioinformatics provides a powerful tool to analyze these data. Bioinformatics studies are essentially based on an assumption that a hierarchical relation exists among the gene structures, their arrangement within the genome, the function of proteins, and protein-protein interactions within the organism. Phylogenetic analysis studies using sequence data are a useful tool for dividing gene families into classes in the absence of anatomical, biochemical, and other functionally relevant data. As more and more genomes become available, the bioinformatics approach can be applied to understand intricate evolutionary patterns of proteins within organisms and thus provide a sounder molecular basis for evolution.

REFERENCES

Arinaminpathy, Y., M. S. P. Sansom, and P. C. Biggin. 2002. Molecular dynamics simulations of the ligand binding domain of the ionotropic glutamate receptor GluR2. *Biophys J.* **82:**676–683.

Armstrong, N. A., and E. Gouaux. 2000. Mechanism for activation and antagonism of an AMPA-sensitive glutamate receptor: crystal structures of the GluR2 ligand-binding core. *Neuron* **28:**164–181.

Armstrong, N. A., Y. Sun, G.-Q. Chen, and E. Gouaux. 1998. Structure of glutamate-receptor ligand-binding core in complex with kainite. *Nature* **395:**913–917.

Ayalon, G., and Y. Stern-Bach. 2001. Functional assembly of AMPA and kainite receptors is mediated by several discrete protein-protein interactions. *Neuron* **13:**103–113.

Bennet, J. A., and R. Dingledine. 1995. Topology profile for a glutamate receptor: three transmembrane domains and a channel-lining re-entrant membrane loop. *Neuron* **14:**373–384.

Bigge, C. F. 1999. Ionotropic glutamate receptors. *Curr. Opin. Chem. Biol.* **3:**441–447.

Bönigk, W., F. Muller, R. Middendorff, I. Wenard, and U. B. Kaup. 1996. Two alternatively spliced forms of the cGMP-gated channel α-subunit from cone photoreceptor are expressed in the chick pineal organ. *J. Neurosci.* **16:**7458–7468.

Brakeman, P. R., A. A. Lanahan, R. O'Brien, K. Roche, C. A. Barnes, R. L. Huganir, and P. F. Worley. 1997. Homer: a protein that selectively binds metabotropic glutamate receptors. *Nature* **386:**284–288.

Burnashev, N., A. Villarroel, and B. Sakmann. 1996. Dimensions and ion selectivity of recombinant AMPA and kainate receptor channels and their dependence on Q/R site residues. *J. Physiol.* **496:**165–176.

Chen, G.-Q., C. Cui, M. L. Mayer, and E. Gouaux. 1999. Functional characterization of a potassium-selective prokaryotuc glutamate receptor. *Nature* **402:**817–821.

Chiu, J., R. DeSalle, H. M. Lam, L. Meisel, and G. M. Coruzzi. 1999. Molecular evolution of glutamate receptors: a primitive signaling mechanism that existed before plants and animals diverged. *Mol. Biol. Evol.* **16:**826–838.

Chiu, J. C., E. D. Brenner, R. DeSalle, M. N. Nitabach, T. C. Holmes, and G. M. Coruzzi. 2002. Phylogenetic and expression analysis of the glutamate-receptor-like gene family in *Arabidopsis thaliana*. *Mol. Biol. Evol.* **19:**1066–1082.

Choi, Y. B. and S. A. Lipton. 1999. Identification and mechanism of action of two histidine residues underlying high-affinity Zn2+ inhibition of the NMDA receptor. *Neuron* **23:**171–180.

Conn, P. J. 2003. Physiological roles and therapeutic potential of metabotropic glutamate receptors. *Ann. N. Y. Acad. Sci.* **1003:**12–21.

Conn, P. J., and J. P. Pin. 1997. Pharmacology and functions of metabotropic glutamate receptors. *Annu. Rev. Pharmacol. Toxicol.* **37:**205–237.

Dingledine, R., K. Borges, D. Bowie, and S. F. Traynelis. 1999. The glutamate receptor ion channels. *Pharmacol. Rev.* **51:**8–61.

Doyle, D. A., C. J. Morais, R. A. Pfuetzner, A. Kuo, J. M. Gulbis, S. L. Cohen, B. T. Chait, and R. MacKinnon. 1998. The structure of the potassium channel: molecular basis of K^+ conduction and selectivity. *Science* **280:**69–77.

Gilbert, W. 1978. Why genes in pieces. *Nature* **271:**501.

Harte, R., and C. A. Ouzounis. 2002. Genome-wide detection and family of clustering of ion channels. *FEBS Lett.* **515:**129–134.

Hille, B. 2001. *Ion Channels of Excitable Membranes*, 3rd ed. Sinauer, Sunderland, Mass.

Hollmann, M., C. Maron, and S. Heinemann. 1994. *N*-glycosylation site tagging suggests a three transmembrane domain topology for the glutamate receptor GluR1. *Neuron* **12:**1331–1343.

Ji, O., P. J. Currie, M. A. Norell, and S.-A. Ji. 1998. Two feathered dinosaurs from northeastern China. *Nature* **393:**753–761.

Kim, S. A., J. M. Kwak, S. K. Jae, M. H. Wang, and H. G. Nam. 2001. Overexpression of the *AtGluR2* gene encoding an Arabidopsis homolog of the mammalian glutamate receptors, impairs calcium utilization and sensitivity to ionic stress in transgenic plants. *Plant Cell Physiol.* **42:**74–84.

Knight, E. D., and L. F. Landweber. 2000. The early evolution of the genetic code. *Cell* **101:**569–572.

Krupp, J. J., B. Vissel, S. F. Heinemann, and G. L. Westbrook. 1998. N-terminal domains in the NR2 subunit control desensitization of NMDA receptors. *Neuron* **20:**317–327.

Kuner, T., P. H. Seeburg, and H. R. Guy. 2003. A common architecture for K+ channels and ionotropic glutamate receptors. *Trends Neurosci.* **26:**27–32.

Lam, H.-M., J. Chiu, M.-H. Hsieh, L. Meisel, I. C. Oliveira, M. Shin, and G. Coruzzi. 1998. Glutamate receptors genes in plants. *Nature* **396:**125–126.

Lipsky, R. H., and D. Goldman. 2003. Genomics and variation of ionotropic glutamate receptors. *Ann. N. Y. Acad. Sci.* **1003:**22–35.

Lomeli, H., J. Mosbacher, T. Melcher, T. Höger, J. R. P. Geiger, T. Kuner, H. Monyer, M. Higuchi, A. Bach, and P. H. Seeburg. 1994. Control of kinetic properties of AMPA receptor channels by nuclear RNA editing. *Science* **261:**1709–1713.

MacKinnon, R. 1991. Determination of the subunit stoichiometry of a voltage-activated potassium channel. *Nature* **350:**232—235.

Madden, D. R. 2002. The structure and function of glutamate receptor ion channels. *Nat. Rev. Neurosci.* **3:**91–101.

Marciq, A. V., E. Peckol, M. Driscoll, and C. I. Bargmann. 1995. Mechanosensory signaling in *C. elegans*, mediated by the GLR-1 glutamate receptor. *Nature* **378:**78–81.

Mayer, M. L., and N. Armstrong. 2004. Structure and functions of glutamate receptor ion channels. *Annu. Rev. Physiol.* **66:**161–181.

Mayer, M. L., R. Olson, and E. Gouaux. 2001. Mechanisms for ligand binding to GluR0 ion channels: crystal structure of the glutamate and serine complexes and a closed apo state. *J. Mol. Biol.* **311:**815-836.

Nakanishi, N., N. A. Shneider, and R. Axel. 1990. A family of glutamate receptor genes: evidence for formation of heteromultimeric receptors with distinct channel properties. *Neuron* **5:**569–581.

Nichols, C. G., and A. N. Lopatin. 1997. Inward rectifier potassium channels. *Annu. Rev. Physiol.* **59:**171–191.

O'Hara, P. J., P. O. Sheppard, H. Thogersen, D. Venezia, B. A. Haldmen, V. McGrane, K. M. Houamed, C. Thomsen, T. L. Gilbert, and E. R. Mulvihill. 1993. The ligand-binding domain in metabotropic glutamate receptors is related to bacterial periplasmic proteins. *Neuron* **11:**41–52.

Osborne, H. B., J. Egeberg, E. Ø. Nielsen, U. Madsen, and P. Krogsgaard-Larsen. 2000. Ligands for glutamate receptors: design and therapeutic prospects. *J. Med. Chem.* **43:**2609–2645.

Paas, Y. 1998. The macro- and microarchitectures of the ligand-binding domain of glutamate receptors. *Trends Neurosci.* **21:**117–125.

Plummer, N. W., and M. H. Meisler. 1999. Evolution and diversity of mammalian sodium channel genes. *Genomics* **57:**323–331.

Quiocho, F. A., and P. S. Ledvina. 1996. Atomic structure and specificity of bacterial periplasmic receptors for active transport and chemotaxis: variation of common themes. *Mol. Microbiol.* **20:**17–25.

Ranganathan, R. 1994. Evolutionary origins of ion channels. *Proc. Natl. Acad. Sci. USA* **91:**3484–3486.

Rosenmund, C., Y. Stern-Bach, and C. F. Stevens. 1998. The tetrameric structure of a glutamate receptor channel. *Science* **280:**1596–1599.

Schoepp, D. D., D. E. Jane, and J. A. Monn. 1999. Pharmacological agents acting at subtypes of metabotropic glutamate receptors. *Neuropharma* **38:**1431–1476.

Schuster, C. M., A. Ultsch, P. Schloss, J. A. Cox, B. Schmitt, and H. Betz. 1991. Molecular cloning of an invertebrate glutamate receptor subunit expressed in Drosophila muscle. *Science* **254:**112–114.

Sommer, B., M. Köhler, R. Sprengel, and P. H. Seeburg. 1991. RNA editing in brain controls a determinant of ion flow in glutamate-gated channel. *Cell* **67:**11–19.

Sprengel, R., and P. H. Seeburg. 1995. Ionotropic glutamate receptors. *In* R. A. North (ed.), *Handbook of Receptors and Channels: Ligand- and Voltage-Gated Ion Channels.* CRC, Boca Raton, Fla.

Sugiyama, H., I. Ito, and C. Hirono. 1987. A new type of glutamate receptor linked to inositol phospholipids metabolism. *Nature* **325:**531–533.

Tikhonov, D. B., J. R. Mellor, P. N. R. Usherwood, and L. G. Magazanik. 2002. Modeling of the pore domain of the GLUR1 channel: homology with K^+ channel and binding of channel blockers. *Biophys. J.* **82:**1884–1893.

Toro, L., M. Wallner, P. Meera, and Y. Tanaka. 1998. Maxi-K_{Ca}, a unique member of the voltage-gated K channel superfamily. *News Physiol. Sci.* **13:**112–117.

Villarroel, A., N. Burnashev, and B. Sakmann. 1995. Dimensions of the narrow portion of a recombinant NMDA receptor channel. *Biophys. J.* **68:**866–875.

Wenthold, R. J., N. Yokotani, K. Doi, and K. Wada. 1992. Immunochemical characterization of the non-NMDA glutamate receptor using subunit-specific antibodies. Evidence for hetero-oligomeric structure in rat brain. *J. Biol. Chem.* **267:**501–507.

Wo, Z. G., and R. E. Oswald. 1994. Transmembrane topology of two kainate receptor subunits revealed by *N*-glycosylation. *Proc. Natl. Acad. Sci. USA* **91:**7154–7158.

Wo, Z. G., and R. E. Oswald. 1995. Unraveling the modular design of glutamate-gated ion channels. *Trends Neurosci.* **18:**161–168.

Wood, M., H. M. A. VanDongen, and A. M. J. VanDongen. 1995. Structural conservation of ion conduction pathways in K channels and glutamate receptors. *Proc. Natl. Acad. Sci. USA* **92:**4882–4886.

Zuo, J., L. P. De Jager, K. A. Takahashi, W. Jiang, D. J. Linden, and N. Heintz. 1997. Neurodegeneration in Lurcher mice caused by mutation in δ2 glutamate receptor gene. *Nature* **388:**769–773.

Bacterial Ion Channels and Their Eukaryotic Homologs
Edited by A. Kubalski and B. Martinac

Chapter 6

Voltage-Gated K^+ Channels

H. Robert Guy and Indira H. Shrivastava

Voltage-gated channels have been studied extensively for over half a century. One might expect the bloom is off the rose by now, but not so. In fact, for those interested in the structure and functional mechanisms of these proteins, or in their molecular evolution and diversity, the fun is just beginning. Biochemical isolation and partial sequencing of a voltage-gated Na^+ channel protein from the electric organ of an electric eel led to the cloning of the first voltage-gated channel (Noda et al., 1984). The introduction of molecular biology into the channel field, coupled with the ability to use patch-clamping techniques to precisely analyze properties of channels from many different types of cells, revitalized ion channel biophysics. The molecular mechanisms for conceptual models of channel gating and selectivity began to be revealed. It soon became apparent that there were many different types of voltage-gated channels and that they were not limited to electrically excitable cells. Even prokaryotes had them. Some prokaryotes have no apparent K^+ channel proteins, and most have only a few; however, homologs of many types of eukaryotic channels have been found in some prokaryotes. Most scientists who study ion channels are unabashed eukaryotic chauvinists; however, this news interested even them because it is usually much easier to solve crystal structures of prokaryotic membrane proteins.

The KcsA K^+ channel from *Streptomyces lividans* was the first crystal structure of a relative of a voltage-gated channel to be solved (Doyle et al., 1998) (Fig. 1). This structure was immediately heralded as a tremendous breakthrough. One of the most satisfying aspects was that it provided a sound and precise molecular basis for many of the concepts about the structure and functional mechanisms that had been developed previously. Ever since the Na^+ channel sequence was first determined, it was suspected that these types of channels would have four homologous domains or subunits that are related by approximate or real fourfold symmetry about the axis of the pore (Guy and Seetharamulu, 1986). This suspicion was confirmed for the Shaker channel by mutagenesis (MacKinnon et al., 1993) and electron microscopy (Sokolova et al., 2001; Sato et al., 1998) studies. The secondary structure and general topology of the M1, P, and M2 helices in the KcsA crystal

H. Robert Guy and Indira H. Shrivastava • Laboratory of Experimental and Computational Biology, Division of Basic Sciences, National Cancer Institute, National Institutes of Health, Bethesda, MD 20892-5567.

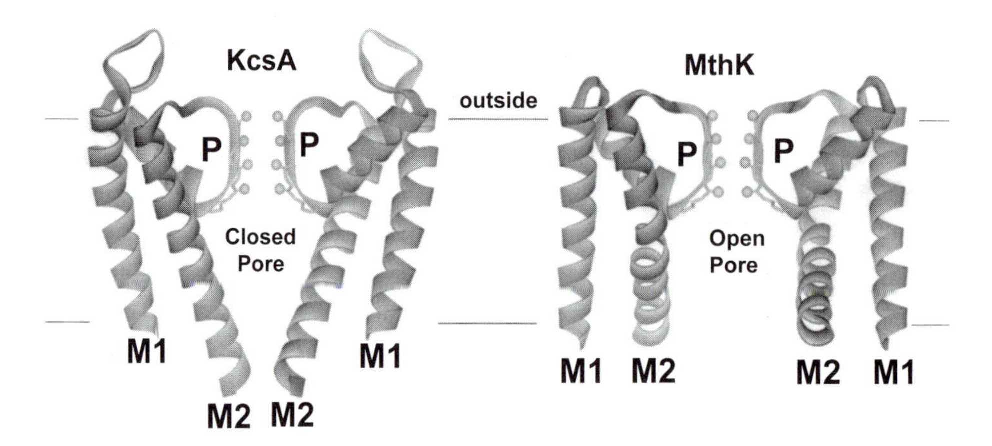

Figure 1. Side view of ribbon representation of two subunits of crystal structures of the transmembrane regions of KcsA (Doyle et al., 1998) and MthK (Jiang et al., 2002). The spheres in the P segment represent backbone oxygen atoms that bind K^+ ions.

structure were similar to those of some models that had been developed (Guy and Durell, 1995; Durell et al., 1998). The hypothesis that the P segment formed a reentrant loop and determined the selectivity of the channel had been proposed for analogous portions of the Na^+ channel shortly after it was first cloned (Guy and Seetharamulu, 1986). This hypothesis was first supported by experiments on Na^+ channels showing that mutations in the four P segments altered the binding of tetrodotoxin (Terlau et al., 1991), which blocks the extracellular entrance to the pore, and later by experiments showing that mutation of a positively charged lysine and an uncharged alanine in two different P segments to a negatively charged glutamate, as occurs in the analogous position in Ca^{2+} channels, converts Na^+ channels to Ca^{2+}-selective channels (Heinemann et al., 1992). During the same period, experiments on the Shaker K^+ channel showed that mutations on each end of the P segment affected blockade of the channel from the outside by tetraethylammonium (TEA), but that mutations near the middle of the P segment affected binding of TEA to a second site that is accessible only from inside the cell (Yellen et al., 1991). It was also shown that mutations of the two glycines in the highly conserved signature sequence of the P segment eliminated the selectivity of the channel for K^+ ions, but that the side chains of most of the other P segment residues could be altered without substantially affecting the pore's ion selectivity (Heginbotham et al., 1994). It was known that K^+ channels possess multiple K^+-binding sites (Neyton and Miller, 1988). To explain the selectivity of K^+ channels, Armstrong (1990) had postulated that the K^+ ions bind between four oxygen atoms that are related by fourfold symmetry about the axis of the pore. Based on these findings and concepts, models were developed for the selectivity filter of voltage-gated and inward rectifying K^+ channels in which backbone carbonyl oxygens of the latter portion of the P segments formed multiple binding sites (Guy and Durell, 1994, 1995). The basic features of these models were confirmed by the KcsA crystal structure; e.g., the root-mean-square deviation of the KcsA structure from the models for the backbone atoms of the TVGY residues of the signature sequence was about 0.5 Å, water and ions in the central portion of the selectivity filter were in essentially the same positions in the models and crystal structures, and the first part of the P segment was a tilted α-helix with the axis of the helix oriented toward the pore (Moczydlowski, 1998). It was also known that positively charged organic cations, such as TEA, can enter and block the pore from the inside and that some of these compounds can be trapped in the inner portion of the pore when the channel closes (Armstrong, 1990). Thus, it had been proposed that the closed pore would possess a relatively hydrophobic cavity that is sufficiently large to hold these compounds, but that is sufficiently electronegative to bind positively charged molecules. The inner portion of the pore of the KcsA crystal structure is in a closed conformation; however, this structure has a relatively hydrophobic cavity in the center of the transmembrane region just below the P segment selectivity filter that has been shown to form the TEA-binding site (Zhou et al., 2001; Zhou and MacKinnon, 2004). The axes of the four P helices are oriented directly toward this cavity and their electronegative C termini facilitate cation binding in the cavity (Doyle et al., 1998). Mutagenesis studies of voltage-gated Shaker K^+ channels had indicated that the S6 segment, which is analogous to M2 of KcsA, forms the primary activation gate (Liu et al., 1997). This appears to be true for KcsA as well. Computational analyses of permeation through the KcsA crystal structure strongly support the concept that it is in a nonconducting conformation (Sansom et al., 2002; Roux et al., 2000). Studies using electroparamagnetic resonance (EPR) of probes attached to cysteines introduced into the M2 segment of KcsA indicate that M2 moves during the pH-dependent gating (Liu et al., 2001). The second prokaryotic 2-TM K^+ channel structure to be solved was that of MthK from *Methanobacterium thermautotrophicum* (Jiang et al., 2002) (Fig. 1). MthK channels are gated by Ca^{2+} that binds to a large cytoplasmic domain. The crystal structure of its transmembrane region is similar to that of KcsA; however, the M2 helices are kinked at a highly conserved glycine residue to produce a structure in which the inner pore is very wide, indicating that it is in an open conformation.

Numerous experiments indicate that the pore-forming domains of these prokaryotic K^+ channels are similar to those of the more intensely studied eukaryotic voltage-gated K^+ channels. The selectivity of KcsA to monovalent inorganic cations is similar to that of the Shaker channel and the K^+ channel of squid axons (LeMasurier et al., 2001). The sequence of the M1-P-M2 region of KcsA can be aligned with the analogous sequence of the S5-P-S6 region of Shaker without introducing insertions or deletions (indels). Mutations of a few residues in the outer vestibule of KcsA to make it more similar to that of Shaker create a high-affinity binding site for a scorpion toxin, agitoxin 2 (AgTx2) (MacKinnon et al., 1998), that blocks the outer entrance of the Shaker pore. The binding of AgTx2 to the outer vestibule of Shaker channels has been studied by extensive thermodynamic cyclic mutagenesis studies in which effects on the toxin-binding affinity of mutations of residues in both the toxin and channel proteins are analyzed (Ranganathan et al., 1996). These studies are designed to identify which toxin residues interact directly with specific channel residues. Simulations of AgTx2 binding to homology models of the Shaker channels constructed from the KcsA structure are consistent with results of these studies (Eriksson and Roux, 2002). Chimeric channels created by replacing the S5-P-S6 segments of Shaker channels with the analogous M1-P-M2 segments of KcsA produce functional channels (Lu et al., 2001), suggesting that the structures of the pore-forming domains are similar in the two proteins.

KvAP CRYSTAL STRUCTURE

While all of this news about KcsA and MthK was exciting, those interested in voltage-dependent gating were still disappointed. The primary Kv channel subunit has four additional segments, S1-S4, that precede the S5-P-S6 segments of the pore-forming domain (S5 is analogous to M1 and S6 is analogous to M2). The S1-S4 segments comprise a voltage-sensing domain, and the positively charged S4 segment acts as the voltage sensor. KcsA and MthK structures provided no information about the voltage-sensing mechanism, since these segments are absent in these 2-TM channels. Thus, the news that the MacKinnon laboratory had solved the crystal structure of a prokaryotic voltage-gated K^+ channel (KvAP) from *Aeropyrum pernix* was greeted enthusiastically (Jiang et al., 2003a). The structure of the pore-forming domain formed by segments S5-P-S6 of the crystal structure of the full-length KvAP protein was similar to that of the MthK crystal, suggesting that it was in an open conformation (Fig. 2a). Unfortunately, interpretation of the structure and functional mechanism of the voltage-sensing domain has been complicated by the likelihood that this domain is distorted in the crystal and does not have a native conformation (Jiang et al., 2003a; Shrivastava et al., 2004). Early models of the Na^+ channel proposed that S1-S4 segments are all transmembrane α-helices with the N termini of S1 and S3 on the cytoplasmic side and the N termini of S2 and S4 on the extracellular side (Noda et al., 1986; Guy and Seetharamulu, 1986). This transmembrane topology has been verified by many mutagenesis experiments performed on both Na^+ and K^+ channels (Durell et al.,

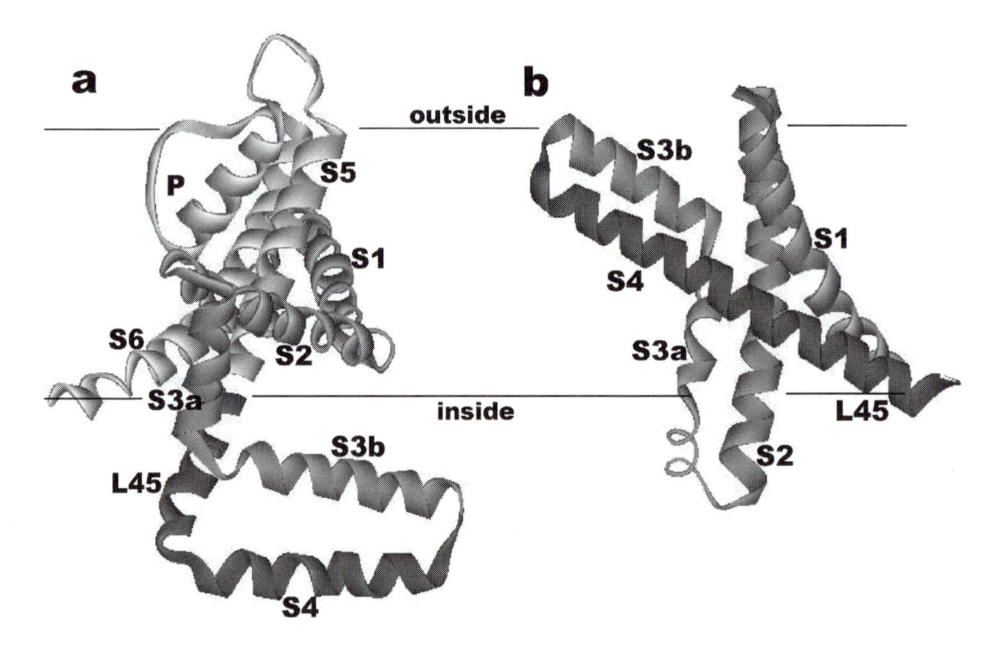

Figure 2. Crystal structures of the KvAP channel. (a) One subunit of the full-length structure. (b) Structure of the isolated voltage-sensing domain in orientation relative to the bilayer used in developing models. The lines represent the boundries of a lipid bilayer: S1-S3 in medium gray, S4 in dark gray, S5-P-S6 in light gray.

1998). Although the S1-S4 segments of the KvAP crystal structure had the anticipated secondary structure, their tertiary structure was different. Using the structure of the pore-forming domain to approximate the location of the lipid bilayer, S1 and S2 encircled the pore-forming domain with their axes approximately parallel to the plane of the membrane, and with the S1-S2 loop near the cytoplasmic surface, S3 began in the transmembrane region and extended into the cytoplasmic phase, and S4 resided in the cytoplasmic phase. Everyone acknowledges that this is unlikely to be a native structure; however, a controversy has arisen about how much it deviates from a native conformation and whether it provides useful information about the native structure and gating mechanism (Ahern and Horn, 2004b; Gandhi et al., 2003; Starace and Bezanilla, 2004; Shrivastava et al., 2004; Durell et al., 2004). Jiang et al. (2003b) proposed that the native conformation of the closed conformation is not very different from the crystal structure. They used a second crystal structure of an isolated voltage-sensing domain (S1-S4) (Fig. 2b) to develop a model of the closed conformation in which S2 is in the same general location as in the full-length crystal structure. Although they acknowledged that this location is inconsistent with the finding that the S1-S2 loop is glycosylated in the Shaker channel, they speculated that this difference could be due to the highly dynamic nature of the protein. In these closed conformation models, the C terminus half of S3, called S3b, and S4 form an α-helical hairpin that is positioned in the transmembrane region near the cytoplasmic surface. They hypothesized that this hairpin, which they called a paddle, remains intact and diffuses through to the lipid phase of the membrane during the activation gating process. To test this hypothesis, they substituted cysteines, one residue at a time, to positions throughout S3b and S4 and then attached biotin adducts to these cysteines. They then analyzed the voltage dependencies of the binding of the biotin adducts to the protein avidin, to which biotin binds virtually irreversibly. Biotin adducts to two adjacent positions, L121C and L122C, near the middle of S4, bound to intracellular avidin at negative voltages and to extracellular avidin at positive voltages, suggesting that this portion of S4 transverses much of the transmembrane region during gating. They also found that the rate at which the biotin adducts to positions in S3b and the initial part of S4 bound to extracellular avidin was much greater at positive voltages than at negative voltages, suggesting that this portion of the protein is relatively inaccessible from the outside at negative voltages. Recently the MacKinnon laboratory has used electron microscopy (EM) to analyze the structure of KvAP channels that have Fab fragments attached to their S3b-S4 paddles (Jiang et al., 2004). They were able to approximate two possible positions for the paddles relative to the pore-forming domain by fitting the X-ray structures of the pore-forming domain and of the Fab fragments with bound isolated paddle peptides to their EM data. In both models, the paddles of this open conformation structure are on the extracellular surface where they interact with the pore-forming domain and with lipids of the extracellular leaflet.

The paddle model has not been widely accepted for several reasons (Shrivastava et al., 2004; Durell et al., 2004): (i) it is inconsistent with many experimental results that have been obtained from Shaker channels, (ii) some consider its postulate that positively charged S4 residues diffuse through the alkyl phase of the bilayer to be energetically unrealistic, (iii) it is based in part on an obviously misfolded crystal structure, and (iv) it is possible that the irreversible binding of the biotin adducts to avidin during long exposures traps the protein in a perturbed conformation that occurs only rarely. Here we will discuss experiments performed on Shaker that we consider to be most informative about the structure and voltage-sensing domain. The similarity of gating properties of the KvAP and Shaker channels and of their sequences makes it probable that both proteins gate by similar mechanisms (Ruta et al., 2003).

TRANSMEMBRANE TOPOLOGY

Contrary to predictions of the paddle model, numerous experiments indicate that the S1-S2 and S3-S4 loops of Shaker channels are in the extracellular aqueous phase in all conformations. Charged sulfhydryl reagents react rapidly with cysteines introduced throughout these loops, in the C-terminal ends of S1 and S3, and in the N terminus region of S2 in a voltage-independent manner (Gandhi et al., 2003). The S1-S2 loop is glycosylated, and the glycosylation does not alter gating properties of the channel substantially (Santacruz-Toloza et al., 1994). Lanthanide-based resonance energy transfer (LRET) and fluorescent resonance energy transfer (FRET) probes attached to some positions within each of these loops detect little or no motion during gating (Cha et al., 1999; Glauner et al., 1999). Disulfide bridges can form at all voltages between cysteines introduced into the S3-S4 loops at positions 353 to 356 that immediately precede S4 and position 416 in S5 near its extracellular C terminus end (Gandhi et al., 2003; Broomand et al., 2003). Spider toxins, such as hanatoxin, that affect activation gating by binding from the outside to the lipid bilayer (Lee and MacKinnon, 2004) and S3b-S4 segments (Li-Smerin and Swartz, 2000) bind with even higher affinity to eukaryotic Kv channels at negative voltages (Lee et al., 2003).

CREVASSE MODELS OF ACTIVATION GATING

Although the S1-S2 and S3-S4 loops of Shaker channels appear to remain relatively static during gating, the S4 segment, which is thought to be the primary voltage sensor, may move substantially. Gating currents can be measured that are due to movement of charges on the voltage-gated channel through the electric field of the membrane (Bezanilla, 2000). The equivalent of about 13 charges (or more than 3 charges per subunit) crosses the electric field during activation of a Shaker channel (Schoppa et al., 1992). S4 segments are atypical of transmembrane helices in that every third residue is positively charged. Most of the gating charge movement of activation in Shaker channels is due to the first four positively charged residues of S4 (Aggarwal and MacKinnon, 1996). Cysteine scanning studies of S4 in both Shaker and a Na^+ channel indicate that the accessibilities of many residues to charged sulfhydryl reagents are voltage dependent: some S4 residues near the N terminus are accessible from the outside only at positive voltages, and other residues near the C terminus are accessible from the inside only at negative voltages (Larsson et al., 1996; Baker et al., 1998; Gandhi et al., 2003; Yang et al., 1996). A few residues near the center of S4 are accessible from the inside at negative voltages and from the outside at positive voltages. The length of the portion of S4 that is inaccessible from either side at a given voltage is much too short to span the entire transmembrane region (the length depends on the channel studied and the probes used). These findings have given rise to the concept that the voltage-sensing domain possesses water-filled crevasses through which charged reagents can diffuse to reach much of the S4 segment (Ahern and Horn, 2004c; Gandhi and Isacoff, 2002; Bezanilla, 2002) (Fig. 3) and that a transition barrier, or gating pore, separates these crevasses. Crevasse models also have been supported by experiments in which the positively

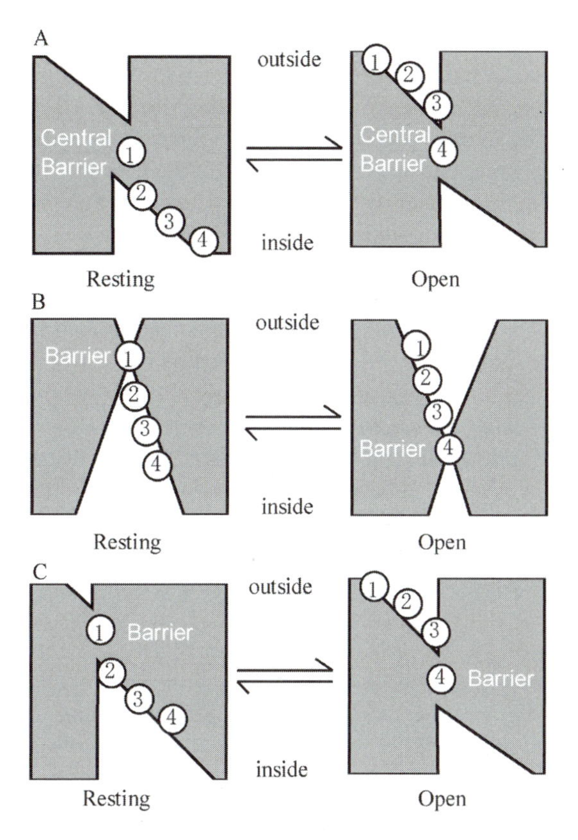

Figure 3. Schematic representations of three crevasse models. Circles with + inside represent positively charged side chains of S4. (A) Model in which S4 moves outwardly relative to a single transition barrier during activation (Shrivastava et al., 2004). (B) Transport model in which S4 does not move much, but the location of the transition barrier changes during activation (Starace and Bezanilla, 2004). (C) Hybrid model in which S4 moves outwardly and the transition barrier moves inwardly during activation.

charged residues of S4 are replaced one at a time by histidine in mutant channels that have normal gating currents but that do not conduct ions through their pore (Starace and Bezanilla, 2001, 2004). In mutants in which the first or fourth S4 arginines are mutated to histidine, a proton pore forms through the voltage-sensing domain at negative or positive voltages, respectively. These results have been interpreted as indicating that the voltage-sensing domain has a transmembrane pathway that is sufficiently polar to conduct protons, but in which proton conduction is blocked by the first arginine in the closed conformation and by the last arginine in the open conformation (Starace and Bezanilla, 2004) (Fig. 3). Mutants in which the second and third arginines are mutated to histidine transport a proton across the membrane whenever the protein undergoes a normal voltage-dependent conformational change. These results suggest that these two residues cross the transition barrier during gating and that the pathway through which these residues move during gating is sufficiently polar and electronegative for the easily titratable histidine side chains to remain protonated throughout the transition. Mutants in which the fifth or sixth positively charged residues are mutated to histidine do not transport protons, suggesting that they never cross the transition barrier to become accessible to protons from the outside. Additional support for crevasse models comes from experiments in which the contribution of charged adducts to S4 on the gating current were measured (Ahern and Horn, 2004a). Positively charged adducts at the positions on S4 normally occupied by the first or second arginine contribute about one charge per subunit to the gating current, even for adducts that are not charged when in a hydrophobic environment. However, adducts at adjacent positions between these arginines or immediately following the second arginine do not. These results suggest that the first and second arginines cross the transition barrier region and enter an intracellular polar crevasse when the channel deactivates, but that adjacent residues on the opposite face of the helix do not. They also suggest that much of the transmembrane voltage change occurs in a relatively short transition barrier region. Note that none of these experimental results that suggest crevasse models are consistent with the paddle model (Jiang et al., 2003b), in which the positively charged S4 residues are postulated to diffuse through the very hydrophobic and relatively thick alkyl phase of the membrane.

Three types of crevasse models can be envisioned. Figure 3A depicts a model in which the transition barrier remains fixed and the positively charged S4 residues move past it during gating (Shrivastava et al., 2004). These models imply that the transmembrane movement of S4 is relatively large and that the rest of the protein remains relatively static. Figure 3B depicts the opposite extreme in which there is little transmembrane movement of S4, but the location of the barrier and the structures of the crevasses change (Starace and Bezanilla, 2004). This has been called a "transporter"-type model due to its resemblance to some models for transporters. The model of Fig. 3C is a hybrid of the other two; S4 charges move outward during activation, but this movement is accompanied by an inward movement of the transition barrier and a change in the structures of the crevasses. This may be the most realistic representation since it is probable that the nature of the crevasses changes during activation. In our recent models of the Shaker channel (Durell et al., 2004), the S1-S3 segments that form the barrier move inward as the S4 segment moves outward during activation. The major differences between the first and second crevasse models are the magnitude of the transmembrane movement of S4 and whether there are one or two barrier regions. The transporter model is supported by studies that suggest little transmembrane movement of S4 (Bezanilla, 2002). Two types of experiments suggest that the movement of S4 may be relatively small. The first type involves the use of FRET (Glauner et al., 1999) and LRET (Cha et al., 1999) measurements. However, most of these data are from probes attached to the loop region that precedes S4; e.g., both techniques indicate only small motions for probes attached to residues 353 to 355. The relatively static nature of these residues is supported by additional experiments that have shown that when residues 353, 354, 355, or 356 of the S3-S4 loop and residue 416 near the end of S5 are mutated to cysteine, a disulfide bridge can form at both positive and negative voltages (Gandhi et al., 2003; Broomand et al., 2003). The Shaker residue that aligns with the first residue of S4 in the KvAP structures is 361. It thus seems unlikely that S4 of Shaker moves 20 Å across the membrane, as suggested for S4 of KvAP by the biotin/avidin experiments (Jiang et al., 2003b). However, these experiments do not demonstrate that the S4 movement is small; e.g., if the linker segment from residue 356 to 361 can have a coiled or helical structure when the channel is open and an extended structure when the channel closes, then S4 could move substantially without affecting the positions of residues preceding 356. Although an LRET measurement to probes attached to residue 363 detects no movement during activation (Cha et al., 1999), this could be an artifact of either the large size of the acceptor probe (8 by 8 by 13 Å) or the fact that only the distance from the axis of the pore is measured. Other experiments suggest substantial movement of this part of S4. Electrostatic interactions between charged adducts to residue position 362 on S4 (the position of the first S4 charge) and residue 418 on S5 appear to occur only when the channel is open (Elinder et al., 2001b; Elinder et al., 2001a), and interactions between a cysteine at position 362 and a second cysteine at either position 416 or 419 on S5 occur only at positive voltages (Laine et al., 2003). The primary advantage of the crevasse model that postulates transmembrane movement of S4 is that it is more consistent with the results of the biotin/avidin experiments on KvAP.

MODELS BASED ON THE CRYSTAL STRUCTURE OF THE ISOLATED VOLTAGE-SENSING DOMAIN OF KvAP

The Voltage-Sensing Domain and Gating

Before determining the crystal structure of the full-length protein, Jiang et al. (2003a) determined a crystal structure for an isolated voltage-sensing domain. The S4-S5 linker, or L45 helix, is an extension of the S4 helix in this structure. This long helix will be called the S4/L45 helix here. This crystal structure has a bundle of four antiparallel α-helices, S1-S4/L45, that are packed in a manner similar to models that had been developed earlier (Durell et al., 1998). Jiang et al. (2003b) used the S2-S4 portion of this structure in developing their paddle model (S1 was not included in their model). In the resting conformation of their model, S2 was positioned in the same general location as in the crystal structure of the full-length structure so that the S2-S4 segments did not span the transmembrane region, and S4 was far from S5 in all conformations. Several groups (Gandhi et al., 2003; Broomand et al., 2003; Shrivastava et al., 2004) have proposed alternative models in which the crystal structure of this domain, or a model similar to this crystal structure, is positioned to be more consistent with the topological data from the Shaker channel; i.e., the S1-S4 helices span the transmembrane region and the N terminus of the S4 helix packs next to the C terminus of the S5 helix of an adjacent subunit. However, these models are not in agreement with the recent EM studies of Jiang et al. (2004); e.g., the S4/L45 helix must break to fit the EM data. Figure 4A illustrates a model that we developed to be consistent with the EM data. The pore-forming domain and S3-S4 paddle were positioned as proposed by Jiang et al. (2004), and S1 and S2 from the crystal structure of the isolated voltage-sensing domain were positioned to span the transmembrane region in a location that is occupied by protein in the EM studies. The position of the L45 segment relative to S1 and S2 in the crystal structure was maintained. Figure 4B illustrates a model that was constructed simply by linking the crystal structure of the isolated voltage-sensing domain to the pore-forming domain of the full-length crystal structure, keeping the link between L45 and S5 very short. The primary difference between these models is the position of the S3-S4 paddle; in both models it is near the extracellular surface. However, in the first model S4 interacts with the pore-forming domain; in the second model, it has swung

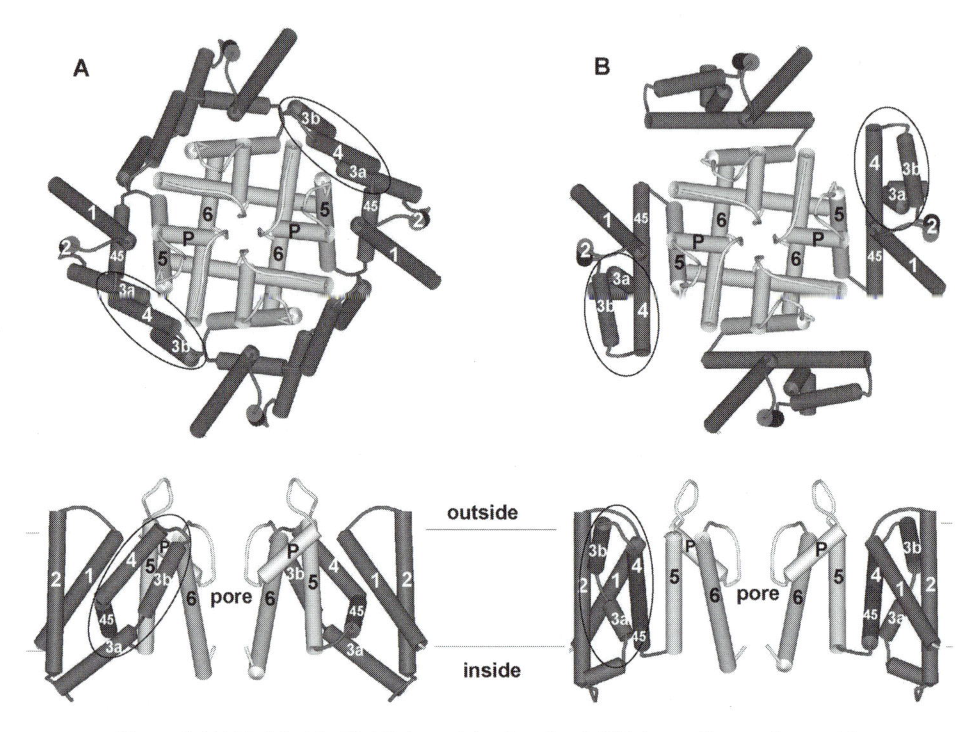

Figure 4. (A) Model of the KvAP channel developed to fit EM data and be consistent with models of Jiang et al. (2004). (B) Model of the open KvAP channel based on the crystal structures of the isolated voltage-sensing domain for S1-S4 and the full-length protein for the pore-forming domain (S5-P-S6). S1, S2, L45, and pore-forming domains are about the same in both models. The voltage-sensing domains are dark and the pore-forming domain is light. The S3-S4 paddles are encircled.

farther into the lipid region of the bilayer and interacts more with the S1 and S2 segments. The first four positively charged residues of S4 are near the extracellular surface in both models, consistent with being in either an open or an inactivated conformation. It is plausible that both types of conformation occur; e.g., Fig. 4B could be an intermediate or open conformation and the paddle could swing to its position in Fig. 4A when the channel opens or inactivates. Unfortunately, homology models of the Shaker channel developed using the backbone structure of either of the KvAP models of Fig. 4 would be inconsistent with experimental results; e.g., in Shaker the N terminus of S4 has been shown to interact with the C terminus of S5 of an adjacent subunit, whereas in Fig. 4 the S4-S5 interactions occur within the same subunit. Thus, there is still much uncertainty about the structure of these channels, and more studies will be required to resolve apparent inconsistencies.

Can models developed using the crystal structure of the isolated voltage-sensing domain of KvAP explain the experimental data from Shaker that led to the concept of crevasses? The composition of the outer surface of the transmembrane region of the pore-forming domain is similar in 2-TM and 6-TM channels; i.e., it is composed almost exclusively of hydrophobic residues. Thus, if the transmembrane region of voltage-gated channels possesses water-filled crevasses that are lined predominantly by hydrophilic residues, these crevasses must reside primarily within the voltage-sensing domains. To examine this possibility, we first considered how the crystal structure of the isolated KvAP voltage-sensing domain would likely be oriented in a lipid bilayer, reasoning that the lipid would mimic reasonably well the hydrophobic exterior of the pore-forming domain (Shrivastava et al., 2004). The orientation that we favor is shown in Color Plate 6A [see color insert]), which is similar to the orientation in Fig. 4B. The hydrophobic alkyl phase of a bilayer is about 25 Å thick. This orientation places the termini of the transmembrane helices and most of the charged residues that are on the surface of the protein outside this region. Some charged residues are in the transmembrane region, but these residues are in the core of the domain where they form salt bridges with oppositely charged residues. If this structure is a native conformation, then it must correspond to the open or inactivated conformation since most of the positively charged S4 residues are on the outer surface, as are L121 and L122 (biotin adducts at positions 121 and 122 bind to extracellular avidin at positive voltages and to intracellular avidin at negative voltages [Jiang et al., 2003a]). The most tightly packed portion of this bundle occurs where the axes of S4 and S2 helices cross. We postulate that this region corresponds to the transition barrier of the crevasse models. The S4/L45 helix is very tilted relative to the membrane's normal orientation, and the regions above and below S4 on the extracellular and intracellular sides of the putative transition barrier are very hydrophilic. We propose that these hydrophilic regions correspond to the hypothesized crevasses.

If the crystal structure of the isolated KvAP voltage-sensing domain is a native open conformation, then what conformational changes occur when the channel closes and how does the protein move between these conformations? The helical screw mechanism is the oldest model for the movement of S4 during activation (Guy and Seetharamulu, 1986). In this model, the S4 helix translocates along and rotates about its axis so that the arginines remain in the same spiral pathway as they traverse the transmembrane region. One inward helical screw step places the backbone of the nth residue in the position occupied by the $(n + 3)$th residue of the previous conformation; e.g., the helix translates by ~4.5 Å along and rotates by ~60° about its axis. Initial helical screw steps from the crystal structure result in an increase in the number of electrostatic interactions between positively and negatively charged residues; e.g., after two helical screw steps each of the first six positively charged residues of S4 bind to negatively charged residues on S1-S3 (Color Plate 6B). Thus, the negatively charged residues of S1, S2, and S3 are positioned in the crystal structure in a manner that complements the spiral of positively charged S4 residues. This arrangement allows positively charged S4 residues to move through the transmembrane region in an energetically realistic manner that never exposes them to lipid alkyl chains (Shrivastava et al., 2004). The proximity of all positively charged S4 side chains to negatively charged side chains explains why the S4 residues remain protonated during activation, even when they are replaced by histidine or adducts that are deprotonated when placed in a hydrophobic environment. In this putative transition conformation, residues of S4 that are highly conserved among distantly related families of voltage-gated channels (K^+, Na^+, Ca^{2+}, cyclic nucleotide-gated, and polycystin) interact with similarly conserved residues of S1-S3 to form a highly conserved cluster in the transition region of the voltage-sensing domain (Shrivastava et al., 2004). This observation supports the supposition that this is a functionally important conformation that occurs in all of these voltage-gated channels. However, a two-helical-screw-step model is insufficient to explain the data in either KvAP or Shaker channels. Six to seven helical screw steps are required to explain the biotin/avidin results for KvAP if S4 remains very tilted throughout the transition. This mechanism requires conformational changes in the segments on each end of S4. We have proposed that the latter half of S3b jackknifes near its midpoint and becomes an N terminus extension of S4 in the resting conformation. Additional movement may involve the hinge point between S3a and S3b. These movements can explain why biotin adducts to S3b do not bind readily to extracellular avidin at negative voltages. Additional conformational changes are proposed to occur at S4/L45 and L45/S5 junctions.

The models for the voltage-sensing domain of KvAP described above are not completely consistent with the data from Shaker. Charge reversal experiments have been interpreted to indicate that in the open conformation, the fourth arginine of S4, *R4*, interacts with a glutamate near the N terminus of S2. Based on the alignment of the Shaker and KvAP sequences of Fig. 5, analogous residues do not interact in the crystal structure of the isolated domain. (Generic designations for charged residues that are found in many voltage-gated channels are presented in italics.) However, if the alignment of the Shaker sequence is shifted three residues to the right, these interactions do occur. They also occur in the model of the KvAP transition conformation in which S4 has moved one helical screw step inward, since this is the modeling equivalent of shifting the alignment by three residues. This arrangement is consistent with the crevasse models for the open conformation of Shaker, i.e., *R4* is positioned in the transition barrier, and regions above and below *R4* are sufficiently polar and open to conduct protons through the membrane (Durell et al., 2004). The crevasse models propose that *R1*, the first arginine of S4, is similarly located in a transition barrier in the resting conformation. This can be accomplished for the single barrier models by moving S4 inward by three helical screw steps during deactivation. Unlike KvAP, S3b and S3-S4 linker residues of Shaker are outwardly accessible in all conformations (Gandhi et al., 2003). This difference may be due to the long S3-S4 linker. Much of the linker can be deleted without altering the voltage dependency of gating substantially (Gonzalez et al., 2001); however, in these mutants, S3b and N terminus S4 residues are outwardly accessible only at positive voltages (Gonzalez et al., 2004). This is consistent with the concept that S4 moves outwardly during activation and that its movement is coupled to a movement of S3b if the linker is short.

Interactions between the Voltage-Sensing and Pore-Forming Domains

Four categories of criteria have been used to constrain models of the interactions between the voltage-sensing and pore-forming domains of Shaker channels. The first category involves studies of the proximity of residues near the N terminus of S4 to residues near the C terminus of S5. The first evidence that the N terminus of S4 is near the C terminus of S5 came from studies of the effects of charged adducts and mutations to the first arginine, R362, of S4 and residue E418, which is near the end of S4 (Elinder et al., 2001b; Elinder et al., 2001a). These residues appear to interact electrostatically only when the channel is open. This finding led several groups to utilize pairwise cysteine substitutions to analyze interactions of these and other residue pairs (Laine et al., 2003; Gandhi et al., 2003;

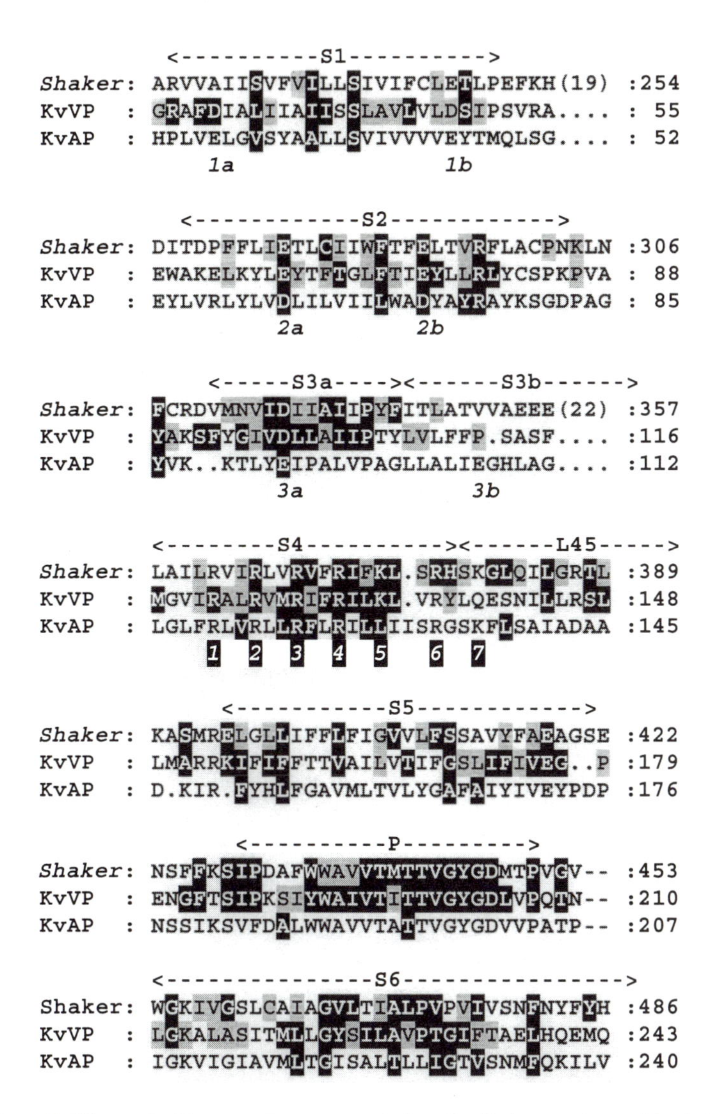

Figure 5. Alignment of transmembrane segments from the Shaker, a category 1 bacterial Kv protein (KvVP) from *V. parahaemolyticus*, and KvAP. The Shaker sequence is shaded according to the mutability of residues calculated from a multisequence alignment of eukaryotic Kv channel sequences; the KvVP sequence is shaded according to the mutability of residues within a prokaryotic family of putative 6-TM channels with sequences intermediate between those of eukaryotic Kv and KvAP, and the KvAP sequence is shaded according to the mutability of residues among numerous distantly related families of 6-TM channels. Color code: black background, highly conserved; gray background, moderately conserved; white background, poorly conserved. Simplified numbers of charge residues in the voltage-sensing domain that interact during the helical screw transitions are indicated in italics below the KvAP sequence. The parentheses indicate insertions of the indicated number of residues in the Shaker S1-S2 and S3-S4 loops.

Broomand et al., 2003; Neale et al., 2003). The rationale is that if residue pairs are proximal in a given conformation, then replacement of these residues by cysteines will lead to the formation of disulfide bridges and Cd^{2+}-binding sites under oxidizing and reducing conditions, respectively. Results of these experiments are summarized and related to our models of Shaker channels in Fig. 6. Residues 353 to 356, which precede the S4 segment, were found to interact with F416C in all conformations (Gandhi et al., 2003; Broomand et al., 2003), suggesting that they do not move much during gating. In contrast, L358C, I359C, and R362C were found to interact with A419C and/or F416C only at positive voltages (Gandhi et al., 2003; Laine et al., 2003), suggesting that these residues do move.

The second approach involves measuring distances between fluorescent probes (FRET [Glauner et al., 1999] or LRET [Cha et al., 1999]) that are bound to introduced cysteine residues of adjacent subunits. Although the absolute values of the distances calculated by these two methods do not agree well, both methods detected only small movements of probes attached to positions 353 to 355, consistent with the disulfide-bridging experiments described above. The LRET distances are consistent with our models, except for R362C in the open conformation (Durell et al., 2004).

The third approach involves attaching TEA analogs to introduced cysteines using tethers of varying lengths (Blaustein et al., 2000). TEA binds to two sites in the K^+ channel: an outwardly accessible site in the outer entrance to the pore and an inwardly accessible

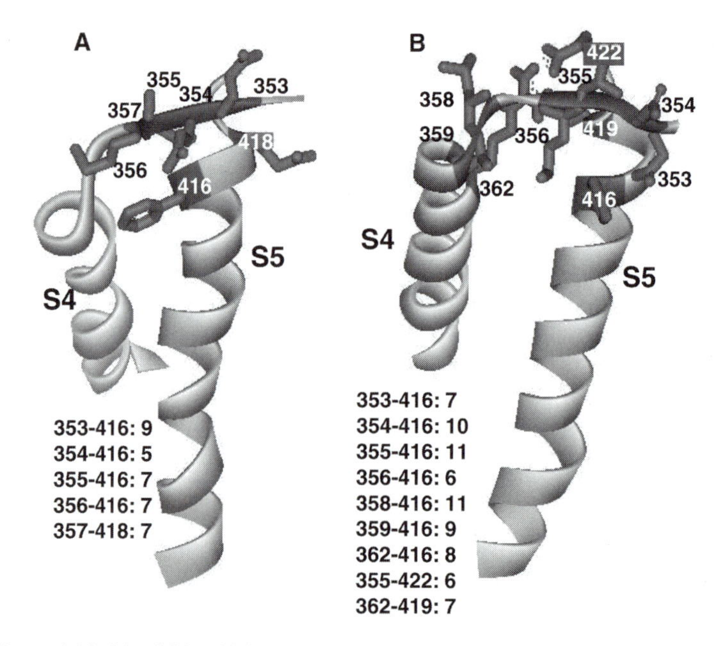

Figure 6. Models of S4 and S5 segments of adjacent subunits of Shaker channels in resting (A) and open (B) conformations illustrating residue pairs that have been shown to be proximal by formation of Cd^{2+}-binding sites and/or disulfide bridges when replaced by cysteines. Distances between β-carbons of specific pairs are indicated in each figure. (Reprinted from Durell et al. [2004] with permission.)

site in the cavity region of the pore. In these experiments, the length of the shortest tether that allowed blockade of the outer entrance was used to approximate the distance from the introduced cysteine to which the probe was attached and the outer TEA-binding site. We used these distances to constrain our Shaker models of the extracellular loops and to demonstrate that our models of the transmembrane regions were consistent with these data (Fig. 7).

The fourth approach involves analyzing the tolerance of residue positions to mutations. Two groups have performed scanning mutagenesis on the S1-S3 segments of eukaryotic Kv channels; one group replaced existing residues with tryptophan (Monks et al., 1999; Hong and Miller, 2000) and the other with alanine (Li-Smerin et al., 2000b). Residues were considered tolerant if the mutation did not alter voltage-dependent gating significantly. Both groups found that S1 and S2 have intolerant faces and opposing tolerant faces; the results on S3 were more complex and less consistent. An alternative approach uses information from multisequence alignments to analyze the tolerance of residues to mutations. We have used this approach for many families of K^+ channels, including both prokaryotic and eukaryotic Kv channels (Shrivastava et al., 2004). The results of these theoretical analyses are similar to those of the experimental findings; both S1 and S2

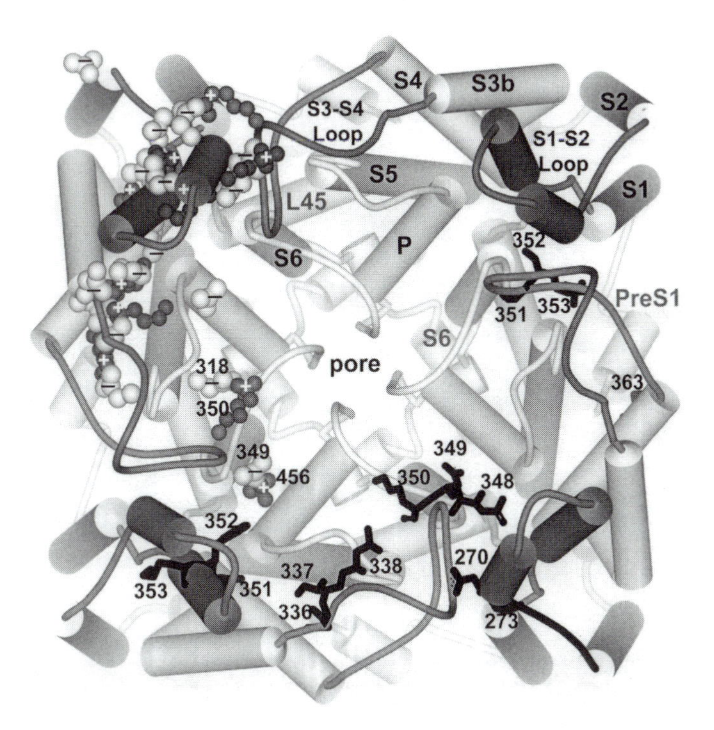

Figure 7. Models of the extracellular loops of the Shaker channel in the open conformation. Residues where LRET probes have been attached are indicated in the subunit on the right, residues where TEA tethers have been attached are indicated in the subunit on the bottom, and charged residues of the loops are indicated in the subunits on the left and top. (Reprinted from Durell et al. [2004] with permission.)

have tolerant and intolerant faces (Fig. 8). These findings have been interpreted to indicate that the tolerant faces of S1 and S2, which are also hydrophobic, are exposed to lipid alkyl chains. The analysis of S3 is complicated by the fact that it is not a single α-helix, but is rather two helices, S3a and S3b. Nonetheless, these results support our contention that portions of S3 are also exposed to lipid alkyl chains. The N-terminal half of S4 also has a poorly conserved hydrophobic face for eukaryotic Kv channels, suggesting that it too has substantial exposure to lipids; however, this pattern was not obtained from the multisequence alignment of the largest subfamily of prokaryotic Kv channels. Many of the predictions of these analyses were already satisfied by the crystal structure of the isolated voltage-sensing domain; i.e., the residues in the core of the domain were highly conserved and intolerant, whereas the poorly conserved tolerant faces of S1 and S2 were oriented away from the core. Alanine substitution experiments of residues on the outer surface of the pore-forming domain also identified intolerant stripes that are tilted by about 40° relative to the pore's axis (Li-Smerin et al., 2000a). We docked our models of the voltage-sensing domain into these intolerant stripes on the pore-forming domain in a manner that exposed the tolerant hydrophobic residues to alkyl chains and that allowed the N terminus of S4 to interact with the C terminus of S5 of an adjacent subunit. The model was also constrained by the requirement that the L45 helix connects to the N terminus of S5.

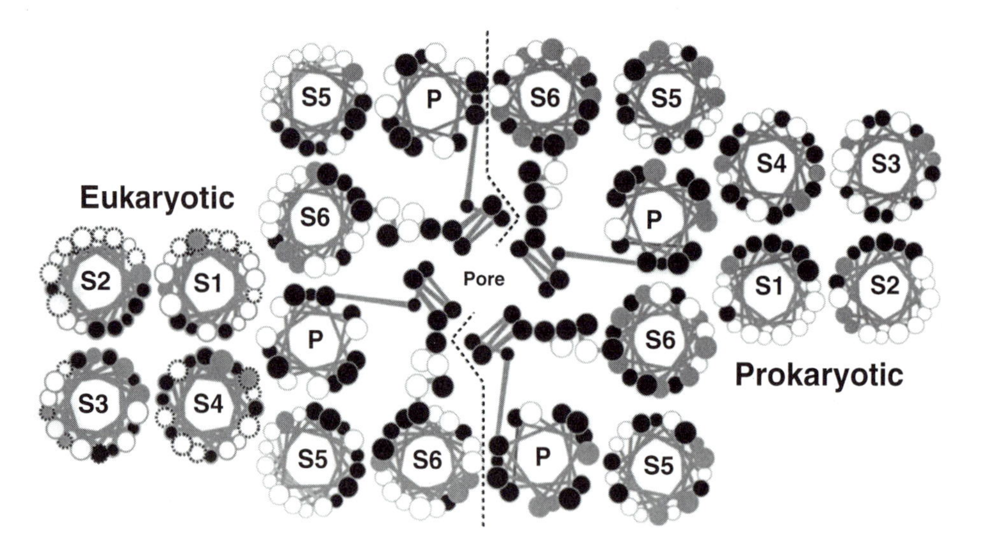

Figure 8. Helical wheel representation of a Kv channel as viewed from outside the cell illustrating residues that are well conserved (black), moderately conserved (gray), or poorly conserved (white) (Shrivastava et al., 2004). S1-S4 residues of eukaryotic channels that are classified as tolerant from results of experimental mutagenesis (Monks et al., 1999; Hong and Miller, 2000; Li-Smerin et al., 2000b) are indicated by dashed lines around the residues.

Coupling the Voltage-Sensing Mechanism to the Voltage Gate

Although the gate formed by the C terminus of S6 and the voltage sensor involving S4 have been analyzed extensively, little is known about how these mechanisms are coupled. The L45 helix that connects S4-S5 is likely to play a key role (Latorre et al., 2003). It forms an amphipathic α-helix in which the hydrophobic face is composed primarily of highly conserved leucines in most voltage-gated channels. Even conservative mutations of some of these leucines alter activation gating dramatically (Judge et al., 2002; McCormack et al., 1991). In our models, L45 is positioned between adjacent S5 helices with its highly conserved face interacting with highly conserved S6 residues that move during gating. The outward movement of S4 during activation pulls the L45 helix away from the axis of the pore, which allows the C terminus half of S6 to swing radially outward to open the pore.

Lipid Exposure of S4

Proponents of crevasse models usually emphasize that most, if not all, positively charged residues of S4 reside in a hydrophilic environment where they interact with water and/or polar, often negatively charged, atoms of other segments or lipid headgroups. However, two-thirds of S4 residues are hydrophobic. There are several reasons to believe that many of these hydrophobic residues are exposed to lipid alkyl chains.

1. EM studies of KvAP with Fab fragments attached to the paddle indicate that S4 is on the periphery of the structure where it should be exposed to lipids.
2. It is difficult to construct atomic scale crevasse models using the pore-forming domain of the full-length KvAP crystal and the voltage-sensing domain of the isolated structure without exposing portions of S4 to lipids.
3. The KvAP channel can still gate when very bulky biotin adducts are attached to positions 121 and 122, even though the studies suggest that biotin adducts transverse most of the transmembrane region during gating (Jiang et al., 2003b). It seems unlikely that this could occur if the hydrophobic faces of S4 were tightly packed next to other protein segments throughout the gating process. In our KvAP model, the L121 and L122 side chains are exposed to lipid alkyl chains in both resting and open conformations.
4. Models with substantial lipid exposure of S4 are more consistent with some modeling criteria; e.g., hydrophobic residues on the N terminus half of S4 segments are predicted to be exposed to lipid alkyl chains because they are poorly conserved in an alignment of eukaryotic Kv sequences and because experimentally induced mutations have little effect on gating (Durell et al., 2004; Li-Smerin et al., 2000b).
5. Exposure of portions of S4 to lipids appears to be energetically favorable in some models. In developing our models, we found that the energies and root-mean-square deviations of the models during molecular

dynamic simulations were lower if some lipids of the cytoplasmic leaflet were sandwiched between the voltage-sensing and pore-forming domains (Durell et al., 2004). Interactions of the positively charged S4 residues with lipid head groups near the membrane surfaces are energetically favorable.

6. Exposure of hydrophobic S4 residues to relatively fluid and dynamic lipid alkyl chains may be important in reducing steric barriers to large movements during gating. Thus, it seems probable that a substantial amount of S4 is exposed to lipids; however, this exposure is not inconsistent with the basic premises of crevasse models.

OTHER TYPES OF PROKARYOTIC Kv CHANNELS

Our searches for prokaryotic 6-TM K^+ channel sequences have so far identified more than 120 sequences. These sequences vary greatly and can be classified into numerous categories or families. Unfortunately, functional properties of very few of these sequences have been analyzed, so we do not know the function of most of these channels or how the sequence changes alter their functional properties.

One criterion that we use to classify sequences is the length of the S2-S3 linker. The first and largest category of sequences has the same length of S2-S3 linkers as most eukaryotic Kv channels; in fact, this category is the most similar to eukaryotic Kv channels (see the KvVP sequence from *Vibrio parahaemolyticus* in Fig. 5). The greatest difference is that eukaryotic Kv channels possess a rather long N terminus domain, T1, that precedes the transmembrane region. We have found no prokaryotic homolog to this domain and suspect that it evolved after the eukaryotics branched from prokaryotics. The transmembrane domains are very similar; almost all of the S1-S6 residues that are highly conserved among eukaryotic Kv channels are also conserved among these sequences. They have a highly conserved classic type of S4 with positively charged residues, usually arginine, at every third position. Unlike eukaryotic Kv sequences, the length of S3-S4 linker does not vary much and is typically about one residue shorter than that of KvAP. Alignments with KvAP indicate that a deletion of one residue and presence of a proline occur near the midregion of the S3b segment of KvAP. This was part of the rationale that we used to propose that S3b hinges at this location. The second category has substantially longer S2-S3 linkers; however, in most other respects these sequences resemble those of the first category and could be classified in the same family. The eukaryotic Kcnq family also has long S2-S3 linkers; however, we find no compelling evidence linking the evolution of Kcnq sequences to this second category; i.e., Kcnq sequences appear to be about equidistant from eukaryotic Kv sequences and these two categories of prokaryotic sequences. The S2-S3 linker of the third category is two residues shorter than that of the first. This category is of interest because it includes KvAP. We have found no close eukaryotic homologs to members of this category. Although there are fewer sequences in this category than in the first two, the sequences are highly diverse, especially for the S3-S4-L45 region. We divide this category into subfamilies according to the length of this region relative to that of KvAP, which has one of the longest S3-S4-L45 sequences in this category (Fig. 9). In one subfamily, this region is 20 residues shorter than in KvAP. This subfamily contains several sequences, so it is unlikely that this large deletion is a sequencing error or corresponds to

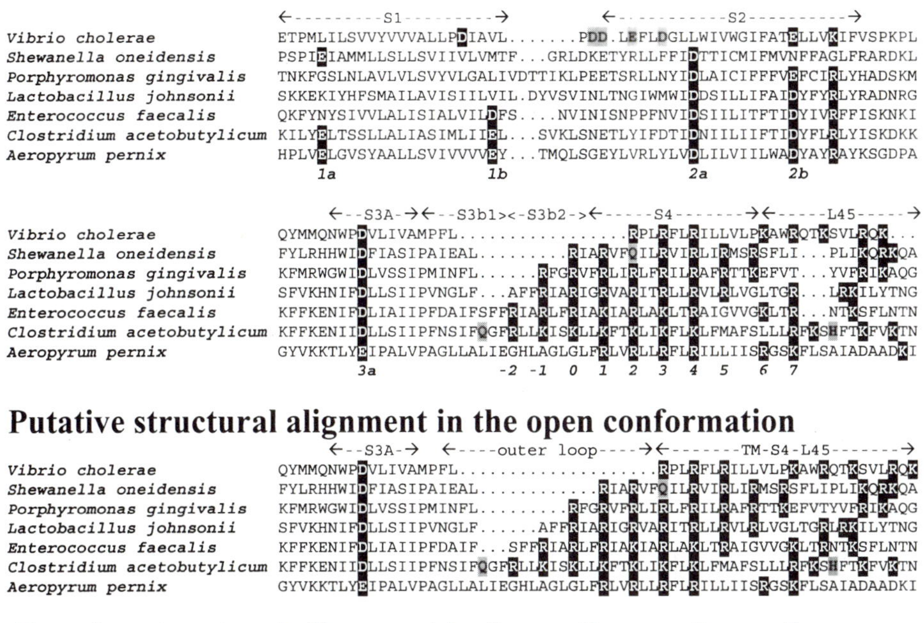

Figure 9. Alignments of the voltage-sensing domain of category 3 prokaryotic Kv channels with varying lengths of S3-S4 and S4-S5 linkers. Most charged residues are highlighted in black.

a nonfunctional gene. Most of the deletions relative to KvAP appear to occur in the S3b-S4 region; however, three residues appear to be deleted at the S4-L45 junction. In spite of these deletions, the sequences of the S4 segments are similar; e.g., a string of eight consecutive residues is identical in the S4 segments of KvAP and the sequence from *Vibrio cholerae*. Implications of the variable lengths of these segments on models of gating are discussed below. The fourth category consists of 6-TM K^+ channels that have long C-terminal domains homologous to cyclic-nucleotide-binding domains. Several families of eukaryotic channels from paramecia, plants, and animals (e.g., the EAG, ERG, AKT, CNG, and HCN families) have homologous domains. Only a few prokaryotic K^+ channels with homologous domains have been identified and most of these do not possess residues in the transmembrane regions that are highly conserved among the eukaryotic homologs; however, a sequence from *Trichodesmium erythraeum* does. Members of the fifth category have a different type of C-terminal cytoplasmic domain, the structure of which has been solved from the *Escherichia coli* homolog (Jiang et al., 2001). Many 2-TM channels, including the MthK channel for which the crystal structure has been solved (Jiang et al., 2002), have homologous cytoplasmic domains. In MthK this domain has a Ca^{2+}-binding site and is apparently responsible for the Ca^{2+} dependency of its gating. Eukaryotic Ca^{2+}-activated K^+ channels also have similar domains (Jiang et al., 2001). This domain is also homologous to putative nucleotide-binding subunits of prokaryotic Trk and Ktr transporters that have been postulated to have evolved from K^+ channels (Durell et al., 1999). The transmembrane segments of this category vary substantially; e.g., the transmembrane sequence from *Aquifex aeolicus* is similar to those of the first category, whereas the S1-S4 segments from the *E. coli* homolog have few of the charged residues that are highly conserved among voltage-gated channels. There are a few other prokaryotic sequences that fit none of these categories. Among these, the 6-TM K^+ channel from *Methanococcus jannaschii* is interesting because its functional properties have been measured (Sesti et al., 2003). This channel is similar to hypolarization-activated HCN channels (Mannikko et al., 2002) and plant KAT1 type channels (Latorre et al., 2003) in that it is opened by hyperpolarization. It has the shortest sequence of any 6-TM channel that we have identified. The facts that this protein has been expressed, has a compact sequence, and comes from a thermophile (which should increase its stability) make it a good candidate for structural studies. The NaChBac channel and its homologs have sequences similar to those of each of four homologous domains of eukaryotic Ca^{2+} channels (Durell and Guy, 2001); however, those that have been studied are selective for Na^+ (Koishi et al., 2004). A few other sequences that are not closely related to NaChBac do not have the K^+ channel signature sequence of the selectivity filter. These additional outliers do not have close eukaryotic homologs.

We attempt to develop models so that same basic model is viable for portions of the protein that can be aligned unambiguously. The least stringent test of this premise is to model closely related homologs. The most closely related sequence to that of KvAP that we have found is the KvCA sequence from *Clostridium acetobutylicum*. While the sequence of its transmembrane region can be aligned with that of KvAP without insertions or deletions, the repetitive pattern of positively charged residues at every third position in S4 begins earlier in the S3b region (Fig. 9). We have labeled the positively charged KvCA residues *R-2*, *K-1*, *K0*, *K1*, *K2*, *K3*, and *K4*. The open conformation of KvCA was modeled from the crystal structure of the isolated domain of KvAP (Jiang et al., 2003a) (Color Plate 7A [see color insert]). Deactivation of the KvCA channel was modeled by moving the S3b-S4 region inwardly via the helical screw mechanism. As S4 moves inward, the positively charged portion of its S3b helix merges with the S4 helix to form a single helix. When this occurs, the additional positively charged residues fall into the classic spiral pattern of S4, supporting our contention that that latter portion of S3b becomes an extension of S4 at negative voltages. S4 may move up to seven helical screw steps inward via this mechanism (Color Plate 7B). This large motion never exposes the S4 charges to lipid alkyl chains, and all charges in central transmembrane regions can form salt bridges throughout the transition. In contrast, in the paddle model, the cluster of seven positive charges of the S3b-S4 paddle shown in Color Plate 7A would diffuse through the alkyl phase, which we consider to be energetically improbable.

In most of the sequences of the third category of prokaryotic K^+ channels, the S3-S4 linker is shorter than that in KvAP or KvCA (Fig. 9). The deletions typically occur in multiples of three residues. We hypothesize that each three-residue deletion reduces the movement of S4 during activation by one helical screw step. This hypothesis is reflected in the putative structural alignments in the lower portion of Fig. 9 for open and resting conformations. The alignment for the open conformation allows the L45 connecting loop to be the same length and thus have similar backbone conformations for all sequences, but the lengths and thus conformations of the S3-S4 loop would vary substantially. Likewise, the alignment for the resting conformation allows the S3-S4 loop to have similar backbone conformations for all sequences except the shortest KvVC sequence, but the lengths of the cytoplasmic loop connecting S4 to S5 vary. Thus, according to this hypothesis, the deletion of three residues from the L45 linker and of nine residues from the S3-S4 linker of the sequence from *Shewanella oneidensis* relative to those of KvAP and KvCA would imply that in the open conformation its S4 segment is positioned one helical screw step (three residues) inward and in the resting conformation three helical screw steps (nine residues) outward relative to that of KvCA. This would still allow S4 of *S. oneidensis* to move up to three helical screw steps during activation. These are precisely the positions for S4 that we propose for the open and resting Shaker channels.

The sequences of the S3-S4 segments of a prokaryotic subfamily that includes the KvVC sequence from *V. cholerae* are extremely short (Fig. 9). To explore whether these very truncated sequences are compatible with our KvAP models, we developed homology models of the KvVC channel in open and resting conformations (Color Plate 7C and D). We selected KvVC because what remains of its S4 sequence is similar to that of KvAP (eight consecutive identical residues), because it is a member of a subfamily with the shortest S3-S4 linker that we have observed (17 residues shorter than that of KvAP) and its L45 linker is three residues shorter, and because the sequence of its pore-forming domain is similar to that of KcsA. In these models, the S3b segment is too short to be modeled as an α-helix and contains two "helix-breaking" prolines, so we modeled it as a random coil extending from M89 through R93. The portion of S4 that contains the arginines was assumed to be helical in all conformations, and the P94 residue was assumed to inhibit a helical conformation for the residues that immediately precede it. S4 of the open conformation model was placed one helical screw step further inward than in the KvAP model (structure 2) to connect S3a to S4 with the very short linker and to account for the three-residue deletion in the L45 linker. In this open conformation model, the first three arginine residues of S4 (*R2-R4*; KvVC has no *R1* residue) form salt bridges with the first three negatively charged residues of the S1-S3 loop, *R4* binds to an aspartate on S1 that is spatially near the position of *E2a* in KvAP, *K6* binds to the S1 aspartate, and *R7* binds to *E2b*. We found that S4 could move inwardly by approximately two helical screw steps without disrupting the S3a or S4 α-helices. In the putative resting conformation, *R2* binds to the S1 glutamate and the third negatively charged residue of the S1-S2 linker, *R3* to *E2b*, *R4* to *D3a*, and *R6* and *R7* extend into the cytoplasm and bind to the head groups of phospholipids located between the two domains.

Our modeling demonstrates how the helical screw mechanism allows different magnitudes of S4 movement in different Kv channels. Differing magnitudes of S4 motion may explain some of the apparent discrepancies obtained from experiments performed on different channels. If this hypothesis is valid, then understanding the intermediate conformations of the S3b-S4-L45 segments will be important in developing homology models of target channels in which the magnitude of motion of S4 may differ from that of the structures of channels that are used as the templates.

CONCLUSION

Prokaryotes provide us with a highly diverse population of 6-TM channels. Unfortunately, we know almost nothing about the function of these channels in prokaryotes, and functional properties of only a few have been analyzed. Nonetheless, these channels are already yielding a great deal of information about the structure, functional mechanisms, and evolution of this important superfamily of proteins. Additional studies that solve more crystal structures and that relate how sequence alterations observed in these proteins affect their functional properties are likely to provide additional insights into the basic functional mechanisms of these proteins. Considering the difficulties in purifying and crystallizing eukaryotic channel proteins, it is indeed fortunate that there are reasonably close prokaryotic homologs of many eukaryotic channel families.

REFERENCES

Aggarwal, S. K., and R. MacKinnon. 1996. Contribution of the S4 segment to gating charge in the Shaker K^+ channel. *Neuron* **16:**1169–1177.

Ahern, C. A., and R. Horn. 2004a. Specificity of charge-carrying residues in the voltage sensor of potassium channels. *J. Gen. Physiol.* **123:**205–216.

Ahern, C. A., and R. Horn. 2004b. Stirring up controversy with a voltage sensor paddle. *Trends Neurosci.* **27:**303–307.

Armstrong, C. M. 1990. Potassium channel architecture and channel blockers. *Prog. Clin. Biol. Res.* **334:**1–15.

Baker, O. S., H. P. Larsson, L. M. Mannuzzu, and E. Y. Isacoff. 1998. Three transmembrane conformations and sequence-dependent displacement of the S4 domain in shaker K^+ channel gating. *Neuron* **20:** 1283–1294.

Bezanilla, F. 2000. The voltage sensor in voltage-dependent ion channels. *Physiol. Rev.* **80:**555–592.

Bezanilla, F. 2002. Voltage sensor movements. *J. Gen. Physiol.* **120:**465–473.

Blaustein, R. O., P. A. Cole, C. Williams, and C. Miller. 2000. Tethered blockers as molecular 'tape measures' for a voltage-gated K^+ channel. *Nat. Struct. Biol.* **7:**309–311.

Broomand, A., R. Mannikko, H. P. Larsson, and F. Elinder. 2003. Molecular movement of the voltage sensor in a K channel. *J. Gen. Physiol.* **122:**741–748.

Cha, A., G. E. Snyder, P. R. Selvin, and F. Bezanilla. 1999. Atomic scale movement of the voltage-sensing region in a potassium channel measured via spectroscopy. *Nature* **402:**809–813.

Doyle, D. A., C. J. Morais, R. A. Pfuetzner, A. Kuo, J. M. Gulbis, S. L. Cohen, B. T. Chait, and R. MacKinnon. 1998. The structure of the potassium channel: molecular basis of K+ conduction and selectivity. *Science* **280:**69–77.

Durell, S. R., and H. R. Guy. 2001. A putative prokaryote voltage-gated $Ca^{(2+)}$ channel with only one 6TM motif per subunit. *Biochem. Biophys. Res. Commun.* **281:**741–746.

Durell, S. R., Y. Hao, and H. R. Guy. 1998. Structural models of the transmembrane region of voltage-gated and other K^+ channels in open, closed, and inactivated conformations. *J. Struct. Biol.* **121:**263–284.

Durell, S. R., Y. Hao, T. Nakamura, E. P. Bakker, and H. R. Guy. 1999. Evolutionary relationship between K(+) channels and symporters. *Biophys. J.* **77:**775–788.

Durell, S. R., I. H. Shrivastava, and H. R. Guy. 2004. Models of the structure and voltage-gating mechanism of the shaker K^+ channel. *Biophys. J.* **87:**2116–2130.

Elinder, F., P. Arhem, and H. P. Larsson. 2001a. Localization of the extracellular end of the voltage sensor S4 in a potassium channel. *Biophys. J.* **80:**1802–1809.

Elinder, F., R. Mannikko, and H. P. Larsson. 2001b. S4 charges move close to residues in the pore domain during activation in a K channel. *J. Gen. Physiol.* **118:**1–10.

Eriksson, M. A. and B. Roux. 2002. Modeling the structure of agitoxin in complex with the Shaker K^+ channel: a computational approach based on experimental distance restraints extracted from thermodynamic mutant cycles. *Biophys. J.* **83:**2595–2609.

Gandhi, C. S., E. Clark, E. Loots, A. Pralle, and E. Y. Isacoff. 2003. The orientation and molecular movement of a $K^{(+)}$ channel voltage-sensing domain. *Neuron* **40:**515–525.

Gandhi, C. S., and E. Y. Isacoff. 2002. Molecular models of voltage sensing. *J. Gen. Physiol.* **120:** 455–463.

Glauner, K. S., L. M. Mannuzzu, C. S. Gandhi, and E. Y. Isacoff. 1999. Spectroscopic mapping of voltage sensor movement in the Shaker potassium channel. *Nature* **402:**813–817.

Gonzalez, C., E. Rosenman, F. Bezanilla, O. Alvarez, and R. Latorre. 2001. Periodic perturbations in Shaker K^+ channel gating kinetics by deletions in the S3-S4 linker. *Proc. Natl. Acad. Sci. USA* **98:**9617–9623.

Gonzalez, C., R. Morera, F. Munoz, E. Rosenman, O. Alvarez, and R. Latorre. 2004. Displacement of S3 segment during activation of *Shaker* K^+ channel with a short S3-S4 linker. *Biophys. J.* **86:**123a.

Guy, H. R., and S. R. Durell. 1995. Structural models of Na+, Ca2+, and K+ channels. *Soc. Gen. Physiol. Ser.* **50:**1–16.

Guy, H. R., and S. R. Durell. 1994. Using sequence homology to analyze the structure and function of voltage-gated ion channel proteins. *Soc. Gen. Physiol. Ser.* **49:**197–212.

Guy, H. R., and P. Seetharamulu. 1986. Molecular model of the action potential sodium channel. *Proc. Natl. Acad. Sci. USA* **83:**508–512.

Heginbotham, L., Z. Lu, T. Abramson, and R. MacKinnon. 1994. Mutations in the K^+ channel signature sequence. *Biophys. J.* **66:**1061–1067.

Heinemann, S. H., H. Terlau, W. Stuhmer, K. Imoto, and S. Numa. 1992. Calcium channel characteristics conferred on the sodium channel by single mutations. *Nature* **356:**441–443.

Hong, K. H., and C. Miller. 2000. The lipid-protein interface of a Shaker $K^{(+)}$ channel. *J. Gen. Physiol.* **115:**51–58.

Jiang, Q. X., D. N. Wang, and R. MacKinnon. 2004. Electron microscopic analysis of KvAP voltage-dependent K^+ channels in an open conformation. *Nature* **430:**806–810.

Jiang, Y., A. Lee, J. Chen, M. Cadene, B. T. Chait, and R. MacKinnon. 2002. Crystal structure and mechanism of a calcium-gated potassium channel. *Nature* **417:**515–522.

Jiang, Y., A. Lee, J. Chen, V. Ruta, M. Cadene, B. T. Chait, and R. MacKinnon. 2003a. X-ray structure of a voltage-dependent K^+ channel. *Nature* **423:**33–41.

Jiang, Y., A. Pico, M. Cadene, B. T. Chait, and R. MacKinnon. 2001. Structure of the RCK domain from the *E. coli* K^+ channel and demonstration of its presence in the human BK channel. *Neuron* **29:**593–601.

Jiang, Y., V. Ruta, J. Chen, A. Lee, and R. MacKinnon. 2003b. The principle of gating charge movement in a voltage-dependent K^+ channel. *Nature* **423:**42–48.

Judge, S. I., J. Z. Yeh, J. E. Goolsby, M. J. Monteiro, and C. T. Bever, Jr. 2002. Determinants of 4-aminopyridine sensitivity in a human brain kv1.4 $K^{(+)}$ channel: phenylalanine substitutions in leucine heptad repeat region stabilize channel closed state. *Mol. Pharmacol.* **61:**913–920.

Koishi, R., H. Xu, D. Ren, B. Navarro, B. W. Spiller, Q. Shi, and D. E. Clapham. 2004. A superfamily of voltage-gated sodium channels in bacteria. *J. Biol. Chem.* **279:**9532–9538.

Laine, M., M. C. Lin, J. P. Bannister, W. R. Silverman, A. F. Mock, B. Roux, and D. M. Papazian. 2003. Atomic proximity between S4 segment and pore domain in Shaker potassium channels. *Neuron* **39:** 467–481.

Larsson, H. P., O. S. Baker, D. S. Dhillon, and E. Y. Isacoff. 1996. Transmembrane movement of the shaker K^+ channel S4. *Neuron* **16:**387–397.

Latorre, R., R. Olcese, C. Basso, C. Gonzalez, F. Munoz, D. Cosmelli, and O. Alvarez. 2003. Molecular coupling between voltage sensor and pore opening in the Arabidopsis inward rectifier K^+ channel KAT1. *J. Gen. Physiol.* **122:**459–469.

Lee, H. C., J. M. Wang, and K. J. Swartz. 2003. Interaction between extracellular hanatoxin and the resting conformation of the voltage-sensor paddle in Kv channels. *Neuron* **40:**527–536.

Lee, S. Y., and R. MacKinnon. 2004. A membrane-access mechanism of ion channel inhibition by voltage sensor toxins from spider venom. *Nature* **430:**232–235.

LeMasurier, M., L. Heginbotham, and C. Miller. 2001. KcsA: it's a potassium channel. *J. Gen. Physiol.* **118:**303–314.

Li-Smerin, Y., D. H. Hackos, and K. J. Swartz. 2000a. A localized interaction surface for voltage-sensing domains on the pore domain of a K^+ channel. *Neuron* **25:**411–423.

Li-Smerin, Y., D. H. Hackos, and K. J. Swartz. 2000b. Alpha-helical structural elements within the voltage-sensing domains of a $K^{(+)}$ channel. *J. Gen. Physiol.* **115:**33–50.

Li-Smerin, Y., and K. J. Swartz. 2000. Localization and molecular determinants of the hanatoxin receptors on the voltage-sensing domains of a $K^{(+)}$ channel. *J. Gen. Physiol.* **115:**673–684.

Liu, Y., M. Holmgren, M. E. Jurman, and G. Yellen. 1997. Gated access to the pore of a voltage-dependent K^+ channel. *Neuron* **19:**175–184.

Liu, Y. S., P. Sompornpisut, and E. Perozo. 2001. Structure of the KcsA channel intracellular gate in the open state. *Nat. Struct. Biol.* **8:**883–887.

Lu, Z., A. M. Klem, and Y. Ramu. 2001. Ion conduction pore is conserved among potassium channels. *Nature* **413:**809–813.

MacKinnon, R., R. W. Aldrich, and A. W. Lee. 1993. Functional stoichiometry of Shaker potassium channel inactivation. *Science* **262:**757–759.

MacKinnon, R., S. L. Cohen, A. Kuo, A. Lee, and B. T. Chait. 1998. Structural conservation in prokaryotic and eukaryotic potassium channels. *Science* **280:**106–109.

Mannikko, R., F. Elinder, and H. P. Larsson. 2002. Voltage-sensing mechanism is conserved among ion channels gated by opposite voltages. *Nature* **419:**837–841.

McCormack, K., M. A. Tanouye, L. E. Iverson, J. W. Lin, M. Ramaswami, T. McCormack, J. T. Campanelli, M. K. Mathew, and B. Rudy. 1991. A role for hydrophobic residues in the voltage-dependent gating of Shaker K^+ channels. *Proc. Natl. Acad. Sci. USA* **88:**2931–2935.

Moczydlowski, E. 1998. Chemical basis for alkali cation selectivity in potassium-channel proteins. *Chem. Biol.* **5:**R291–R301.

Monks, S. A., D. J. Needleman, and C. Miller. 1999. Helical structure and packing orientation of the S2 segment in the Shaker K^+ channel. *J. Gen. Physiol.* **113:**415–423.

Neale, E. J., D. J. Elliott, M. Hunter, and A. Sivaprasadarao. 2003. Evidence for intersubunit interactions between S4 and S5 transmembrane segments of the Shaker potassium channel. *J. Biol. Chem.* **278:**29079–29085.

Neyton, J., and C. Miller. 1988. Discrete Ba^{2+} block as a probe of ion occupancy and pore structure in the high-conductance Ca^{2+}-activated K^+ channel. *J. Gen. Physiol.* **92:**569–586.

Noda, M., T. Ikeda, T. Kayano, H. Suzuki, H. Takeshima, M. Kurasaki, H. Takahashi, and S. Numa. 1986. Existence of distinct sodium channel messenger RNAs in rat brain. *Nature* **320:**188–192.

Noda, M., S. Shimizu, T. Tanabe, T. Takai, T. Kayano, T. Ikeda, H. Takahashi, H. Nakayama, Y. Kanaoka, N. Minamino, and S. Numa. 1984. Primary structure of *Electrophorus electricus* sodium channel deduced from cDNA sequence. *Nature* **312:**121 127.

Ranganathan, R., J. H. Lewis, and R. MacKinnon. 1996. Spatial localization of the K^+ channel selectivity filter by mutant cycle-based structure analysis. *Neuron* **16:**131–139.

Roux, B., S. Berneche, and W. Im. 2000. Ion channels, permeation, and electrostatics: insight into the function of KcsA. *Biochemistry* **39:**13295–13306.

Ruta, V., Y. Jiang, A. Lee, J. Chen, and R. MacKinnon. 2003. Functional analysis of an archaebacterial voltage-dependent K+ channel. *Nature* **422:**180–185.

Sansom, M. S., I. H. Shrivastava, J. N. Bright, J. Tate, C. E. Capener, and P. C. Biggin. 2002. Potassium channels: structures, models, simulations. *Biochim. Biophys. Acta* **1565:**294–307.

Santacruz-Toloza, L., Y. Huang, S. A. John, and D. M. Papazian. 1994. Glycosylation of shaker potassium channel protein in insect cell culture and in *Xenopus* oocytes. *Biochemistry* **33:**5607–5613.

Sato, C., M. Sato, A. Iwasaki, T. Doi, and A. Engel. 1998. The sodium channel has four domains surrounding a central pore. *J. Struct. Biol.* **121:**314–325.

Schoppa, N. E., K. McCormack, M. A. Tanouye, and F. J. Sigworth. 1992. The size of gating charge in wild-type and mutant Shaker potassium channels. *Science* **255:**1712–1715.

Sesti, F., S. Rajan, R. Gonzalez-Colaso, N. Nikolaeva, and S. A. Goldstein. 2003. Hyperpolarization moves S4 sensors inward to open MVP, a methanococcal voltage-gated potassium channel. *Nat. Neurosci.* **6:**353–361.

Shrivastava, I. H., S. R. Durell, and H. R. Guy. 2004. A model of voltage gating developed using the KvAP channel crystal structure. *Biophys. J.* **87:**2255–2270.

Sokolova, O., L. Kolmakova-Partensky, and N. Grigorieff. 2001. Three-dimensional structure of a voltage-gated potassium channel at 2.5 nm resolution. *Structure* (Cambridge) **9:**215–220.

Starace, D. M., and F. Bezanilla. 2001. Histidine scanning mutagenesis of basic residues of the S4 segment of the shaker K^+ channel. *J. Gen. Physiol.* **117:**469–490.

Starace, D. M., and F. Bezanilla. 2004. A proton pore in a potassium channel voltage sensor reveals a focused electric field. *Nature* **427:**548–553.

Terlau, H., S. H. Heinemann, W. Stuhmer, M. Pusch, F. Conti, K. Imoto, and S. Numa. 1991. Mapping the site of block by tetrodotoxin and saxitoxin of sodium channel II. *FEBS Lett.* **293:**93–96.

Yang, N., A. L. George, Jr., and R. Horn. 1996. Molecular basis of charge movement in voltage-gated sodium channels. *Neuron* **16:**113–122.

Yellen, G., M. E. Jurman, T. Abramson, and R. MacKinnon. 1991. Mutations affecting internal TEA blockade identify the probable pore-forming region of a K+ channel. *Science* **251:**939–942.

Zhou, M., J. H. Morais-Cabral, S. Mann, and R. MacKinnon. 2001. Potassium channel receptor site for the inactivation gate and quaternary amine inhibitors. *Nature* **411:**657–661.

Zhou, Y., and R. MacKinnon. 2004. Ion binding affinity in the cavity of the KcsA potassium channel. *Biochemistry* **43:**4978–4982.

Bacterial Ion Channels and Their Eukaryotic Homologs
Edited by A. Kubalski and B. Martinac

Chapter 7

Inward Rectifier K^+ Channels

Indira H. Shrivastava and H. Robert Guy

A crucial function of K^+ channels is to stabilize the membrane potential by facilitating the flow of K^+ ions across the membrane. Inward rectifier K^+ channels (Kirs) act as a valve or a diode, allowing an inward current upon hyperpolarization but not allowing exit of K^+ ions upon depolarization. These channels are thus effectively unidirectional, allowing greater influx than efflux, thereby controlling the resting potential without causing massive loss of K^+ ions (Hille, 2001). The degree of rectification in the Kir channels is correlated with the binding affinity of the channel for blocking cations.

Kirs are unique with respect to the stimuli that cause their gating, and they are gated by a variety of mechanisms. For example, blockade of the pore from the intracellular side by Mg^{2+} (Vandenberg, 1987; Matsuda et al., 1987) and polyamines such as spermine and spermidine (Lopatin et al., 1994; Nichols and Lopatin, 1997) are some of the gating stimuli. Some Kir channels are gated by G-protein-coupled receptors (North, 1989) whereas others are modulated by GTP-binding proteins, by the membrane phospholipids phosphatidylinostol-4,5-bisphosphate, and by ATP and Na^+ ions (Reimann and Ashcroft, 1999). Some Kir channels are inhibited by intracellular acidification (Jiang et al., 2002; Mao et al., 2003). The functional roles of Kirs are varied due to the diversity of the gating signals. Heart muscles have a variety of K^+ channels, having the property of rapid inward rectification or rapid inactivation. Kir channels are involved in transepithelial membrane transport in the kidney and in mediation of the electrical activity in cardiac, neuronal, and neurosecretory cells (Yamada et al., 1998). They are also important in the regulation of insulin secretion and the control of vascular smooth muscle tone. Malfunctioning in ion channels due to mutations in genes encoding channel proteins is implicated in the pathogenesis of a growing number of diseases, termed channelopathies (George, 1995; Keating and Sanguinetti, 1996). So far, three major diseases are linked to mutations in Kir channels or an associated protein in *Homo sapiens* (Abraham et al., 1999). These diseases are persistent hyperinsulinemic hypoglycemia of infancy, which affects the function of pancreatic B cells; Bartter syndrome, which is characterized by hypokalemic alkalosis, hypercalciurea, and increased serum aldosterone

---

Indira H. Shrivastava and H. Robert Guy • Laboratory of Experimental and Computational Biology, Division of Basic Sciences, National Cancer Institute, National Institutes of Health, Bethesda, MD 20892-5567.

and plasma renin activity; and Andersen's syndrome, characterized by periodic paralysis, cardiac arrhythmias, and dysmorphic features. Point mutations in the signature sequence of K^+ channels (TXGYG or TXGFG) abolish K^+ selectivity and are linked to the movement disorder observed in weaver mice (Patil et al., 1995; Navarro et al., 1996). Kir channels thus serve diverse and important roles throughout the human body and pose major challenges in the recognition of the molecular basis of Kir-mediated channelopathies (Abraham et al., 1999).

MOLECULAR DIVERSITY

The Kirs are a functionally diverse and heterogeneous group, with varying strengths of rectification and different gating mechanisms. The genes for this family of potassium channels encode proteins ranging from ~360 to 500 amino acids. The Kir gene family consists of at least 16 genes with the official nomenclature of Kir*m.x*, where *m* denotes the subfamily and *x* the subtype (Doupnik and Davidson, 1995). The Kir channels are complexes with tetrameric subunits that are sometimes associated with auxillary subunits (Hille, 2001). There are seven Kir subfamilies (Fig. 1), which are classified on the basis of their strength of rectification and responses to cellular signals. The Kirs thus range from Kir1.*x* to Kir7.*x*, with *x* taking values between 1 and 4, depending on the subfamily. The multimeric association of subunits from different Kir subfamilies and splicing variants gives rise to functional and molecular diversity in the Kir family. Kir1 and Kir6 are weakly rectifying Kir channels, influenced by ATP/ADP ratio, whereas Kir2 and Kir3 are strongly rectifying Kir channels, activated by the G-protein subunit, Gβγ (Bichet et al., 2003). Alternative splicing at the 5′ end in Kir1.1 (ROMK1) generates multiple variants (Yano et al., 1994; Zhou et al., 1994). Weak inward rectifying property and lack of macroscopic time-dependent gating distinguish Kir1.1 from other Kir channels (Lu and MacKinnon, 1997). Also, unlike other members of the Kir family, Kir1.1 contains a putative phosphate-binding loop in the C-terminal region (Ho et al., 1993). In the Kir2 subfamily, four subtypes (Kir2.1 to Kir2.4 or IRK1 to 4) have been

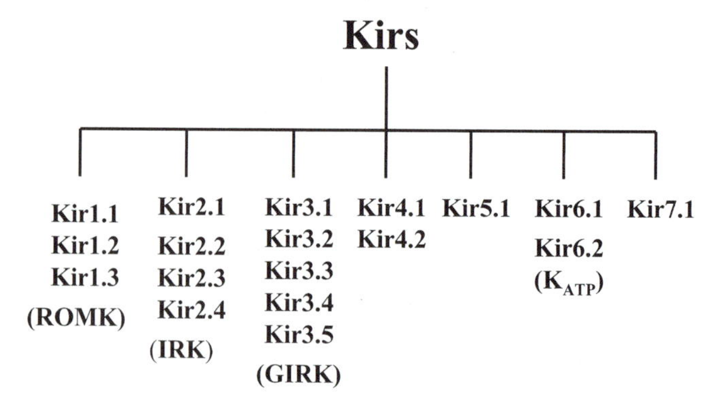

Figure 1. Classification and nomenclature of major human Kir channels are shown. Also indicated are the alternative names in parentheses.

cloned to date, all encoding strong inward rectifiers. The channels differ in the single channel conductance and in sensitivity to phosphorylation and other secondary messengers (Fakler et al., 1994) and are ubiquitously expressed in the heart and nervous system (Wible et al., 1995). The Kir3 (GIRK) family expresses G-protein-activated, strongly rectifying K^+ channels (Lesage et al., 1994). So far, four mammalian members in these G-protein-gated K^+ channels (Kir3.1 to Kir3.4 or GIRK1 to 4) have been cloned (Dascal et al., 1993; Chan et al., 1996; Krapivinsky et al., 1995). The GIRK complex is suggested to be a heterotetramer, consisting of two GIRK1 and two GIRK4 subunits (Corey et al., 1998; Silverman et al., 1996). The GIRK channels in atria are heteromultimers of Kir3.1 and Kir3.4. These channels are responsible for acetylcholine-activated K^+ current in atrial muscle. The neurons contain functional heteromultimeric coassembly of Kir3.1 and Kir3.2 or Kir3.4. In contrast to Kir3 subunits, there is less experimental evidence for heteromultimeric assembly within other Kir subfamilies (Coetzee et al., 1999). However, there is indirect experimental evidence for heteromultimerization between subunits from different Kir subfamilies. Kir4 and Kir5 were discovered in brain (Takumi et al., 1995; Bond et al., 1994). Kir4.1 forms a weakly rectifying K^+ channel when expressed by itself, but when coexpressed with Kir5.1, they form novel channels and tandem dimers and tetramers in a specific 4-5-4-5 arrangement (Pessia et al., 1996). Kir5.1, identified through a database search, was found to be expressed significantly in human kidney, pancreas, and thyroid gland (Liu et al., 2000). The Kir6.*x* channels are unique in that they coassemble with a regulatory subunit, a sulfonyl urea receptor (SUR), to form an octameric channel (Clement et al., 1997; Ashcroft and Gribble, 1998), although Kir6 principal subunits (C-terminal truncated) can also be expressed independently in heterologous expression systems (Coetzee et al., 1999). Kir6.1 was first isolated by Inagaki (1995). Kir6.1 (or Kir6.2) forms the channel pore while the SUR subunit provides nucleotide sensitivity (Nichols et al., 1996). SUR belongs to the ATP-binding cassette transporter family (Higgins, 1995). Speculation that Kirs may also couple to the ATP-binding cassette transporters remains controversial, in the absence of experimental evidence. Kir7.1 was first cloned and isolated by Krapivinsky et al. (1998). Kir7.1 is only ~38% identical to its closest relative (Kir1.3) and displays none of the functional properties unique to the Kir1.1 family of channels (Krapivinsky et al., 1998). Unlike other inwardly rectifying K^+ channels, the low conductance of Kir7.1 shows K^+ permeability that is poorly dependent on external K^+ concentration and low sensitivity to blocking by Ba^{2+} (Doring et al., 1998).

There is no apparent consistency in the coassembly of Kir subfamilies, either within or between subfamilies (Fig. 2). In some Kir subfamilies, heteromultimerization of subunits from the same subfamily is essential for formation of functional channels (for example, Kir3.*x*) while in others coassembly of subunits from different subfamilies is necessary (Kir4.1 and Kir5.1) (Coetzee et al., 1999). This combinatorial diversity in assembly leads to a great deal of functional diversity. Alternative splicing in genes also leads to diversity in channel functions. The functional consequences of K^+ channel gene splicing include changes in electrophysiological properties, gene expression, tissue distribution, subcellular localization, and modulation. For example, alternative splicing in Kir1.1 and Kir6.1 results in alternative 5′ untranscribed regions and alternative N termini, whereas alternative splicing in Kir3.1 results in alternative C termini isoforms (Coetzee et al., 1999).

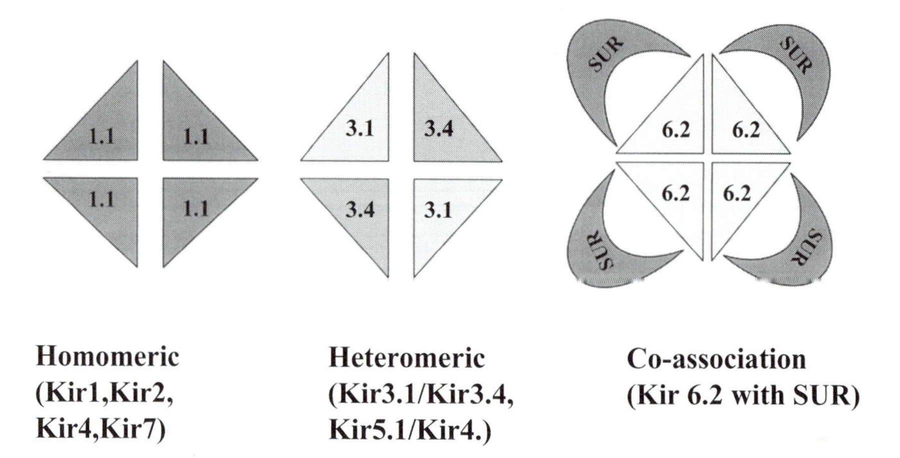

Figure 2. Cartoon representation of multimeric association of Kir subunits. (left) Homomeric association of tetramers; (middle) heteromeric association; (right) coassembly of Kir subunits with SURs.

MOLECULAR STRUCTURE

The first Kir channels that were cloned (Ho et al., 1993; Kubo et al., 1993) showed a membrane topology characteristic of the entire family: each subunit is composed of a cytoplasmic β-domain attached to a two-transmembrane (TM) α-domain, linked by a pore loop (this region is now called the P segment, but it was also known as the H5 segment), consisting of the selectivity filter selective for K^+ ions. Kir channels are tetramers with four identical (homomeric) or related (heteromeric) subunits (Fig. 3). Each subunit consists of two domains, the TM domain (α-subunit) and the intracellular cytoplasmic domain (β-subunit). The α-subunit consists of two putative TM helices (outer helix [M1] and inner helix [M2]) linked by a short P-loop and an extracellular loop. The TM domain is analogous to the channel-forming pore domain S5-P-S6 of the six TM K^+ channels. The structural topology of the hydrophobic TM domain of Kirs is similar to that of bacterial KcsA from *Streptomyces lividans* (Doyle et al., 1998) and other known K^+ channels. Recently, a high-resolution structure of a prokaryote inward rectifier K^+ channel, KirBac1.1 from *Burkholderia pseudomallei* (Kuo et al., 2003), has provided a major breakthrough in understanding the structure and functional mechanism of the K^+ channel family of proteins. This structure contains both the TM domain and the cytoplasmic domain. In KirBac1.1, the selectivity filter, containing the K^+ channel signature motif, GYG; the pore cavity; and the two TM helices (M1 and M2) are structurally homologous to the 2-TM bacterial K^+ channel, KcsA. However, the P-loop helices in KirBac1.1 are oriented less directly toward the pore cavity as in KcsA. The volume of the pore cavity and the size of the portion of the pore formed by the M2 segments are smaller, leading to the speculation that the structure of KirBac1.1 represents the closed form (Kuo et al., 2003). Surprisingly, sequences of the TM domain of KirBac channels are more similar to those of some other prokaryotic 2-TM K^+ channels than to those of eukaryotic Kir channels (Durell and Guy, 2001a). However,

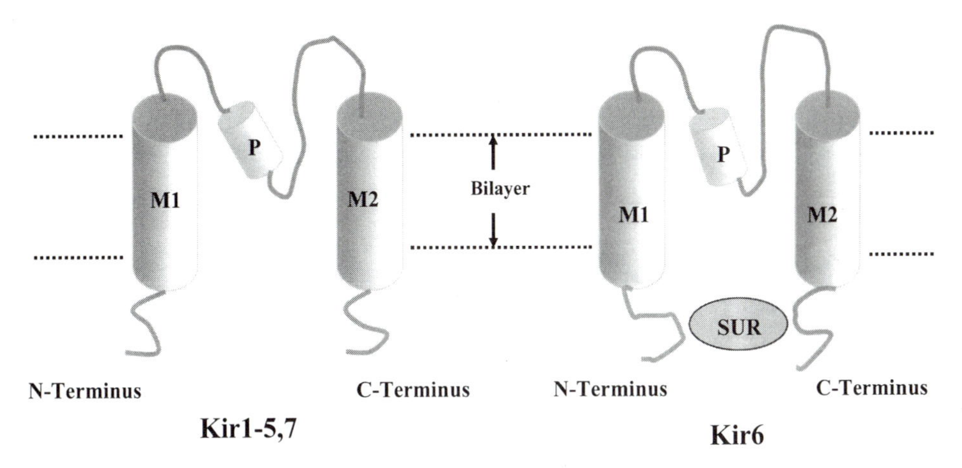

Figure 3. General topology of TM domain of Kirs. The dotted lines indicate approximate positions of the bilayer interfaces. (left) The TM helices are denoted as M1-P-M2; (right) association of Kir with intracellular SUR.

the structure of KirBac1.1 is in general agreement with the structural information derived from multiple sequence alignments of Kir channels, suggesting that KirBac1.1 may indeed be a structural representative of the eukaryotic channel family (Kuo et al., 2003).

There is a structural homology of the TM domain of Kirs with other 2-TM K^+ channels such as KcsA and MthK, both of which have been crystallized (Doyle et al., 1998; Nishida and MacKinnon, 2002). However, there are considerable differences in the sequences of the eukaryotic Kirs and the prokaryotic Kirs. The sequence of the TM domain of KirBac1.1 is intermediate between that of KcsA and MthK and eukaryotic Kirs. Also, mutagenesis studies suggest that the TM domains of Kir2.1 have an organization different from that of KcsA (Minor et al., 1999). A sequence alignment of representative members from each subfamily of Kirs identified some restricted residue motifs strongly conserved in M1 and M2 helices (Minor et al., 1999). The conserved motifs are FXXSWXXFGXXA and A(V/I)XX(V/L)XXQXXXG, respectively, in M1 and M2 of Kir2.1, which is conserved in all eukaryotic Kir families. The prevalence of these conserved patterns within the Kir superfamily probably reflects the stereochemical requirements for a structural framework common to the TM region of all eukaryotic Kir channels (Minor et al., 1999). These hallmark patterns are conspicuously absent in all other known K^+ channels, including KirBacs, supporting an assertion that Kir channels are structurally unique. The P-loop of the TM region that has the highest sequence homology among potassium channels (Heginbotham et al., 1994) is notably different in Kir channels as evidenced by tetraethylammonium and peptide toxin blocker studies (Lu and MacKinnon, 1997; Imredy et al., 1998; Jin and Lu, 1998). A salt bridge between conserved charged residues in adjacent subunits anchors and stabilizes the signature sequence motif in K^+ channels. This salt bridge cannot be formed in KirBac1.1 or Kir7.1, due to absence of one of the charged residues (Yang et al., 1997; Shieh et al., 1999; Bichet et al., 2003). This further supports the idea that eukaryotic Kir channels are structurally different from other K^+ channels.

The features that most clearly establish the relationship between prokaryotic and eukaryotic Kirs and distinguish them from other K^+ channels are their unique intracellular cytoplasmic N and C termini domains. Studies with chimeras of weakly rectifying Kir1.1 and strongly rectifying Kir2.1 indicate that the C-terminal region, beyond the M2 helix, contains residues necessary for strong inward rectification and high-affinity Mg^{2+} block (Taglialatela et al., 1994; Pessia et al., 1995). The residues in KirBac1.1 that are equivalent to the residues important in rectification in Kir2.1 and Kir1.1 also form part of the lining of the C-terminal vestibule. The cytoplasmic domain of KirBac1.1, consisting chiefly of β-sheets, is structurally homologous to the intracellular domain of G protein coupled to the inwardly rectifying K^+ channel, Kir3.1 (GIRK1) (Nishida and MacKinnon, 2002). The crystal structure of the cytoplasmic domain of GIRK1 revealed the cytoplasmic pore formed on the axis of symmetry of the four identical domains to be wide enough to permit diffusion of hydrated K^+ ions from the TM channel into the intracellular solution (Nishida and MacKinnon, 2002). The structure also revealed the cytoplasmic pore to be lined with negatively charged residues interspersed with hydrophobic residues. These residues provide an electrostatic surface to polyamines, which bind effectively to the cytoplasmic pore and thereby modulate inward rectification (Nishida and MacKinnon, 2002). The GIRK1 crystal structure was proposed to correspond to an open channel conformation, whereas the KirBac1.1 crystal structure appears to be in a closed conformation.

The KirBac1.1 crystal structure also possesses an α-helix, called the slide helix, that precedes the M1 TM helix. The amphipathic slide helices from the four subunits lie on the cytoplasmic surface and encircle the pore-forming TM domain. Eukaryotic Kirs have similar sequences preceding M1 that also can form amphipathic α-helices. Thus this structural feature is likely common to all Kirs.

MOLECULAR EVOLUTION

Analysis of the evolutionary relationships among K^+ channels and their homologs led to the discovery of prokaryotic KirBac1.1 in the genome of *B. pseudomallei* (Durell and Guy, 2001a). KirBacs were found to be the only other family of K^+ channels that possessed domains homologous to the cytoplasmic domains of eukaryotic Kirs. However, when only the M1-P-M2 motif was aligned, KirBac1.1 was found to be more homologous to the K^+ channel families than to eukaryotic Kirs. The M1-P-M2 region of eukaryotic Kirs has unique features, distinguishing them from other K^+ channel families (Color Plate 8 [see color insert]). These features include a DXXTTXXDXXWR sequence, immediately preceding M1, highly conserved tryptophans in M1, an insertion in the M1-loop containing the CVXX(V/I) sequence, and an RXXTXXCP sequence in the P-M2 loop. None of these features are present in KirBac1.1, suggesting that the common ancestor to KirBac1.1 and eukaryotic Kirs did not have these features and that they developed after the divergence of KirBac from other Kir subfamilies (Durell and Guy, 2001a).

Gene transfer between organisms often complicates the interpretation of their evolution. In most of the cases that we have studied (i.e., glutamate-activated K^+ channel GluR0 from *Synechocystis* sp. strain PCC 6803 [Chen et al., 1999], the NaChBac channel from *Bacillus halodurans* [Durell and Guy, 2001b], and KirBac1.1 from *B. pseudomallei* [Durell and Guy, 2001a]), the nature of the prokaryotic sequence supports the hypothesis that the gene evolved first in prokaryotes rather than being transferred to prokaryote from eukaryote. The latter hypothesis is inconsistent with several findings vis-à-vis KirBac1.1 and Kirs, i.e., all eukaryotic Kirs are more closely related to each other than to KirBac1.1. The M1-P-M2 region of KirBac1.1 is more similar to other K^+ channels than to the Kirs, and the M1-P-M2 region of Kirs has some features that do not occur in other K^+ channels or KirBac1.1 (Color Plate 9 [see color insert]). These findings suggest that Kirs probably evolved first in prokaryotes and that KirBac1.1 has diverged less from the ancestor common to KirBac1.1 and eukaryotic Kirs than have the eukaryotic Kirs.

A comprehensive sequence analysis of K^+ channels in *Caenorhabditis elegans* (CE), *Drosophila melanogaster* (DM), and mammalian genomes has been performed (Moulton et al., 2003). Although the fundamental pore structure is the same in all members of the K^+ channel family, other parts of the sequence indicate significant structural diversity. Phylogenetic analysis of the Kir genes grouped CE together, excluding the genes from the other two species, indicating that gene duplication occurred after the divergence of CE from the lineage-leading mammals and DM. The human Kir genes formed two clusters, one with a single DM gene and the other with the remaining DM and CE gene sequences, indicating that independent gene duplications also occurred in the human lineage, suggesting there are two ancestral genes in the most recent common ancestor, one for each group (Moulton et al., 2003). A genealogical analysis of the K^+ channel principal subunits of Kirs was performed (Coetzee et al., 1999) in which the genes of the subfamilies Kir2, Kir3, Kir5, and Kir6 each formed their own respective groups while the genes of Kir1, Kir4, and Kir7 formed a separate group. A threshold identity of >55% exists between members of individual Kir families, except for the subfamily that includes Kir1, Kir4, and Kir7. For this latter group of Kir subfamilies, the identity score was as low as 36%, suggesting that some of these genes may belong to different subfamilies. The genes of the Kir1/Kir4/Kir7 group are more closely related to each other than to genes of other Kir subfamilies.

Inward rectifying properties of K^+ channels appear to have evolved independently at least twice. Plants possess K^+ channels that conduct primarily at negative voltages. However, these currents are due to 6-TM-type K^+ channels that have sequences more closely related to Eag-type K^+ channels and cyclic-nucleotide-gated channels than to the 2-TM-type Kirs. The inward rectification of these K^+ channels in plants is independent of intracellular magnesium, thus differing mechanistically from gating of Kir channels in the animal kingdom. Moreover, chimeras of plant and animal Kirs that contain the S1 to S4 segments of plants are activated by hyperpolarization (Cao et al., 1995), suggesting that plant Kirs have a membrane topology similar to that of eukaryotic Kv channels. There are a few prokaryotic 6-TM K^+ channel gene sequences that, like the plant K^+ channels, appear to possess a cytoplasmic nucleotide-binding domain. However, the functional properties of these channels have not been determined.

REFERENCES

Abraham, M. R., A. Jahangir, A. E. Alekseev, and A. Terzic. 1999. Channelopathies of inwardly rectifying potassium channels. *FASEB. J.* **13:**1901–1910.

Ashcroft, F. M., and F. M. Gribble. 1998. Correlating structure and function in ATP sensitive K^+ channels. *Trends Neurosci.* **21:**288–294.

Bichet, D., F. A. Haas, and L. Y. Jan. 2003. Merging functional studies with structures of inward rectifier K^+ channels. *Nat. Rev. Neurosci.* **4:**957–967.

Bond, C. T, M. Pessia, X. M. Xia, A. Lagrutta, M. P. Kavanaugh, and J. P. Adelman. 1994. Cloning and expression of a family of inward rectifier potassium channels. *Recept. Channels* **2:**183–191.

Cao, Y., N. M. Crawford, and J. I. Schroeder. 1995 Amino terminus and the first four membrane spanning segments of the Arabidopsis K^+ channel KAT1 confer inward-rectification property of plant-animal chimeric channels. *J. Biol. Chem.* **270:**17697–17701.

Chan, K. W., J. L. Sui, M. Vivadou, and D. E. Logothetis. 1996. Control of channel activity through a unique amino acid residue of a G protein-gated inwardly rectifying K^+ channel subunit. *Proc. Natl. Acad. Sci. USA* **93:**14193–14198.

Chen, G. Q., C. Cui, M. L. Mayer, and E. Gouaux. 1999. Functional characterization of a potassium-selective prokaryotic glutamate receptor. *Nature* **402:**817–821.

Clement, J. P., K. Kunjilwar, G. Gonzalez, M. Schwanstecher, U. Panten, B. L. Aguilar, and J. Bryan. 1997. Association and stoichiometry of K(ATP) channel subunits. *Neuron* **18:**827–838.

Coetzee, W. A., Y. Amarillo, J. Chiu, A. Chow, T. McCormack, H. Moreno, M. Nadal, A. Ozaita, D. Pountney, E. Vega-Saenz de Miera, and B. Rudy. 1999. Molecular diversity of K^+ channels. *In* B. Rudy and P. Seeburg (ed.), *Molecular and Functional Diversity of Ion Channels and Receptors*. New York Academy of Sciences, New York, N.Y.

Corey, S., and D. E. Clapham. 1998. Identification of native atrial G-protein-regulated inwardly rectifying K^+ (GIRK4) channel homomultimers. *J. Biol. Chem.* **273:**27499–27504.

Dascal, N., W. Schreibmayer, N. F. Lim, W. Wang, C. Chavkin, L. DiMagno, C. Labarca, B. L. Kieffer, C. Gaveriaux-Ruff, D. Trollinger, H. A. Lester, and N. Davidson. 1993. Atrial G protein-activated K^+ channel: expression cloning and molecular properties. *Proc. Natl. Acad. Sci. USA* **90:**10235–10239.

Doring F., C. Derst, E. Wischmeyer, C. Karschin, R. Schneggenburger, J. Daut, and A. Karschin. 1998. The epithelial inward rectifier channel Kir7.1 displays unusual K^+ permeation properties. *J. Neurosci.* **18:** 8625–8636.

Doupnik, C. A., and N. L. Davidson. 1995. The inward rectifier potassium channel family. *Curr. Opin. Neurobiol.* **5:**268–277.

Doyle, D. A., J. M. Cabral, R. A. Pfuetzner, A. Kuo, J. M. Gulbis, S. L. Cohen, B. T. Chait, and R. MacKinnon. 1998. The structure of potassium channel: molecular basis of K^+ conduction and selectivity. *Science* **280:**69–76.

Durell, S. R., and H. R. Guy. 2001a. A family of putative Kir potassium channels in prokaryotes. *BMC Evol. Biol.* **1:**14.

Durell, S. R., and H. R. Guy. 2001b. A putative prokaryote voltage-gated Ca^{2+} channel with only one 6TM motif per subunit. *Biochem. Biophys. Res. Commun.* **281:**741–746.

Fakler, B., U. Brandle, E. Glowatzki, H. P. Zenner, and J. P. Ruppersberg. 1994. Kir2.1 inward rectifier K^+ channels are regulated independently by protein kinases and ATP hydrolysis. *Neuron* **13:**1413–1420.

George, A. L. 1995. Molecular genetics of ion channel diseases. *Kidney Int.* **48:**1180–1190.

Heginbotham, L., Z. Lu, T. Abramson, and R. MacKinnon. 1994. Mutations in the K^+ channel signature sequence. *Biophys. J.* **66:**1061–1067.

Higgins, C. 1995. The ABC of channel regulation. *Cell* **82:**693–696.

Hille, B. 2001. *Ion Channels of Excitable Membranes*, 3rd ed., Sinauer, Sunderland, Mass.

Ho, K., C. G. Nichols, W. J. Lederer, J. Lytton, P. M. Vassilev, M. V. Kanazirska, and S. C. Hebert. 1993. Cloning and expression of an inwardly rectifying ATP-regulated potassium channel. *Nature* **362:**31–38.

Imredy, J. P., C. Chen, and R. MacKinnon. 1998. A snake toxin inhibitor of inward rectifier potassium channel ROMK1. *Biochemistry* **37:**14867–14874.

Inagaki, N., Y. Tsuura, N. Namba, K. Masuda, H. T. Gonoi, M. Horie, Y. Seino, M. Mizuta, and S. Seino. 1995. Cloning and functional characterization of a novel ATP sensitive potassium channel, ubiquitously expressed in rat tissues, including pancreatic islets, pituitary, skeletal muscle and heart. *J. Biol. Chem.* **270:**5691–5694.

Jiang, C., Z. Qu, and H. Xu. 2002. Gating of inward rectifier K^+ channels by proton-mediated interactions of intracellular protein domains. *Trends Cardiovasc. Med.* **12:**5–13.

Jin, W., and Z. Lu. 1998. A novel high-affinity inhibitor for inward rectifier K^+ channels. *Biochemistry* **37:** 13291–13299.

Keating, M. T., and M. C. Sanguinetti. 1996. Pathophysiology of ion channel mutations. *Curr. Opin. Genet. Dev.* **6:**326–333.

Krapivinsky, G., E. A. Gordon, K. Wickman, B. Velimirovic, L. Krapivinsky, and D. E. Clapham. 1995. The G-protein-gated atrial K^+ channel IKACh is a heteromultimer of two inwardly rectifying K^+ channel proteins. *Nature* **374:**135–141.

Krapivinsky, G., I. Medina, L. Eng, L. Krapivinsky, Y. Yang, and D. E. Clapham. 1998. A novel inward rectifier K^+ channel with unique pore properties. *Neuron* **20:**995–1005.

Kubo, Y., T. J. Baldwin, Y. N. Jan, and L. Y. Jan. 1993. Primary structure and functional expression of a mouse inward rectifier potassium channel. *Nature* **362:**127–133.

Kuo, A., J. M. Gulbis, J. F. Antcliff, T. Rahman, E. D. Lowe, J. Zimmer, J. Cuthbertson, F. M. Ashcroft, T. Ezaki, and D. A. Doyle. 2003. Crystal structure of the potassium channel KirBac1.1 in the closed state. *Science* **300:**1922–1926.

Lesage, F., F. Duprat, M. Fink, E. Guillemare, T. Coppola, M. Lazdunski, and J.-P. Hugnot. 1994. Cloning provides evidence for a family of inward rectifier and G-protein coupled K^+ channels in the brain. *FEBS Lett.* **353:**37–42.

Liu, Y., E. McKenna, D. J. Figueroa, R. Blevins, C. P. Austin, P. B. Bennett, and R. Swanson. 2000. The human inward rectifier K^+ channel subunit. Kir5.1 (KCNJ16) maps to chromosome 17q25 and is expressed in kidney and pancreas. *Cytogenet. Cell Genet.* **90:**60–63.

Lopatin, A. N., E. N. Makhina, and C. G. Nichols. 1994. Potassium channel block by cytoplasmic polyamines as the mechanism of intrinsic rectification. *Nature* **372:**366–369.

Lu, Z., and R. MacKinnon. 1997. Purification, characterization, and synthesis of an inward-rectifier K^+ channel inhibitor from scorpion venom. *Biochemistry* **36:**6936–6940.

Mao, J., J. Wu, F. Chen, X. Wang, and C. Jiang. 2003. Inhibition of G-protein coupled inward rectifying K^+ channels by intracellular acidosis. *J. Biol. Chem.* **278:**7091–7098.

Matsuda, H., A. Saigusa, and H. Irisawa. 1987. Ohmic conductance through the inwardly rectifying K^+ channel and blocking by internal Mg^{2+}. *Nature* **325:**156–159.

Minor, D. L., S. J. Masseling, Y. N. Jan, and L. Y. Jan. 1999. Transmembrane structure of an inwardly rectifying potassium channel. *Cell* **96:**879–891.

Moulton, G., T. K. Attwood, D. J. Parry-Smith, and J. C. L. Packer. 2003. Phylogenomic analysis and evolution of the potassium gene family. *Recept. Channels* **9:**363–377.

Navarro, B., S. A. Corey, M. Kennedy, and D. E. Clapham. 1996. Nonselective and G-βγ-insensitive weaver K^+ channels. *Science* **272:**1950–1953.

Nichols, C. G., and A. N. Lopatin. 1997. Inward rectifier potassium channels. *Annu. Rev. Physiol.* **69:**171–191.

Nichols, C. G., S.-L. Shyng, A. Nestorowicz, A. Glaser, J. P. Clement IV, G. Gonzalez, A. Aguilar-Bryan, M. A. Permutt, and J. B. Stable. 1996. ADP as the intracellular regulator of insulin secretion. *Science* **272:**1785–1787.

Nishida, M., and R. MacKinnon. 2002. Structural basis of inward rectification: cytoplasmic pore of the G protein-gated inward rectifier at 1.8 Å resolution. *Cell* **111:**957–965.

North, R. A. 1989. Drug receptors and the inhibition of nerve cells. *Br. J. Pharmacol.* **98:**13–28.

Patil, N., D. R. Cox, D. Bhat, M. Faham, R. M. Meyers, and A. S. Peterson. 1995. A potassium channel mutation in *weaver* mice implicates membrane excitablility in granule cell differentiation. *Nat. Genet.* **11:**126–129.

Pessia, M., C. T. Bond, M. P. Kavanaugh, and J. P. Adelman. 1995. Contributions of the C-terminal domain to gating properties of inward rectifier potassium channels. *Neuron* **14:**1039–1045.

Pessia, M., S. J. Tucker, K. Lee, C. T. Bond, and J. P. Adelman. 1996. Subunit positional effects revealed by novel heteromeric inwardly rectifying K^+ channels. *EMBO J.* **15:**2980–2987.

Reimann, F., and F. M. Ashcroft. 1999. Inwardly rectifying potassium channels. *Curr. Opin. Cell. Biol.* **11:**503–508.

Shieh, R. C., J. C. Chang, and C. C. Kuo. 1999. K^+ binding sites and interactions between permeating K^+ ions at the external pore mouth of an inward rectifier K^+ channel (Kir2.1). *J. Biol. Chem.* **274:**17424–17430.

Silverman, S. K., H. A. Lester, and D. A. Dougherty. 1998. Asymmetrical contributions of subunit pore regions to ion selectivity in an inward rectifier K^+ channel. *Biophys. J.* **75:**1330–1339.

Taglialatela, M., B. A. Wible, R. Caporaso, and A. M. Brown. 1994. Specification of pore properties by the carboxyl terminus of inwardly rectifying K^+ channels. *Science* **264:**844–847.

Takumi, T., T. Ishii, Y. Horio, K.-I. Morishige, N. Takahashi, M. Yamada, T. Yamashita, H. Kiyama, K. Sohmiya, S. Nakanishi, and Y. Kurachi. 1995. A novel ATP-dependent inward rectifier potassium channel expressed predominantly in glial cells. *J. Biol. Chem.* **270:**16339–16346.

Vandenberg, C. A. 1987. Inward rectification of a potassium channel in cardiac ventricular cells depends on internal magnesium ions. *Proc. Natl. Acad. Sci. USA* **84:**2560–2564.

Wible, B. A., M. De Biasi, K. Majumder, M. Taglialatela, and A. M. Brown. 1995. Cloning and functional expression of an inwardly rectifying K^+ channel from human atrium. *Circ. Res.* **76:**343–350.

Yamada, M., A. Ianobe, and Y. Kurachi. 1998. G protein regulation of potassium channels. *Pharmacol. Rev.* **50:**723–757.

Yang, J., M. Yu, Y. N. Jan, and L. Y. Jan. 1997. Stabilization of ion selectivity filter by pore loop ion pairs in inwardly rectifying potassium channel. *Proc. Natl. Acad. Sci. USA* **94:**1568–1572.

Yano, H., L. H. Philipson, J. L. Kugler, Y. Tokuyama, E. M. Davis, M. M. Le Beau, D. J. Nelson, G. I. Bell, and J. Takeda. 1994. Alternative splicing of human inwardly rectifying K^+ channel ROMK1 mRNA. *Mol. Pharmacol.* **45:**854–860.

Zhou, H., S. S. Tate, and L. G. Plamer. 1994. Primary structure and functional properties of an epithelial K channel. *Am. J. Physiol.* **266:**C809–C824.

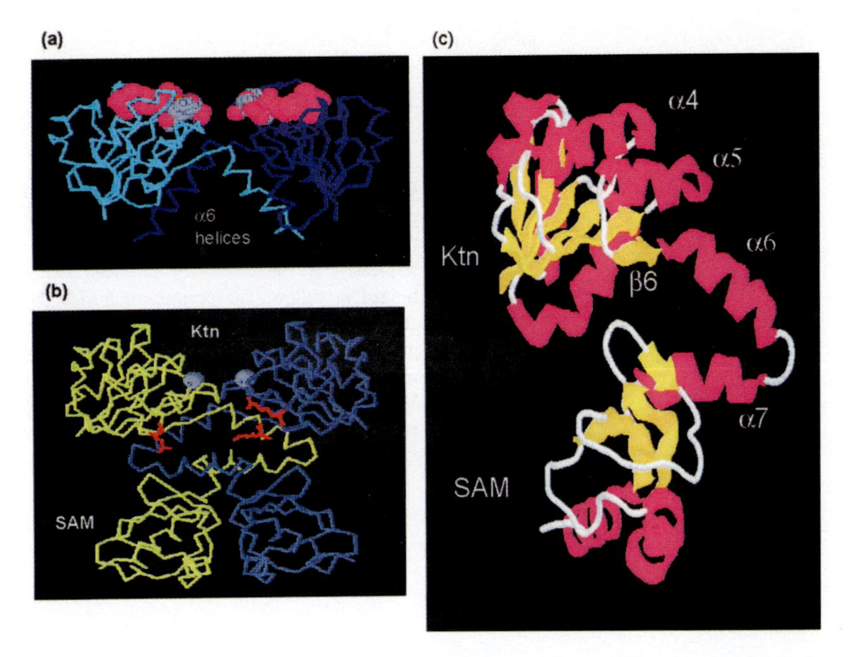

Color Plate 1 (chapter 2). Molecular cartoons of Ktn domains. (a) The Ktn domain of the *B. subtilis* KtrAB system (Roosild et al., 2002) showing the bound NADH in the Rossman fold. (b) The Ktn-SAM domains of the MthK channel (Jiang et al., 2002a) showing the bound Ca^{2+} (gray spheres) and the positions of the three critical Arg residues in the KefC protein (red). The two Ktn domains are oriented in the same way so that their relatedness can be discerned. (c) A single Ktn-SAM domain from MthK showing the positions of α-helices and β-sheets described in the text. The figure was drawn with CHIME (Martz, 2002).

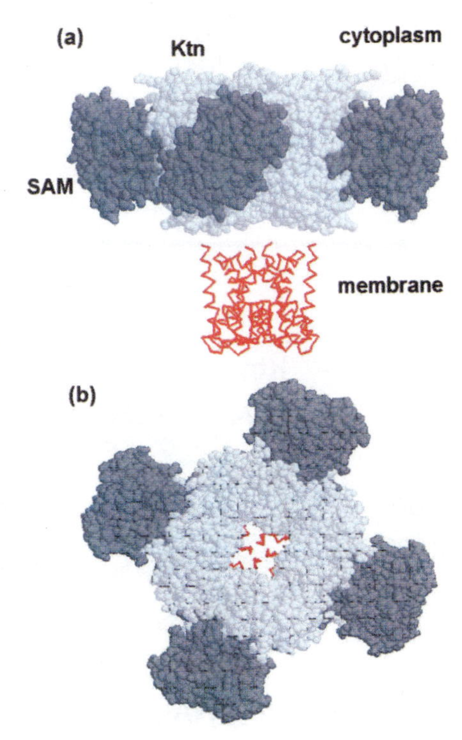

Color Plate 2 (chapter 2). The organization of Ktn and SAM domains in the MthK channel. The channel is shown in backbone (red) and is visible below (a) and through the central "pore" (b) of the Ktn-SAM domains. Ktn domains (pale gray) and SAM domains (dark gray) are each dimers constructed from one subunit attached to each channel-forming domain and one soluble domain synthesized from an intragenic translation start site. The protein is displayed viewed from the side and from the cytoplasmic face. In the original crystal structure another channel-forming domain is located above the Ktn-SAM complex, but this is omitted from this representation to indicate the probable structure of the MthK channel with an octameric assembly of Ktn-SAM domains. The figure was drawn with CHIME (Martz, 2002).

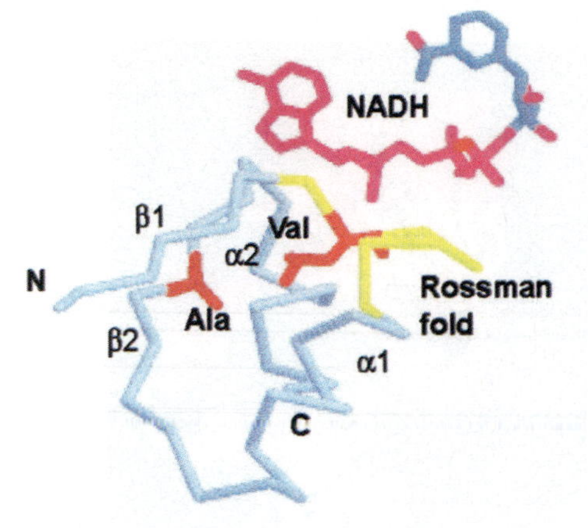

Color Plate 3 (chapter 2). The V427A/F407L mutation pair indicated using the *B. subtilis* Ktn structure (Roosild et al., 2002). The positions of the β1-α1-β2-α2 sequences are indicated; the Rossman fold is highlighted in yellow. The position of NADH binding is shown to indicate the potential proximity of the V427A mutation in KefC to the Rossman fold. The residues in red are L14 and A34, which are the equivalents of the V427A mutation and its F407L suppresser. The figure was drawn with CHIME (Martz, 2002).

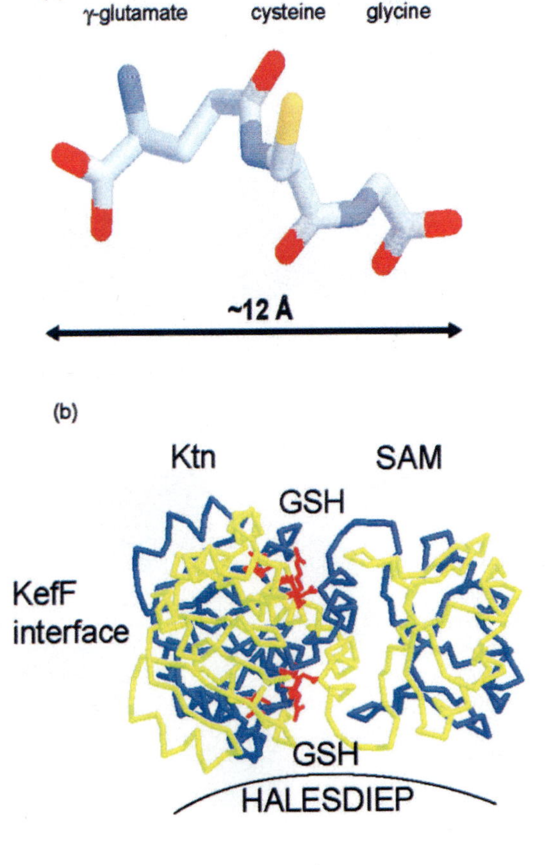

Color Plate 4 (chapter 2). Glutathione and gating of KefC. (a) The structure of glutathione is depicted and the length of the molecule determined using the CHIME program (Martz, 2002). (b) The KefC Ktn-SAM domain organization with respect to the HALESDIEP sequence in the membrane domain, to glutathione, and to the KefF protein based on the MthK structure. A pair of Ktn-SAM domains (each domain colored either blue or yellow) are depicted in contact with the membrane surface, such that one set of the Arg clusters (red) (see text) is placed in close opposition to the HALESDIEP sequence in accordance with genetic data (Roosild et al., 2002). The other set of Arg residues face the cytoplasm. The bridge between Ktn-SAM and HALESDIEP may be a direct salt bridge between R527 and E262 or may be created via GSH acting as an intermediary. The latter may account for the extreme sensitivity of the E262K and E262R mutants to the removal of GSH. The figure was drawn with CHIME (Martz, 2002).

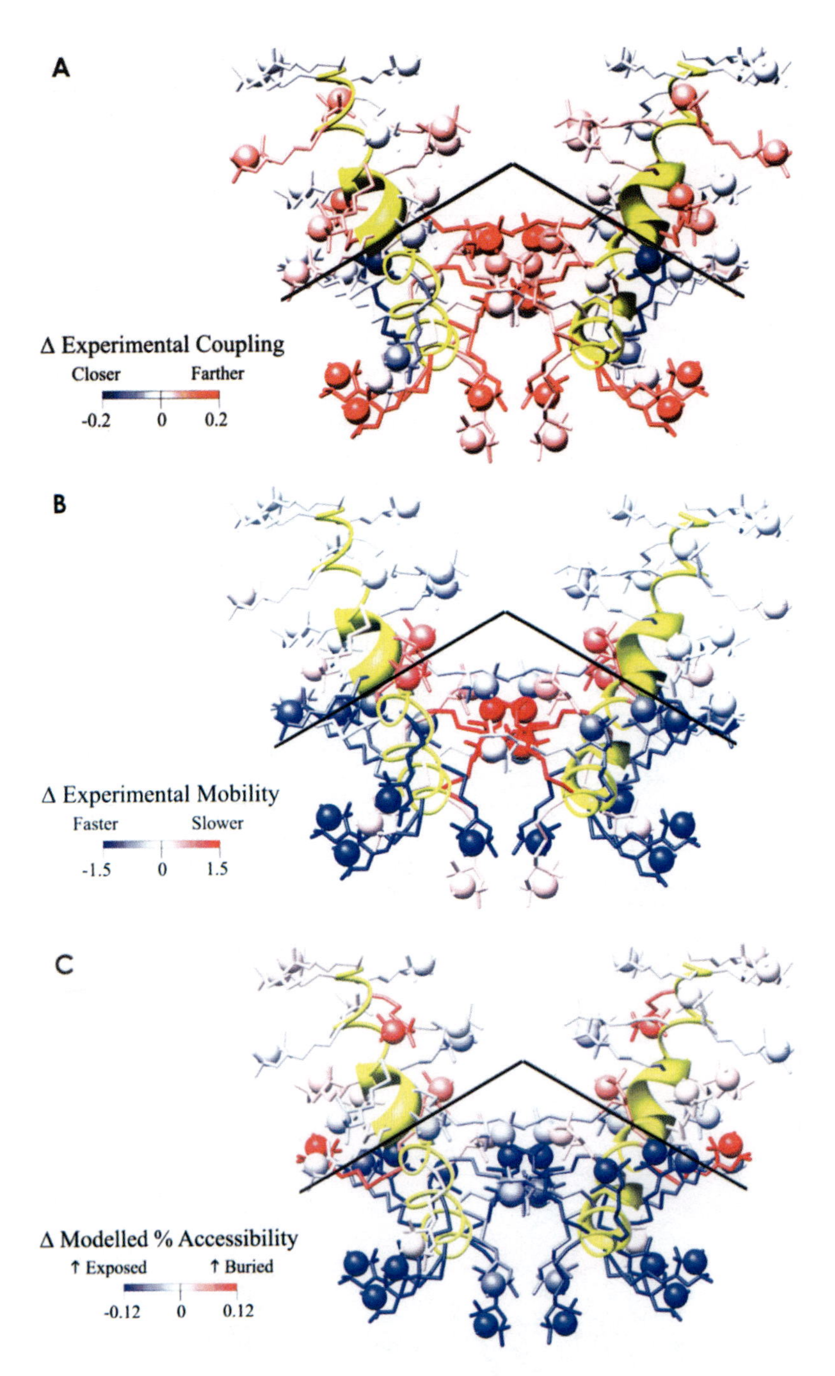

Color Plate 5 (chapter 4). Detection of gating movements. Changes in EPR parameters (coupling [A] and mobility [B]) for KcsA between open (pH4) and closed (pH7) are mapped onto the opposite inner helix of MthK. Residues have been replaced with spin-label side chains. The lines separate the helix before and after the conserved glycine (G99). (C) Changes in side-chain surface accessibility from the KcsA crystal structure to an open KcsA model based on MthK are mapped.

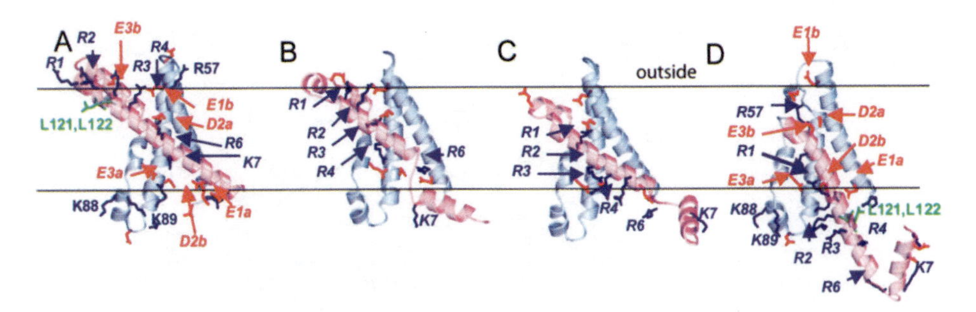

Color Plate 6 (chapter 6). Models of the voltage-sensing domain. (A) The crystal structure of the isolated voltage-sensing domain oriented in the bilayer (dashed lines) as predicted for the open channel. (B and C) Model in which S4 has moved inwardly by two (B) and four (C) helical screw steps. (D) Model of the resting conformation. S1-S3a segments are gray; S3b-S4-L45 segments are magenta. Positively charged residues of S4 are blue, negatively charged residues of S1-S3 are red, and L121 and L122 are green. The italicized residue numbers are generic and apply to all voltage-gated channel sequences. (Reprinted from Shrivastava et al. [2004] with permission.)

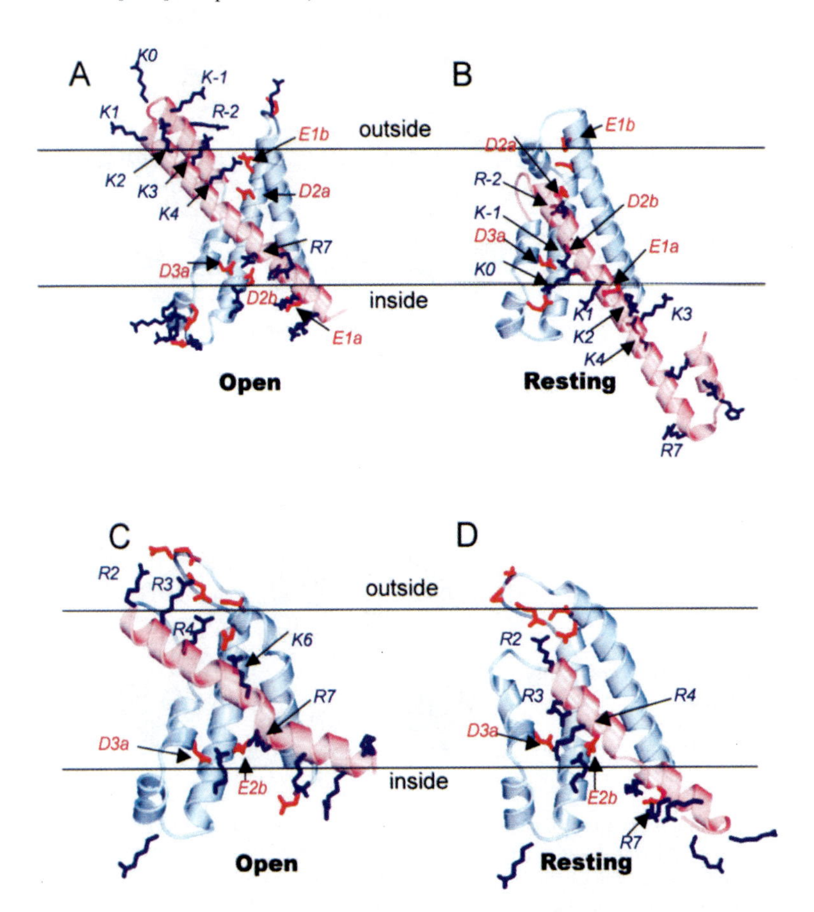

Color Plate 7 (chapter 6). Models of the voltage-sensing domain of the KvCA and KvVC channels. (A) Open KvCa, (B) resting KvCa, (C) open KvVC, and (D) resting KvVC conformations are shown. S1-S3a segments are gray; S3b-S4-S45 segments are magenta. Positively and negatively charged residues are blue and red.

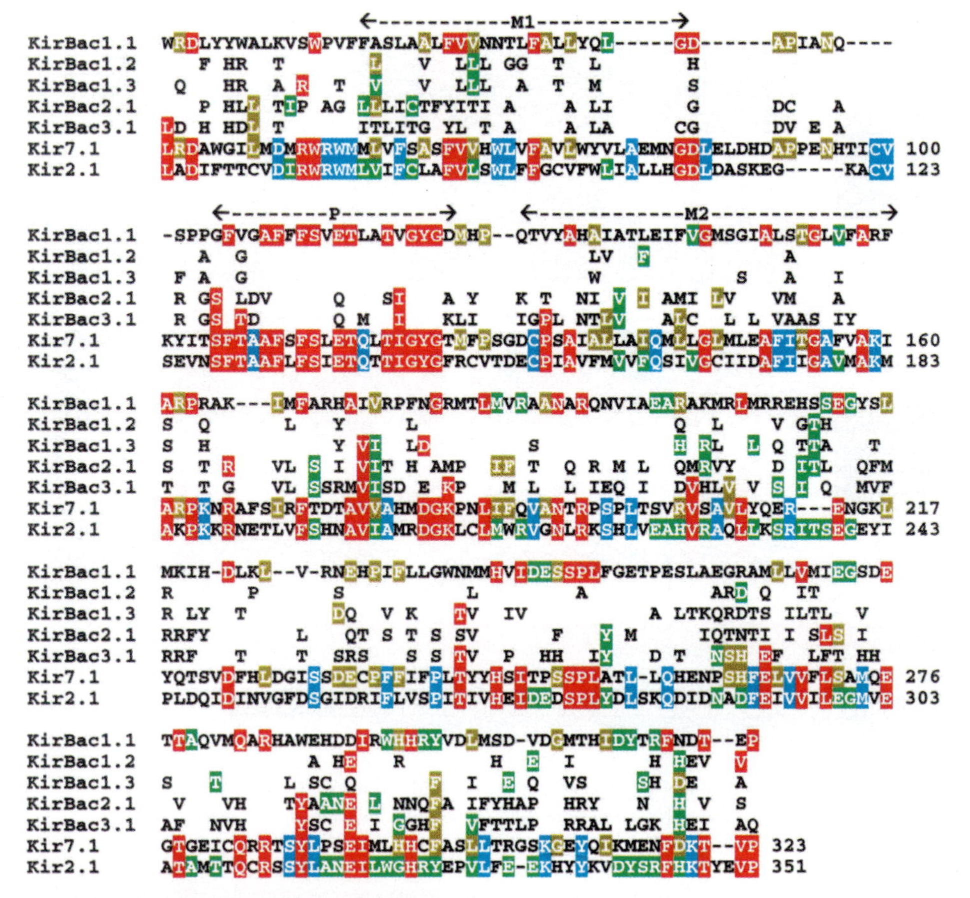

Color Code: number and percentage of residue positions in each category:

Identical in both Kir's and at least one KirBac

Identical in only Kir7.1 and at least one KirBac

Identical in only Kir2.1and at least one KirBac

Identical in only Kir7.1 and Kir2.1

Color Plate 8 (chapter 7). Alignment of KirBac sequence with Kir7.1 and Kir2.1. Locations of the M1-P-M2 segments are indicated above the sequences. The numbers under the Kir2.1 sequence indicate features that are well conserved among most eukaryotic Kirs but do not occur in KirBacs or other K^+ channels. Only residues that differ from KirBac1.1 are shown for other KirBac sequences. The KirBac sequences are from *B. pseudomallei* (KirBac1.1), *Burkholderia cepacia* (KirBac1.2), *Burkholderia fungorum* strain LB400 (KirBac1.3), *Magnetospirillum magnetotacticum* (KirBac2.1), and *Nostoc punctiforme* (KirBac3.1).

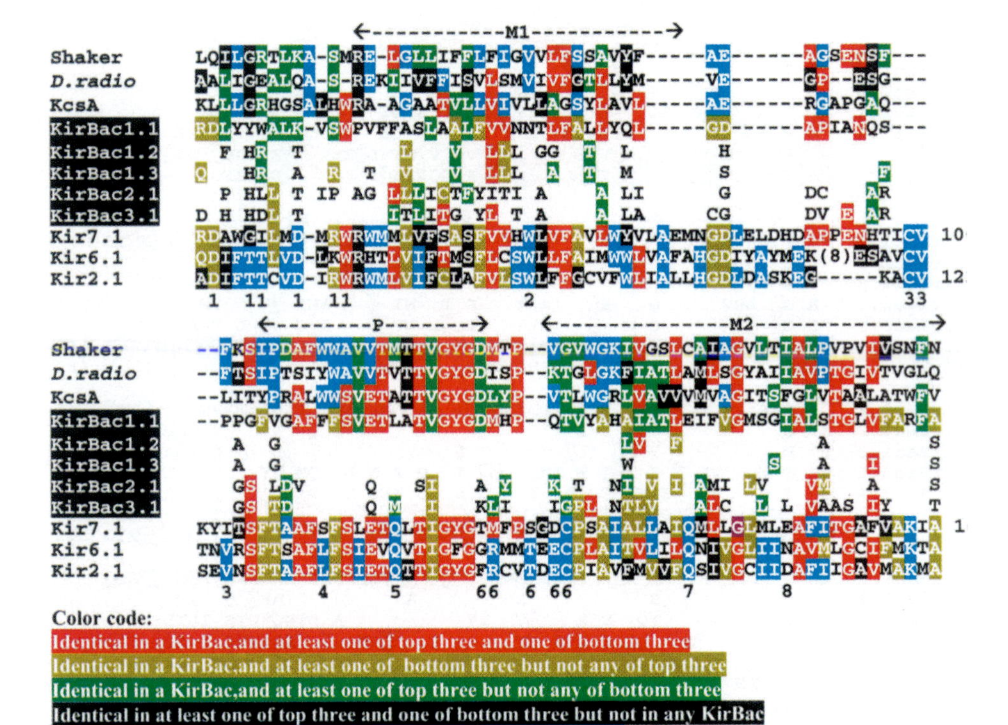

Color Plate 9 (chapter 7). Alignment of the M1-P-M2 region of the KirBac family with three non-Kir (Shaker Kv, a sequence from *Deinococcus radiodurans* that belongs to the Kbac 6TM1 family, and KcsA) and with three eukaryotic Kirs (Kir7.1, Kir4.1, and Kir2.1). Only those residues that differ from KirBac1.1 are shown for the other KirBacs. The relatively large number of residues in the first and last three sequences are identical to at least one residue in the KirBac sequences. Features unique to Kirs are indicated by numbers under the sequences.

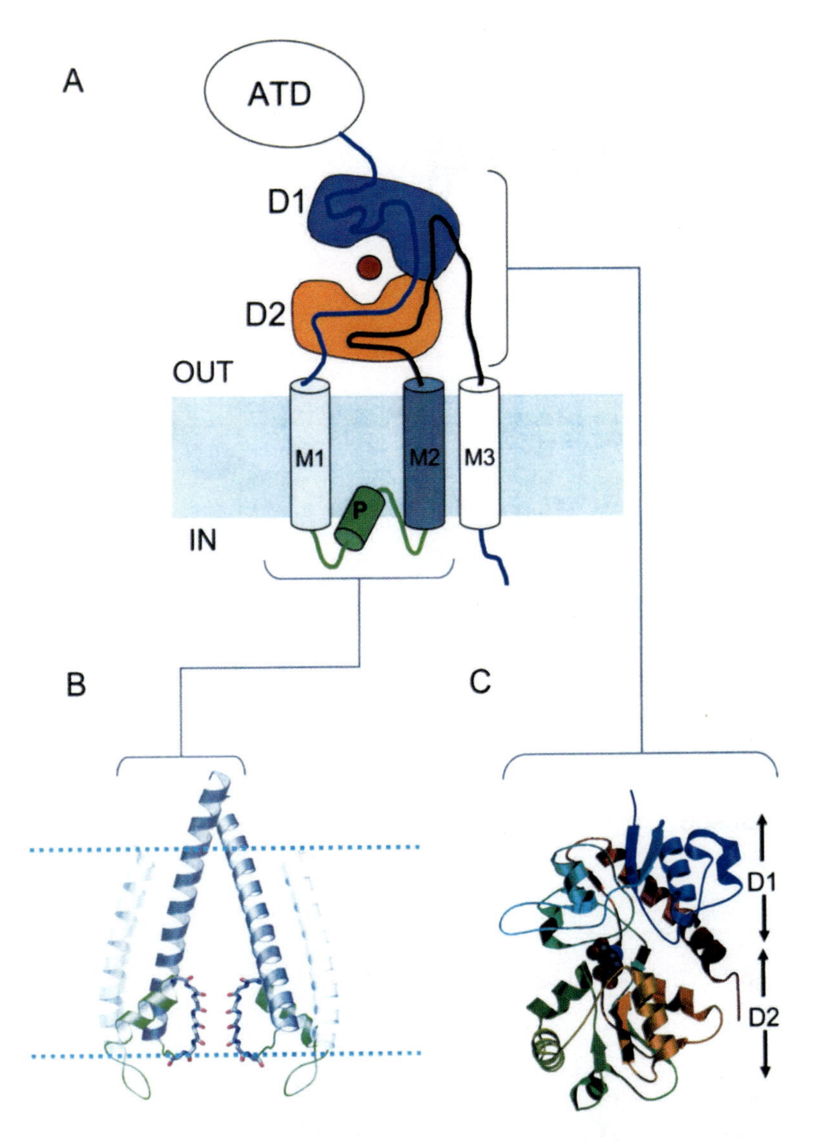

Color Plate 10 (chapter 8). Structure of glutamate receptors. (A) Schematic of a single subunit from the (tetrameric) GluR. (B) Homology model of TM domain of GluR0 (for clarity only two of the four subunits are shown). (C) Ligand-binding domain structure of GluR0.

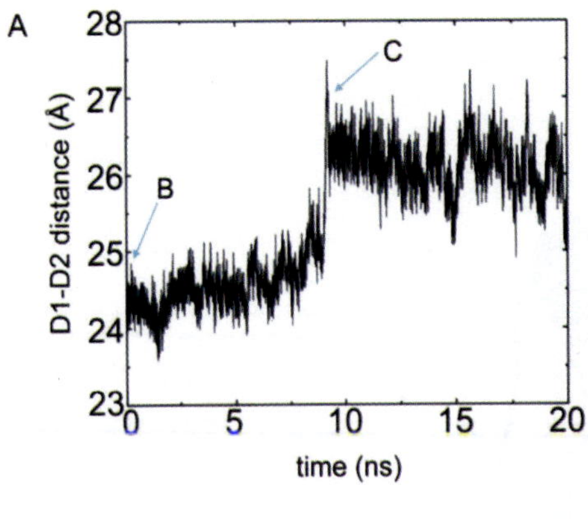

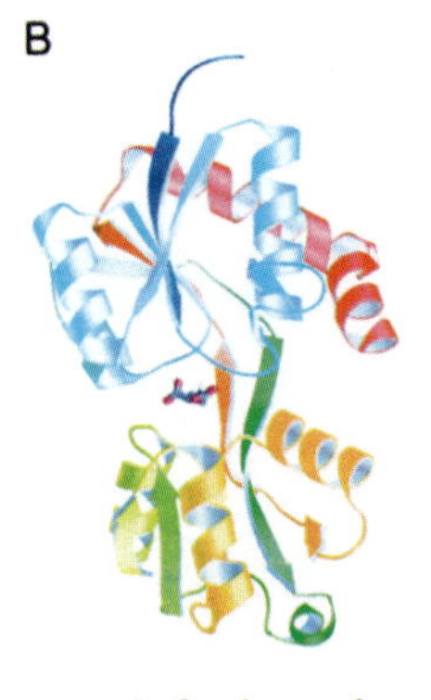

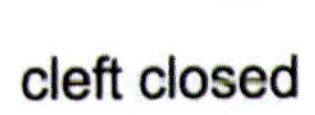

Color Plate 11 (chapter 8). Intersubdomain motions in a 20-ns simulation of GluR0 (in the presence of bound glutamate). (A) The D1-D2 intersubdomain distance is shown as a function of time. The arrows indicate the times at which snapshots of structures, illustrated in panels B and C, were taken. (B and C) Snapshots along the simulation at $t \sim 0$ and $t \sim 9$ ns, revealing the closed and open conformations of the D1/D2 cleft.

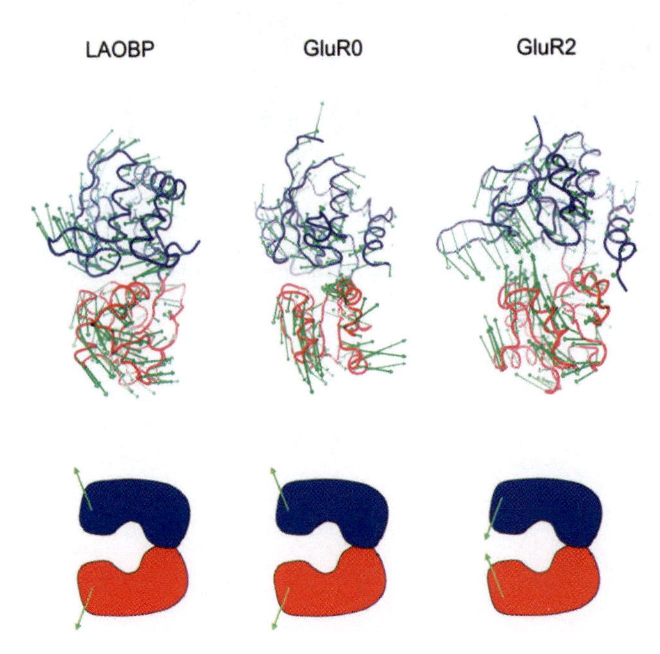

Color Plate 12 (chapter 8). Comparison of the hinge-bending motion of LAOBP, GluR0, and GluR2 as revealed by principal components analysis of simulations. The green bars indicate the direction of motion of the Cα atoms corresponding to the hinge-bending motion as revealed by principal components analysis of 10-ns MD simulations of the corresponding proteins. The schematic diagrams below the structures indicate the nature of the hinge-bending motion revealed.

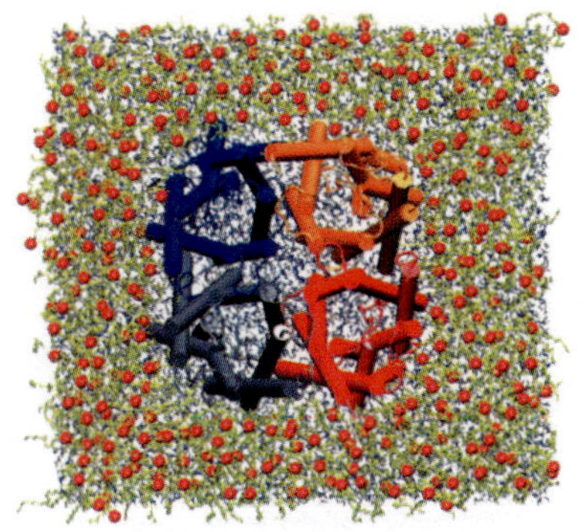

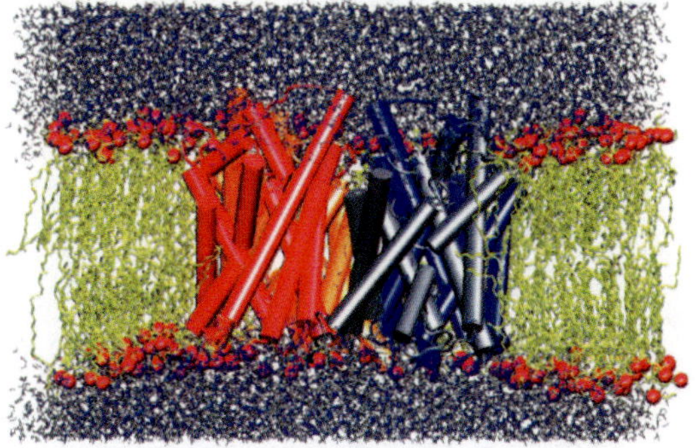

Color Plate 13 (chapter 9). Top (left) and side (right) views of an AQP tetramer in a fully hydrated lipid bilayer (Zhu et al., 2001; Jensen et al., 2001; Tajkhorshid et al., 2002). The model includes 106,000 atoms. The monomers are shown using cartoon representations. The phosphate groups of lipids are shown using vdW representation to illustrate the variation of membrane thickness at its interface to the protein. Four water pores each formed by a monomer are discernible from the top view. A fifth pore is formed in the middle of the tetramer.

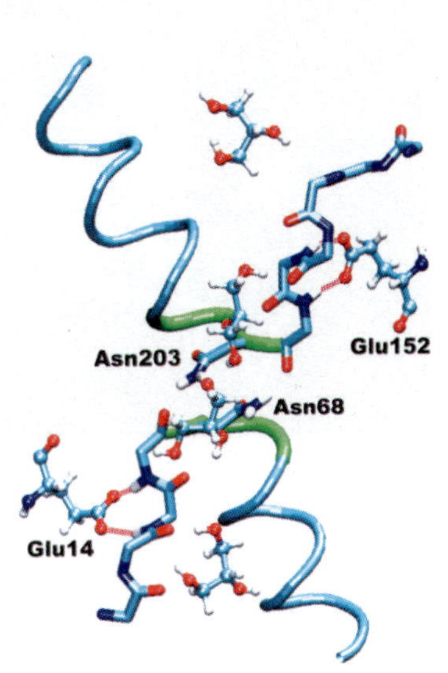

Color Plate 14 (chapter 9). The pathway for glycerol transport in GlpF is formed by the nonhelical halves (shown in red tube) of the two reentrant loops and the two conserved asparagine side chains (shown in licorice representation) of the NPA motifs in the middle of the channel.

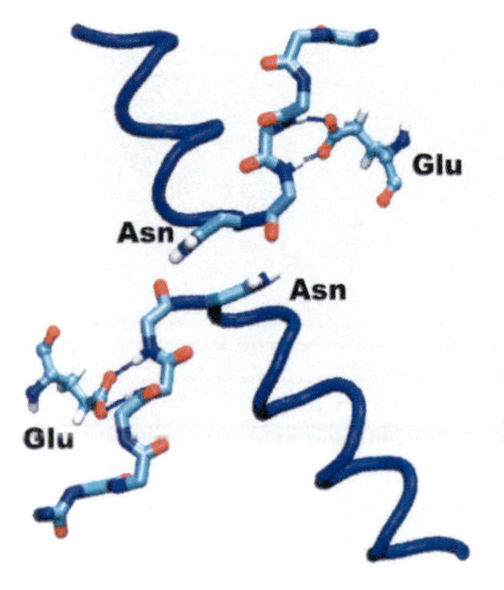

Color Plate 15 (chapter 9). Substrate conduction pathway in AQPs formed by backbone oxygen atoms of the reentrant loops. Carbonyl oxygens are shown in red. The peculiar secondary structure of the nonhelical halves of the loops is stabilized by specific hydrogen bonds with two highly conserved glutamates shown in CPK representation. Asparagine side chains of the NPA motifs are also shown.

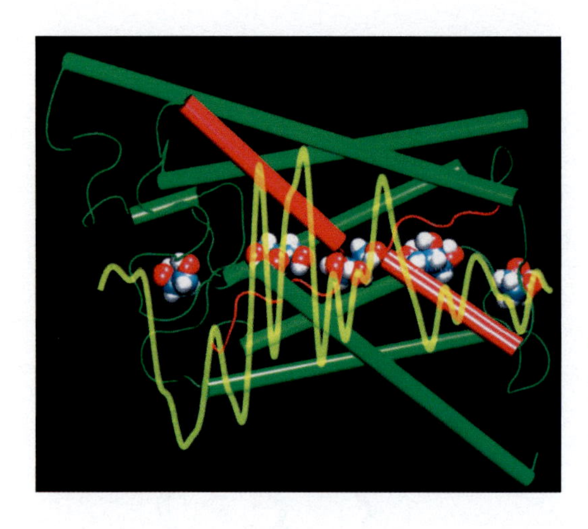

Color Plate 16 (chapter 9). PMF along the pathway of glycerol conduction in GlpF, constructed from SMD simulations (Jensen et al., 2002). The PMF is superimposed on a GlpF channel, in which several positions adopted by glycerol along its transport are highlighted.

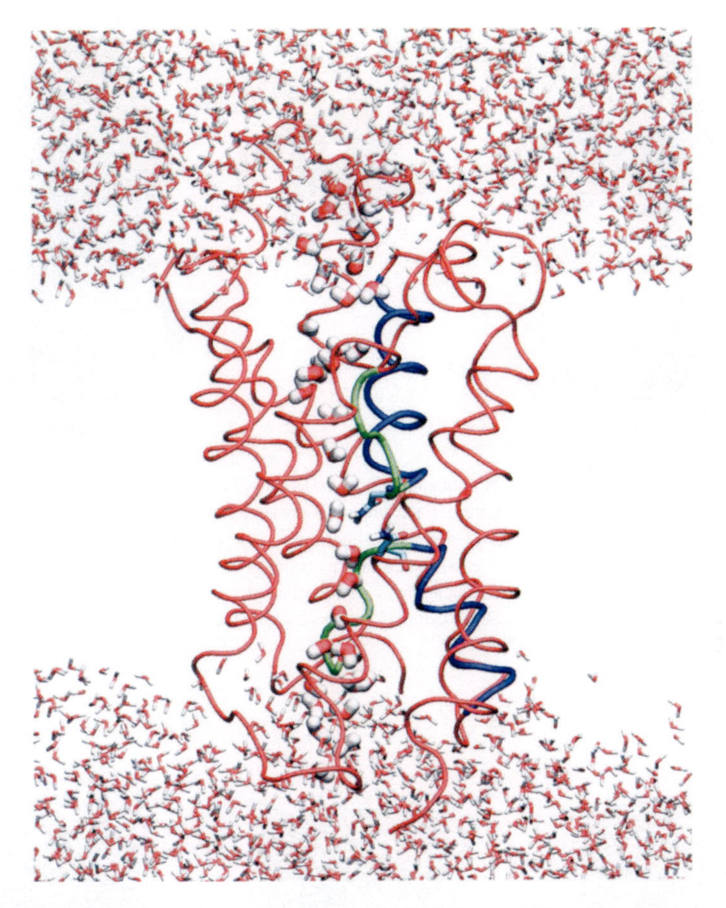

Color Plate 17 (chapter 9). A snapshot from an MD simulation of GlpF. Only one monomer is shown. A single file of water forms in the channel during the simulation. The orientation of water molecules in the single file is reversed in the two halves of the channel due to the electric field of the protein. This bipolar configuration of water prevents proton conduction (Tajkhorshid et al., 2002; Jensen et al., 2003).

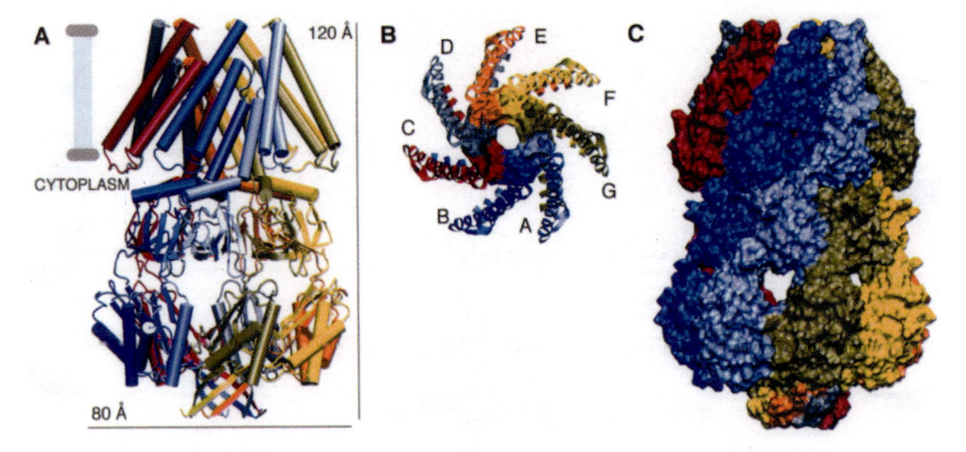

Color Plate 18 (chapter 9). Homoheptameric architecture of MscS. (A) Cartoon representation of the MscS crystal structure. Each subunit is presented in a different color. The suggested membrane position (Bass et al., 2002) is shown as a gray bar. (B) Top view of the transmembrane domain (residues 27 to 128). Pore residues (96 to 113) are shown in surface representation. Each subunit is labeled by a color and a letter. (C) "Solvent-excluded" surface representation of MscS. Seven cytoplasmic openings are located on the side (one visible in front) and one at the bottom.

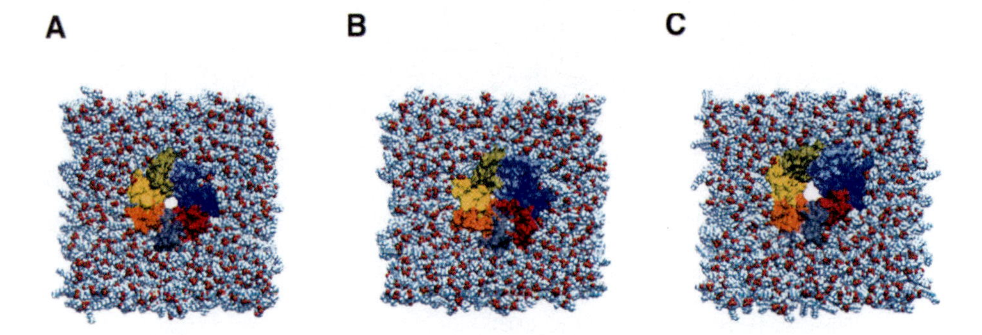

Color Plate 19 (chapter 9). Simulated conformations of MscS. Top views of final states of simulations sim1 (A), sim2a (B), and sim2b (C). The transmembrane domain is shown in surface representation and the lipid bilayer in space-filling representation. The color code for each protein subunit is the same as in Color Plate 18B. The simulations were carried out on a protein, lipid, water, and ion system with 224,340 atoms.

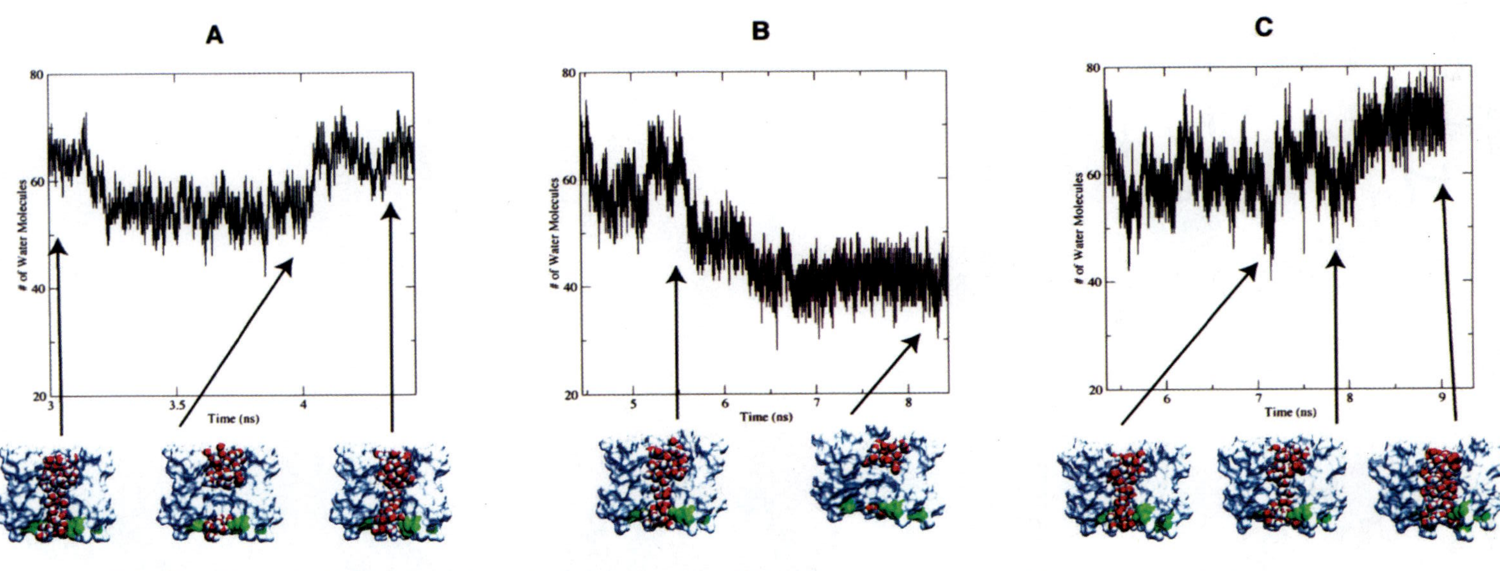

Color Plate 20 (chapter 9). Water permeation of MscS in open and closed forms. Shown is the occupancy of the hydrophobic (transmembrane) pore of MscS by water molecules as a function of time for simulations (A) sim1, (B) sim2a, and (C) sim2b. The number of water molecules inside the pore was monitored every picosecond assuming the positions of C_α atoms of residues 96 and 113 as pore ends. Snapshots of the pore (shown in surface representation and colored by residue type) with water molecules are shown for representative states.

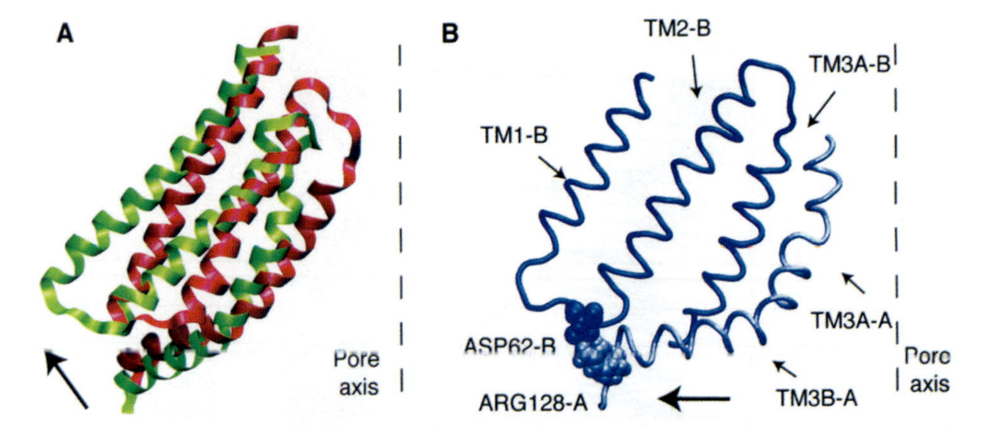

Color Plate 21 (chapter 9). Gating of MscS. (A) Movement of transmembrane helices. Transmembrane domains are shown in ribbon representation. Red shows the state of helices in simulation sim2a (open pore) and green shows the state of helices in simulation sim2b (closed pore). (B) Contacts between TM1-TM2 loop and the TM3B helix of adjacent subunits. Segments of helices TM1, TM2, and TM3 of subunit B are shown in blue along with helix TM3 of subunit A shown in light blue, all in tube representation. Residue Asp62 of subunit B and residue Arg128 of subunit A are shown in space-filling representation. The interaction of both residues is clearly discernible. The large arrow indicates the suggested movement of transmembrane domains during channel opening.

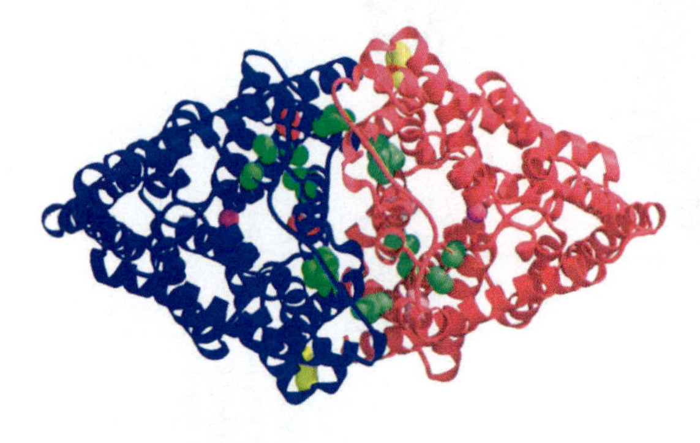

Color Plate 22 (chapter 11). Ribbon representation of the ClC-Set1 dimer viewed from outside the membrane. Monomers are in blue and red with the Cl^- ions indicated as magenta CPK spheres. The positions mutated in the study of Duffield et al. (2003) are indicated by CPK spheres colored as follows: green, A179, A188, A189, I200, I223, P405, and M425; yellow, F438; red/pink, S225 and T407; numbered according to ClC-Set1.

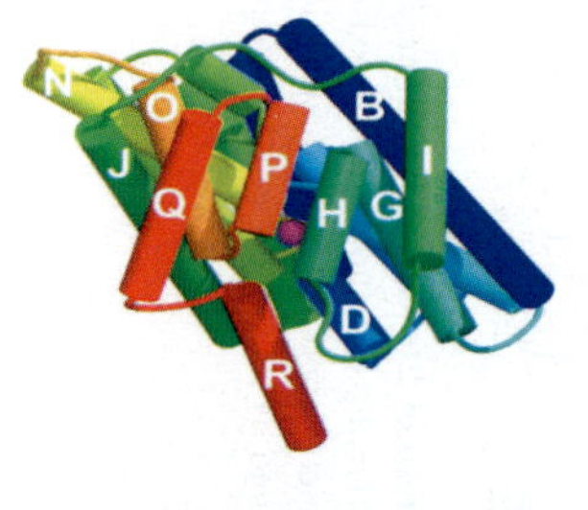

Color Plate 23 (chapter 11). Cartoon of a ClC-Set1 monomer viewed from within the plane of the membrane from the direction of the opposing monomer with the extracellular face at the top. Helices are indicated by cylinders, colored from blue (N terminus) to red (C terminus), with labels according to Dutzler et al. (2002). The Cl^- ion is indicated as a magenta CPK sphere.

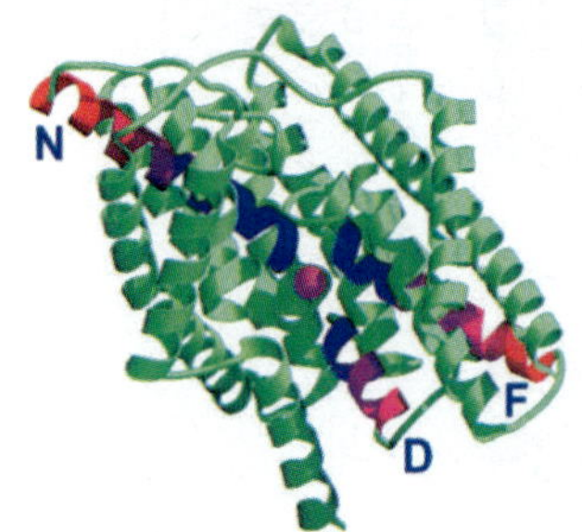

Color Plate 24 (chapter 11). Ribbon representation of a ClC-Set1 monomer as for Color Plate 23. Helices whose dipoles stabilize the Cl^- ion at the simple gate are colored from blue (N terminus) to red (C terminus) and labeled. For clarity, helices P and H and their connecting loops have been omitted.

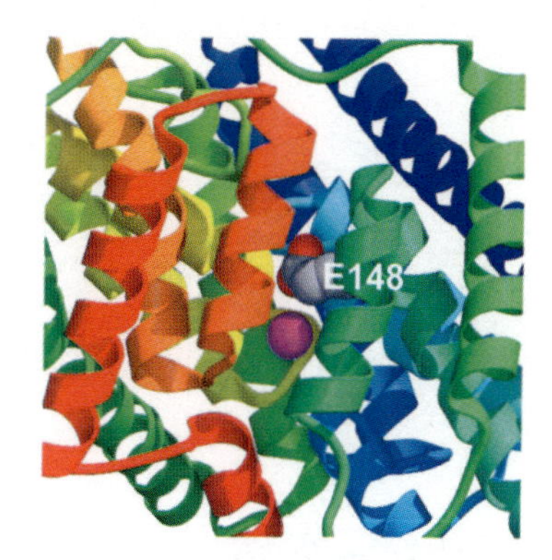

Color Plate 25 (chapter 11). Ribbon representation of ClC-Set1 selectivity filter. The side chain of Glu^{148} is shown with the epsilon oxygen atoms indicated in red.

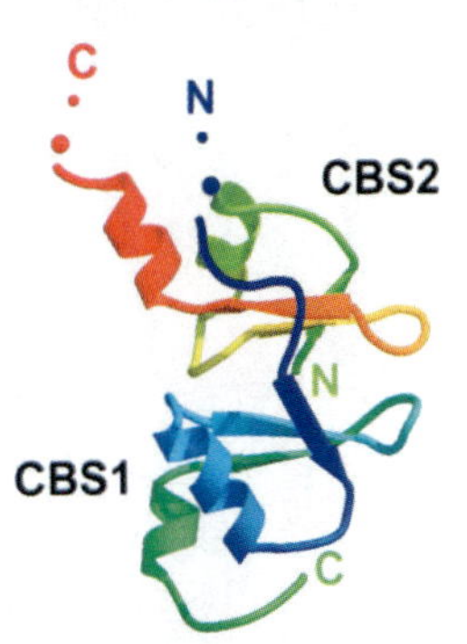

Color Plate 26 (chapter 11). Ribbon representation of the CBS domains of ClC-Hs1 based on the structure of the CBS domains of inosine monophosphate dehydrogenase (PDB code 1zfj). The sequence is colored from blue (N terminus) to red (C terminus). Sequences between CBS1 and CBS2 are not included in the model.

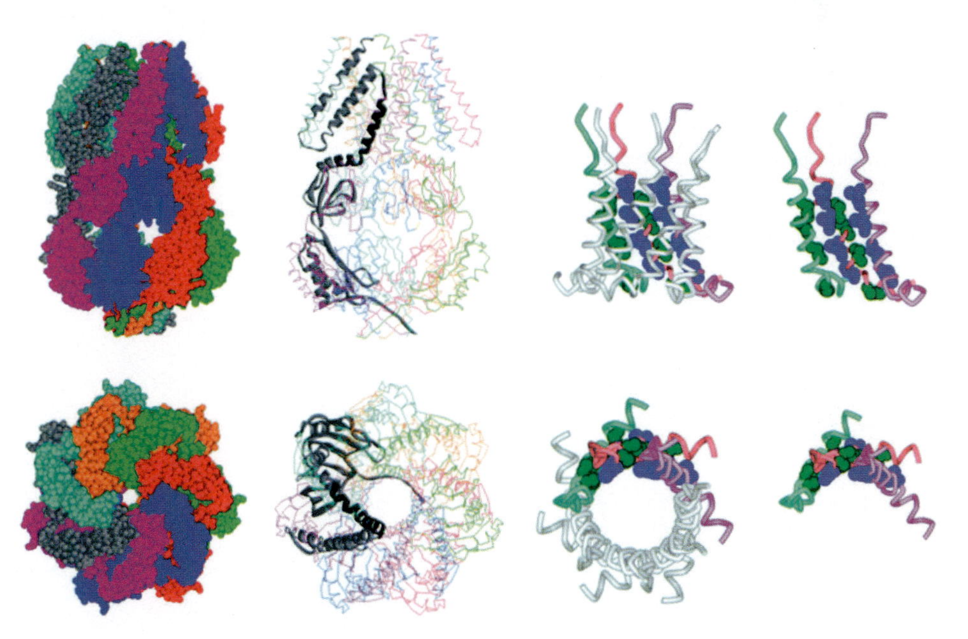

Color Plate 27 (chapter 12). The structure and pore design of the MscS channel. The model shown is derived from X-ray crystallographic data (Bass et al., 2002). The side (top panels) and top (bottom panels) views are shown. Each of the seven subunits in the complex is shown in a different color in a space-filling model (left) or in a wire diagram with a single subunit shown as a ribbon diagram in gray with the pore domain highlighted in black (left center). The pore domain is enlarged to show the heptameric (right center) and simple trimeric (right; the obstructing subunit structures have been removed) structures. Note the tight packing of the glycines (green CPK residues) and alanines (blue CPK residues) of the three subunits in the right panel.

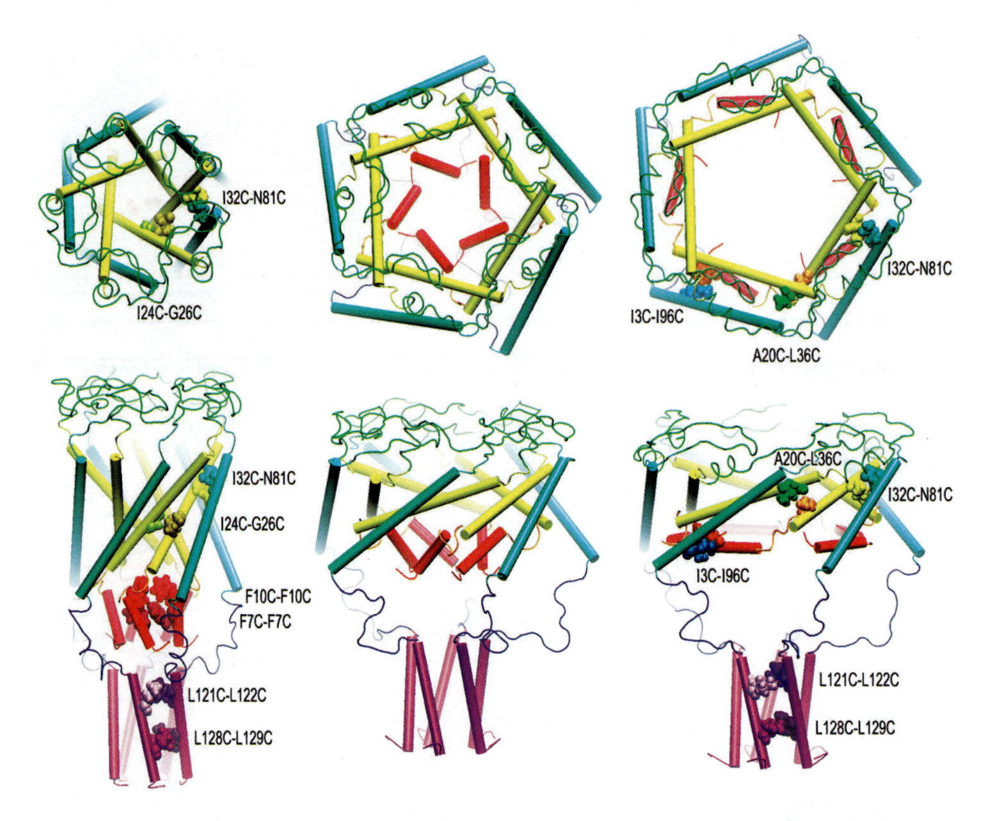

Color Plate 28 (chapter 13). The hypothetical gating transition in MscL modeled as a two-stage process, proceeding from the closed (left) to expanded (center) and then to the open state (right). The first stage is opening of the main M1 gate associated with major expansion of the barrel. In the intermediate expanded state S1 helices hold together, thus representing the low-conducting substate. Separation of S1 domains associated with minor expansion leads to the fully open state. The domains are color coded: S1 (red), M1 (yellow), S2 loops (green line), M2 (blue-green), and S3 (purple). Pairs of cysteine positions shown to form disulfide bonds in cross-linking experiments are mapped on the closed- and open-state models (see text).

Bacterial Ion Channels and Their Eukaryotic Homologs
Edited by A. Kubalski and B. Martinac

Chapter 8

From Prokaryotes to Eukaryotes: Molecular Modeling and Simulation Studies of Ion Channels

Philip C. Biggin, Alessandro Grottesi, and Mark S. P. Sansom

Ion channels provide a paradigm for ongoing attempts to relate the structures of membrane proteins to their physiological function. Recent years have seen considerable progress in the determination of ion channel structures, mainly by X-ray diffraction studies of bacterial ion channels. However, to understand the relationship between structure and function of human ion channels we must extrapolate beyond static snapshots of bacterial ion channels. This is an area where computational studies, including molecular modeling and simulation methods, can play a vital role.

It is important to define what is meant by modeling and simulation at a molecular level. By modeling, we imply prediction of the structure of one ion channel based on the structure of a related channel. This is usually done by homology modeling (Fiser et al., 2000; Sali and Blundell, 1993), although other techniques may also be required, especially when, for example, assembling the structure of a complex multidomain channel protein from structures of fragments of related channel proteins (Fig. 1). By simulation, we mean the use of the static structure of a channel protein as the starting point for simulations, at the molecular level, of its dynamic physiological function. This may be achieved via a number of techniques, including molecular dynamics (MD) (Karplus and McCammon, 2002) and Brownian dynamics simulations (Chung and Kuyucak, 2002).

When considering simulation studies, it must be recalled that a crystal structure is a static (spatial and temporal average) structure of a channel protein in a nonmembrane environment. The aim of simulation studies is to understand the conformational dynamics of a channel protein, its interactions with ions and water molecules, and its interactions with the membrane bilayer environment. Typically these interactions take place on an ~1-ns timescale, which can be readily addressed by MD simulations. However, as will be discussed below, longer timescale events such as channel gating have to be addressed by less direct simulation approaches.

Philip C. Biggin, Alessandro Grottesi, and Mark S. P. Sansom • Department of Biochemistry, University of Oxford, South Parks Road, Oxford OX1 3QU, United Kingdom.

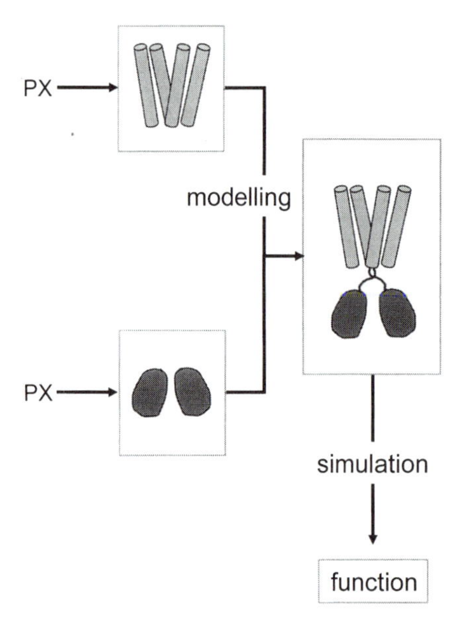

Figure 1. Schematic diagram of the role of modeling and simulation in studying the structure and function of ion channels. Protein crystallography (PX) yields structures of bacterial channels and their constituent domain/subunits. Molecular modeling may be used to integrate these structures within a complete model of a mammalian ion channel. The latter model may then be used as the starting point for simulation studies of channel structure-function relationships.

Modeling and simulation may be combined to enable us to study the conformational dynamics of mammalian (e.g., human) ion channels in relation to their function. In this respect, a major challenge for current research is to understand the strengths and limitations of computational methodologies. In particular, we need to establish the extent to which such computational approaches may be used to understand, for example, the effects of channel mutations or the nature of drug-protein interactions in human channel proteins. This is of considerable biomedical importance as it remains uncertain to what extent high-throughput structural studies of mammalian membrane proteins and their mutants will become a reality in the near future.

In this study we focus on two (distantly) related classes of channel for which there has been some progress in the application of modeling and simulation techniques. These are the potassium channels, for which a number of crystal structures are available, and glutamate receptor channels (GluRs). The GluRs share some distant homology in their transmembrane (TM) domains with K channels (Fig. 2) but possess distinct extracellular ligand-binding domains for which several structures, of both bacterial and mammalian homologs, are known.

POTASSIUM CHANNELS

Potassium channels are a ubiquitous family of integral membrane proteins, whose role is mediating the passive flux of K^+ ions across cell membranes. This function is of fundamental importance for a wide range of biological processes including cell volume regulation, movement, and electrical signal generation (Hille, 2001). K^+ ion permeation takes place on an $\sim 10^{-8}$-s timescale. K channels are selective for potassium ions; thus Na^+ ions are not able to pass through these channels. K^+ channel activity is regulated on an ~1-ms timescale by conformational changes that allow the protein to switch between a closed

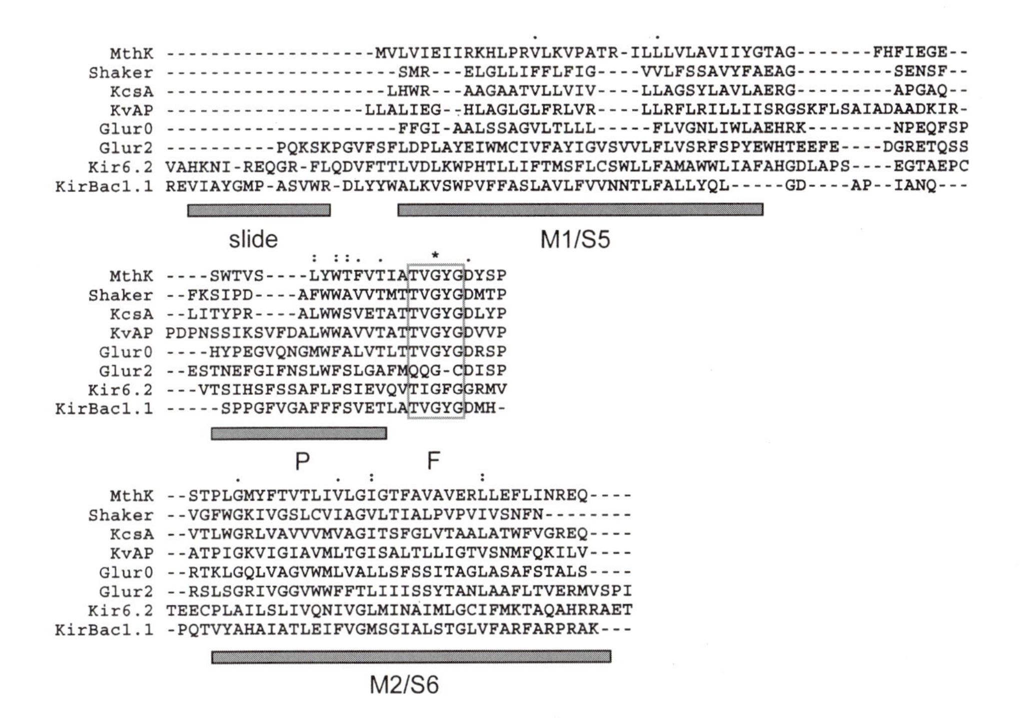

Figure 2. Sequence alignment of the core TM pore domain of K channels and GluRs. The locations of the slide helix (in Kir channels), the M1/S5 helix, the P helix and filter, and the M2/S6 helix are indicated.

(resting) and an open (activated) conformation, a process known as gating. There are a number of major classes of potassium channels, including voltage-gated (Kv) channels (Yellen, 2002) and inward rectifier (Kir) channels (Reimann and Ashcroft, 1999). The structures of bacterial homologs of both have been determined (see below). Furthermore, sequence alignment and structural studies indicate that the core pore-forming domain architecture is likely to be conserved across all K channels (Shealy et al., 2003). Thus, studies of bacterial K channels are able to inform us of structure-function relationships in their mammalian homologs.

Structural Studies of K Channels

The elucidation of the structures of several K^+ channels (MacKinnon, 2003) has shed light on the structural basis of the mechanisms of ion selectivity and permeation (Doyle et al., 1998; Jiang et al., 2002a, 2002b; Jiang et al., 2003; Kuo et al., 2003; Morais-Cabral et al., 2001; Zhou and MacKinnon, 2003). All of these structures are of bacterial K channels. They share a common architecture for their core pore-forming TM domain (Fig. 3A). The differences in their structures correspond to their different gating mechanisms. KcsA is gated by low pH; MthK is gated by Ca^{2+} ions; KvAP is gated by transmembrane voltage; and KirBac, for which the gating mechanism remains unknown, is presumed to be activated by an intracellular ligand. All K channels are tetrameric, with

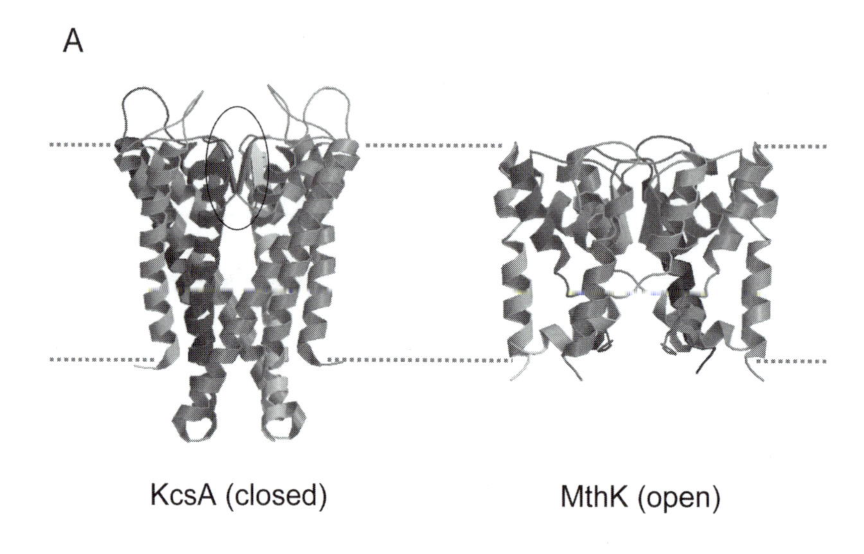

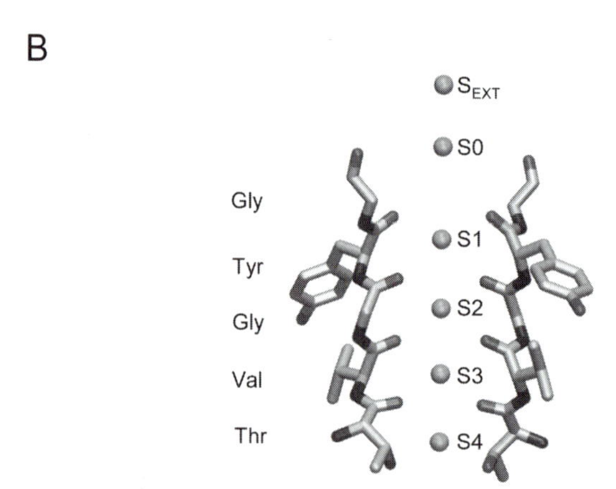

Figure 3. (A) Closed and open conformations of a K channel pore domain as revealed by the X-ray structures of KcsA and MthK, respectively. In both cases, only the structure of the TM domain is shown. The vertical ellipse indicates the location of the selectivity filter in the KcsA structure. The approximate location of the lipid bilayer is indicated by the horizontal gray broken lines. (B) Structure of the filter region of KcsA (for clarity only two of the four polypeptide chains are shown). The locations of the possible binding sites (S_{EXT} to S4) for the potassium ions are shown as spheres.

the monomers surrounding a central pore. The pore-forming domain is formed of four M1-P-M2 motifs (or S5-P-S6 motifs in Kv channels), where M1/S5 and M2/S6 are TM helices and the short P-helix and extended filter (F) region form a reentrant loop in between the two TM helices. In the KirBac structure an additional helix, referred to as the "slide" helix (Kuo et al., 2003), runs parallel to the cytoplasmic face of the membrane. In Kv channels each subunit also contains a voltage sensor domain composed of four

TM helices (S1 to S4), and thus each Kv subunit contains a total of six TM helices. The other K channels contain two TM helices per subunit. A number of K channels also have intracellular domains that are associated with control of channel gating: the structures of a number of these domains, both prokaryotic and eukaryotic, have been determined (Jiang et al., 2002a; Jiang et al., 2001; Nishida and MacKinnon, 2002; Zagotta et al., 2003).

Selectivity Filter

The selectivity filter is the structural element responsible for the conduction of K^+ ions at rates approaching 10^8 ions s^{-1}. It is formed by a highly conserved sequence motif, TVGYG, the glycines of which enable the polypeptide to adopt an unusual extended conformation in which all of the carbonyl oxygen atoms point toward the same side of the backbone. The crystal structure of KcsA reveals that the selectivity filter contains multiple discrete K^+-binding sites (Fig. 3B). Under physiological conditions the flow of potassium ions through the pore occurs in single-file mode, with two K^+ ions and an intervening water molecule undergoing a one-dimensional diffusion along the filter (Morais-Cabral et al., 2001). A recent study suggests that the filter occupancy is determined by a complex balance between charge distribution and subtle protein conformational changes (Zhou and MacKinnon, 2003).

Cytoplasmic Gate

Closure of K channels occurs by occluding the pore pathway at the intracellular mouth, localized in the channel structure at the crossing point of the four pore-lining M2/S6 helices. On the basis of comparing the crystal structures of KcsA (in a closed conformation) and of the bacterial Ca^{2+}-activated potassium channel MthK (in an open conformation), it was suggested that the crucial step in the activation mechanism of K channels is a bending or kinking of the pore-lining M2 helices such that the cytoplasmic segments of these helices move outward with respect to the pore axis and thus enlarge the cytoplasmic pore radius. This model is supported by the observed sequence conservation of the Gly residue that appears to form the hinge region in the M2 helices of KcsA and MthK (Jiang et al., 2002b).

For Kir channels, two structures provide some insight into possible gating mechanisms. The structure of the intact bacterial homolog KirBac reveals the presence of two slide helices, which may act to couple conformational changes in the intracellular domain to channel activation via motion of the TM M1 and M2 helices (Kuo et al., 2003). The crystal structure of the intracellular domain of a mammalian Kir channel (Kir3.1) (Nishida and MacKinnon, 2002) provides possible insights into the nature of conformational changes regulating channel activation (see below).

In Kv channels, the nature of the conformational change in the S6 helices underlying channel activation may be more complex. In addition to a conserved Gly in S6, there is also a PVP motif downstream of the glycine that may act as a further molecular hinge point. Data from Yellen and coworkers on Shaker Kv channels suggest that the PVP motif may act as a flexible hinge or swivel that keeps the upper portion of the S6 helix static relative to the remainder of the channel-forming domain but allows the lower (i.e., cytoplasmic) half of S6 to move, opening the pore by swiveling the blocking residues away from the central axis (Webster et al., 2004). Unfortunately, the role of the PVP motif cannot be resolved via direct examination of the structure of the bacterial Kv homolog KvAP, as the PVP motif is absent from the S6 helix of KvAP.

Gating and the Filter

In addition to activation of K channels via opening of the gate at the intracellular mouth, there is increasing evidence that conformational changes in the filter region may also be related to channel gating, underlying phenomena such as C-type inactivation of Kv channels. In particular, crystallographic studies of KcsA in the presence of lowered concentrations of K^+ ions have demonstrated that the selectivity filter is not rigid but rather is deformable and able to switch between an open conformation (observed in the crystal in the presence of a high concentration of K^+ ions) and a closed conformation (observed in the crystal in the presence of a low concentration of K^+) (Zhou et al., 2001).

Kv Channels and Voltage Sensing

Activation gating of Kv channels is the conformational response to a change in voltage across the cell membrane. It is thought that the principal structural element responsible for sensing the change in voltage is the positively charged S4 helix. The structure of the bacterial KvAP channel (Jiang et al., 2003) has revealed one conformation of the voltage sensor (S1 to S4) domain, which has been interpreted in terms of a "gating paddle" model of channel activation. However, the conformation of the isolated voltage sensor domain is different from that in the intact channel, providing structural evidence for conformational flexibility. There is currently some controversy concerning the structure of KvAP and how it informs our models of Kv channel activation (Starace and Bezanilla, 2004). Although computational models and simulations have been used to inform this debate (Monticelli et al., 2004), such studies are perhaps premature until we have a better understanding of the native conformation Kv channels and how Kv channel conformation may respond to changes in experimental environment such as the presence of detergent molecules.

Molecular Dynamics Simulations of Channels

Ion Permeation

A number of simulation approaches have been employed in computational studies of ion permeation through channels and related membrane proteins, helping to bridge the divide between static structure and dynamic function of ion channels (Chung and Kuyucak, 2002; Compoint et al., 2004; Domene et al., 2003a; Giorgetti and Carloni, 2003; Roux, 2002; Roux et al., 2000; Sansom et al., 2000; Tieleman et al., 2001a). In particular, MD simulations enable us to model the nanosecond timescale dynamics of channels embedded in an atomistic model of a cell membrane (Fig. 4), and they have been applied to a range of ion channels, including KcsA (Allen et al., 2000; Allen and Chung, 2001; Allen et al., 1999; Åqvist and Luzhkov, 2000; Bernèche and Roux, 2000, 2001, 2003; Biggin et al., 2001; Capener and Sansom, 2002; Capener et al., 2000; Guidoni and Carloni, 2002; Guidoni et al., 1999, 2000; Jordan, 2000; Luzhkov and Åqvist, 2000, 2001; Mashl et al., 2001; Ranatunga et al., 2001a; Ranatunga et al., 2001b; Sansom et al., 2002; Sansom et al., 2000; Shrivastava and Sansom, 2000, 2002; Tieleman et al., 2001a), and to related channel-like proteins such as the aquaporins (de Groot and Grubmuller, 2001; Jensen et al., 2001; Tajkhorshid et al., 2002; Zhu et al., 2001). Simulation studies of KcsA have provided information on the conformational dynamics of the selectivity filter and on the dynamics and energetics of ion permeation.

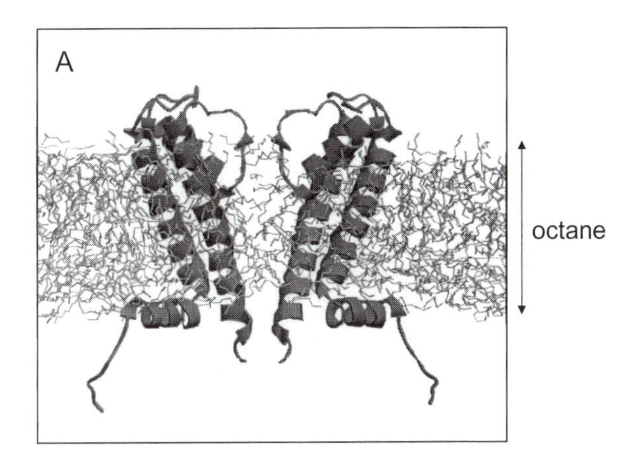

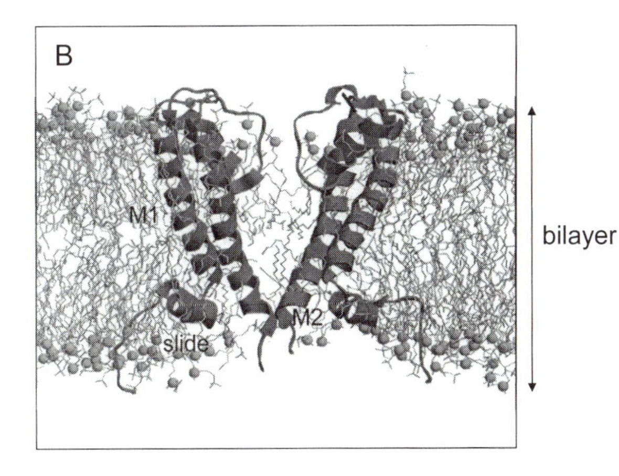

Figure 4. K channel simulation systems illustrated using KirBac. For clarity only two subunits of the protein are shown, and all water molecules and ions are omitted. KirBac is shown embedded in a membrane-mimetic octane slab (A) and a 1-palmitoyl-2-oleoyl-phosphatidylcholine (POPC) lipid bilayer (B). In the latter diagram, the helices are labeled and the phosphorus atoms of the lipid molecules are shown as gray spheres.

In terms of detailed atomistic models of ion permeation through K channels, such simulations have revealed that K^+ ions move in a single-file mode in the filter via a concerted translocation of ions between the corresponding binding sites. Thus, K^+ ions are able to alternately occupy the five binding sites along the filter pathway. Simulations also suggest that the conformational flexibility of the selectivity filter plays an important role in enabling rapid translocation of K^+ ions between adjacent binding sites. Simulations of ion movement through the pore have confirmed that two K^+ ions can occupy the filter simultaneously and that the timescale of ion movement is compatible with experimentally measured conductivities (Compoint et al., 2004; Domene and Sansom, 2003; Shrivastava and Sansom, 2000). A number of detailed studies support the view of a concerted motion of ions along the pore and provide an atomistic interpretation of potassium-binding energies in the filter (Åqvist and Luzhkov, 2000; Bernèche and Roux, 2001; Garofoli and Jordan, 2003; Guidoni and Carloni, 2002; Luzhkov and Åqvist, 2000).

Filter Flexibility and Distortion

Several simulations have provided evidence for limited (< 0.1 nm) filter flexibility during the concerted motion of ions and water molecules within the filter (e.g., Domene et al.,

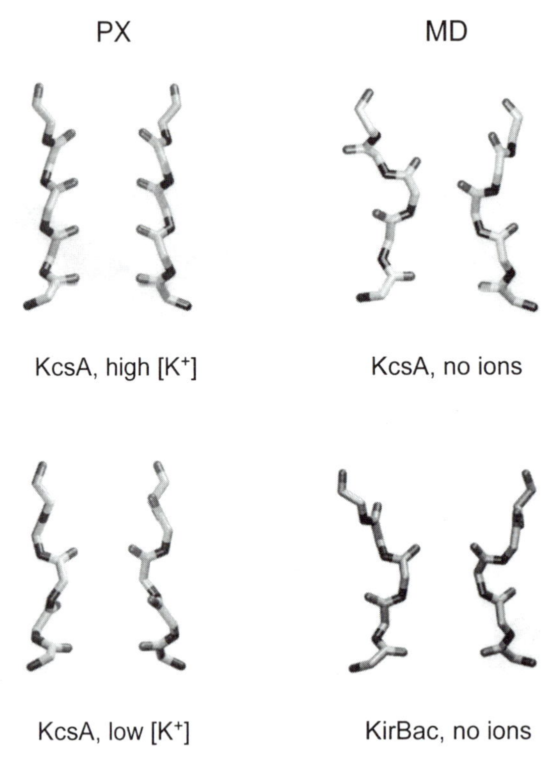

Figure 5. Structures of the selectivity filter in crystal structures (PX) and simulations (MD) are compared. In each case the backbones of two subunits of the filter are shown. The PX structures are of KcsA, crystallized in the presence of a high concentration of K^+ ions (PDB code 1k4c) and in the presence of a low concentration of K^+ ions (PDB code 1k4d). The MD structures are of KcsA, from a simulation in which all K^+ ions have left the filter (Holyoake et al., 2003), and of KirBac at the end (10 ns) of a simulation in the absence of K+ ions (Domene et al., 2004).

2004). These local conformational changes occur on an ~1-ns timescale. Furthermore, in the absence of K^+ ions, both the KcsA (Shrivastava and Sansom, 2000) and KirBac (Domene et al., 2004) selectivity filters undergo more substantial distortions (Fig. 5). Similar conformational distortions have been seen in KcsA-based homology models of mammalian Kir channels (Capener et al., 2003; Capener and Sansom, 2002) (see below). These conformational distortions of the filter resemble those seen in the low [K^+] structure of KcsA and are thought to lead to functional closure of the channel. This suggests comparable filter distortions, albeit on a longer timescale, may provide a mechanism of K channel gating in addition to changes in the hydrophobic gate formed at the intracellular mouth by the crossing point of the M2 helices.

Simulations and Gating

Current simulation techniques do not enable us to address the intracellular gating mechanism of K channels directly. The maximum times achievable by atomistic simulation of such large systems (~10 to 100 ns) fall short of the expected timescales (>1 μs) of channel gating. Nevertheless, MD simulations can suggest possible dynamic patterns that form the basis of larger conformational changes underlying gating. Indeed, early simulation studies of KcsA (Shrivastava and Sansom, 2000, 2002) suggested that the motion of the M2 helix to widen the intracellular gate might correspond to a possible gating mechanism. A similar model was suggested on the basis of normal mode analysis of KcsA (Shen et al., 2002). This was in agreement with spectroscopic (site-directed spin labeling) studies (Liu et al., 2001; Perozo et al., 1998, 1999) and confirmed by subsequent comparison of the KcsA (closed) and MthK (open) crystal structures (Jiang et al., 2002b). Simple model calculations (Beckstein et al., 2003; Beckstein et al., 2001; Beckstein and Sansom, 2004) indicate that relatively small changes in pore radius may switch a hydrophobic gate between closed and open. Models of the open state of KcsA may be constructed on the basis of the MthK structure and demonstrate that the hydrophobic barrier to ion permeation is removed and that the open-state channel model is stable during nanosecond duration simulations (Holyoake et al., 2003). Steered MD simulation may be used to probe possible transition pathways between the closed and open states (Biggin and Sansom, 2002). Taken together, such approaches offer the prospect of modeling the open state of a K channel pore domain once the structure of the closed state has been determined.

Both structural and simulation studies suggest a key role for M2/S6 helix flexibility in enabling gating of K channels. The intrinsic flexibility of TM helices may be studied by simulation of such helices in isolation, embedded either in a lipid bilayer or a bilayer mimetic octane slab (Bright and Sansom, 2003; Shrivastava et al., 2000; Tieleman et al., 2001b). Such simulations have been used to define the flexibility possible about the molecular hinge(s) within the S6 helix of Kv channels, both in isolation (Bright et al., 2002) and as part of a homology model of the pore-forming domain (Beckstein et al., 2003; Bright and Sansom, 2004). More recently it has been possible to compare these motions with those seen in long (~10 ns) simulations of the TM domain of the KirBac channel (A. Grottesi et al., unpublished data). Simulations of both KirBac and the model TM domain of Kv reveal kinked geometries of the M2/S6 helix that are consistent with an open form of the channel. Significantly, these simulation-derived structures for M2 and S6 are comparable to the degree of kink observed in the crystal structures of M2 and S6 in MthK and KvAP, respectively (Fig. 6).

It is important to examine the relationship of these simulation results to channel gating, especially given the difference in timescales noted above. It would seem that simulations of the isolated TM domain of K channels, especially in the lower viscosity environment provided by an octane slab, can reveal aspects of the intrinsic short-timescale flexibility of the pore-lining helices that are associated with channel gating. However, in the intact channel, these motions are restricted by the presence of either intracellular regulatory domains (in KcsA and Kir) or an outer ring of voltage sensor domains (in Kv) channels. One may think of this in terms of a channel gate per se, which is regulated by a gatekeeper domain. Thus, to understand the overall mechanism of gating we must understand the conformational dynamics of the gate, the gatekeeper, and the coupling mechanism between these two domains.

The conformational dynamics of a gatekeeper domain may be approached via direct simulation of that domain. For example, the crystal structure of the intracellular domain of Kir3.1 (Nishida and MacKinnon, 2002) reveals a stable tetrameric assembly around a central pore. Simulations of the Kir3.1 intracellular domain tetramer reveal "symmetry-breaking" fluctuations in the packing of the subunits in a dimer-of-dimers fashion (Haider et al., in press). Such motions of the gatekeeper domain may be related to the mechanism of channel gating. Dimer-of-dimers structures have been implicated in the gating mechanisms of Kir channels (Kuo et al., 2003) and of HCN channels (Ulens and Siegelbaum, 2003). Furthermore, recent simulations of a homology model of the

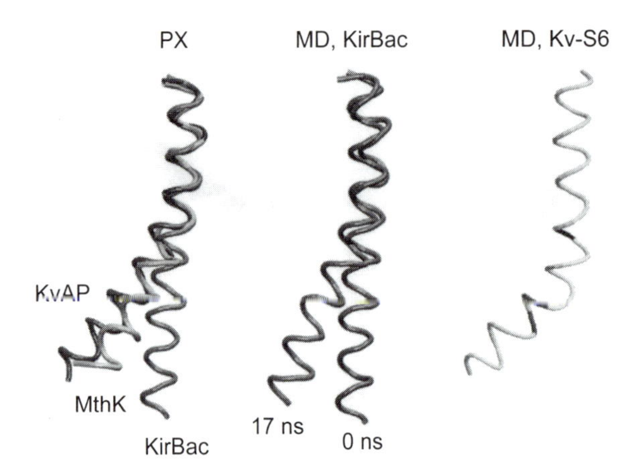

Figure 6. Kinking of the pore-lining M2 or S6 helix of K channels associated with channel activation. PX compares the M2 helices from the crystal structures of KirBac (closed) and MthK (open) with the S6 helix from KvAP. MD, KirBac shows selected structures of M2 helices from a simulation of KirBac (Domene et al., 2004). MD, Kv-S6 shows a selected structure of the S6 helix from a simulation of a model of the pore domain of the Shaker Kv channel (Bright and Sansom, 2004).

ligand-binding domain of the nicotinic acetylcholine receptor suggest that symmetry-breaking motions may be a more general feature of channel gating mechanisms (Henchman et al., 2003).

Channel-Membrane Interactions

The interactions between membrane proteins and their lipid bilayer environment play important roles in the stability and function of such proteins. Extended (>10 ns) MD simulations have been used to explore the interactions of KcsA with phosphatidylcholine bilayers (Deol et al., 2004; Domene et al., 2003b). These simulations enable analysis of a number of aspects of lipid-protein interactions, including the interactions of aromatic amphipathic side chains (i.e., Trp and Tyr) with lipid headgroups, and "snorkeling" interactions of basic side chains (i.e., Lys and Arg) with lipid phosphate groups. There are two clear bands of interacting residues on the surface of KcsA. A large number of Arg-phosphate interactions are seen for KcsA. Overall, these simulations present a dynamic picture of nonspecific lipid-channel interactions. More recent studies (S. S. Deol, C. Domene, and M. S. P. Sansom, unpublished results) have also revealed the details of specific interactions of lipid molecules with KcsA that are required for structural and functional integrity of the channel (Demmers et al., 2003; Valiyaveetil et al., 2002).

Homology Modeling

As discussed above, modeling plays a key role in attempts to extrapolate from bacterial channel structure to mammalian channel function (Capener et al., 2002). Considerations of space prevent us from undertaking a complete survey of all homology models of K channels. Instead, we will focus on a single example, namely, modeling of mammalian Kir channels.

Before considering the results of homology modeling studies, it is useful to consider the methodology in a little more detail. The main steps used to build a homology model are as follows:

1. Sequence alignment of target sequence onto a suitable template structure.

2. Identification of TM helices both by means of prediction methods and key residues from analysis of conservation in multiple sequence alignments.
3. Threading the sequence onto the structural template and optimizing the conformations of side chains. This is usually done using a homology modeling program such as Modeller (Sali and Blundell, 1993) (also see http://salilab.org/modeller/modeller.html).
4. Evaluation and exploration of the resultant model using structural bioinformatics and MD simulation-based methods.

The first step in this procedure is crucial, as the quality of the sequence alignment dominates the resultant model. In evaluating the model, considerations from the structural bioinformatics of membrane proteins can be useful. For example, one may consider the presence or absence of bands of Trp and Tyr side chains on the surface of the TM domain of a model at the interface between the lipid bilayer and the surrounding water. Such bands of side chains are frequently found on the surface of membrane proteins and play a key role in anchoring them into a bilayer (de Planque and Killian, 2003; Schiffer et al., 1992; Yau et al., 1998), as has been revealed by MD simulations of membrane proteins (Domene et al., 2003b; Petrache et al., 2002). Once one is satisfied with a given model from a structural bioinformatics perspective, its conformational stability in a lipid bilayer or bilayer-mimetic environment may be assessed via MD simulations.

Kir6.2 Channels

Homology modeling has been used by a number of investigators to study mammalian Kir channels, both from a perspective of aiding experimental studies (Loussouarn et al., 1999; Loussouarn et al., 2000) and exploring the conformational dynamics of Kir channels in relationship to biological function. In particular, Kir6.2 has been intensively studied. Kir6.2 is a component of K_{ATP}, which is regulated by intracellular ATP and ADP (Trapp et al., 1998; Tucker et al., 1998) and controls, for example, insulin release from pancreatic β-cells.

The first generation of models of the Kir6.2 TM domain used KcsA as a structural template. These studies were among the first to use MD simulations to assess the conformational stability of a channel homology model (Capener et al., 2000) and to explore the intrinsic flexibility of the selectivity filter in that model (Capener and Sansom, 2002). The model was subsequently used to aid interpretation of the effects of mutations in the vicinity of the selectivity filter on channel conductance and on intraburst gating kinetics (Proks et al., 2001). More extensive simulations of some of the mutants suggested that their effects on conductance could be correlated with changes in the local conformation of the selectivity filter of the channel (Capener et al., 2003).

More recently, direct structural information on Kir channels has become available (see above) with the crystal structures of the intact KirBac molecule and of the intracellular domain of Kir3.1. The Kir3.1 intracellular domain is highly homologous with that of Kir6.2 and so has been used to construct a homology model of the Kir6.2 intracellular domain. This model has been used to aid interpretation of mutational data (Trapp et al., 2003) relating to ATP binding to Kir6.2. More recently, simulations of the Kir6.2 intracellular domain model have revealed a pattern of dimer-of-dimers motion similar to that observed in simulations of the Kir3.1 intracellular domain (Haider et al., in press) (see above), thus providing possible insights into the mechanism of channel gating. By combining structural data from KirBac and the Kir3.1 intracellular domain, it is possible to build a model of the complete Kir6.2 tetramer, thus enabling simulations that will provide insights into the interactions of the gate and the gatekeeper domains (S. Haider and M. S. P. Sansom, unpublished results).

GLUTAMATE RECEPTORS

The Structure of GluRs

As mentioned above, GluRs share a distant homology in their TM domains with K channels. Ionotropic glutamate receptors (iGluRs) are ligand-gated ion channels that mediate fast excitatory neurotransmission in the brain. Glutamate is released from the presynaptic neuron, diffuses across the synaptic cleft, and then binds to a GluR on the postsynaptic membrane. This binding event leads to a change in conformation within the receptor such that cations flow through it across the membrane. iGluRs have been classified according to their sequences and the pharmacology of their responses to various ligands (Hollmann et al., 1994). Those receptors (GluR1-4) that show the greatest response to α-amino-3-hydroxy-5-methyl-4-isoxazole propionic acid (AMPA) are termed AMPA receptors (Borges and Dingledine, 1998). Similarly, those receptors that show the greatest sensitivity to kainate (gluR5-7, KA 1-2) are referred to as kainate receptors. Finally, receptors that are activated by *N*-methyl-D-aspartate (NMDAR1, NMDAR2a-d) are called NMDA receptors. In vivo, NMDA receptors require both glutamate and glycine to bind for activation.

The overall architecture of an iGluR subunit is complex (Color Plate 10A [see color insert]), containing two extracellular domains and a TM domain. The complete channel is formed by a tetramer of such subunits. The first extracellular domain (the amino-terminal domain) is thought to play a role in subtype-specific interactions and shares ~16% identity with the bacterial leucine/isoleucine/valine-binding protein. The second extracellular domain is the ligand-binding domain, for which several crystallographic structures now exist (Armstrong and Gouaux, 2000; Armstrong et al., 1998; Hogner et al., 2003; Hogner et al., 2002; Jin et al., 2003; Jin and Gouaux, 2003; Jin et al., 2002; Kasper et al., 2002; Lunn et al., 2003; Sun et al., 2002). The ligand-binding domain shares a common fold with the periplasmic binding protein family (Wilkinson and Verschueren, 2003), which includes the glutamine-binding protein (GlnBP) and the lysine/arginine/ornithine-binding protein (LAOBP). In both the bacterial-binding proteins and the ligand-binding domain of mammalian GluRs, the ligand-binding pocket is located between two lobes or subdomains (D1 and D2 in Color Plate 10C). In the case of the GluRs, the ligand-binding domain is formed by a discontinuous polypeptide chain, with part of the TM domain inserted in the middle of the ligand-binding domain sequence. Sequence comparisons (Fig. 2) have revealed a similarity between the TM domains of GluRs and K channels (Kuner et al., 2003; Wo and Oswald, 1995; Wood et al., 1995). This suggests that the topology of the glutamate receptor TM domain resembles that of an inverted K channel (Color Plate 10B).

Modeling the TM Domain

Further support for the distant homology of K channels and GluRs was provided by the discovery and identification of a bacterial glutamate receptor, GluR0 (Chen et al., 1999). This receptor is a glutamate-gated channel that is selective for potassium ions. There is high sequence identity in the TM region between GluR0 and the bacterial potassium channel, KcsA, especially within the P helix and M2 regions (44% sequence identity). GluR0 even possesses the potassium channel signature sequence, TVGYG. The subsequent identification of bacterial sequences with M2 regions even closer to eukaryotic glutamate receptors makes the link between the two classes of channel rather convincing (Kuner et al., 2003). Molecular simulation has been used to evaluate the conformational stability of a homology model (based on KcsA) of the TM domain of GluR0 (Arinaminpathy et al., 2003). Two alternative homology models of the TM domains of GluR0, differing in whether a sequence insertion was placed either in the P helix or in the M1-P-helix loop, were compared by running 6-ns duration simulations in a membrane-mimetic octane slab. The model with the insertion between the M1 and P helix regions was preferred as it showed lower structural drift and the conformation of the P helix was maintained for the duration of the simulation. Furthermore, this model was also stabilized by the presence of an intersubunit salt bridge, a feature not observed in the alternative model. The simulations also showed concerted ion movement, similar to that reported in numerous K channel simulations (see above).

Analysis of the GluR0 TM domain model and simulations suggested that mobility of residue Ala232 might correlate with changes in the dimensions of the intracellular pore mouth, i.e., the gate in K channels. Residue Ala232 of GluR0 corresponds to Gly99 in KcsA, i.e., the conserved glycine that forms the M2 hinge in K channels. Thus the glycine in K channel M2 helices is not preserved in GluR0. Indeed, there appears not to be an obvious "gating hinge" in the M2 helices of glutamate receptors. In K channels mutating the gating hinge glycine to alanine yields nonfunctional channels (Ding and Horn, 2003; Yifrach and MacKinnon, 2002). In contrast, corresponding mutations in GluR-A were indistinguishable from wild-type channels in their behavior. These results, combined with the fact that in some subunits there is no glycine at all in the M2 segment, led to the suggestion that during gating of GluRs the M2 helix may behave more like a rigid rod (Sobolevsky et al., 2003, 2004).

Attempts to extend homology modeling to the TM domain of mammalian GluRs have proved more challenging. Mammalian iGluRs do not possess the canonical TVGYG motif. For example, in the case of GluR2, the corresponding sequence is QQGCD. This suggests a substantial difference in the conformation of the filter region in mammalian GluRs. Indeed, MD simulation of GluR2 homology models based on KcsA leads to rapid structural collapse of the filter (Y. Arinaminpathy, P. C. Biggin, and M. S. P. Sansom, unpublished data). Thus the GluRs provide examples of both the strength (GluR0) and limitations (GluR2) of homology modeling applied to channel TM domains. A further problem in GluR TM domain modeling is how to include the additional TM helix (M3 in Color Plate 10).

There is considerable evidence that the overall stoichiometry of GluRs is tetrameric (Rosenmund et al., 1998; Safferling et al., 2001). However, there is growing evidence that the structural symmetry, at least of the ligand-binding domain, may be better represented by a dimer-of-dimers (Armstrong and Gouaux, 2000). As discussed above, this may also be the case for Kir intracellular domains (as is seen in the X-ray structure of KirBac). This raises the interesting question of the consequences, in terms of conformational dynamics, of a switch in channel symmetry from tetrameric in the TM domain to dimer-of-dimers in the extramembraneous domain.

Dynamics of the Ligand-Binding Domain

Structures

Several structures of GluR ligand-binding domains are known (for GluR2, the NMDA receptor NR1 subunit, and GluR0) (see Gouaux [2003] and Mayer and Armstrong [2004] for reviews of the structural studies). These proteins have been crystallized without bound ligands, with bound agonists (both full and partial), and with bound antagonists. Comparison of the structures shows differences in the degree of closure of the two subdomains, D1 and D2. This has led to formulation of a hypothesis in which activation of the receptor was associated with closure of D1 against D2 around the agonist bound at the cleft between the two subdomains. Partial activation (e.g., of GluR2 by the partial agonist kainate) was associated with incomplete closure, while antagonists such as 6,7-dinitro-2,3-quinoxalinedione in the case of GluR2 and 5,7-dichlorokynurenic acid were proposed to hold the D1 and D2 subdomains apart in an "open cleft" conformation (Armstrong and Gouaux, 2000; Furukawa and Gouaux, 2003). This static structural model was complicated by simulation studies (Arinaminpathy et al., 2002) that suggested a more dynamic model in which the resting-state receptor was in a dynamic equilibrium between open-cleft and closed-cleft states, with agonist binding leading to preferential stabilization of the closed-cleft state and consequent reduction of intersubdomain mobility. Subsequent structural studies of GluR2 ligand-binding domains resulted in a still more complex model, with differences in conformation and intersubunit interactions between the resting, activated, and desensitized states (Sun et al., 2002).

GluR0

Structural studies of the ligand-binding domain of the bacterial homolog GluR0 suggested that the equilibrium between the activated (D1-D2 cleft closed) and resting (D1-D2 cleft open) states is finely balanced and that crystal packing forces may bias the sampling of this equilibrium. For example, of the series of structures solved (Mayer et al., 2001), one corresponded to a D1-D2 cleft-closed conformation even though no ligand was bound to the protein. This supports the model of a preexisting equilibrium between cleft-open and cleft-closed conformations that may be perturbed by ligand binding. Indeed, such equilibria have been discussed in a more general sense for a number of proteins (Goh et al., 2004) in the context of protein-protein (rather than protein-ligand) interactions.

Extended (20 ns) MD simulations of the ligand-binding domain of GluR0 have provided direct support for this model (Y. Arinaminpathy, P. C. Biggin, and M. S. P. Sansom, unpublished results). Simulations of the apo (i.e., no ligand-bound) state of GluR0 reveal fluctuations in the intersubdomain (i.e., D1 to D2) distance on an ~5-ns timescale. Other simulations of GluR0 have revealed a direct cleft-opening event even in the presence of glutamate (Color Plate 11 [see color insert]). It should be noted that these simulations are probing the intrinsic flexibility of the ligand-binding domain and its modulation by ligands on an ~10-ns timescale. Of course, this does not imply that activation of the receptor channel takes place on this timescale. The overall process of modulation of the cleft-open ↔ cleft-closed equilibrium by agonist binding and of coupling this to the gate-closed ↔ gate-open equilibrium of the channel will doubtless be a much slower process. However, this study illustrates how molecular simulations enable us to dissect out the elements of a complex gating process.

Comparative MD Simulations

One way to test the robustness of a model of conformational change derived from MD simulations of proteins is to see whether the conformational dynamics behavior is conserved across different members of the protein family. For the GluR simulations, this has been possible because of the presence of a similar architecture and ligand-induced conformational change in the bacterial periplasmic binding proteins. To this end, we have performed extended multiple simulations of two periplasmic binding proteins, GlnBP and LAOBP, and compared their conformational dynamics with those of the ligand-binding domains of GluR2 and of GluR0.

Multinanosecond MD simulations on GlnBP (Pang et al., 2003) revealed a complex intersubdomain motion that was modulated by ligand (Gln) binding to the protein. Principal component analysis (also known as essential dynamics) (Amadei et al., 1993) enabled us to dissect out the molecular motions into two main components, namely, (i) a hinge-bending motion (as discussed above in the context of GluR structures) and (ii) a twisting motion, whereby one lobe twists with respect to the other. (The latter motion may be relevant in higher-order complexes such as in the tetrameric arrangement of the GluRs, where the interface between monomers has been implicated in desensitization kinetics.) Subsequent simulation studies (Y. Arinaminpathy, A. Pang, M. S. P. Sansom, and P. C. Biggin, unpublished observations) extended this analysis to LAOBP, revealing similar principal motions. Comparison of the hinge-bending motion in particular shows that this is similar for GlnBP, LAOBP, GluR2, and GluR0 (Color Plate 12 [see color insert]). This suggests that the cleft-open ↔ cleft-closed equilibrium is an intrinsic property of the binding protein/GluR fold that has been exploited by evolution both in transport systems (using ligand-binding proteins) and in GluR activation.

CONCLUSIONS

It can be seen that molecular modeling and simulations can contribute to studies of ion channels in two respects. Modeling studies enable extrapolation from experimental structures of prokaryotic ion channels to molecular models of eukaryotic homologs, thus aiding design and interpretation of, for example, mutation experiments for dissecting structure-function relationships. Ion channel structures and ion channel models may also be used as the basis of multinanosecond MD simulations. Such simulations can be used to explore a number of aspects of channel function, including interactions with ions, conformational mobility in relationship to ion channel gating, and interactions of channel proteins with their membrane environment.

Structural and simulation studies are starting to provide insights into channel gating mechanisms. In particular, it would seem the breaking of strict tetrahedral symmetry to give local dimer-of-dimers symmetry may be a common feature of the gating mechanisms of both K channels and GluRs. This may reflect the similar folds of the TM channel domains of these two classes of protein.

It is valuable to reflect on future applications of computational methods to channel proteins and related systems. As more structures of channels and of channel fragments and domains emerge, molecular modeling will play an increasingly important role in allowing extrapolation of structural data to systems of biomedical importance. From a simulation perspective, there is a need to develop methods that will enable us to simulate longer timescales and larger-scale conformational changes in channel proteins. Finally, it will become increasingly important to run multiple simulations on multiple channels to allow comparative analysis of simulation results, which in turn will enable the formulation of more general hypotheses concerning the relationship between the conformational dynamics of channel proteins and their physiological functions.

REFERENCES

Allen, T. W., A. Bliznyuk, A. P. Rendell, S. Kuyucak, and S. H. Chung. 2000. The potassium channel: structure, selectivity and diffusion. *J. Chem. Phys.* **112:**8191–8204.

Allen, T. W., and S. H. Chung. 2001. Brownian dynamics study of an open-state KcsA potassium channel. *Biochim. Biophys. Acta* **1515:**83–91.

Allen, T. W., S. Kuyucak, and S. H. Chung. 1999. Molecular dynamics study of the KcsA potassium channel. *Biophys. J.* **77:**2502–2516.

Amadei, A., A. B. M. Linssen, and H. J. C. Berendsen. 1993. Essential dynamics of proteins. *Proteins.* **17:** 412–425.

Åqvist, J., and V. Luzhkov. 2000. Ion permeation mechanism of the potassium channel. *Nature* **404:**881–884.

Arinaminpathy, Y., P. C. Biggin, I. H. Shrivastava, and M. S. P. Sansom. 2003. A prokaryotic glutamate receptor: homology modelling and molecular dynamics simulations of GluR0. *FEBS Lett.* **553:**321–327.

Arinaminpathy, Y., M. S. P. Sansom, and P. C. Biggin. 2002. Molecular dynamics simulations of the ligand binding domain of the ionotropic glutamate receptor, GluR2. *Biophys. J.* **82:**676–683.

Armstrong, N., and E. Gouaux. 2000. Mechanisms for activation and antagonism of an AMPA-sensitive glutamate receptor: crystal structures of the GluR2 ligand binding core. *Neuron* **28:**165–181.

Armstrong, N., Y. Sun, G.-Q. Chen, and E. Gouaux. 1998. Structure of a glutamate-receptor ligand-binding core in complex with kainate. *Nature* **395:**913–917.

Beckstein, O., P. C. Biggin, P. J. Bond, J. N. Bright, C. Domene, A. Grottesi, J. Holyoake, and M. S. P. Sansom. 2003. Ion channel gating: insights via molecular simulations. *FEBS Lett.* **555:**85–90.

Beckstein, O., P. C. Biggin, and M. S. P. Sansom. 2001. A hydrophobic gating mechanism for nanopores. *J. Phys. Chem. B* **105:**12902–12905.

Beckstein, O., and M. S. P. Sansom. 2004. The influence of geometry, surface character and flexibility on the permeation of ions and water through biological pores. *Phys. Biol.* **1:**42–52.

Bernèche, S., and B. Roux. 2000. Molecular dynamics of the KcsA K^+ channel in a bilayer membrane. *Biophys. J.* **78:**2900–2917.

Bernèche, S., and B. Roux. 2001. Energetics of ion conduction through the K^+ channel. *Nature* **414:**73–77.

Bernèche, S., and B. Roux. 2003. A microscopic view of ion conduction through the K^+ channel. *Proc. Natl. Acad. Sci. USA* **100:**8644–8648.

Biggin, P. C., and M. S. P. Sansom. 2002. Open-state models of a potassium channel. *Biophys. J.* **83:**1867–1876.

Biggin, P. C., G. R. Smith, I. H. Shrivastava, S. Choe, and M. S. P. Sansom. 2001. Potassium and sodium ions in a potassium channel studied by molecular dynamics simulations. *Biochim. Biophys. Acta* **1510:**1–9.

Borges, K., and R. Dingledine. 1998. AMPA receptors: molecular and functional diversity. *Prog. Brain Res.* **116:**153–170.

Bright, J. N., and M. S. P. Sansom. 2003. The flexing/twirling helix: exploring the flexibility about molecular hinges formed by proline and glycine motifs in transmembrane helices. *J. Phys. Chem. B* **107:**627–636.

Bright, J. N., and M. S. P. Sansom. 2004. The Kv channel S6 helix as a molecular switch: simulation studies. *IEE Proc. Nanobiotechnol.* **151**:17–27.

Bright, J. N., I. H. Shrivastava, F. S. Cordes, and M. S. P. Sansom. 2002. Conformational dynamics of helix S6 from Shaker potassium channel: simulation studies. *Biopolymers* **64:**303–313.

Capener, C. E., H. J. Kim, Y. Arinaminpathy, and M. S. P. Sansom. 2002. Ion channels: structural bioinformatics and modelling. *Hum. Mol. Genet.* **11:**2425–2433.

Capener, C. E., P. Proks, F. M. Ashcroft, and M. S. P. Sansom. 2003. Filter flexibility in a mammalian K channel: models and simulations of Kir6.2 mutants. *Biophys. J.* **84:**2345–2356.

Capener, C. E., and M. S. P. Sansom. 2002. MD simulations of a K channel model—sensitivity to changes in ions, waters and membrane environment. *J. Phys. Chem. B* **106:**4543–4551.

Capener, C. E., I. H. Shrivastava, K. M. Ranatunga, L. R. Forrest, G. R. Smith, and M. S. P. Sansom. 2000. Homology modelling and molecular dynamics simulation studies of an inward rectifier potassium channel. *Biophys. J.* **78:**2929–2942.

Chen, G. Q., C. H. Cui, M. L. Mayer, and E. Gouaux. 1999. Functional characterization of a potassium-selective prokaryotic glutamate receptor. *Nature* **402:**817–821.

Chung, S. H., and S. Kuyucak. 2002. Ion channels: recent progress and prospects. *Eur. Biophys. J.* **31:**283–293.

Compoint, M., P. Carloni, C. Ramseyer, and C. Girardet. 2004. Molecular dynamics study of the KcsA channel at 2.0-Å resolution: stability and concerted motions within the pore. *Biochim. Biophys. Acta* **1661:**26–39.

de Groot, B. L., and H. Grubmuller. 2001. Water permeation across biological membranes: mechanism and dynamics of aquaporin-1 and GlpF. *Science* **294:**2353–2357.

Demmers, J. A. A., A. van Dalen, B. de Kruijiff, A. J. R. Heck, and J. A. Killian. 2003. Interaction of K channel KcsA with membrane phospholipids as studied by ESI mass spectrometry. *FEBS Lett.* **541:**28–32.

Deol, S. S., P. J. Bond, C. Domene, and M. S. P. Sansom. 2004. Lipid-protein interactions of integral membrane proteins: a comparative simulation study. *Biophys. J.* **87:**3737–3749.

de Planque, M. R. R., and J. A. Killian. 2003. Protein-lipid interactions studied with designed transmembrane peptides: role of hydrophobic matching and interfacial anchoring. *Mol. Membr. Biol.* **20:**271–284.

Ding, S., and R. Horn. 2003. Glycine scan of Shaker's gating hinge. *Biophys. J.* **84:**73A.

Domene, C., P. Bond, and M. S. P. Sansom. 2003a. Membrane protein simulation: ion channels and bacterial outer membrane proteins. *Adv. Protein Chem.* **66:**159–193.

Domene, C., P. J. Bond, S. S. Deol, and M. S. P. Sansom. 2003b. Lipid-protein interactions and the membrane/water interfacial region. *J. Am. Chem. Soc.* **125:**14966–14967.

Domene, C., A. Grottesi, and M. S. P. Sansom. 2004. Filter flexibility and distortion in a bacterial inward rectifier K+ channel: simulation studies of KirBac1.1. *Biophys. J.* **87:**256–267.

Domene, C., and M. S. P. Sansom. 2003. A potassium channel, ions and water: simulation studies based on the high resolution X-ray structure of KcsA. *Biophys. J.* **85:**2787–2800.

Doyle, D. A., J. M. Cabral, R. A. Pfuetzner, A. Kuo, J. M. Gulbis, S. L. Cohen, B. T. Cahit, and R. MacKinnon. 1998. The structure of the potassium channel: molecular basis of K^+ conduction and selectivity. *Science* **280:**69–77.

Fiser, A., R. K. G. Do, and A. Sali. 2000. Modeling of loops in protein structures. *Protein Sci.* **9:**1753–1773.

Furukawa, H., and E. Gouaux. 2003. Mechanisms of activation, inhibition and specificity: crystal structures of the NMDA receptor NR1 ligand-binding core. *EMBO J.* **22:**2873–2885.

Garofoli, S., and P. C. Jordan. 2003. Modelling permeation energetics in the KcsA potassium channel. *Biophys. J.* **84:**2814–2830.

Giorgetti, A., and P. Carloni. 2003. Molecular modeling of ion channels: structural predictions. *Curr. Opin. Chem. Biol.* **7:**150–156.

Goh, C. S., D. Milburn, and M. Gerstein. 2004. Conformational changes associated with protein-protein interactions. *Curr. Opin. Struct. Biol.* **14:**104–109.

Gouaux, E. 2003. Structure and function of AMPA receptors. *J. Physiol.* **554.2:**249–253.

Guidoni, L., and P. Carloni. 2002. Potassium permeation through the KcsA channel: a density functional study. *Biochim. Biophys. Acta* **1563:**1–6.

Guidoni, L., V. Torre, and P. Carloni. 1999. Potassium and sodium binding in the outer mouth of the K^+ channel. *Biochemistry* **38:**8599–8604.

Guidoni, L., V. Torre, and P. Carloni. 2000. Water and potassium dynamics in the KcsA K^+ channel. *FEBS Lett.* **477:**37–42.

Haider, S., A. Grottesi, F. M. Ashcroft, and M. S. P. Sansom. Conformational dynamics of the ligand-binding domain of inward rectifier K channels as revealed by MD simulations: towards an understanding of Kir channel gating. *Biophys. J.*, in press.

Henchman, R. H., H. L. Wang, S. M. Sine, P. Taylor, and J. A. McCammon. 2003. Asymmetric structural motions of the homomeric a7 nicotinic receptor ligand binding domain revealed by molecular dynamics simulation. *Biophys. J.* **85:**3007–3018.

Hille, B. 2001. *Ionic Channels of Excitable Membranes*. Sinauer Associates, Inc. Sunderland, Mass.

Hogner, A., J. R. Greenwood, T. Liljefors, M.-L. Lunn, J. Egebjerg, I. K. Larsen, E. Gouaux, and J. S. Kastrup. 2003. Competitive antagonism of AMPA receptors by ligands of different classes: crystal structure of ATPO bound to the GluR2 ligand-binding core, in comparison with DNQX. *J. Med. Chem.* **46:**214–221.

Hogner, A., J. S. Kastrup, R. Jin, T. Liljefors, M. L. Mayer, J. Egebjerg, I. K. Larsen, and E. Gouaux. 2002. Structural basis for AMPA receptor activation and ligand selectivity: crystal structures of five agonist complexes with the GluR2 ligand-binding core. *J. Mol. Biol.* **322:**93–109.

Hollmann, M., C. Maron, and S. Heinemann. 1994. *N*-glycosylation site tagging suggests a three transmembrane domain topology for the glutamate receptor GluR1. *Neuron* **13:**1331–1343.

Holyoake, J., C. Domene, J. N. Bright, and M. S. P. Sansom. 2003. KcsA closed and open: modelling and simulation studies. *Eur. Biophys. J.* **33:**238–246.

Jensen, M. O., E. Tajkhorshid, and K. Schulten. 2001. The mechanism of glycerol conduction in aquaglyceroporins. *Structure* **9:**1083–1093.

Jiang, Y., A. Lee, J. Chen, M. Cadene, B. T. Chait, and R. MacKinnon. 2002a. Crystal structure and mechanism of a calcium-gated potassium channel. *Nature* **417:**515–522.

Jiang, Y., A. Lee, J. Chen, M. Cadene, B. T. Chait, and R. MacKinnon. 2002b. The open pore conformation of potassium channels. *Nature* **417:**523–526.

Jiang, Y., A. Lee, J. Chen, V. Ruta, M. Cadene, B. T. Chait, and R. Mackinnon. 2003. X-ray structure of a voltage-dependent K^+ channel. *Nature* **423:**33–41.

Jiang, Y. X., A. Pico, M. Cadene, B. T. Chait, and R. MacKinnon. 2001. Structure of the RCK domain from the *E. coli* K+ channel and demonstration of its presence in the human BK channel. *Neuron* **29:**593–601.

Jin, R., T. G. Banke, M. L. Mayer, S. F. Traynelis, and E. Gouaux. 2003. Structural basis for partial agonist action at ionotropic glutamate receptors. *Nat. Neurosci.* **6:**803–810.

Jin, R., and E. Gouaux. 2003. Probing the function, conformational plasticity, and dimer-dimer contacts of the GluR2 ligand-binding core: studies of 5-substituted willardines and gluR2 S1S2 in the crystal. *Biochemistry* **42:**5201–5213.

Jin, R., M. Horning, M. L. Mayer, and E. Gouaux. 2002. Mechanism of activation and selectivity in a ligand-gated ion channel: structural and functional studies of GluR2 and quisqualate. *Biochemistry* **41:**15635–15643.

Jordan, P. 2000. Ionic energetics in narrow channels. *In* H. E. Layton and A. M. Weinstein (ed.), *Membrane Transport and Renal Physiology*. Springer-Verlag, NewYork, N.Y.

Karplus, M. J., and J. A. McCammon. 2002. Molecular dynamics simulations of biomolecules. *Nat. Struct. Biol.* **9:**646–652.

Kasper, C., M.-L. Lunn, T. Liljefors, E. Gouaux, J. Egebjerg, and J. S. Kastrup. 2002. GluR2 ligand-binding core complexes: importance of the isoxazolol moiety and 5-substituent for the binding mode of AMPA-type agonists. *FEBS Letters* **531:**173–178.

Kuner, T., P. H. Seeburg, and H. R. Guy. 2003. A common architecture for K+ channels and ionotropic glutamate receptors? *Trends Neurosci.* **26:**27–32.

Kuo, A., J. M. Gulbis, J. F. Antcliff, T. Rahman, E. D. Lowe, J. Zimmer, J. Cuthbertson, F. M. Ashcroft, T. Ezaki, and D. A. Doyle. 2003. Crystal structure of the potassium channel KirBac1.1 in the closed state. *Science* **300:**1922–1926.

Liu, Y., P. Sompornpisut, and E. Perozo. 2001. Structure of the KcsA channel intracellular gate in the open state. *Nat. Struct. Biol.* **8:**883–887.

Loussouarn, G., E. N. Amkhina, and C. G. Nichols. 1999. Structure of the second transmembrane domain of Kir6.2 revealed by the substituted cysteine accessibility method. *Biophys. J.* **76:**A75.

Loussouarn, G., E. N. Makhina, T. Rose, and C. G. Nichols. 2000. Structure and dynamics of the pore of inwardly rectifying KATP channels. *J. Biol. Chem.* **275:**1137–1144.

Lunn, M.-L., A. Hogner, T. B. Stensbøl, E. Gouaux, J. Egebjerg, and J. S. Kastrup. 2003. Three-dimensional structure of the ligand-binding core of GluR2 in complex with the agonist (*S*)-ATPA: implications for receptor subunit selectivity. *J. Med. Chem.* **46:**872–875.

Luzhkov, V. B., and J. Åqvist. 2000. A computational study of ion binding and protonation states in the KcsA potassium channel. *Biochim. Biophys. Acta* **1481:**360–370.

Luzhkov, V. B., and J. Åqvist. 2001. Mechanisms of tetraethylammonium ion block in the KcsA potassium channel. *FEBS Lett.* **495:**191–196.

MacKinnon, R. 2003. Potassium channels. *FEBS Lett.* **555:**62–65.

Mashl, R. J., Y. Z. Tang, J. Schnitzer, and E. Jakobsson. 2001. Hierarchical approach to predicting permeation in ion channels. *Biophys. J.* **81:**2473–2483.

Mayer, M. L., and N. Armstrong. 2004. Structure and function of glutamate receptor ion channels. *Annu. Rev. Physiol.* **66:**161–181.

Mayer, M. L., R. Olson, and E. Gouaux. 2001. Mechanisms for ligand binding to GluR0 ion channels: crystal structures of the glutamate and serine complexes and a closed apo state. *J. Mol. Biol.* **311:**815–836.

Monticelli, L., K. M. Robertson, J. L. MacCallum, and D. P. Tieleman. 2004. Computer simulation of the KvAP voltage-gated potassium channel: steered molecular dynamics of the voltage sensor. *FEBS Lett.* **564:**325–332.

Morais-Cabral, J. H., Y. Zhou, and R. MacKinnon. 2001. Energetic optimization of ion conduction by the K^+ selectvity filter. *Nature* **414:**37–42.

Nishida, M., and R. MacKinnon. 2002. Structural basis of inward rectification: cytoplasmic pore of the G protein-gated inward rectifier GIRK1 at 1.8 Å resolution. *Cell* **111:**957–965.

Pang, A., Y. Arinaminpathy, M. S. P. Sansom, and P. C. Biggin. 2003. Interdomain dynamics and ligand binding: molecular dynamics simulations of glutamine binding protein. *FEBS Lett.* **550:**168–174.

Perozo, E., D. M. Cortes, and L. G. Cuello. 1998. Three-dimensional architecture and gating mechanism of a K^+ channel studied by EPR spectroscopy. *Nat. Struct. Biol.* **5:**459–469.

Perozo, E., D. M. Cortes, and L. G. Cuello. 1999. Structural rearrangements underlying K^+-channel activation gating. *Science* **285:**73–78.

Petrache, H. I., D. M. Zuckerman, J. N. Sachs, J. A. Killian, R. E. Koeppe, and T. B. Woolf. 2002. Hydrophobic matching mechanism investigated by molecular dynamics simulations. *Langmuir* **18:**1340–1351.

Proks, P., C. E. Capener, P. Jones, and F. Ashcroft. 2001. Mutations within the P-loop of Kir6.2 modulate the intraburst kinetics of the ATP-sensitive potassium channel. *J. Gen. Physiol.* **118:**341–353.

Ranatunga, K. M., I. H. Shrivastava, G. R. Smith, and M. S. P. Sansom. 2001a. Sidechain ionisation states in a potassium channel. *Biophys. J.* **80:**1210–1219.

Ranatunga, K. M., G. R. Smith, R. J. Law, and M. S. P. Sansom. 2001b. Electrostatics and molecular dynamics of a homology model of the *Shaker* K^+ channel pore. *Eur. Biophys. J.* **30:**295–303.

Reimann, F., and F. M. Ashcroft. 1999. Inwardly rectifying potassium channels. *Curr. Opin. Cell Biol.* **11:**503–508.

Rosenmund, C., Y. Stern-Bach, and C. F. Stevens. 1998. The tetrameric structure of a glutamate receptor channel. *Science* **280:**1596–1599.

Roux, B. 2002. Theoretical and computational models of ion channels. *Curr. Opin. Struct. Biol.* **12:**182–189.

Roux, B., S. Bernèche, and W. Im. 2000. Ion channels, permeation and electrostatics: insight into the function of KcsA. *Biochemistry* **39:**13295–13306.

Safferling, M., W. Tichelaar, G. Kummerle, A. Jouppila, A. Kuusinen, K. Keinanen, and D. R. Madden. 2001. First images of a glutamate receptor ion channel: oligomeric state and molecular dimensions of GluRB homomers. *Biochemistry* **40:**13948–13953.

Sali, A., and T. L. Blundell. 1993. Comparative protein modelling by satisfaction of spatial restraints. *J. Mol. Biol.* **234:**779–815.

Sansom, M. S. P., I. H. Shrivastava, J. N. Bright, J. Tate, C. E. Capener, and P. C. Biggin. 2002. Potassium channels: structures, models, simulations. *Biochim. Biophys. Acta* **1565:**294–307.

Sansom, M. S. P., I. H. Shrivastava, K. M. Ranatunga, and G. R. Smith. 2000. Simulations of ion channels—watching ions and water move. *Trends Biochem. Sci.* **25:**368–374.

Schiffer, M., C. H. Chang, and F. J. Stevens. 1992. The functions of tryptophan residues in membrane proteins. *Protein Eng.* **5:**213–214.

Shealy, R. T., A. D. Murphy, R. Ramarathnam, E. Jakobsson, and S. Subramaniam. 2003. Sequence-function analysis of the K^+-selective family of ion channels using a comprehensive alignment and the KcsA channel structure. *Biophys. J.* **84:**2929–2942.

Shen, Y. F., Y. F. Kong, and J. P. Ma. 2002. Intrinsic flexibility and gating mechanism of the potassium channel KcsA. *Proc. Natl. Acad. Sci. USA* **99:**1949–1953.

Shrivastava, I. H., C. Capener, L. R. Forrest, and M. S. P. Sansom. 2000. Structure and dynamics of K^+ channel pore-lining helices: a comparative simulation study. *Biophys. J.* **78:**79–92.

Shrivastava, I. H., and M. S. P. Sansom. 2000. Simulations of ion permeation through a potassium channel: molecular dynamics of KcsA in a phospholipid bilayer. *Biophys. J.* **78:**557–570.

Shrivastava, I. H., and M. S. P. Sansom. 2002. Molecular dynamics simulations and KcsA channel gating. *Eur. Biophys. J.* **31:**207–216.

Sobolevsky, A. I., M. V. Yelshansky, and L. P. Wollmuth. 2003. Different gating mechanisms in glutamate receptor and K+ channels. *J. Neurosci.* **23:**7559–7568.

Sobolevsky, A. I., M. V. Yelshansky, and L. P. Wollmuth. 2004. The outer pore of the glutamate receptor channel has 2-fold rotational symmetry. *Neuron* **41:**367–378.

Starace, D. M., and F. Bezanilla. 2004. A proton pore in a potassium channel voltage sensor reveals a focused electric field. *Nature* **427:**548–553.

Sun, Y., R. Olson, M. Horning, N. Armstrong, M. L. Mayer, and E. Gouaux. 2002. Mechanism of glutamate receptor desensitization. *Nature* **417:**245–253.

Tajkhorshid, E., P. Nollert, M. O. Jensen, L. J. W. Miercke, J. O'Connell, R. M. Stroud, and K. Schulten. 2002. Control of the selectivity of the aquaporin water channel family by global orientational tuning. *Science* **296:**525–530.

Tieleman, D. P., P. C. Biggin, G. R. Smith, and M. S. P. Sansom. 2001a. Simulation approaches to ion channel structure-function relationships. *Q. Rev. Biophys.* **34:**473–561.

Tieleman, D. P., I. H. Shrivastava, M. B. Ulmschneider, and M. S. P. Sansom. 2001b. Proline-induced hinges in transmembrane helices: possible roles in ion channel gating. *Proteins* **44:**63–72.

Trapp, S., S. Haider, M. S. P. Sansom, F. M. Ashcroft, and P. Jones. 2003. Identification of residues contributing to the ATP binding site of Kir6.2. *EMBO J.* **22:**2903–2912.

Trapp, S., P. Proks, S. J. Tucker, and F. M. Ashcroft. 1998. Molecular analysis of ATP-sensitive K channel gating and implications for channel inhibition by ATP. *J. Gen. Physiol.* **112:**333–349.

Tucker, S. J., F. M. Gribble, P. Proks, S. Trapp, T. J. Ryder, T. Haug, F. Reimann, and F. M. Ashcroft. 1998. Molecular determinants of K_{ATP} channel inhibition by ATP. *EMBO J.* **17:**3290–3296.

Ulens, C., and S. A. Siegelbaum. 2003. Regulation of hyperpolarization-activated HCN channels by cAMP through a gating switch in binding domain symmetry. *Neuron* **40:**959–970.

Valiyaveetil, F. I., Y. Zhou, and R. MacKinnon. 2002. Lipids in the structure, folding and function of the KcsA channel. *Biochemistry* **41:**10771–10777.

Webster, S. M., D. del Camino, J. P. Dekker, and G. Yellen. 2004. Intracellular gate opening in Shaker K^+ channels defined by high-affinity metal bridges. *Nature* **428:**864–868.

Wilkinson, A., and K. H. G. Verschueren. 2003. Crystal structures of periplasmic solute-binding proteins in ABC transport complexes illuminate their function, p. 187–208. *In* B. I. Holland, S. P. C. Cole, K. Kuchler, and C. F. Higgins (ed.), *ABC Proteins: From Bacteria to Man*. Academic Press, London, United Kingdom.

Wo, Z. G., and R. E. Oswald. 1995. Unraveling the modular design of glutamate-gated ion channels. *Trends Neurosci.* **18:**161–168.

Wood, M. W., H. M. A. VanDongen, and A. M. J. VanDongen. 1995. Structural conservation of ion conduction pathways in K channels and glutamate receptors. *Proc. Natl. Acad. Sci. USA* **92:**4882–4886.

Yau, W. M., W. C. Wimley, K. Gawrisch, and S. H. White. 1998. The preference of tryptophan for membrane interfaces. *Biochemuistry* **37:**14713–14718.

Yellen, G. 2002. The voltage-gated potassium channels and their relatives. *Nature* **419:**35–42.

Yifrach, O., and R. MacKinnon. 2002. Energetics of pore opening in a voltage-gated K^+ channel. *Cell* **111:** 231–239.

Zagotta, W. N., N. B. Ollvier, K. D. Black, E. C. Young, R. Olson, and E. Gouaux. 2003. Structural basis for modulation and agonist specificity of HCN pacemaker channels. *Nature* **425:**200–205.

Zhou, Y., and R. MacKinnon. 2003. The occupancy of ions in the K^+ selectivity filter: charge balance and coupling of ion binding to a protein conformational change underlie high conduction rates. *J. Mol. Biol.* **333:**965–975.

Zhou, Y., J. H. Morais-Cabral, A. Kaufman, and R. MacKinnon. 2001. Chemistry of ion coordination and hydration revealed by a K^+ channel-Fab complex at 2.0 Å resolution. *Nature* **414:**43–48.

Zhu, F. Q., E. Tajkhorshid, and K. Schulten. 2001. Molecular dynamics study of aquaporin-1 water channel in a lipid bilayer. *FEBS Lett.* **504:**212–218.

Bacterial Ion Channels and Their Eukaryotic Homologs
Edited by A. Kubalski and B. Martinac

Chapter 9

Towards an Understanding of Membrane Channels

Emad Tajkhorshid, Jordi Cohen, Aleksij Aksimentiev, Marcos Sotomayor, and Klaus Schulten

Membranes are the most conspicuous components of the cell body, and it is estimated that 30% of all human genes code for membrane proteins. These proteins perform many different tasks, ranging from mediating the exchange of materials and information across the membrane to converting between electro-osmotic, mechanical, and chemical energy. Membrane receptors are major sites of action for hormones and neurotransmitters, and, for this reason, they are the most important targets for pharmacological intervention. Membrane channels, on the other hand, are extremely critical for the transmembrane traffic of a wide range of compounds and for the establishment, maintenance, and control of the membrane's electrical properties.

The functions of membrane proteins have been studied for a long time. In particular, membrane channels have been characterized in great detail, e.g., by the celebrated patch-clamp method. Membrane proteins have been resistant to systematic attempts at crystallization. As a result, there are few available membrane protein structures and this situation impedes the discovery of the physical mechanisms underlying their functions. The situation is finally beginning to change, with dramatic consequences for our understanding of membrane processes. Today, 84 membrane protein structures are available. These proteins come from the inner and outer membranes of bacteria, from eukaryotic and mitochondrial membranes, and from the membranes of other organelles (http://www.mpibp-frankfurt.mpg.de/michel/public/memprotstruct.html; http://blanco.biomol.uci.edu/Membrane_Proteins_xtal.html). The recent increase in the rate of success at solving the structures of membrane proteins and their great medical relevance make the study of membrane proteins one of the most exciting areas of biomedicine today.

Molecular modeling has developed over the past decade into the key research methodology that links structural features to the corresponding biological mechanisms. The application of the methodology to membrane proteins has been reviewed in recent publications (Saiz et al., 2002; Roux, 2002; Ash et al., 2004; Roux et al., 2004; Domene et al., 2003; Bond and Sansom, 2004).

Emad Tajkhorshid, Jordi Cohen, Aleksij Aksimentiev, Marcos Sotomayor, and Klaus Schulten • Theoretical and Computational Biophysics Group, Beckman Institute for Advanced Science and Technology, University of Illinois at Urbana-Champaign, Urbana, IL 61801.

During the past decade, our own group completed several modeling projects focused on processes in biological membranes, including studies of pure lipid bilayers (Heller et al., 1993; Zhou and Schulten, 1995; Gullingsrud and Schulten, 2004), glycophorin A transmembrane helices in micelles (Braun et al., 2004), artificial membranes (Sheng et al., 1995), rhodopsin (Saam et al., 2002), bacteriorhodopsin (Tajkhorshid et al., 2000; Hayashi et al., 2001; Hayashi et al., 2003; Hayashi et al., 2004) and the purple membrane (Baudry et al., 2001), phospholipase A_2 (Zhou and Schulten, 1996; Stepaniants et al., 1997), the mechanosensitive channel of large conductance MscL (Gullingsrud et al., 2001; Gullingsrud and Schulten, 2003), the cytochrome bc_1 complex (Izrailev et al., 1999), F_o ATP synthase (Aksimentiev et al., 2004a), light harvesting complex 2 (Damjanović et al., 2002), and other proteins in the photosynthetic unit of purple bacteria as reviewed by Hu et al. (2002). These studies used molecular dynamics (MD) simulations along with other computational methodologies such as quantum chemistry calculations. In the remainder of this chapter we present exemplary studies of the structure-function relationship of four membrane channels of diverse function that illustrate the recent advance in membrane protein modeling: aquaporin water channels, the chloride channel, hemolysin, and the mechanosensitive channel of small conductance. The simulations describe the proteins in integral membrane environments; the main focus of each study is the investigation of the physical mechanisms underlying function.

METHODS

Two obstacles stand in the way of the application of MD to membrane channels: the large size of systems to be simulated and the short timescale to which the method traditionally applies. Fortunately, for some exciting membrane processes, these obstacles have been overcome. In regard to the system size, our own group has completed and published MD simulations of membrane proteins in lipid bilayers and water that last up to tens of nanoseconds, include up to 360,000 atoms, avoid boundary effects, control pressure, and account for full electrostatic forces (reviewed by Tajkhorshid et al. [2003]). These advances were made possible by the decade-long development of NAMD, a computer science engineered simulation program that runs efficiently on parallel computers with over one thousand processors (Kalé et al., 1999). The timescale problem can be addressed, on the one hand, by focusing on processes whose natural dynamics are in the nanosecond range, e.g., water conduction (Tajkhorshid et al., 2002), and on the other hand, by using steered molecular dynamics (SMD) (Izrailev et al., 1998; Gullingsrud et al., 1999; Isralewitz et al., 2001a, 2001b), a method of applying external forces to accelerate a thermally activated process without significantly altering the process pathway. The latter methodology requires that a pathway be known and is particularly suitable for transport in membrane channels, where the pathway is often obvious (see, for example, Jensen et al., 2002; Amaro et al., 2003).

The simulations employ classical MD for conformational sampling and simulation of dynamical processes in biomolecular systems. Atomic coordinates of the proteins are obtained from known crystallographic structures; lipid configurations are based on experimentally determined densities; and the simulations embed the proteins in native-like environments. The protein-membrane systems are solvated with explicit (TIP3P) water molecules, as well as ions to neutralize the entire system. The CHARMM22 parameter set for proteins (MacKerell et al., 1992, 1998) and the CHARMM27 set for lipids (Schlenkrich et al., 1996) were used in all simulations. MD simulations were performed in a constant temperature and pressure ensemble (Feller et al., 1995) with periodic boundary conditions minimizing finite-size effects in the simulation. Full long-range electrostatic interactions were calculated using the particle mesh Ewald method (Darden et al., 1993; Essmann et al., 1995), which has been shown to be important for accurate representation of the highly polarized interfacial region of lipid bilayers (Feller et al., 1996).

As stated above, the simulations employed NAMD (http://www.ks.uiuc.edu/Research/namd/) (Nelson et al., 1996; Kalé et al., 1999), a publicly available MD program developed within our group. NAMD employs several methods to reduce the amount of calculation required by the MD algorithm. First, the complexity of the long-range electrostatic force evaluation is reduced from $O(N^2)$ to $O(N \log N)$ via the particle-mesh Ewald (Darden et al., 1993) algorithm, which combines a cutoff direct calculation with a fast Fourier transform-based mesh calculation. Second, nonbonded interactions are split into a smooth long-range component and a complementary short-range component; the frequency of the long-range (electrostatic) evaluation is then reduced from every 1 fs to every 4 fs via an impulse-based symplectic multiple timestepping method. Finally, either the frequency of the short-range nonbonded force evaluation is reduced to 2 fs via multiple timestepping or the timestep itself is increased to 2 fs by constraining the lengths of the bonds vibrating with the highest frequencies. Timestep limits are imposed by the requirement of energy conservation (larger values result in energy drift); the use of a symplectic integrator allows us to interpret energy conservation as an indicator of accuracy. For the analysis of the results, the program VMD (http://www.ks.uiuc.edu/Research/vmd/) (Humphrey et al., 1996), which can handle very large trajectory data sets, is used. In our investigations, we also employed the communication and collaboration program BioCoRE (http://www.ks.uiuc.edu/Research/biocore/) (Bhandarkar et al., 1999).

Our MD simulations follow a basic protocol consisting of minimization (to eliminate initial sterical overlaps that would destabilize the integrator), equilibration to a temperature of 300 or 310 K and a pressure of 1 atm, and simulation in an isobaric, isothermal ensemble for 1 to 10 ns. This protocol is sufficient to test the stability of a biomolecular aggregate or to observe the relaxation of the system into a more favorable conformation. However, if the goal is to study events that would not spontaneously occur during the time span of the simulation, the system must be forced to undergo a transition via the application of forces in an SMD protocol. Important conclusions may be drawn from such a simulation of nonequilibrium events, even though the simulated timescale is much shorter than the natural one (Izrailev et al., 1998; Gullingsrud et al., 1999; Isralewitz et al., 2001a; Israłewitz et al., 2001b; Park and Schulten, 2004).

PERMEATION AND SELECTIVITY IN AQUAPORIN WATER CHANNELS

Our first case study focuses on aquaporin (AQP) water channels. These channels are particularly amenable to MD investigations due to their rather simple function, their great structural rigidity, and the short timescale of the elementary conduction process. These fortunate characteristics were by no means certain when investigations were initiated, but they became obvious very quickly, explaining why AQPs attracted a great deal of attention from computational biologists. This section covers various computational studies of AQPs, in particular the experimental-computational study by Tajkhorshid et al. (2002).

Living cells need to constantly exchange water with their environment. A substantial amount of water can be conducted through pure lipid bilayers and, in fact, such transport plays an important role in cellular water homeostasis. For example, in mammalian cells water, driven by osmotic forces, can pass indiscriminately through the cell lipid membrane. However, this conduction is too slow for an active regulation of water. Membrane channels that permit a rapid water flux had been previously postulated to explain unusually high water permeabilities of certain cellular membranes. However, it was not until 1991 that Peter Agre and his colleagues (Preston and Agre, 1991) demonstrated in their experiments that expression of a protein significantly increases the membrane permeability of a cell to water. They termed the protein aquaporin, or water channel, and their discovery promised insight into the fundamental physiology of water balance and the pathophysiology of water balance disorders.

AQPs are now known to be abundantly present in all forms of life, including mammals, amphibia, insects, plants, and bacteria (Agre et al., 1998; Borgnia et al., 1999a; Heymann and Engel, 1999). In humans, 11 different AQPs have been identified and characterized in various organs that are either directly involved in handling large volumes of water, such as the kidneys, or need to precisely and/or quickly respond to osmotic conditions of the environment, e.g., red blood cells and the central nervous system. Impaired function of AQPs has been associated with diseases like nephrogenic diabetes insipidus and congenital cataract (Agre et al., 1998; Deen and van Os, 1998; Borgnia et al., 1999a; Li and Verkman, 2001).

AQPs are open channels by design; no gating mechanism has been reported so far for water. However, the water pores can be blocked by mercurial compounds (Verkman and Mitra, 2000). AQPs are highly selective channels. Despite their high permeability to water, they completely exclude protons, a property that puzzled researchers since protons can be readily transported through a hydrogen-bonded chain of water molecules, which was expected to exist in AQPs. Proton exclusion is a critical aspect of the biological function of AQPs, enabling them to transport high volumes of water without dissipating the electrochemical potential across cell membranes.

Water permeation remains the best-characterized physiological function of AQPs. However, as researchers continue to investigate these channels, it becomes more evident that AQPs participate in rather diverse cellular functions. In fact, permeability to small substrates other than water has been very well characterized for some members of the AQP family. A subfamily of AQPs, known as aquaglyceroporins, for instance, permit a highly stereoselective passage of small sugar molecules, such as glycerol (Heller et al., 1980; Borgnia and Agre, 2001; Grayson et al., 2003). The *Escherichia coli* glycerol uptake facilitator, GlpF, is a prominent member of the subfamily, partly because it was the first AQP for which a high-resolution structure was solved by X-ray crystallography (Fu et al., 2000). GlpF provides the cell with efficient access to extracellular carbon sources and is particularly important for the cell metabolism under low sugar concentration conditions (Rickey and Lin, 1972).

During the past few years the structures of several AQPs have been solved at high resolution, a development that presented computational biologists with a great opportunity to investigate the structure-function relationship of these channels at an atomic level. Water channels are a particularly attractive subject for MD simulations, as their main biological function, namely water permeation, is known to happen on a nanosecond timescale, i.e., a timescale that can be simulated effectively with classical MD simulations.

In this section we present the results of MD simulations of AQPs and the lessons learned from these simulations. The questions to be addressed are typical for all channels, namely, how channels have been designed to permit rapid passage of the desired molecules, yet prevent with high selectivity the transit of undesirable molecules without getting clogged. Equilibrium simulations have been used first to describe water permeation through AQPs. The simulations were extended with hydrostatic pressure applied, both simulations yielding permeabilities in close agreement with observation. The simulations focused then on the problem of selectivity against protons, revealing an astonishing selection mechanism that is closely related to the overall architecture of AQPs. The simulations finally investigated the selective conductivity of sugar molecules, offering an energetic picture of the conduction process and how the channel distinguishes between different sugar molecules. Obviously, the simulations address the most fundamental questions about channel function, the answers found being of value not only for AQPs but also for other channels.

For methodological details, we refer readers to articles recently published on the subject (Zhu et al., 2001, 2002, 2004b; de Groot et al., 2001; Jensen et al., 2001, 2003; de Groot and Grubmüller, 2001; Jensen et al., 2002; Tajkhorshid et al., 2002; Lu et al., 2003; Grayson et al., 2003; Chakrabarti et al., 2004; Ilan et al., 2004), to a recent review (Roux and Schulten, 2004), and to an AQP web page (additional information on AQPs including several images and animations describing the function of these channels can be found at http://www.ks.uiuc.edu/Research/aquaporins/).

To describe AQPs under physiological conditions, simulations were carried out on membrane-embedded AQP tetramers fully solvated by two layers of water molecules (Color Plate 13 [see color insert]). This setting results in a large system size of about 100,000 atoms, which is unavoidable, as simulation of the main biological function of the channel, namely transmembrane water transport, requires lipids and water molecules to be explicitly represented in the model. Simulating the system at ambient pressure (1 atm) and physiological temperature (310 K) ensured a proper density of water, which can influence the kinetics of water permeation through the channel. Since water permeation occurs on a nanosecond timescale in AQPs, multinanosecond simulations were needed for sufficient sampling of the event. With the program NAMD (Kalé et al., 1999), it took about 30 h to complete 1 ns of the needed 106,000-atom simulation on a 48-processor (3-GHz Athlon) cluster.

Architectural Design of Water Channels

AQPs form tetramers in the membrane (Walz et al., 1997; de Groot et al., 2000). A model of an AQP tetramer in a lipid bilayer taken from an MD simulation (Zhu et al., 2001; Jensen et al., 2001; Tajkhorshid et al., 2002) is shown in Color Plate 13. Each monomer contains a single water channel, the four channels apparently functioning completely independently of each other (Jung et al., 1994; Shi et al., 1994). The four monomers form at their common center a fifth pore, which supposedly functions as a gated ion channel (Yool and Weinstein, 2002). Functional characterization of the central channel is also a goal of simulation studies that are still ongoing. Presently, atomic resolution structures are available for four AQPs: AQP1 (Sui et al., 2001) and AQP0 (Gonen et al., 2004) from mammalian cells, as well as GlpF (Fu et al., 2000) and AqpZ (Savage et al., 2003) from *E. coli*.

A side view of an AQP monomer with the described structural motif is shown in Fig. 1. Each monomer consists of six transmembrane α-helices that surround the actual water pore. The pore is lined largely by hydrophobic side groups of these helices. The remaining part of the lining is realized through very unusual structural elements that are found likewise in potassium channels (Roux and Schulten, 2004). The elements, which are known as the reentrant loops (shown in dark shading in Fig. 1), are each composed of two halves: a short helical half (dark cylinders) and an inverted helical half (shown as dark tubes) pointing with its carboxyl groups into the channel and with its amino groups away from it. The inverted α-helices, which are unfavorable energetically, are fixed in the channel center by the two half-membrane spanning helical halves that protrude from both membrane surfaces to meet in the channel center. One can recognize that the helix ends with their connections to the inverted helices are juxtaposed in the protein's middle and held together through strong hydrogen bonds involving two strictly conserved NPA (Asn-Pro-Ala) motifs (Walz and Ghosh, 1997; Verkman and Mitra, 2000; Zhu et al., 2001).

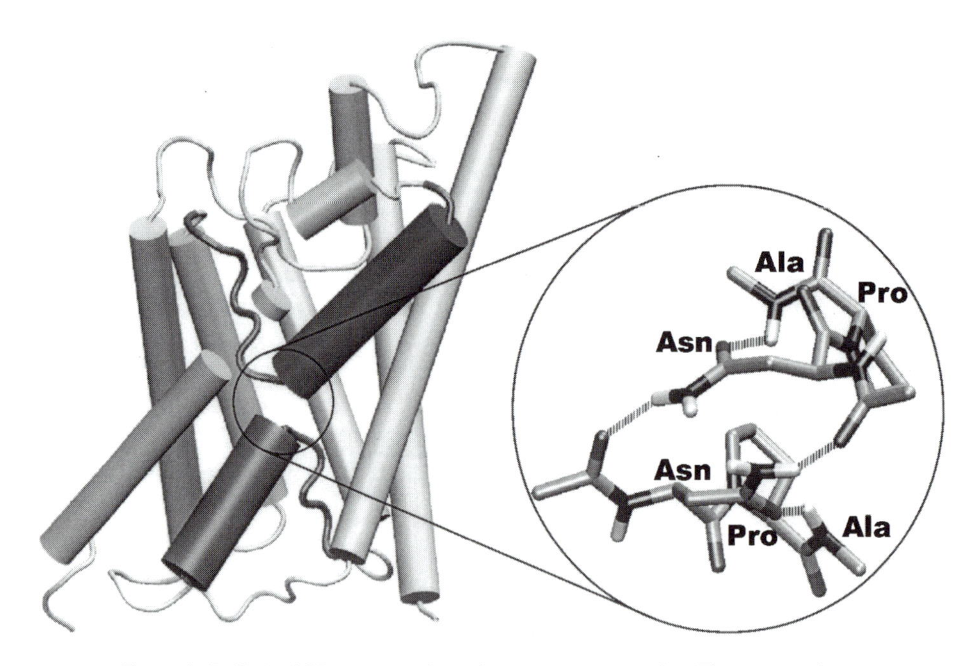

Figure 1. (Left) An AQP monomer shown in cartoon representation. The reentrant loops described in the text are very important structural elements in AQP's architecture. They are shown in a darker shade and are surrounded by the helical bundle formed by transmembrane helices of the protein. (Right) Hydrogen bonds between the two NPA motifs are essential for the stability of the structure and for the function of AQPs. Due to two stable hydrogen bonds between the amino group of the asparagine side chain of each NPA motif with neighboring side chains, one of the amino hydrogens of the asparagine is restrained to be fully exposed toward the interior of the channel, where it forms hydrogen bonds with the permeating substrate. Hydrogen bonds between the two NPA motifs are also important for the stability of the two reentrant loops.

To investigate the functional role of the structural motifs described in Fig. 1, a tetrameric model of GlpF, an aquaglyceroporin, was embedded in a palmitoyl-oleyl-phosphatidylethanolamine bilayer that is a good substitute for an *E. coli* cell membrane. The model was based on a crystallographic structure obtained from a solution with a high concentration of glycerol (Fu et al., 2000), the structure containing three glycerols in each channel. The simulation reported by Jensen et al. (2001) kept all glycerol molecules in the channels (a total of 12 in four monomers) and monitored their conduction. During the simulation, thermal fluctuations promoted indeed significant movements of the glycerol molecules, some of the glycerol molecules even diffusing to the outside of the channel. By combining displacements of individual glycerol molecules, the complete permeation pathway was determined (Jensen et al., 2001) and was found to involve the inverted α-helices as seen in Color Plate 14 (see color insert) that shows snapshots of glycerol during the conduction process.

Further analysis of interactions between glycerol and water molecules and the channel lining shed light onto the architectural design of the protein. Glycerol and water molecules form a hydrogen-bonded single file in the pore, their translocation along the channel axis being strictly correlated (Jensen et al., 2001). Water molecules accompanying glycerol in the channel proved to be very important for the conduction process, lubricating the movement of glycerol through competition for the hydrogen-binding sites (Jensen et al., 2001). Interestingly, almost all of the protein-substrate (glycerol, water) hydrogen bonds in the pore region are mediated by the backbone carbonyl groups from the two inverted helices and by the two asparagines of the conserved NPA side groups mentioned above. These hydrogen-bonding groups act in an otherwise hydrophobic environment as stepping stones used by glycerol and water to move along the channel.

Therefore, adopting an inverted helical structure in this region is essential to the function of the protein, as it ensures a maximal participation of carbonyl groups in substrate permeation (Jensen et al., 2001). It is interesting to note that the two inverted helices are further stabilized through hydrogen bonds between the backbone amino groups pointing away from the channel and two conserved glutamates (Color Plate 15 [see color insert]). The glutamate residues are the only negatively charged residues buried in the transmembrane region of the protein, which clearly reflects their important role in the structure and function of the channel. Given the common fold of all AQPs, the results of this computational study have provided an explanation for the functional implication of the protein architecture in the whole family of AQPs.

Sugar Permeation through GlpF

The inner lining of GlpF consists of a hydrophobic face complemented by hydrogen-bonding groups on the opposite side, matching closely the amphiphilic structure of glycerol and other linear sugar molecules. The narrowest region of the channel is located close to the periplasmic mouth of the channel and is proposed to function as its selectivity filter (Fu et al., 2000; Nollert et al., 2001; Law and Sansom, 2002; Fujiyoshi et al., 2002).

The permeation of glycerol through GlpF is controlled in large part by energetics. The molecule needs to be attracted to the channel and as it transits the channel it must be subjected to interactions that select it, rejecting other compounds that may also fit in the channel. One expects that the energetics can be cast into an energy profile that exhibits attractive wells at the channel ends and barriers in the selectivity regions of the channel, the barriers being surmountable by glycerol and related molecules but too high for other compounds. The energy profile must take into account the thermal motion of the channel protein, i.e., average overall degrees of freedom other than the channel axis coordinate z. Such a profile, termed the potential of mean force (PMF), can indeed be determined, namely through SMD simulations as described elsewhere (Isralewitz et al., 2001b; Park et al., 2003; Park and Schulten, 2004).

SMD was applied to GlpF embedded in a membrane as described above; the simulations pulled glycerol through the channel by applying an external force, $f(t)$, along the channel axis (Jensen et al., 2002). The simulations monitored the work done on the glycerol, i.e., W, and, employing the Jarzynski identity (Jarzynski, 1997a, 1997b), $\langle \exp(-\beta W) \rangle = \exp(-\beta F)$, with $\beta = k_B T$, sampled over trials and calculated the free-energy difference, $F(z)$, between bulk water and positions z inside the channel (for methodological details and references, see Park and Schulten [2004]). The resulting energy profile, shown in Color Plate 16 (see color insert), captured major features of the glycerol-channel interaction and confirmed the expectations, exhibiting an energy well on the periplasmic side of the channel, a pair of barriers in a region that was suggested to function as the selectivity filter (Fu et al., 2000), and a third barrier at the NPA motifs. The positions of crystallographically observed glycerol molecules (Fu et al., 2000) closely correspond to minima in the energy profile (Jensen et al., 2003). Additional minima, due to multiple hydrogen bonds between glycerol and the channel, were also discovered.

One of the most interesting aspects of the constructed PMF (Color Plate 16) is its clear asymmetry in regard to the periplasmic well and the locations of the main barriers. Although GlpF is a passive channel and conduction of nutrient molecules is facilitated equally in either direction by the channel under equilibrium conditions, the asymmetry reflected by the periplasmic energy well, referred to also as the attractive vestibule, might be of importance for the dominant physiological role of the channel under nonequilibrium conditions, namely uptake of glycerol from the periplasmic space. After entering the cytoplasmic region, glycerol becomes phosphorylated and cannot exit the cell. Interestingly, the asymmetric PMF corresponds to the asymmetric shape of the channel, when one compares the periplasmic and cytosolic segments of the channel. The significantly bigger protrusion of GlpF to the periplasmic region provides an attractive site for nutrient molecules, which are usually not abundant. Kinetic rate models built up on the basis of the calculated PMF indicate that the attractive well in the periplasmic vestibule of GlpF may increase the efficiency of the channel for glycerol uptake, especially at low concentrations of the substrate (Lu et al., 2003). In keeping with this suggestion, it is interesting to note that the asymmetry of the structure is much less pronounced in AQP1, a pure water channel, where substrate (water) is always present in the environment at a high concentration. The asymmetry of the channel energetics may have the surprising consequence that it can turn GlpF from a passive into an active transporter of glycerol (Kosztin and Schulten, 2004). The only attribute required for this dramatic change is a symmetrically fluctuating or oscillating force also acting on the glycerol. On the basis of the theory of ratchets Kosztin and Schulten (2004) argued that GlpF driven by volume fluctuations of *E. coli* in the megahertz-kilohertz range can transport glycerol into the cell as well as protect the cell against poisoning through large doses of glycerol.

Definitely the most intriguing aspect of sugar transport in GlpF is the mechanism of selectivity. Experimental measurements of permeation of various linear sugar molecules through GlpF have clearly established that substrate permeation through GlpF is a stereoselective process (Fu et al., 2000). To investigate what structural elements discriminate between stereoisomers, Grayson et al. (2003) studied the permeation of ribitol and arabitol, two five-carbon sugar molecules (pentaols) that differ only in regard to the position of one of their hydroxyl groups. Experiments had shown that there is an almost 10-fold difference between the permeability of GlpF to these diastereoisomers (Fu et al., 2000). Since the pentaols are significantly longer than glycerol, unfavorable steric interactions or hydrogen-bonding mismatches are amplified. As a result, one expects the mechanisms of stereoselectivity to surface more clearly for the pentaols than for glycerol.

To investigate the stereoselectivity in GlpF, the method of interactive molecular dynamics (IMD) (Stone et al., 2001; Grayson et al., 2003) was employed in which external forces were applied to the pentaols and continuously readjusted through visual and mechanical feedback during an ongoing MD simulation. A single molecule of ribitol or arabitol was positioned in the periplasmic space and then pulled interactively through the channel, monitoring the conformation of the molecule. At key moments the molecule was allowed to relax with forces switched off. Potentially important configurations found in IMD simulations were simulated for 100 ps, each without external forces. The method permits one to not only probe quickly the passage of sugar molecules through the channel but also to manipulate the conformation of the sugar at each point during its permeation, thus significantly increasing the likelihood of obtaining an optimal configuration for the sugar inside the channel.

Selectivity was found to be achieved incrementally during several stages of conduction, the key stage arising at the location with the tightest fit around the molecule, namely at the selectivity filter (Fu et al., 2000). Here the protein presents hydrophobic and hydrophilic linings that match the amphiphilic structure of the sugar; the IMD simulations revealed that the selectivity is due to an induced fit and to optimal hydrogen bonding. The interactions of the two pentaols in the selectivity filter are shown in Fig. 2. Unfavorable steric interactions resulting from the tightness of the selectivity filter are found to be balanced by

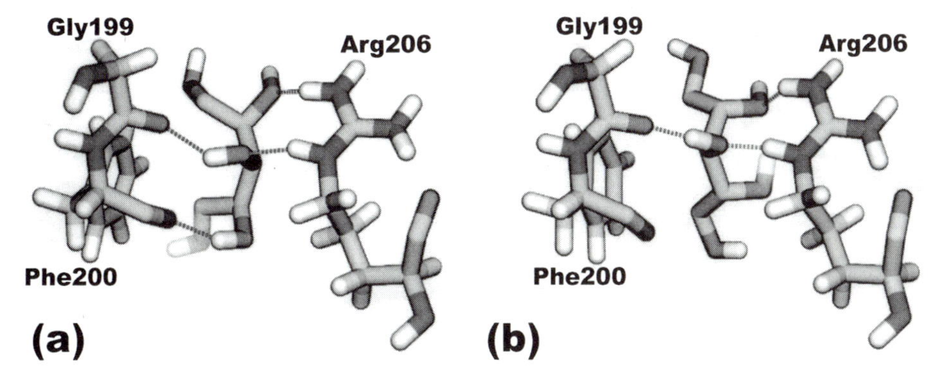

Figure 2. Comparison of ribitol (a) and arabitol (b) at the selectivity filter of GlpF. Hydrogen bonds are shown as dotted lines. Ribitol is able to form an optimal number of hydrogen bonds with the channel without losing its linear conformation in this region. Formation of the same number of hydrogen bonds for arabitol requires a tilted conformation of the molecule that is unfavorable due to strong steric hindrance of the filter region.

multiple hydrogen bonds to accessible sites. Fu et al. (2000) had indeed suggested earlier that the multiple hydrogen-binding sites found in the selectivity filter of GlpF hold multiple parts of a passing sugar in place and that this locking mechanism is responsible for the selectivity of GlpF. Figure 2 shows that ribitol can form an optimal number of synchronous hydrogen bonds to the selectivity filter of GlpF, but that arabitol cannot engage in hydrogen bonding to the same degree with the result of a much lower conduction rate (Grayson et al., 2003).

Water Conduction and Proton Exclusion

AQPs facilitate fast water transport across cellular membranes, and one wishes to understand how the rapidity is achieved despite the narrowness of the channels. More importantly, water channels must prevent leakage of protons whose concentration gradient across the membrane plays a key role in a cell's life. AQPs achieve the needed selectivity admirably, but the mechanism behind this ability posed a great conceptual challenge due to the apparent inseparability of water and proton transport. Protons are known to be able to transfer from one end of a hydrogen-bonded chain to the other end without needing the molecules forming the chain to move. In other words, provided the right arrangement of water molecules and protein groups, protons can be transfered from one side of the channel to the other much faster than diffusing water. This mechanism allows protons to transfer across several water molecules, using them as stepping stones, on timescales of the order of picoseconds, i.e., a thousand times faster than the nanosecond timescale needed for water itself to fully permeate the channel in AQPs. Definitely, AQPs must prevent this fast proton conduction, but they must also prevent the slower extreme of conduction in which hydronium ions (H_3O^+) ride through the channels along with water molecules.

MD simulations of GlpF show that a dehydrated channel fills within 100 to 200 ps with a single file of seven or eight water molecules (Tajkhorshid et al., 2002). The water molecules remain single file during diffusive transport. Analyses of simulations in terms of conduction rates or osmotic and diffusive permeabilities (de Groot and Grubmüller, 2001; Tajkhorshid et al., 2002; Zhu et al., 2004b; Tajkhorshid et al., in press) yield excellent agreement with the observed values reported (Borgnia et al., 1999a; Heymann and Engel, 1999; Borgnia et al., 1999b). The simulations of AQP1 (Zhu et al., 2004b) yield an effective diffusion coefficient of a single water molecule in the channel of 20 Å^2/ns, which is about 1/10 of the coefficient for self-diffusion of bulk water, demonstrating that the channel despite its small size permits fast membrane crossing. It is illustrative to compare the situation with that at a strictly controlled border crossing where a need to slow one's speed to 1/10 of the regular cruising speed seems an outstanding solution.

Atomic resolution MD simulations permit one to monitor the dynamics and behavior of each and every atom in the system, including hydrogen atoms. In this regard, simulations are superior to crystallography. In the experimental-computational study of water transport in GlpF reported by Tajkhorshid et al. (2002), the authors compared the predicted and crystallographically observed occupancy of oxygen atom positions of water in the channel, finding excellent agreement. However, the simulations could resolve also the positions of water hydrogens and, thereby, the orientation and hydrogen bonding of water in the channel. The pattern of orientation found, as shown in Color Plate 17 (see color insert), was quite astonishing: water molecules realize throughout their conduction a strictly bipolar orientation with oxygens entering the channel first and leaving it last, the reorientation occurring at the middle of the channel in the vicinity of the NPA motifs.

The electric field of the protein inside the channel dictates the peculiar configuration of water molecules that was not reported in any other system before (Tajkhorshid et al., 2002; Jensen et al., 2003). This configuration prevents the hopping of protons through the channel as argued by Tajkhorshid et al. (2002) and Zhu and Schulten (2003). Quantitative analyses of the effective energy barrier that AQPs erect against proton passage (de Groot et al., 2003; Chakrabarti et al., 2004; Ilan et al., 2004) confirmed the mechanism proposed by Tajkhorshid et al. (2002). All of these studies found that the largest barrier against proton transfer is indeed located at the NPA region of the channel. A second barrier against protons was found at the selectivity filter of the channel where a conserved positive Arg is located (shown in Color Plate 14). Interestingly, despite the presence of the Arg positive charge, the barrier against protons at the selectivity filter is lower than the one at the NPA motifs. These results are in keeping with the electrostatics analysis (Jensen et al., 2003) showing that the NPA region carries the strongest electrostatic potential along the channel.

Putting Water Channels under Pressure

As described above, equilibrium MD simulations can be used to describe permeation of water molecules through the channel via diffusion, where the net transmembrane water flux is zero. While random diffusion of water is always happening, a physiologically more important event, which is routinely measured experimentally for water channels, is the net flow of water induced by an osmotic or hydrostatic pressure gradient across the membrane. To study such events, MD simulations are needed in which bulk water molecules located on the two sides of the membrane experience different osmotic or hydrostatic pressures.

A method to apply such pressure differences to determine computationally the osmotic permeability of water channels by MD simulations is described by Zhu et al. (2002, 2004b), Zhu and Schulten (2003), and Tajkhorshid et al. (in press). Through application of external forces to bulk water molecules, a hydrostatic pressure gradient is generated across the membrane that promotes water transport (Fig. 3). Through adjustment of the force applied to individual water molecules and the number of water molecules affected, a wide range of pressure gradients can be realized. This method induces a net water flux through the channel, and one can measure the channel's osmotic permeability from the flux/pressure gradient ratio (Zhu et al., 2002, 2004b; Tajkhorshid et al., in press).

Conclusions

The molecular mechanisms by which membrane channels facilitate a highly selective transport of materials across biological membranes are of great importance in all cells. Such mechanisms are due to the specific architecture of the channel-forming protein(s), but even when the structures are known, the mechanisms are not obvious. Often only MD simulations reveal enough detail to recognize the mechanisms of selectivity and conduction as exemplified above for the AQP channels GlpF and AQP1.

MD simulations of AQPs have significantly contributed to our understanding of the molecular basis of function and selectivity in this important family of membrane channels.

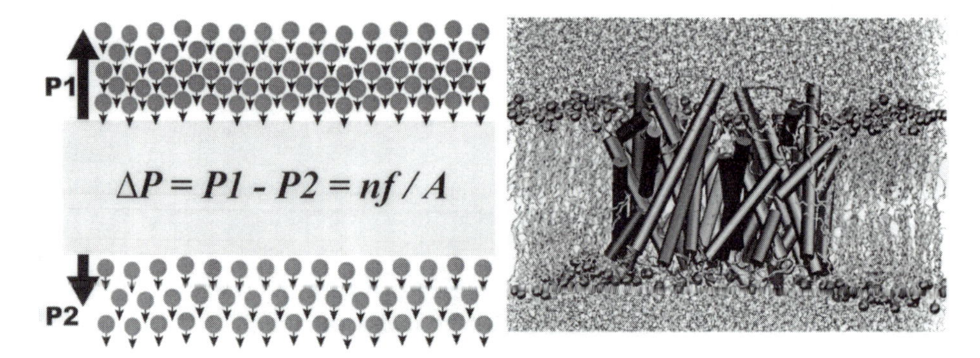

Figure 3. Illustration of the method to produce a pressure gradient across the membrane through forces, shown by small arrows, applied to individual water molecules in the bulk region. Either all or only some water molecules may be selected for force application. The total pressure difference, ΔP, is determined by the number of water molecules, n; the applied force on each water molecule, f; and the area of the membrane, A. In this case, ΔP drives the water downward.

The results presented in this section showcase the wealth of information that can be obtained from atomic resolution simulation of macromolecules.

Future simulations should investigate the possible involvement of the central tetrameric pore of AQP in ion transport. Comparison of GlpF and AqpZ, both from *E. coli*, one being a glycerol channel and one a pure water channel, can provide further insights into the mechanisms underlying substrate selectivity in AQPs. A recent structure of mammalian AQP0 can be used to understand the effect of disease related to mutations of this protein. Three-dimensional structures of other AQPs could be modeled based on their high degree of similarity with structurally known AQPs.

ION PERMEATION IN ClC CHLORIDE CHANNELS

In the case study of ClC chloride channels, MD simulation is used to determine how Cl^- conducts through the channel pore. Here, the possibility that Cl^- ions diffuse one at a time across the channel is ruled out, and following closely Cohen and Schulten (2004), a coordinated two-ion mechanism is suggested instead.

The ClC chloride channel family was discovered by C. Miller in 1982 during investigations of the torpedo ray electroplax membrane (Miller, 1982). Since then, various members of the ClC family have been isolated in a wide variety of organisms, ranging from animals and plants to yeast and almost all bacteria except for a few species with small genomes. All ClCs have in common a selectivity for small inorganic anions (e.g., Cl^-, NO_3^-, Br^-, I^-, SCN^-, and some larger hydrophobic anions), though they tend to discriminate rather poorly between these different anions. Despite their poor interanion selectivity, ClCs are called "chloride channels" because Cl^- is the only inorganic anion with a significant presence at physiological conditions.

Many roles for ClCs have been identified in higher organisms: they play vital cellular functions, such as the regulation of blood pressure, cell volume, organelle pH, and membrane excitability (Maduke et al., 2000; Jentsch et al., 1999; Valverde, 1999; Estévez and

Jentsch, 2002). In prokaryotes, however, the ClCs' various roles are only now emerging from obscurity. First, Iyer et al. (2002) found that ClC is essential for *E. coli* to survive extreme acid shock. Then, in an unexpected discovery, Accardi and colleagues concluded that the *E. coli* ClC was not a passive channel, as had been assumed for all ClCs, but behaved like an active Cl^--H^+ antiporter (Accardi and Miller, 2004; Accardi et al., 2004). This finding is particularly interesting because most other eukaryotic ClC homologs are known to be passive channels, and the *E. coli* ClC has conduction rates (<0.2 pS) that are intermediate between typical rates observed in channels and those observed in transporters, prompting the suggestion that the evolutionary distance between channel and transporter proteins is less than previously thought.

While most electrophysiological measurements have been performed on eukaryotic ClC homologs (Fahlke et al., 1997; Fahlke, 2001), such as the torpedo ray ClC-0 and the human homologs ClC-1 and ClC-2, there are believed to be many similarities between the core structures of prokaryotic and eukaryotic ClCs (Dutzler et al., 2002; Miller, 2003). All ClCs are believed to share a "double-barreled" architecture, in which each channel consists of two identical monomers, each monomer consisting of two heterogeneous but structurally similar segments arranged in an antiparallel fashion, and each monomer containing its own independent water-filled pore (Dutzler et al., 2002; Miller, 1982; Mindell et al., 2001) (Fig. 4a). ClCs share a very high sequence similarity for the selectivity filter: the central region of the pore that has been found to coordinate the permeating Cl^- ions according to X-ray crystallographic structures (Dutzler et al., 2002; Fahlke et al., 1997). On the other hand, bacterial and eukaryotic ClCs differ considerably in size. Bacterial ClCs are much smaller (typically 395 to 492 residues), whereas eukaryotic homologs are longer (687 to 988 residues), with most of the extra residues lying in cytoplasmic dangling ends or in the periplasmic regions responsible for regulation and gating functions. And while most, if not all, eukaryotic ClCs are voltage-gated in some way or another (Fahlke, 2001; Pusch et al., 1995), it is still unclear as to whether the recently characterized bacterial *E. coli* ClC possesses a voltage-gating mechanism at all.

In addition to its biological relevance, the study of ClC is important because it is one of only two channel families of known atomic structure that are specifically selective to ions, along with the potassium channel (Dutzler et al., 2002; Dutzler et al., 2003; Doyle et al., 1998; Jiang et al., 2002, 2003). The preceding discovery of the potassium channel structure (Doyle et al., 1998) and subsequent MD studies (Bernèche and Roux, 2000, 2001; Shrivastava and Sansom, 2000; Morais-Cabral et al., 2001; Åqvist and Luzhkov, 2000; Noskov et al., 2004a) revealed for the first time the detailed mechanism of ion conduction through an ion channel. Still, little is currently known about the mechanism of ion conduction in other ion channels such as ClC.

Cl^- Conduction Pathway and the Selectivity Filter

Examining the general shape of ClC, one immediately discovers the first strategy used by the channel to permit anion conduction across the membrane: to a first approximation, the ClC channel fulfills its role by reducing the effective thickness of the membrane seen by the ions. Looking at the channels' cross-section in Fig. 4b, it can be seen that ClC allows water to significantly enter the membrane in two vestibules on either side of each of the ClC's narrow pores. This way, the lipid membrane's dielectric barrier can more easily

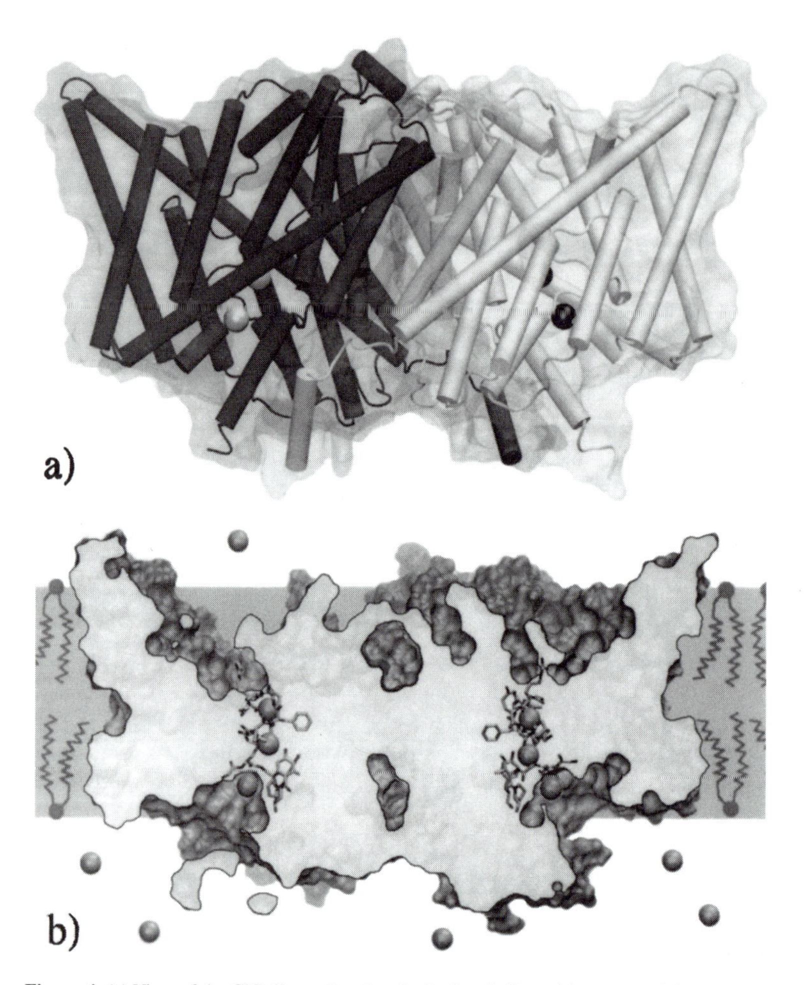

Figure 4. (a) View of the ClC dimer showing the broken helix architecture and the position of the Cl^- ions in the crystal structure. Each monomer and pair of ions is displayed in a different shade. (b) Vertical cross-section of the solvent-accessible surface of the ClC protein embedded in a lipid bilayer. The simulated model comprises 97,000 atoms. In the narrowest part of the protein, where the Cl^- ions permeate, the residues that define the selectivity filter are shown.

be overcome by the conducting ions since the thickness of the membrane to be crossed has been reduced and its dielectric constant has been increased by the substitution of lipids for protein. Following this chain of thought, the remainder of ClC's function can be attributed to the region of the pore between the two vestibules, the selectivity filter. This region, highlighted in Fig. 5, is the thinnest part of the channel and allows anions to pass but blocks all other particles. From here on, the focus will be primarily on ClC's selectivity filter.

Here, we will set aside the issues related to ClC gating and the Cl^--H^+ coupling mechanism and pretend for one instant that the channel is frozen into a perfectly conducting

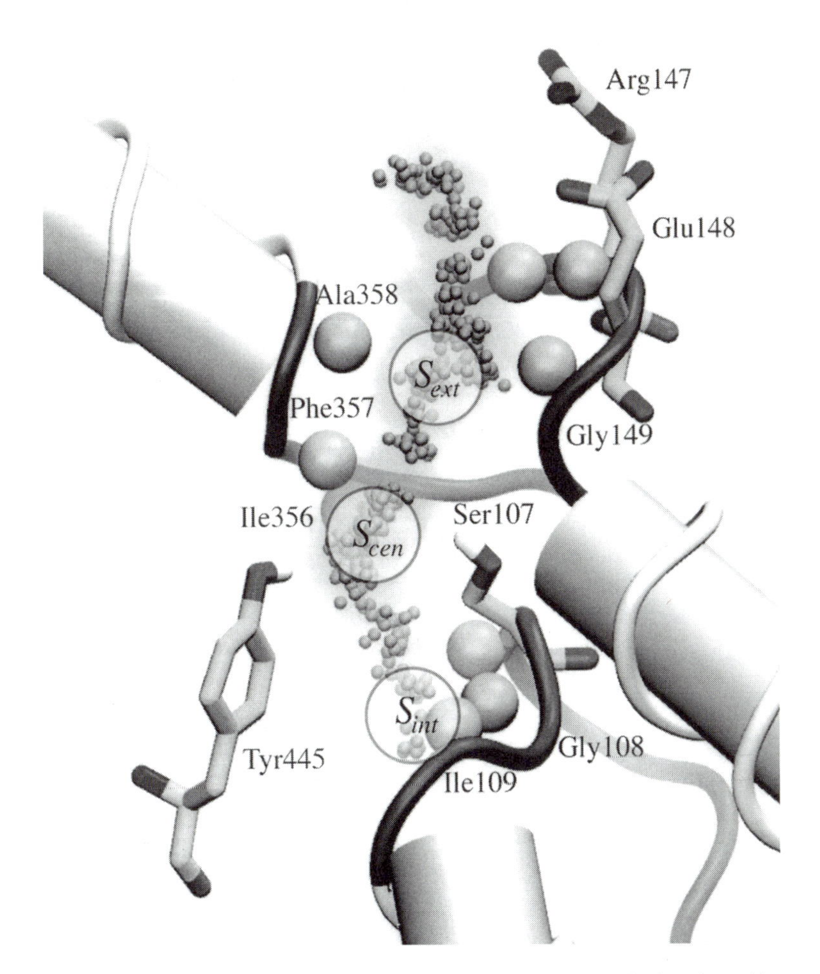

Figure 5. Detailed view of residues forming the ClC selectivity filter. The locations of hydrophobic residues are indicated as balls, whereas polar and charged residues are drawn explicitly. The positions of the Cl^- ions as they permeate across the selectivity filter are plotted as small spheres, and the locations of the three Cl^--binding sites (S_{int}, S_{cen}, and S_{ext}) identified by X-ray crystallography (Dutzler et al., 2003) are indicated as large circles.

state: all gates are open, and the channel's titration state is just right for Cl^- conduction. MD allows us to study conduction across such a system and can provide many clues about how conduction really occurs. Additional effects, such as gating or interactions with transiently protonated side chains, will, of course, temporarily interrupt the mechanism observed in the perfectly conductive model described above. But even though a global theory of conduction needs to take into account such events as gate opening and closing, these transient effects can be omitted and the perfectly conductive conduction mechanism will accurately represent ion conduction in the time interval between the gating events. The same applies to the poorly understood Cl^--H^+ coupling mechanism in ClC, which presumably can be described by involving a second H^+/Cl^--controlled gate.

Computing a Slow Process in Little Time

Our goal is to characterize the way in which Cl^- passes through the ClC channel under extremely favorable conditions: open gates and no proton-coupling to slow the dynamics. For this, the energy profile of Cl^- ions going through the selectivity filter will be measured. When one wishes to understand a long timescale process using computer simulation, MD trajectories of the atomic motions, by themselves, are of limited interest. In the present case, we know that a Cl^- ion will conduct across *E. coli* ClC, in a favorable electrochemical gradient, within roughly 1 ms (Accardi et al., 2004). This timescale is out of the reach of contemporary atomic-level simulations running on the best currently available hardware. But even if we could simulate the complete translocation of a Cl^- ion across one of ClC's pores, the computed trajectory would represent only a single event (or a handful at most). As we know, nature allows many paths between two end states, all with different probabilities. Computing just one of these paths leaves us with no information about its statistical relevance and prevents us from reaching any meaningful conclusion about the studied mechanism.

Fortunately, one does not require a full description of the atomic trajectories to understand the translocation process. What is really needed is the energy profile—the energy mountains and valleys—experienced by anions as they conduct through ClC. A complete description of the free-energy profile would provide us with all the statistical information about the ion translocation process, since the free energy for a given ion state (i.e., the set of positions along the channel occupied by ions at a given instant) can be interpreted as a measure of the probability of occurrence of that ion state. Now, to compute the energy profile for a conduction event, we will need to carry out calculations of simulation trajectories. There are, however, two superfluous aspects of real-time simulation that can be taken advantage of to allow the calculation of the relevant energy profile using minimal computational requirements: (i) not all degrees of freedom are important to the problem at hand and (ii) not all parts of a trajectory are sampled equally by equilibrium simulations. In fact, most slow processes spend the overwhelming majority of the time in a small favorable region of phase space. If the system's time evolution can be biased such that states with high energy are sampled as often as the more favorable low-energy states, we could calculate the energy profile of the system projected along a chosen reaction coordinate much faster than it would take to perform a full equilibrium simulation.

The PMF is the desired quantity that describes the overall free-energy profile experienced by the system as it evolves along one or more reaction coordinates, averaged over all other degrees of freedom. In our case, this means that the PMF will describe energy as a function of the position of ions in the channel, averaged over all possible conformations that the ClC channel itself can take to accommodate the permeant ions.

Energetics of Cl^- Permeation

For the case of ClC, if one tries to pull a single ion across the channel, one encounters extremely large energy barriers, rendering the calculation of the PMF impossible to perform. A single Cl^- ion inside ClC is bound too tightly to be easily separated from it by pulling with a realistic force, and a separate mechanism is needed to be able to explain how it is dislodged. A similar behavior has been observed in the KcsA potassium channel, and for that channel, it was shown that conduction requires two to three K^+ ions simultaneously in the pore (Bernèche and Roux, 2001). If such is also the case for ClC permeation, notably, that conduction can only be possible if the ClC pore is occupied by more than one Cl^- ion, then if we calculate the PMF of conduction across ClC for two simultaneous ions and observe low-energy barriers for a conduction event, we will know that ClC requires two ions for conduction. This hypothesis is further supported by experimental evidence suggesting that more than one anion are able to simultaneously occupy the pores of eukaryotic ClC homologs (Pusch et al., 1995).

The PMF for the coordinated motion of two Cl^- ions through ClC has been computed by Cohen and Schulten (2004), using umbrella sampling (Roux, 1995; Gullingsrud et al., 1999), an alternative to the sampling method used for the GlpF channel above, but which can be extended more readily to sample along more than one reaction coordinate at a time. In the case of ion conduction through a linear pore, the reaction coordinates can be chosen to be the position of the permeating ions along the axis of the pore. Thus, a series of sampling simulations were performed, in which two Cl^- ions per ClC pore were separately tethered by means of virtual springs acting along the pore axis. The role of the virtual spring is to constrain the ions to sample a region of space that might otherwise not be sampled adequately by an equilibrium simulation. The tethering points for the umbrella potentials were never distributed more than 1 Å apart for any ion, requiring 92 simulations of 370 ps each to sample the range of motion of the ions needed to describe conduction through the selectivity filter. The spatial distributions of the ions along the z axis for each simulation were then combined using the weighted histogram analysis method (Kumar et al., 1992) applied in two dimensions to obtain the full two-ion distribution of Cl^-, which was inverted to obtain the PMF (Cohen and Schulten, 2004).

To prepare the protein for the PMF calculation, a model was constructed from the *Salmonella enterica* serovar Typhimurium ClC X-ray structure (Dutzler et al., 2002), which was then embedded in a POPE lipid membrane and solvated. This system was equilibrated for 5 ns, during which the Cl^- ions present in the selectivity filter of the crystal structure did not budge, due to the very strong binding energy of Cl^- with the ClC pore. Each pore of the X-ray ClC structure was found to be blocked by a charged glutamic acid residue (Glu148) at the periplasmic end (Dutzler et al., 2002; Dutzler et al., 2003). Recent electrophysiological measurements performed on the torpedo ray ClC-0 show that both the substitution of the pore-blocking glutamic acid (Glu166 in ClC-0) with small noncharged residues (E166G, E166A, E166V, or E166Q) and the protonation of this glutamate strongly reduce the voltage dependence of the fast gate, allowing the pore to remain open for a wider range of conditions (Dutzler et al., 2003; Friedrich et al., 1999). Similar mutation studies reached similar conclusions for the human ClC-4 (E224A) and ClC-5 (E211A) homologs (Friedrich et al., 1999). In addition, crystal structures of *E. coli* ClC mutants E148A and E148Q (electrophysiological measurements are not currently possible on the native protein) showed them to be virtually identical to the wild-type structures except that the mutant pores were unobstructed by the Glu148 side chain. To enable the conduction of Cl^-, the pore-blocking Glu148 side chain has been displaced so that it points away from the pore, while still being hinged at the pore entrance, in agreement with the structure of the E148Q mutant (Dutzler et al., 2003). Other computational work has dealt with Glu148 by either pushing it aside (Cohen and Schulten, 2004; Corry et al., 2004) or protonating or neutralizing it (Bostick and Berkowitz, 2004; Yin et al., 2004; Miloshevsky and Jordan, 2004). In both cases, conduction has been shown to be possible by computational studies; however, it is unlikely that Glu148 would spontaneously move out of the pore in its unprotonated state due to strong electrostatic interactions. Because of this, the protonation of Glu148 is almost certainly a necessary step for conduction and is an extremely likely candidate for one of the gating mechanisms necessarily associated with the *E. coli* ClC transporter mechanism. After Glu148 was displaced, the pore was solvated with a single file of water and equilibrated for 0.5 ns and a second Cl^- ion was added near the pore entrance to start measuring the PMF for the two-ion conduction process.

The final computed PMF for the two-ion process in the selectivity filter of one of the two pores is shown in Fig. 6. The map describes the energy landscape involved in the evolution of the channel-ion system between an initial state I and a final state II. In state I, the first Cl^- is bound to the channel's central binding site (determined from the crystal structure)

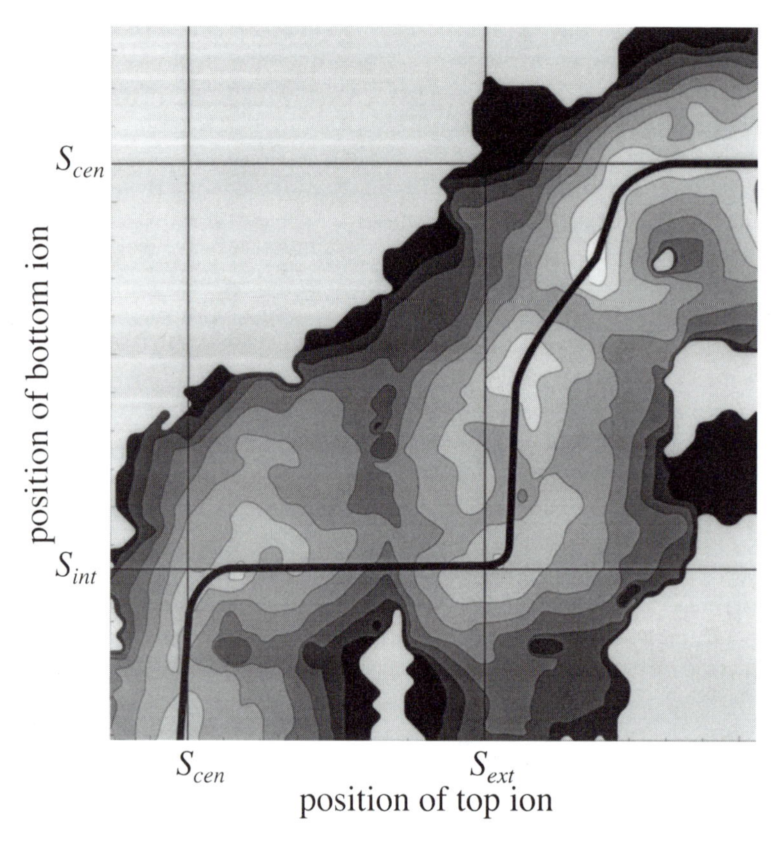

Figure 6. Map of the PMF for a pair of Cl^- ions moving across one of ClC's two pores as a function of the independent positions of the top and bottom Cl^- ion. The thick line represents the coordinated motion of the pair of ions that follows the path of minimum energy (i.e., the most probable path). Each contour represents an energy difference of 1 kcal/mol. The axes correspond to the distance of the top and bottom permeating ions along a line perpendicular to the membrane, with the three binding sites from the crystal structure indicated as S_{in}, S_{cen}, and S_{ext} for reference.

and the second Cl^- is positioned in the channel's cytoplasmic entrance; in state II, the first Cl^- is positioned in the periplasmic exit and the second Cl^- is bound to the central binding site. This process effectively describes a conduction event since states I and II share identical pore configurations except that in state I, one Cl^- is at the top of the pore, and in state II, it is at the bottom. For a new translocation to occur, one simply has to wait for a third ion to diffuse from the bulk solution into either of the pore's entrances, after which the transition I → II or II → I can repeat. By following the path that minimizes the PMF (drawn as a thick line), one can deduce the most probable sequence of events in the coordinated permeation of two Cl^- ions. Along this path, the anions move in turn in the quasi-stepwise manner detailed in Fig. 7; the maximum barrier height experienced by this two-ion process is measured to be 4 to 4.5 kcal/mol (Cohen and Schulten, 2004).

Looking at the PMF, one sees that the second Cl^- enters or exits while the first Cl^- is tightly bound to the central crystal-binding site (S_{cen}) in the selectivity filter. When only one anion is present, the interaction energy between an anion and the pore is very attractive (estimated at 75 to 100 kcal/mol [Corry et al., 2004; Cohen and Schulten, 2004] in absolute terms or 5.5 to 9 kcal/mol with respect to a water solvent [Faraldo-Gómez and Roux, 2004]). It is very likely that such a strong attraction ensures that an anion is present in the pore at all times. To dislodge the central Cl^-, its binding energy needs to be considerably reduced. The presence of additional anions in the pore is thus required so that the channel-Cl^- attraction and the Cl^--Cl^- repulsion could balance each other. The competition between the ion-channel attraction and the ion-ion repulsion is a crucial feature of ion channels. The ion-channel attraction determines which particle can enter the channel and accounts for anion versus cation selectivity. In the case of ClC, only anions are strongly attracted into the channel. A strong ion-channel attraction, however, also implies that to conduct efficiently, such a channel would need a mechanism to dislodge the ions caught in the channel's attraction basin, otherwise the ions that the channel wants to conduct would simply remain bound to the channel. In the case of ClC, this mechanism is provided by the ion-ion repulsive forces between many anions simultaneously present in the pore, as

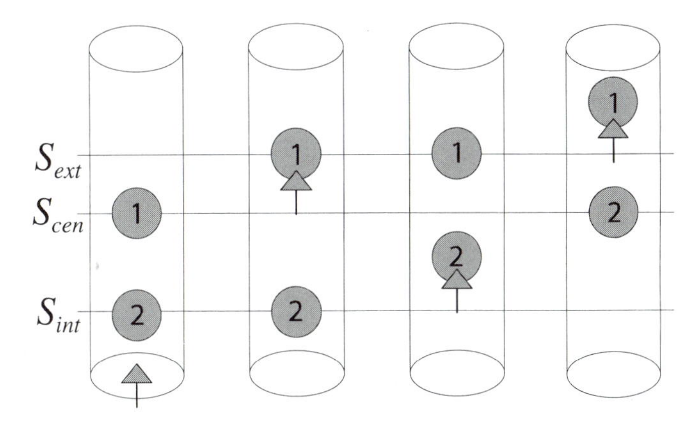

Figure 7. Sequence of motion of two ions in the ClC pore, resulting in the conduction of one Cl^- ion across ClC, as inferred from the calculated PMF.

suggested by the PMF shown in Fig. 6. The same mechanism of the ion-channel attraction being counterbalanced by an ion-ion repulsion is also used by the potassium channel, in which two to three simultaneous K^+ ions are needed in the pore for conduction to occur (Bernèche and Roux, 2001).

With a simple tool, notably the calculation of the PMF using biased sampling simulations, we are able to fully characterize a multi-ion (in this case, two-ion) permeation process across a channel pore. This technique can be applied to almost any system that can be described by a few important degrees of freedom (in this case, the positions of two ions projected along a straight line). The PMF then provides us with a complete story of the sequence of steps observed during conduction, inferred by following the path that minimizes the energy differences in the PMF between two end states.

ION AND WATER PERMEATION IN ALPHA-HEMOLYSIN

In the case study of alpha-hemolysin we demonstrate how MD simulations are used to determine the current-voltage dependence of ion conductance and the osmotic permeability of water across a large membrane protein. This section follows closely the report by Aksimentiev and Schulten (in press).

In its fight for resources, *Staphylococcus aureus* secretes alpha-hemolysin monomers to bind to the plasma membrane of susceptible cells. Upon binding, the monomers oligomerize to form a water-filled transmembrane channel (Bhakdi and Tranum-Jensen, 1991; Gouaux, 1998) that facilitates uncontrolled permeation of water, ions, and small organic molecules. Rapid discharge of vital molecules, such as ATP, or irreversible osmotic swelling leading to the cell wall rupture (lysis), can cause death of the host cell. This pore-forming property of alpha-hemolysin has been identified as a major mechanism by which proteinaceous exotoxins can damage cells. Secretion of alpha-hemolysin is believed to contribute to hemostasis disturbances, thrombocytopenia, and pulmonary lesions during severe staphylococcal infections (Bhakdi and Tranum-Jensen, 1991).

The crystallographic structure of the assembled alpha-hemolysin reveals a heptametric organization (Song et al., 1996). The protein has a mushroom-like shape, with a 50-Å β-barrel stem protruding from the cap domain through the lipid bilayer into the cell's interior. The cap of the protein conceals a large vestibule connected to the cell's exterior through a large opening at the top of the cap. The narrowest (1.4 nm in diameter) part of the channel is located at the base of the stem, where the β-barrel pore connects to the vestibule. Figure 8 illustrates the atomic-level model of alpha-hemolysin in its native environment used in our study.

Apart from being one of the most widespread toxic agents of bacterial origin encountered by the human organism (Bhakdi and Tranum-Jensen, 1991), alpha-hemolysin is the principal component of several biotechnological applications, including systems for controlled delivery of ions or small organic compounds across biological or synthetic lipid membranes (Bhakdi et al., 1993; Bayley, 1995; Russo et al., 1997), stochastic sensors for small solutes (Gu et al., 1999; Braha et al., 2000; Kasianowicz et al., 2001), and an alternative to DNA microarray technology for sequencing DNA (Howorka et al., 2001; Deamer and Branton, 2002; Vercoutere and Akeson, 2002).

Permeation of water, ions, and other solutes through the transmembrane pore of alpha-hemolysin is central to both the natural function and biotechnological applications of alpha-hemolysin. Indeed, permeation of ions, ATP, and water through alpha-hemolysin is

Figure 8. Microscopic model of the alpha-hemolysin channel in its native environment, a lipid bilayer membrane. The channel is drawn as a molecular surface separating the protein from the membrane and water. This surface is cut by the plane normal to the lipid bilayer passing through the geometrical center of the protein. All atoms but phosphorus of the dipalmitoyl phosphatidylcholine lipid bilayer are shown as lines; the phosphorus atoms are shown as spheres. Water and ions are not shown. The model comprises 288,680 atoms.

the major mechanism of alpha-hemolysin toxicity; electromigration of ions and analytes enables detection of analytes' concentrations in stochastic sensors (Gu et al., 1999); simultaneous transport of DNA and ions through the alpha-hemolysin pore may be used to decipher the chemical composition of DNA strands (Kasianowicz et al., 1996; Akenson et al., 1999).

Measurements have indicated a high sensitivity of the channel conductance to the atomic details of the pore and to the structure of the conducted solutes (Menestrina, 1986; Krasilnikov and Sabirov, 1989; Kasianowicz et al., 1996; Akenson et al., 1999; Gu et al., 1999; Braha et al., 2000). The alpha-hemolysin channel was found to undergo a dose- and voltage-dependent inactivation in the presence of divalent and trivalent cations (Menestrina, 1986). Its ionic conductance as well as anion selectivity is sensitive to pH and the chemical composition of the lipid bilayer (Krasilnikov and Sabirov, 1989; Korchev et al., 1995). Understanding these phenomena is a prerequisite for developing successful antitoxin treatments. Atomic details are even more important for biotechnological applications of alpha-hemolysin, as fluctuations of the ionic conductance through the transmembrane pore identify the presence of analytes that can be as small as protons (Kasianowicz and Bezrukov, 1995), whereas the absolute value of the modulated ionic current indicates the type of analyte, with a sensitivity that allows one to distinguish different types of metal ions (Braha et al., 2000). Finally, as A, C, G, and T nucleotides differ from each other by only a few atoms, achieving the goal of sequencing DNA requires relating the atomic structure of nucleotides to the measured ionic currents.

Currently, no instrument exists that can visualize permeation of ions and small solutes through transmembrane channels at atomic detail. The permeation process, however, can be investigated at atomic resolution through microscopic simulations. Theoretical models describing ion conductance through alpha-hemolysin supplemented early experimental studies of the channel (Menestrina, 1986; Krasilnikov and Sabirov, 1989). After the discovery of the crystallographic structure (Song et al., 1996), quantitative studies relying on the atomic structure became possible (Smart et al., 1998; Misakian and Kasianowicz, 2003; Noskov et al., 2004b; Cozmuta et al., 2005). Until recently, only small parts of the alpha-hemolysin channel could be simulated with all-atom MD (Shilov and Kurnikova, 2003). Advances in computational technology permit one today to use MD simulations as a computational microscope to obtain dynamic images of biomolecular systems. Bacterial membrane channels now can be investigated as integral units, i.e., in their native environment, and without losing atomic precision. The large scale of the simulations, however, brings the concern about the applicability of the force fields describing interactions between atoms, as these force fields were not tested at the time of their development to reproduce physical properties of biomolecular systems of such size. Using results of our previous study to be described in detail elsewhere (Aksimentiev and Schulten, in press), we shall test below the accuracy of our 300,000-atom simulation by computing the ionic conductance and the water permeability of alpha-hemolysin in its native environment. As the results will show, although our simulations are very large, results are highly accurate and compare favorably with experiments.

Ion Permeation

A typical MD study starts from building a microscopic model of the system of interest. In the case of alpha-hemolysin, the minimal system included one copy of the protein, a patch of lipid bilayer (dipalmitoyl phosphatidylcholine in our case), and water and ions in desired proportions. Once all components are assembled, the system is set free to evolve in time according to the laws of classical mechanics, moving toward an equilibrium state. The interactions between atoms are described by a molecular force field that has been developed and calibrated to reproduce physical properties of the simulated system. The duration of equilibration depends on the type of system, but is typically from 1 to 5 ns. In the present case the simulations included 288,680 atoms and required, for one nanosecond simulation period, 10 h on 256 processors of an Itanium 1.5-GHz cluster.

The first property that we are going to investigate is the current-voltage (I-V) curve of ion conduction through alpha-hemolysin. In a typical experiment, a membrane containing a single alpha-hemolysin channel divides an electrolyte solution into two compartments. A voltage bias is applied across the membrane through two Ag-AgCl electrodes submerged into the electrolyte solution on the two sides of the membrane. The ionic current induced by the applied voltage bias is registered by an amplifier. Typically, when no solutes other than KCl are added to the system, the amplifier records a steady-state current.

To reproduce this experimental setup in an MD simulation, a uniform electric field is applied to all atoms in the system. This uniform electric field induces, at the beginning of the simulation, a rearrangement of the ions and water that focuses the electric field to the vicinity of the membrane and the protein, abolishing the field gradient in the bulk. The resulting voltage bias, V, depends both on the magnitude of the applied field, E, and on the dimension, L_E, of the system in the direction of the applied field (Crozier et al., 2001; Aksimentiev et al., 2004b). If the electric field is directed along the z axis, the resulting transmembrane potential is $V = -EL_z$. Note that although the applied external field is uniform, the resulting total electrostatic potential around the protein is nonuniform, complying with the local dielectric properties of the protein and the membrane.

As in a typical experiment, in MD simulations a gradient of electrostatic potential induces electromigration of ions through the transmembrane pore of alpha-hemolysin whereas the lipid membrane itself is not permeable to ions if the applied voltage bias does not exceed 2 V. The ionic current is computed as

$$I(t) = \frac{1}{\Delta t L_z} \sum_{i=1}^{N} [q_i(z_i(t+\Delta t) - z_i(t))], \tag{1}$$

where z_i and q_i are the z coordinate and the charge of atom i, respectively; L_z is the length of the simulated system in the z direction; the sum runs over all ions. The coordinates of all atoms were recorded every $\Delta t = 1$ ps.

Simulations were performed for several transmembrane potentials varying from −1.2 to 1.2 V. The resulting I-V curve is shown in Fig. 9. Due to the time limitation of MD, the smallest ionic current that can be sampled sufficiently with presently available computer resources is about 100 pA, which, in the case of alpha-hemolysin, corresponds to the

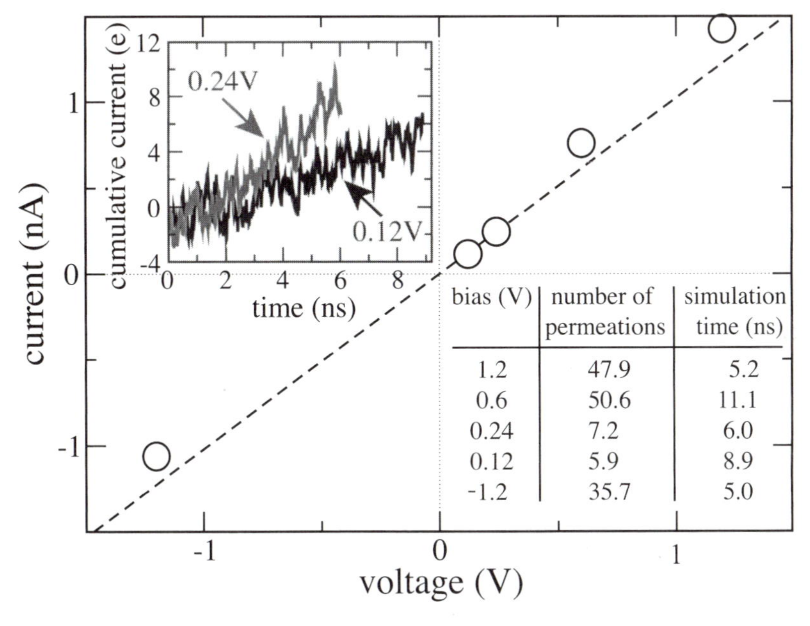

bias (V)	number of permeations	simulation time (ns)
1.2	47.9	5.2
0.6	50.6	11.1
0.24	7.2	6.0
0.12	5.9	8.9
-1.2	35.7	5.0

Figure 9. Current-voltage characteristics of alpha-hemolysin computed with MD. Each data point is derived from a 288,680-atom simulation of the system shown in Fig. 8. The dashed line indicates the linear fit through the data points at 120 and 240 mV and the origin. In accordance with experimental studies (Menestrina, 1986; Krasilnikov and Sabirov, 1989), the I-V curve is sublinear at $V < 0$. The absolute value of the ionic current at 120 mV is also in good agreement with experiment (Meller and Branton, 2002). The inset shows cumulative currents through alpha-hemolysin at 120 and 240 mV. The table shows the number of ion permeations computed by dividing the total charge transported through the alpha-hemolysin pore by e, the unitary charge (1.6×10^{-19} C).

smallest applied voltage bias of about 100 mV. The reason is that a statistically meaningful simulation of ionic permeation must include at least several full ion permeations through the channel during the timescale of the MD run. We ran our simulations typically for 5 to 11 ns and observed tens of full permeations. The table in Fig. 9 provides the simulation times and the respective number of ion permeations for each run used to determine the I-V curve.

The molecular force field describing interatomic interactions in our simulations (CHARMM27) is somewhat approximate and, therefore, before comparing the results of our simulations to experiment, we first have to assess the level of accuracy with which our simulations reproduce conductivity of KCl in the bulk. In a separate simulation, a 1 M solution of KCl was subject to an electric field that induced electromigration of K^+ and Cl^- ions. Computing the resulting ionic current with equation 1 yielded a bulk conductivity of 12.2 S/m for 1 M KCl, which is close to the measured conductivity of 11.0 S/m. We note, however, that this rather good agreement between experiment and simulation is likely due to the cancellation of the imperfections of the force field and is not due to an absolute accuracy of the latter, as the TIP3P model of water used with CHARMM27 in our simulations does not reproduce quantitatively transport properties of bulk water (Lamoureux et al., 2003; Yeh and Hummer, 2004) (see also next section).

The calculated I-V curve (Fig. 9) is in good agreement with experiment. The I-V curve is asymmetric, indicating that alpha-hemolysin partially rectifies the ionic current at negative voltage biases (a negative bias is defined here as the one that drives positive ions from the vestibule of alpha-hemolysin through the transmembrane pore across the membrane). This rectification is caused by the asymmetric distribution of charged residues along the transmembrane pore (Henrickson et al., 2000; Misakian and Kasianowicz, 2003). The absolute values of the ionic currents are also close to experimental values: at 120 mV and 21°C, the simulations predicted ionic currents of 130 ± 10 pA, which are in close agreement with the experimental value of 112 ± 3 pA (Meller and Branton, 2002).

Water Permeation

When concentrations of an impermeable solute on the two sides of a membrane are not equal, water flows from the low concentration side to the other side (Zhu et al., 2004b). In dilute solutions, the flux of water is linearly proportional to the solute concentration difference, which defines the osmotic permeability as the ratio of the net water flux to the concentration difference. Water permeation has been described for aquaporins above; the computational approach applied a pressure gradient across the channel. Rather than following the same route for alpha-hemolysin and computing the osmotic permeability directly, i.e., using an osmotic pressure gradient, we trace diffusion of water through the channel in an equilibrium MD simulation and then, using an elegant theory described by Zhu et al. (2004a) and Tajkhorshid et al. (in press), relate the properties of the freely diffusing water to the osmotic permeability of the channel.

To compute the osmotic permeability of alpha-hemolysin, we can use trajectories generated previously during the initial equilibration of the system and the simulations of the ionic conductance. To identify the volume occupied by water inside the channel, we divided the interior of the channel into disk-like segments centered around the symmetry axis of the alpha-hemolysin pore. Each disk was assigned a radius that confines 95% of all water inside that particular segment. The top and the bottom boundaries of the channel were defined by residues Gly134 and Ser16, respectively.

We define a collective coordinate of all water molecules inside the channel, n, using (Zhu et al., 2004a):

$$\mathrm{d}n = \sum_{i \in S(t)} \mathrm{d}z_i / L, \tag{2}$$

where $S(t)$ denotes the set of water molecules in the channel at time t, L is the length of the channel, and we assumed that the displacement of water molecule i in the z direction during $\mathrm{d}t$ is $\mathrm{d}z_i$. By demanding $n = 0$ at $t = 0$, $n(t)$ can be determined by integrating $\mathrm{d}n$. If a water molecule enters or exits the channel within our sampling interval $\mathrm{d}t$, only the portion of its displacement within the channel contributes to the sum.

Figure 10 (top) illustrates spontaneous diffusion of water through the alpha-hemolysin channel, where the collective coordinate $n(t)$ is plotted versus time. As n quantifies the net amount of water permeation, we learn from Fig. 10 (top) that about 50 water molecules diffuse through the channel one way or another within a 10-ns interval.

To quantify diffusion of water through the alpha-hemolysin channel, we concatenated all $n(t)$ trajectories shown in Fig. 10 (top) into one and then divided equally the resulting trajectory into M short pieces. Each subtrajectory $n_j(t)$ ($j = 1, \ldots, M$) had length t_M and was treated as an independent subtrajectory. After each subtrajectory was shifted to

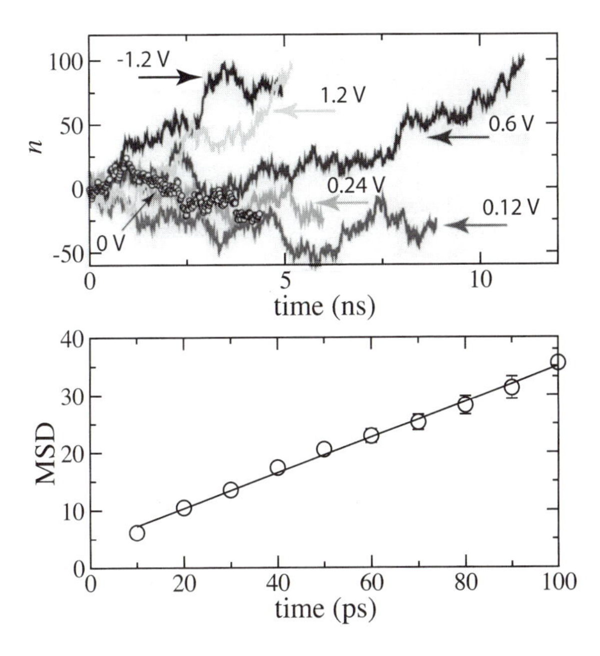

Figure 10. Computing the osmotic permeability of alpha-hemolysin with MD. (Top) Collective coordinate of all water molecules inside the channel $n(t)$ (equation 2) versus time; $n(t)$ quantifies the net amount of water permeation through the channel (see the text). Neither the sign nor the magnitude of the transmembrane potential has a noticeable deterministic effect on water permeability. (Bottom) Mean square displacement of $n(t)$ versus time. The slope of the curve yields the collective diffusion constant of water at 310/ns, which gives, after taking into account a correction for the low viscosity of TIP3P water, the osmotic permeability for alpha-hemolysin of $1.9 \times 10^{-12}\,\mathrm{cm^3/s}$ (see text).

provide $n_j(t)|_{t=0} = 0$, the mean square displacement (MSD) of $n(t)$ over the t_M interval was computed as

$$\text{MSD}(t_M) = \frac{1}{M}\sum_{j=1}^{M}[n_j(t_M)]^2. \tag{3}$$

In Fig. 10 (bottom) we plotted MSDs of the collective coordinate $n(t)$ versus t_M varying from 10 to 100 ps. At equilibrium, $n(t)$ can be described as a one-dimensional unbiased random walk, with a diffusion coefficient, D_n, that obeys

$$\langle n^2(t)\rangle = 2D_n t. \tag{4}$$

A linear regression fit to the plot of MSD(t_M) versus t_M yielded the collective diffusion coefficient, D_n, of water inside the channel of 310/ns. The osmotic permeability of a channel, p_f, is related to D_n (Zhu et al., 2004a) as

$$p_f = v_W D_n \tag{5}$$

where v_W is the average volume of a single water molecule (18 cm^3/mol). From this results the osmotic permeability of 5.6×10^{-12} cm^3/s.

Experimentally, the average single-channel permeability of alpha-hemolysin for water was found to be in the range of 1.3×10^{-12} to 1.5×10^{-12} cm^3/s (Paula et al., 1999), depending on pH. The factor 3 discrepancy between these results and our simulations originates from the properties of the TIP3P model of water that is known to underestimate the viscosity of water by a factor of 2.87 (Yeh and Hummer, 2004). After scaling down the computed osmotic permeability to account for the low viscosity of TIP3P water, we obtain a value of 1.9×10^{-12} cm^3/s, which is much closer to the experimental value.

We have demonstrated that the current-voltage dependence and the osmotic permeability of a transmembrane channel can be determined accurately from large-scale MD simulations. After taking into account known deficiencies of the molecular force field, results of simulations are in excellent agreement with experiment. The results endorse the methodology described above for investigating passive or electrophoretically driven permeation through other structurally known membrane channels, in particular, through porins and toxins, as they often have rather large (at least 1.0-nm-diameter) pores.

PERMEATION AND GATING OF MECHANOSENSITIVE CHANNELS

This final case study, on the mechanosensitive channel of small conductance, demonstrates how MD simulations elucidate the mechanisms underlying channel permeation and gating.

Mechanosensitive channels are ubiquitous membrane proteins in prokaryotes and eukaryotes, playing a fundamental role in hearing, touch, cardiovascular regulation, and volume regulation. Several mechanosensitive channels gated by mechanical strain in the cellular membrane have been identified in bacteria (Martinac et al., 1987; Berrier et al., 1989; Sukharev et al., 1994). These channels are essential for maintaining the viability of the cell through adaptation to osmotic stress (Levina et al., 1999; Blount and Moe, 1999; Sukharev, 2002).

Mechanosensitive channels of large (MscL) and small (MscS) conductance, found in the inner membrane of *E. coli* (Martinac et al., 1987; Berrier et al., 1989; Sukharev et al., 1994), have been extensively studied. Their crystal structures are available at 3.5- and 3.9-Å resolutions, respectively (Chang et al., 1998; Bass et al., 2002). MscL, with a conductance of about 3.2 nS and an activation midpoint of 11 dynes/cm (see Sukharev and Corey [2004] and references therein), forms a homopentameric structure approximately 50 Å wide in the plane of the membrane and 85 Å tall. On the other hand, MscS, with a conductance of about 1 nS and an activation midpoint of about 5 dynes/cm (Sukharev, 2002; Bezanilla and Perozo, 2002), forms a large homoheptameric structure approximately 80 Å wide in the plane of the membrane and 120 Å tall, extending largely into the cytoplasmic region of the cell.

Extensive studies have been carried out to explore and model gating mechanisms of MscL that involve large rearrangements of its transmembrane helices (Sukharev et al., 2001; Betanzos et al., 2002; Perozo et al., 2002a, 2002b; Gullingsrud et al., 2001; Gullingsrud and Schulten, 2003; Elmore and Dougherty, 2003; Kong et al., 2002; Colombo et al., 2003). Here we will focus on MscS. This channel has recently become the subject of experimental and theoretical investigations that seek to reveal its permeation properties and gating mechanism (Levina et al., 1999; Anishkin and Sukharev, 2004; Shapovalov and Lester, 2004; Schumann et al., 2004; Edwards et al., 2004).

The gating of MscS has been characterized by patch-clamp experiments in which the conductance and midpoint activation values mentioned above have been determined. Experiments have also revealed that MscS is reversibly inactivated at low pH (B. Martinac, personal communication; Cui et al., 1995). In addition, MscS shows slight anionic preference and is voltage modulated (Martinac et al., 1987; Sukharev, 2002). The estimated energy of opening, ΔG, is 11.4 ± 0.5 kT and the transition-related area change of the pore, ΔA, is $8.4 \pm 0.4\ nm^2$ (Sukharev, 2002).

Experimental results involving deletions in the C-terminal and N-terminal regions, cross-linking of the C termini, and cross-linking of site-specific cysteines suggest that the protein is very flexible, that its C terminus is structurally highly relevant, and that there are significant structural differences between the closed and open states (Miller, 2003; Koprowski and Kubalski, 2003; Miller et al., 2003; Schumann et al., 2004). Random mutagenesis experiments have identified some functionally relevant residues (Miller, 2003; Okada et al., 2002), whereas recent experiments on channel gating at 3-μs resolution have shown the existence of a substate for MscS in *E. coli* with a conductance corresponding to 2/3 of the conductance of the fully open state (Shapovalov and Lester, 2004).

The overall picture of MscS gating determined by experiments does not identify yet the molecular changes involved in MscS opening and closing. In this section we will show how MD simulations have contributed to the elucidation of gating in MscS. This section follows closely the report by Sotomayor and Schulten (2004).

Architectural Design of MscS

The overall architecture of MscS is shown in Color Plate 18 (see color insert). The channel is formed by seven identical subunits of 286 amino acids each. The transmembrane domain has three α-helices per subunit. The first two transmembrane helices (TM1 and TM2) are tilted 27 to 35 degrees with respect to the membrane normal and have positive residues (arginines) that may act as voltage sensors (Bass et al., 2002; Bezanilla and Perozo, 2002). The third transmembrane helix (TM3) is divided by a pronounced kink into two parts: TM3A and TM3B. The kink occurs at residue Gly113 and marks the end of the pore's cytoplasmic side. The highly hydrophobic transmembrane pore, which is formed by TM3A, appears to be in an open state in the crystal structure (Color Plate 18B) with a diameter of 6.5 Å in its narrowest section, lined by seven Leu105 side chains. The second part of the third transmembrane helix, TM3B, is almost parallel to the membrane plane (Color Plate 18A).

The cytoplasmic domain of MscS is formed by a large chamber (Color Plate 18C) thought to act as a molecular filter or docking site for regulatory proteins (but otherwise of unknown function), with seven openings on the sides and one distal opening at the bottom or intracellular side. Polar and charged residues are abundant in the cytoplasmic domain of MscS.

The crystal structure allows one to view the MscS channel at atomic resolution and to identify domains and residues relevant for MscS function. However, despite the availability of the structure, key questions remain unanswered: Does the structure explain the physiological role of MscS, i.e., is MscS only a safety valve? What residues are relevant for gating? Does the crystal structure show the open form of the channel or a conductance substate? What does the closed state of MscS look like? How do pH and transmembrane voltage control channel gating? What is the role of the large cytoplasmic domain? Some of these questions have been addressed by the MD simulations described below.

Permeation of Water Molecules in an Open State of MscS

In the MD simulations reported by Sotomayor and Schulten (2004), the crystal structure of MscS was embedded in a 1-palmitoyl-2-oleoyl-sn-glycero-3-phosphocholine lipid bilayer and the protein-membrane complex was then solvated, the combined system encompassing 224,340 atoms. Using the program NAMD (Kalé et al., 1999), it takes about 60 h to complete 1 ns of simulation on a 48-processor (3-GHz Athlon) cluster. After an equilibration of 3 ns with the protein restrained to the crystal structure, a close packing of lipids against the protein was achieved. Water permeation was then monitored for 1.45 ns of dynamics with the backbone of the protein restrained (this simulation will be referred to as sim1; see Sotomayor and Schulten [2004] for details). The transmembrane pore of MscS remained sufficiently open during sim1 for water permeation to occur, and in fact, five full permeation events of water molecules were observed (Color Plates 19A and 20A [see color insert]). Interestingly, we found two different states for the channel in this set of simulations: the pore transited from being completely filled with water to being partially empty. Such intermittent liquid-vapor oscillations of water in hydrophobic environments have been reported earlier (Anishkin and Sukharev, 2004; Beckstein and Sansom, 2003, 2004). Despite constraining the backbone of the protein, the radius of the transmembrane pore in its narrowest part decreased from an initial value of 3.2 Å in the crystal structure to a value of 2.8 Å, indicating a strong tendency of the channel to close in the membrane environment.

Spontaneous Closure of the Pore

The transmembrane channel of MscS is observed in an open state in the crystal structure. However, the channel at native physiological conditions should be closed, preventing leakage of solutes and dissipation of the cellular potential. Simulations in which restraints on the backbone of the protein were eliminated allowed us to investigate whether MscS relaxes to a closed state in a membrane without tension. Indeed, gradual elimination of backbone restraints resulted in the spontaneous closure of the transmembrane pore within 4 ns. A top view of the final state of this simulation (sim2a) is shown in Color Plate 19B. Comparing the state with the one obtained in sim1 (Color Plate 19A), one can immediately recognize the complete closure of the transmembrane pore. The pore radius at the end of the simulation was 1.27 Å in its narrowest section, i.e., at Leu105. However, the value of the radius fluctuated between 0.5 and 2.4 Å. No flow of water could be observed through the closed channel (Color Plate 20B). An analysis of the pore dynamics in simulation sim2a revealed that the main motion of the pore closure is an asymmetric displacement of the seven Leu105 side chains along with the appearance of a kink in the α-helical structure of TM3A in some subunits. The motion causing the kink involved breaking and transferring α-helical hydrogen bonds. Two possible causes were identified for the deformation of pore-lining structures: direct interaction with the lipid bilayer and tension transmitted through TM1 and TM2, the latter being the dominant cause as explained below.

Channel Opening under Surface Tension

The activity of MscS is recorded in patch-clamp experiments where curvature in the membrane is generated by a small pressure. The induced curvature and the resulting tension mimic the effect of an osmotic downshock, where cell volume increases due to the intracellular/extracellular difference in ion concentration (osmotic pressure). We simulated a similar external condition (simulation sim2b, 4 ns, surface tension of 20 dynes/cm) and explored the ensuing dynamics of MscS. The initial MscS configuration in this simulation was the restrained equilibrated crystal structure of MscS (see above). A top view of the final state of sim2b is shown in Color Plate 19C. An increase in the transmembrane pore size is evident; the pore radius is 3.75 Å at its narrowest part, a value that is larger than the value observed for the crystal structure (3.26 Å). Intermediate values of the pore radius at its narrowest section during sim2b were found to be about 4 Å and, accordingly, the conduction of water molecules through the transmembrane channel was the largest of all three systems simulated (Color Plate 20C).

The surface tension applied in sim2b (20 dynes/cm) is approximately four times higher than what is needed to experimentally open MscS (Sukharev, 2002), and even higher than the rupture limit for a membrane. However, the high surface tension permitted us to induce the pore opening on the short timescale of MD simulations. Moreover, the force field parameters and simulation conditions (NPT) underestimate the experimental area per lipid, which may justify the strong tension applied (Feller and Pastor, 1999; Gullingsrud and Schulten, 2004). Simulations of MscL under surface tension have also utilized unphysiological tension values of 50 to 500 dynes/cm (Gullingsrud et al., 2001; Colombo et al., 2003). Opening of MscL in a hydrated membrane environment was observed only at 500 dynes/cm, which is more than an order of magnitude larger than the experimentally required surface tension for MscL opening. In none of our simulations did we observe membrane rupture.

The in-plane protein areal expansion during channel opening, ΔA, has been estimated from experiments to be 8.4 nm^2 (Sukharev, 2002). According to our simulations (sim2a and sim2b), the change in area in the *xy* plane of the simulated cell was 8.6 nm^2. However, the change in area occupied by the protein was estimated to be 2.8 nm^2, leaving some room for further expansion of the protein in the membrane plane.

Gating of the Pore

The results of simulations sim2a and sim2b described above suggest a general picture for MscS gating. In our simulations, the TM1 and TM2 helices of some subunits moved in a hinge-like fashion, with the TM1-TM2 loop "swinging" and the TM2-TM3 loop acting as the hinge (Color Plate 21A), in agreement with the gating mechanism suggested by Bass et al. (2002). The movement of the transmembrane helices was found to be caused by the interaction of the lipid headgroups with charged and polar residues of TM1 and TM2. These interactions were favored by a rearrangement of the membrane at the protein interface, where side chains of charged residues reoriented and were not facing the hydrophobic region of the membrane as suggested by the crystal structure, but rather interacted with lipid headgroups, as mentioned above. From these results we infer that the stated interactions explain the movement of the transmembrane helices and are a key element in the gating mechanism. However, a concerted, symmetric movement of all transmembrane subunits was not seen, likely due to the overall displacement of the protein in an irregular environment of lipids.

In all simulations, we observed the formation of one to four salt bridges between residues Asp62 and Arg128 of adjacent subunits (Color Plate 21B). Both residues are conserved in 65% or more of 25 homologs of MscS in prokaryotes and archaea (Bass et al., 2002). Counting in the number of analogs in which Asp is replaced by a Glu (both negative), or Arg is replaced by a Lys (both positive), the conservation of the salt bridge becomes more prominent (~80%), reflecting its functional importance. Although the specific role of these salt bridges in the gating mechanism is not completely clear, they may provide direct links between the movement of the TM1-TM2 domains described above and the pore-lining helices. Therefore, sensitivity of the channel to membrane tension could be increased due to their presence.

The MD simulations summarized above and reported by Sotomayor and Schulten (2004) have shown that MscS is a very flexible protein. Indeed, the transmembrane pore closes in a relaxed membrane and remains open in a membrane subject to surface tension. Simulations also suggest that the movement of transmembrane helices, coupled to lipid headgroups and the cytoplasmic domain of the protein, controls the gating of the MscS transmembrane pore.

CONCLUDING REMARKS

Permeation of materials across cellular membranes is a dynamic process. MD simulations are in a unique position to describe such processes in the increasing number of cases in which the structures of channel proteins are known. MD simulations can describe channel permeability and gating in the native environment of lipid bilayer and water, even in cases of large channels that require simulations of 300,000 atoms. Timescale limitations, currently about 50 ns, can be circumvented through steered and interactive molecular dynamics or computation of potentials of mean force.

We have presented in this chapter four case studies that demonstrate the power of MD simulations in unraveling the mechanisms underlying the function of membrane channels. Simulations of AQP water channels present a prime example of the effectiveness and usefulness of MD in simulating conduction and explaining selectivity. A detailed energetic analysis of ion permeation through chloride channels proposes a two-ion permeation mechanism that can reconcile naturally structural and physiological data. The study of ion and water permeation through hemolysin exemplifies how accurately one can simulate today even very large membrane channel systems. Finally, the study of mechanical gating of MscS shows that molecular dynamics can account even for complex gating motions of channels.

In the past, the explanation of protein mechanisms through structures was often delegated to the last page of reports on the new structures and involved a combination of prior knowledge, a close look at the structure, and intuitive arguments. Rare is the paper reporting a new structure along with an advanced theoretical or computational analysis of the structure's implications. But a true explanation of the mechanisms behind a function of a protein can only be derived from a simulation of the protein's function based on the structure together with a priori knowledge, e.g., force fields, from physics and chemistry; the simulation provides an extrapolation to a noncrystalline environment and to the dynamic domain. There is no shortcut to a sound explanation. The impressive achievements of protein crystallographers in resolving the structures of membrane proteins today and of other experimental researchers to characterize functional and other properties of these proteins need to be complemented by sound computational modeling. No method is perfect in this regard, and computational modeling including MD simulation is far from it, but as past examples suggest, with computational modeling one can reach more quickly to the goal of understanding membrane channels than without.

Acknowledgments. This work was possible only through the advanced computational resources of the National Science Foundation (NSF) supercomputer centers in Urbana and Pittsburgh as well as our group's computer laboratory. We also acknowledge supercomputer time provided via the NSF's National Resources Allocation Committee grant MCA93S028. The reported studies were supported by the National Institutes of Health grants PHS-5-P41-RR05969 and R01-GM67887. Molecular images in the chapter were created with the program VMD (Humphrey et al., 1996).

We thank the staff of the national centers and of our own group.

REFERENCES

Accardi, A., L. Kolmakova-Partensky, C. Williams, and C. Miller. 2004. Ionic currents mediated by a prokaryotic homologue of CLC Cl^- channels. *J. Gen. Physiol.* **123:**109–119.

Accardi, A., and C. Miller. 2004. Secondary active transport mediated by a prokaryotic homologue of ClC Cl^- channels. *Nature* **427:**803–807.

Agre, P., M. Bonhivers, and M. J. Borgnia. 1998. The aquaporins, blueprints for cellular plumbing systems. *J. Biol. Chem.* **273:**14659–14662.

Akenson, M., D. Branton, J. J. Kasianowicz, E. Brandin, and D. W. Deamer. 1999. Microsecond time-scale discrimination among polycytidylic acid, polyadenylic acid, and polyuridylic acid as homopolymers or as segments within single RNA molecules. *Biophys. J.* **77:**3227–3233.

Aksimentiev, A., I. A. Balabin, R. H. Fillingame, and K. Schulten. 2004a. Insights into the molecular mechanism of rotation in the F_0 sector of ATP synthase. *Biophys. J.* **86:**1332–1344.

Aksimentiev, A., J. B. Heng, G. Timp, and K. Schulten. 2004b. Microscopic kinetics of DNA translocation through synthetic nanopores. *Biophys. J.* **87:**2086–2097.

Aksimentiev, A., and K. Schulten. Imaging the permeability of alpha-hemolysin with molecular dynamics. *Biophys. J.*, in press.

Amaro, R., E. Tajkhorshid, and Z. Luthey-Schulten. 2003. Developing an energy landscape for the novel function of a $(\beta/\alpha)_8$ barrel: ammonia conduction through HisF. *Proc. Natl. Acad. Sci. USA* **100:** 7599–7604.

Anishkin, A., and S. Sukharev. 2004. Water dynamics and dewetting transitions in the small mechanosensitive channel MscS. *Biophys. J.* **86:**2883–2895.

Åqvist, J., and V. Luzhkov. 2000. Ion permeation mechanism of the potassium channel. *Nature* **404:**881–884.

Ash, W. L., M. R. Zlomislic, E. O. Oloo, and D. P. Tieleman. 2004. Computer simulations of membrane proteins. *Biochim. Biophys. Acta* **1666:**158–189.

Bass, R. B., P. Strop, M. Barclay, and D. C. Rees. 2002. Crystal structure of *Escherichia coli* MscS, a voltage-modulated and mechanosensitive channel. *Science* **298:**1582–1587.

Baudry, J., E. Tajkhorshid, F. Molnar, J. Phillips, and K. Schulten. 2001. Molecular dynamics study of bacteriorhodopsin and the purple membrane. *J. Phys. Chem. B* **105:**905–918.

Bayley, H. 1995. Pore-forming proteins with built-in triggers and switches. *Bioorg. Chem.* **23:**340–354.

Beckstein, O., and M. S. P. Sansom. 2003. Liquid-vapor oscillations of water in hydrophobic nanopores. *Proc. Natl. Acad. Sci. USA* **100:**7063–7068.

Beckstein, O., and M. S. P. Sansom. 2004. The influence of geometry, surface character, and flexibility on the permeation of ions and water through biological pores. *Phys. Biol.* **1:**42–52.

Bernèche, S., and B. Roux. 2000. Molecular dynamics of the KcsA K^+ channel in a bilayer membrane. *Biophys. J.* **78:**2900–2917.

Bernèche, S., and B. Roux. 2001. Energetics of ion conduction through the K^+ channel. *Nature* **414:**73–77.

Berrier, C., A. Coulombe, C. Houssin, and A. Ghazi. 1989. A patch-clamp study of ion channels of inner and outer membranes and of contact zones of *E. coli*, fused into giant liposomes. *FEBS Lett.* **259:**27–32.

Betanzos, M., C.-S. Chiang, H. R. Guy, and S. Sukharev. 2002. A large iris-like expansion of a mechanosensitive channel protein induced by membrane tension. *Nat. Struct. Biol.* **9:**704–710.

Bezanilla, F., and E. Perozo. 2002. Force and voltage sensors in one structure. *Science* **298:**1562–1563.

Bhakdi, S., U. Weller, I. Walev, E. Martin, D. Jonas, and M. Palmer. 1993. A guide to the use of pore-forming toxins for controlled permeabilization of cell membranes. *Med. Microbiol. Immunol.* **182:**167–175.

Bhakdi, S., and J. Tranum-Jensen. 1991. Alpha-toxin of *Staphylococcus aureus. Microbiol. Rev.* **55:**733–751.

Bhandarkar, M., G. Budescu, W. Humphrey, J. A. Izaguirre, S. Izrailev, L. V. Kalé, D. Kosztin, F. Molnar, J. C. Phillips, and K. Schulten. 1999. BioCoRE: a collaboratory for structural biology, p. 242–251. *In* A. G. Bruzzone, A. Uchrmacher, and E. H. Page (ed.), *Proceedings of the SCS International Conference on Web-Based Modeling and Simulation*. Society for Modeling and Simulation, San Francisco, Calif.

Blount, P., and P. C. Moe. 1999. Bacterial mechanosensitive channels: integrating physiology, structure and function. *Trends Microbiol.* **18:**420–424.

Bond, P. J., and M.S.P. Sansom. 2004. The simulation approach to bacterial outer membrane proteins. *Mol. Mem. Biol.* **21:**151–161.

Borgnia, M., S. Nielsen, A. Engel, and P. Agre. 1999a. Cellular and molecular biology of the aquaporin water channels. *Annu. Rev. Biochem.* **68:**425–458.

Borgnia, M. J., and P. Agre. 2001. Reconstitution and functional comparison of purified GlpF and AqpZ, the glycerol and water channels from *Escherichia coli. Proc. Natl. Acad. Sci. USA* **98:**2888–2893.

Borgnia, M. J., D. Kozono, G. Calamita, P. C. Maloney, and P. Agre. 1999b. Functional reconstitution and characterization of AqpZ, the *E. coli* water channel protein. *J. Mol. Biol.* **291:**1169–1179.

Bostick, D. L., and M. L. Berkowitz. 2004. Exterior site occupancy infers chloride-induced proton gating in a prokaryotic homolog of the ClC chloride channel. *Biophys. J.* **87:**1686–1696.

Braha, O., L. Q. Gu, L. Zhou, X. Lu, S. Cheley, and H. Bayley. 2000. Simultaneous stochastic sensing of divalent metal ions. *Nat. Biotechnol.* **18:**1005–1007.

Braun, R., D. M. Engelman, and K. Schulten. 2004. Molecular dynamics simulations of micelle formation around dimeric glycophorin A transmembrane helices. *Biophys. J.* **87:**754–763.

Chakrabarti, N., E. Tajkhorshid, B. Roux, and R. Pomès. 2004. Molecular basis of proton blockage in aquaporins. *Structure* **12:**65–74.

Chang, G., R. H. Spencer, A. T. Lee, M. T. Barclay, and D. C. Rees. 1998. Structure of the MscL homolog from *Mycobacterium tuberculosis*: a gated mechanosensitive ion channel. *Science* **282:**2220–2226.

Cohen, J., and K. Schulten. 2004. Mechanism of anionic conduction across ClC. *Biophys. J.* **86:**836–845.

Colombo, G., S. J. Marrink, and A. E. Mark. 2003. Simulation of MscL gating in a bilayer under stress. *Biophys. J.* **84:**2331–2337.

Corry, B., M. O'Mara, and S.-H. Chung. 2004. Conduction mechanisms of chloride ions in ClC-type channels. *Biophys. J.* **86:**846–860.

Cozmuta, I., J. T. O'Keeffe, D. Bose, and V. Stolc. 2005. Hybrid MD-Nernst-Planck model of α-hemolysin conductance properties. *Mol. Sim.* **31:**79–93.

Crozier, P. S., D. Henderson, R. L. Rowley, and D. D. Busath. 2001. Model channel ion currents in NaCl-extended simple point charge water solution with applied-field molecular dynamics. *Biophys. J.* **81:**3077.

Cui, C., D. O. Smith, and J. Adler. 1995. Characterization of mechanosensitive channels in *Escherichia coli* cytoplasmic membrane by whole-cell patch clamp recording. *J. Mol. Biol.* **44:**31–42.

Damjanović, A., I. Kosztin, U. Kleinekathofer, and K. Schulten. 2002. Excitons in a photosynthetic light-harvesting system: a combined molecular dynamics, quantum chemistry and polaron model study. *Phys. Rev. E* **65:**031919.

Darden, T., D. York, and L. Pedersen. 1993. Particle mesh Ewald. An N log(N) method for Ewald sums in large systems. *J. Chem. Phys.* **98:**10089–10092.

Deamer, D., and D. Branton. 2002. Characterization of nucleic acids by nanopore analysis. *Acc. Chem. Res.* **35:**817–825.

Deen, P. M. T., and C. H. van Os. 1998. Epithelial aquaporins. *Curr. Opin. Cell Biol.* **10:**435–442.

de Groot, B. L., A. Engel, and H. Grubmüller. 2001. A refined structure of human aquaporin-1. *FEBS Lett.* **504:**206–211.

de Groot, B. L., T. Frigato, V. Helms, and H. Grubmüller. 2003. The mechanism of proton exclusion in the aquaporin-1 water channel. *J. Mol. Biol.* **333:**279–293.

de Groot, B. L., and H. Grubmüller. 2001. Water permeation across biological membranes: mechanism and dynamics of aquaporin-1 and GlpF. *Science* **294:**2353–2357.

de Groot, B. L., J. B. Heymann, A. Engel, K. Mitsuoka, Y. Fujiyoshi, and H. Grubmüller. 2000. The fold of human aquaporin 1. *J. Mol. Biol.* **300:**987–994.

Domene, C., P. J. Bond, and M. S. P. Sansom. 2003. Membrane protein simulations: ion channels and bacterial outer membrane proteins. *Adv. Protein Chem.* **66:**159–193.

Doyle, D. A., J. M. Cabral, R. A. Pfuetzer, A. Kuo, J. M. Gulbis, S. L. Cohen, B. T. Chait, and R. MacKinnon. 1998. The structure of the potassium channel: molecular basis of K^+ conduction and selectivity. *Science* **280:**69–77.

Dutzler, R., E. B. Campbell, M. Cadene, B. T. Chait, and R. MacKinnon. 2002. X-ray structure of a ClC chloride channel at 3.0 Å reveals the molecular basis of anion selectivity. *Nature* **415:**287–294.

Dutzler, R., E. B. Campbell, and R. MacKinnon. 2003. Gating the selectivity filter in ClC chloride channels. *Science* **300:**108–112.

Edwards, M. D., I. R. Booth, and S. Miller. 2004. Gating the bacterial mechanosensitive channels: Mscs a new paradigm? *Curr. Opin. Microbiol.* **7:**163–167.

Elmore, D. E., and D. A. Dougherty. 2003. Investigating lipid composition effects on the mechanosensitive channel of large conductance (MscL) using molecular dynamics simulations. *Biophys. J.* **85:**1512–1524.

Essmann, U., L. Perera, M. L. Berkowitz, T. Darden, H. Lee, and L. G. Pedersen. 1995. A smooth particle mesh Ewald method. *J. Chem. Phys.* **103:**8577–8593.

Estévez, R., and T. J. Jentsch. 2002. ClC chloride channels: correlating structure with function. *Curr. Opin. Struct. Biol.* **12:**531–539.

Fahlke, C. 2001. Ion permeation and selectivity in ClC-type chloride channels. *Am. J. Physiol. Ren. Physiol.* **280:**F748–F757.

Fahlke, C., H. Yu, C. L. Beck, T. R. Rhodes, and J. A. L. George. 1997. Pore-forming segments in voltage-gated chloride channels. *Nature* **390:**529–532.

Faraldo-Gómez, J. D., and B. Roux. 2004. Electrostatics of ion stabilization in a ClC chloride channel homologue from *Escherichia coli*. *J. Mol. Biol.* **339:**981–1000.

Feller, S. E., and R. W. Pastor. 1999. Constant surface tension simulations of lipid bilayers: the sensitivity of surface areas and compressibilities. *J. Chem. Phys.* **111:**1281–1287.

Feller, S. E., R. W. Pastor, A. Rojnuckarin, S. Bogusz, and B. R. Brooks. 1996. Effect of electrostatic force truncation on interfacial and transport properties of water. *J. Phys. Chem.* **100:**17011–17020.

Feller, S. E., Y. H. Zhang, R. W. Pastor, and B. R. Brooks. 1995. Constant pressure molecular dynamics simulation—the Langevin piston method. *J. Chem. Phys.* **103:**4613–4621.

Friedrich, T., T. Breiderhoff, and T. J. Jentsch. 1999. Mutational analysis demonstrates that ClC-4 and ClC-5 directly mediate plasma membrane currents. *J. Biol. Chem.* **274:**896–902.

Fu, D., A. Libson, L. J. W. Miercke, C. Weitzman, P. Nollert, J. Krucinski, and R. M. Stroud. 2000. Structure of a glycerol conducting channel and the basis for its selectivity. *Science* **290:**481–486.

Fujiyoshi, Y., K. Mitsuoka, B. L. de Groot, A. Philippsen, H. Grubmüller, P. Agre, and A. Engel. 2002. Structure and function of water channels. *Curr. Opin. Struct. Biol.* **12:**509–515.

Gonen, T., P. Sliz, J. Kistler, Y. Cheng, and T. Walz. 2004. Aquaporin-0 membrane junctions reveal the structure of a closed water pore. *Nature* **429:**193–197.

Gouaux, E. 1998. α-Hemolysin from *Staphylococcus aureus*: an archetype of β-barrel, channel-forming toxins. *J. Struct. Biol.* **121:**110–122.

Grayson, P., E. Tajkhorshid, and K. Schulten. 2003. Mechanisms of selectivity in channels and enzymes studied with interactive molecular dynamics. *Biophys. J.* **85:**36–48.

Gu, Q., O. Braha, S. Conlan, S. Cheley, and H. Bayley. 1999. Stochastic sensing of organic analytes by a pore-forming protein containing a molecular adapter. *Nature* **398:**686–690.

Gullingsrud, J., R. Braun, and K. Schulten. 1999. Reconstructing potentials of mean force through time series analysis of steered molecular dynamics simulations. *J. Comp. Phys.* **151:**190–211.

Gullingsrud, J., D. Kosztin, and K. Schulten. 2001. Structural determinants of MscL gating studied by molecular dynamics simulations. *Biophys. J.* **80:**2074–2081.

Gullingsrud, J., and K. Schulten. 2003. Gating of MscL studied by steered molecular dynamics. *Biophys. J.* **85:**2087–2099.

Gullingsrud, J., and K. Schulten. 2004. Lipid bilayer pressure profiles and mechanosensitive channel gating. *Biophys. J.* **86:**3496–3509.

Hayashi, S., E. Tajkhorshid, H. Kandori, and K. Schulten. 2004. Role of hydrogen-bond network in energy storage of bacteriorhodopsin's light-driven proton pump revealed by ab initio normal mode analysis. *J. Am. Chem. Soc.* **126:**10516–10517.

Hayashi, S., E. Tajkhorshid, E. Pebay-Peyroula, A. Royant, E. M. Landau, J. Navarro, and K. Schulten. 2001. Structural determinants of spectral tuning in retinal proteins—bacteriorhodopsin vs sensory rhodopsin II. *J. Phys. Chem. B* **105:**10124–10131.

Hayashi, S., E. Tajkhorshid, and K. Schulten. 2003. Molecular dynamics simulation of bacteriorhodopsin's photoisomerization using ab initio forces for the excited chromophore. *Biophys. J.* **85:**1440–1449.

Heller, H., M. Schaefer, and K. Schulten. 1993. Molecular dynamics simulation of a bilayer of 200 lipids in the gel and in the liquid crystal-phases. *J. Phys. Chem.* **97:**8343–8360.

Heller, K. B., E. C. Lin, and T. H. Wilson. 1980. Substrate specificity and transport properties of the glycerol facilitator of *Escherichia coli*. *J. Bacteriol.* **144:**274–278.

Henrickson, S., M. Misakian, B. Robertson, and J. J. Kasianowicz. 2000. Driven DNA transport into an asymmetric nanometer-scale pore. *Phys. Rev. Lett.* **85:**3057–3060.

Heymann, J. B., and A. Engel. 1999. Aquaporins: phylogeny, structure, and physiology of water channels. *News Physiol. Sci.* **14:**187–193.

Howorka, S., S. Cheley, and H. Bayley. 2001. Sequence-specific detection of individual dna strands using engineered nanopores. *Nat. Biotechnol.* **19:**636–639.

Hu, X., T. Ritz, A. Damjanović, F. Autenrieth, and K. Schulten. 2002. Photosynthetic apparatus of purple bacteria. *Q. Rev. Biophys.* **35:**1–62.

Humphrey, W., A. Dalke, and K. Schulten. 1996. VMD—visual molecular dynamics. *J. Mol. Graph.* **14:**33–38.

Ilan, B., E. Tajkhorshid, K. Schulten, and G. A. Voth. 2004. The mechanism of proton exclusion in aquaporin channels. *Proteins* **55:**223–228.

Isralewitz, B., J. Baudry, J. Gullingsrud, D. Kosztin, and K. Schulten. 2001a. Steered molecular dynamics investigations of protein function. *J. Mol. Graph. Model.* **19:**13–25.

Isralewitz, B., M. Gao, and K. Schulten. 2001b. Steered molecular dynamics and mechanical functions of proteins. *Curr. Opin. Struct. Biol.* **11:**224–230.

Iyer, R., T. M. Iverson, A. Accardi, and C. Miller. 2002. A biological role for prokaryotic ClC chloride channels. *Nature* **419:**715–718.

Izrailev, S., A. R. Crofts, E. A. Berry, and K. Schulten. 1999. Steered molecular dynamics simulation of the Rieske subunit motion in the cytochrome bc_1 complex. *Biophys. J.* **77:**1753–1768.

Izrailev, S., S. Stepaniants, B. Isralewitz, D. Kosztin, H. Lu, F. Molnar, W. Wriggers, and K. Schulten. 1998. Steered molecular dynamics, p. 39–65. *In* P. Deuflhard, J. Hermans, B. Leimkuhler, A. E. Mark, S. Reich, and R. D. Skeel (ed.), *Lecture Notes in Computational Science and Engineering*, vol. 4. *Computational Molecular Dynamics: Challenges, Methods, Ideas.* Springer-Verlag, Berlin, Germany.

Jarzynski, C. 1997a. Equilibrium free-energy differences from nonequilibrium measurements: a master equation approach. *Phys. Rev. E* **56:**5018–5035.

Jarzynski, C. 1997b. Nonequilibrium equality for free energy differences. *Phys. Rev. Lett.* **78:**2690–2693.

Jensen, M. Ø., S. Park, E. Tajkhorshid, and K. Schulten. 2002. Energetics of glycerol conduction through aquaglyceroporin GlpF. *Proc. Natl. Acad. Sci. USA* **99:**6731–6736.

Jensen, M. Ø., E. Tajkhorshid, and K. Schulten. 2001. The mechanism of glycerol conduction in aquaglyceroporins. *Structure* **9:**1083–1093.

Jensen, M. Ø., E. Tajkhorshid, and K. Schulten. 2003. Electrostatic tuning of permeation and selectivity in aquaporin water channels. *Biophys. J.* **85:**2884–2899.

Jentsch, T. J., T. Friedrich, A. Schriever, and H. Yamada. 1999. The CLC chloride channel family. *Pflugers Arch.* 437:783–795.

Jiang, Y., A. Lee, J. Chen, M. Cadene, B. T. Chait, and R. MacKinnon. 2002. Crystal structure and mechanism of a calcium-gated potassium channel. *Nature* **417:**515–522.

Jiang, Y., A. Lee, J. Chen, M. Cadene, B. T. Chait, and R. MacKinnon. 2003. X-ray structure of a voltage-dependent K^+ channel. *Nature* **423:**33–41.

Jung, J. S., G. M. Preston, B. L. Smith, W. B. Guggino, and P. Agre. 1994. Molecular structure of the water channel through aquaporin CHIP—the hourglass model. *J. Biol. Chem.* **269:**14648–14654.

Kalé, L., R. Skeel, M. Bhandarkar, R. Brunner, A. Gursoy, N. Krawetz, J. Phillips, A. Shinozaki, K. Varadarajan, and K. Schulten. 1999. NAMD2: greater scalability for parallel molecular dynamics. *J. Comp. Phys.* **151:**283–312.

Kasianowicz, J. J., and S. M. Bezrukov. 1995. Protonation dynamics of the α-toxin ion channel from spectral analysis of pH dependent current fluctuations. *Biophys. J.* **69:**94–105.

Kasianowicz, J. J., E. Brandib, D. Branton, and W. Deamer. 1996. Characterization of individual polynucleotide molecules using a membrane channel. *Proc. Natl. Acad. Sci. USA* **93:**13770–13773.

Kasianowicz, J. J., S. E. Henrickson, H. H. Weetall, and B. Robertson. 2001. Simultaneous multianalyte detection with a nanometer-scale pore. *Anal. Chem.* **73:**2268–2272.

Kong, Y., Y. Shen, T. E. Warth, and J. Ma. 2002. Conformational pathways in the gating of *Escherichia coli* mechanosensitive channel. *Proc. Natl. Acad. Sci. USA* **99:**5999–6004.

Koprowski, P., and A. Kubalski. 2003. C termini of the *Escherichia coli* mechanosensitive ion channel (MscS) move apart upon the channel opening. *J. Biol. Chem.* **278:**11237–11245.

Korchev, Y. E., G. M. Alder, A. Bakhramov, C. L. Bashford, B. S. Joomun, E. V. Sviderskaya, P. N. R. Usherwood, and C. A. Pasternak. 1995. *Staphylococcus aureus* alpha-toxin: channel-like behavior in lipid bilayers and patch clamped cells. *J. Membr. Biol.* **143:**143–151.

Kosztin, I., and K. Schulten. 2004. Fluctuation-driven molecular transport through an asymmetric membrane channel. *Phys. Rev. Lett.* **93:**238102.

Krasilnikov, O. V., and R. Z. Sabirov. 1989. Ion transport through channels formed in lipid bilayers by *Staphylococcus aureus* alpha-toxin. *Gen. Physiol. Biophys.* **8:**213–222.

Kumar, S., D. Bouzida, R. H. Swendsen, P. A. Kollman, and J. M. Rosenberg. 1992. The weighted histogram analysis method for free-energy calculations on biomolecules. I. The method. *J. Comp. Chem.* **13:**1011–1021.

Lamoureux, G., A. D. MacKerell, Jr., and B. Roux. 2003. A simple polarizable model of water based on classical drude oscillators. *J. Chem. Phys.* **119:**5185–5197.

Law, R. J., and M. S. P. Sansom. 2002. Water transporters: how so fast yet so selective? *Curr. Biol.* **12:**R250–R252.

Levina, N., S. Totemeyer, N. R. Stokes, P. Louis, M. A. Jones, and I. Booth. 1999. Protection of *Escherichia coli* cells against extreme turgor by activation of MscS and MscL mechanosensitive channels: identification of genes required for MscS activity. *EMBO J.* **18:**1730–1737.

Li, J., and A. S. Verkman. 2001. Impaired hearing in mice lacking aquaporin-4 water channels. *J. Biol. Chem.* **276:**31233–31237.

Lu, D., P. Grayson, and K. Schulten. 2003. Glycerol conductance and physical asymmetry of the *Escherichia coli* glycerol facilitator GlpF. *Biophys. J.* **85:**2977–2987.

MacKerell, A. D., Jr., D. Bashford, M. Bellott, et al. 1992. Self-consistent parameterization of biomolecules for molecular modeling and condensed phase simulations. *FASEB J.* **6:**A143.

MacKerell, A. D., Jr., D. Bashford, M. Bellott, et al. 1998. All-hydrogen empirical potential for molecular modeling and dynamics studies of proteins using the CHARMM22 force field. *J. Phys. Chem. B* **102:** 3586–3616.

Maduke, M., C. Miller, and J. A. Mindell. 2000. A decade of CLC chloride channels: structure, mechanism, and many unsettled questions. *Annu. Rev. Biophys. Biomol. Struct.* **29:**411–438.

Martinac, B., M. Buechner, A. H. Delcour, J. Adler, and C. Kung. 1987. Pressure-sensitive ion channel in *Escherichia coli*. *Proc. Natl. Acad. Sci. USA* **84:**2297–2301.

Meller, A., and D. Branton. 2002. Single molecule measurements of DNA transport through a nanopore. *Electrophoresis* **23:**2583–2591.

Menestrina, G. 1986. Ionic channels formed by *Staphylococcus aureus* alpha-toxin: voltage-dependent inhibition by divalent and trivalent cations. *J. Membr. Biol.* **90:**177–190.

Miller, C. 1982. Open-state substructure of single chloride channels from Torpedo electroplax. *Philos. Trans. R. Soc. Lond. B* **299:**401–411.

Miller, C. 2003. Reading eukaryotic function through prokaryotic spectacles. *J. Gen. Physiol.* **122:**129–131.

Miller, M. K., M. L. Bang, C. C. Witt, D. Labeit, C. Trombitas, K. Watanabe, H. Granzier, A. S. McElhinny, C. C. Gregorio, and S. Labelt. 2003. The muscle ankyrin repeat proteins: CARP, ankrd2/Arpp and DARP as a family of titin filament-based stress response molecules. *J. Mol. Biol.* **333:**951–964.

Miloshevsky, G. V., and P. C. Jordan. 2004. Anion pathway and potential energy profiles along curvilinear bacterial ClC Cl^- pores: Electrostatic effects of charged residues. *Biophys. J.* **86:**825–835.

Mindell, J. A., M. Maduke, C. Miller, and N. Grigorieff. 2001. Projection structure of a ClC-type chloride channel at 6.5 Å resolution. *Nature* **409:**219–223.

Misakian, M., and J. J. Kasianowicz. 2003. Electrostatic influence of ion transport through the αHL channel. *J. Membr. Biol.* **195:**137–146.

Morais-Cabral, J. H., Y. Zhou, and R. MacKinnon. 2001. Energetic optimization of ion conduction rate by the K^+ selectivity filter. *Nature* **414:**37–41.

Nelson, M., W. Humphrey, A. Gursoy, A. Dalke, L. Kalé, R. D. Skeel, and K. Schulten. 1996. NAMD—a parallel, object-oriented molecular dynamics program. *Int. J. Supercomput. Appl.* **10:**251–268.

Nollert, P., W. E. C. Harries, D. Fu, L. J. W. Miercke, and R. M. Stroud. 2001. Atomic structure of a glycerol channel and implications for substrate permeation in aqua(glycero)porins. *FEBS Lett.* **504:**112–117.

Noskov, S., S. Berneche, and B. Roux. 2004a. Control of ion selectivity in potassium channels by electrostatic and dynamic properties of carbonyl ligands. *Nature* **431:**830–834.

Noskov, S. Y., W. Im, and B. Roux. 2004b. Ion permeation through the α-hemolysin channel: theoretical studies based on Brownian dynamics and Poisson-Nernst-Plank electrodiffusion theory. *Biophys. J.* **87:**2299–2309.

Okada, K., P. C. Moe, and P. Blount. 2002. Functional design of bacterial mechanosensitive channels. *J. Biol. Chem.* **277:**27682–27688.

Park, S., F. Khalili-Araghi, E. Tajkhorshid, and K. Schulten. 2003. Free energy calculation from steered molecular dynamics simulations using Jarzynski's equality. *J. Chem. Phys.* **119:**3559–3566.

Park, S., and K. Schulten. 2004. Calculating potentials of mean force from steered molecular dynamics simulations. *J. Chem. Phys.* **120:**5946–5961.

Paula, S., M. Akeson, and D. Deamer. 1999. Water transport by the bacterial channel α-hemolysin. *Biochim. Biophys. Acta* **1418:**117–126.

Perozo, E., D. M. Cortes, P. Sompornpisut, A. Kloda, and B. Martinac. 2002a. Open channel structure of MscL and the gating mechanism of mechanosensitive channels. *Nature* **418:**942–948.

Perozo, E., A. Kloda, D. M. Cortes, and B. Martinac. 2002b. Physical principles underlying the transduction of bilayer deformation forces during mechanosensitive channel gating. *Nat. Struct. Biol.* **9:**696–703.

Preston, G. M., and P. Agre. 1991. Isolation of the cDNA for erythrocyte integral membrane-protein of 28-kD—member of an ancient channel family. *Proc. Natl. Acad. Sci. USA* **88:**11110–11114.

Pusch, M., U. Ludewig, A. Rehfeldt, and T. J. Jentsch. 1995. Gating of the voltage-dependent chloride channel ClC-0 by the permeant anion. *Nature* **373:**527–531.

Rickey, D. P., and E. E. C. Lin. 1972. Importance of facilitated diffusion for effective utilization of glycerol by *Escherichia coli*. *J. Bacteriol.* **112:**784–790.

Roux, B. 1995. The calculation of the potential of mean force using computer simulations. *Comput. Phys. Commun.* **91:**275–282.

Roux, B. 2002. Computational studies of the gramicidin channel. *Acc. Chem. Res.* 35:366–375.

Roux, B., T. Allen, S. Berneche, and W. Im. 2004. Theoretical and computational models of biological ion channels. *Q. Rev. Biophys.* **37:**15–103.

Roux, B., and K. Schulten. 2004. Computational studies of membrane channels. *Structure* **12:**1343–1351.

Russo, M. J., H. Bayley, and M. Toner. 1997. Reversible permeabilization of plasma membranes with an engineered switchable pore. *Nat. Biotechnol.* **15:**278–282.

Saam, J., E. Tajkhorshid, S. Hayashi, and K. Schulten. 2002. Molecular dynamics investigation of primary photoinduced events in the activation of rhodopsin. *Biophys. J.* **83:**3097–3112.

Saiz, L., S. Bandyopadhyay, and M. L. Klein. 2002. Towards an understanding of complex biological membranes from atomistic molecular dynamics simulations. *Biosci. Rep.* **22:**151–173.

Savage, D. F., P. F. Egea, Y. Robles-Colmenares, J. D. O'Connell III, and R. M. Stroud. 2003. Architecture and selectivity in aquaporins: 2.5 Å x-ray structure of aquaporin z. *PLoS Biol.* **1:**334–340.

Schlenkrich, M., J. Brickmann, A. D. MacKerell Jr., and M. Karplus. 1996. Empirical potential energy function for phospholipids: criteria for parameter optimization and applications, p. 31–81. *In* K. M. Merz and B. Roux (ed.), *Biological Membranes: a Molecular Perspective from Computation and Experiment.* Birkhauser, Boston, Mass.

Schumann, U., M. Edwards, C. Li, and I. R. Booth. 2004. The conserved carboxy-terminus of the mscs mechanosensitive channel is not essential but increases stability and activity. *FEBS Lett.* **572:**233–237.

Shapovalov, G., and H. A. Lester. 2004. Gating transitions in bacterial ion channels measured at 3 μs resolution. *J. Gen. Physiol.* **124:**151–161.

Sheng, Q., K. Schulten, and C. Pidgeon. 1995. A molecular dynamics simulation of immobilized artificial membranes. *J. Phys. Chem.* **99:**11018–11027.

Shi, L. B., W. R. Skach, and A. S. Verkman. 1994. Functional independence of monomeric CHIP28 water channels revealed by expression of wild-type-mutant heterodimers. *J. Biol. Chem.* **269:**10417–10422.

Shilov, I. Y., and M. G. Kurnikova. 2003. Energetics and dynamics of a cyclic oligosaccharide molecule in a confined protein pore environment: a molecular dynamics study. *J. Phys. Chem. B* **107:**7189–7201.

Shrivastava, I. H., and M. S. P. Sansom. 2000. Simulations of ion permeation through a potassium channel: molecular dynamics of KcsA in a phospholipid bilayer. *Biophys. J.* **78:**557–570.

Smart, O. S., G. M. P. Coates, M. S. P. Sansom, G. M. Alder, and C. L. Bashford. 1998. Structure-based prediction of the conductance properties of ion channels. *Faraday Discuss.* **111:**185–199.

Song, L., M. R. Hobaugh, C. Shustak, S. Cheley, H. Bayley, and J. E. Gouaux. 1996. Structure of staphylococcal α-hemolysin, a heptameric transmembrane pore. *Science* **274:**1859–1866.

Sotomayor, M., and K. Schulten. 2004. Molecular dynamics study of gating in the mechanosensitive channel of small conductance MscS. *Biophys. J.* **87:**3050–3065.

Stepaniants, S., S. Izrailev, and K. Schulten. 1997. Extraction of lipids from phospholipid membranes by steered molecular dynamics. *J. Mol. Mod.* **3:**473–475.

Stone, J., J. Gullingsrud, P. Grayson, and K. Schulten. 2001. A system for interactive molecular dynamics simulation, p. 191–194. *In* J. F. Hughes and C. H. Séquin (ed.), *2001 ACM Symposium on Interactive 3D Graphics*. ACM SIGGRAPH, New York, N.Y.

Sui, H., B.-G. Han, J. K. Lee, P. Walian, and B. K. Jap. 2001. Structural basis of water-specific transport through the AQP1 water channel. *Nature* **414:**872–878.

Sukharev, S. 2002. Purification of the small mechanosensitive channel of *Escherichia coli* (MscS): the subunit structure, conduction, and gating characteristics in liposomes. *Biophys. J.* **83:**290–298.

Sukharev, S., M. Betanzos, C.-S. Chiang, and H. R. Guy. 2001. The gating mechanism of the large mechanosensitive channel MscL. *Nature* **409:**720–724.

Sukharev, S., and D. P. Corey. 2004. Mechanosensitive channels: multiplicity of families and gating paradigms. *Sci. STKE* **2004:**re4.

Sukharev, S. I., P. Blount, B. Martinac, F. R. Blattner, and C. Kung. 1994. A large-conductance mechanosensitive channel in *E. coli* encoded by MscL alone. *Nature* **368:**265–268.

Tajkhorshid, E., A. Aksimentiev, I. Balabin, M. Gao, B. Isralewitz, J. C. Phillips, F. Zhu, and K. Schulten. 2003. Large scale simulation of protein mechanics and function, p. 195–247. *In* F. M. Richards, D. S. Eisenberg, and J. Kuriyan (ed.), *Advances in Protein Chemistry*, vol. 66. Elsevier Academic Press, New York, N.Y.

Tajkhorshid, E., J. Baudry, K. Schulten, and S. Suhai. 2000. Molecular dynamics study of the nature and origin of retinal's twisted structure in bacteriorhodopsin. *Biophys. J.* **78:**683–693.

Tajkhorshid, E., P. Nollert, M. Ø. Jensen, L. J. W. Miercke, J. O'Connell, R. M. Stroud, and K. Schulten. 2002. Control of the selectivity of the aquaporin water channel family by global orientational tuning. *Science* **296:**525–530.

Tajkhorshid, E., F. Zhu, and K. Schulten. Kinetic theory and simulation of single-channel water transport. *In* E. Sip (ed.), *Encyclopedia of Materials Modeling*, vol. 1. *Fundamental Models and Methods*, in press. MIT Press, Cambridge, Mass.

Valverde, M. A. 1999. ClC channels: leaving the dark ages on the verge of a new millenium. *Curr. Opin. Cell Biol.* **11:**509–516.

Vercoutere, W., and M. Akeson. 2002. Biosensors for DNA sequence detection. *Curr. Opin. Chem. Biol.* **6:**816–822.

Verkman, A. S., and A. K. Mitra. 2000. Structure and function of aquaporin water channels. *Am. J. Physiol. Ren. Physiol.* **278:**F13–F28.

Walz, T., and R. Ghosh. 1997. Two-dimensional crystallization of the light-harvesting I reaction centre photounit from *Rhodospirillum rubrum*. *J. Mol. Biol.* **265:**107–111.

Walz, T., T. Hirai, K. Murata, J. B. Heymann, K. Mitsuoka, Y. Fujiyoshi, B. L. Smith, P. Agre, and A. Engel. 1997. The three-dimensional structure of aquaporin-1. *Nature* **387:**624–627.

Yeh, I. C., and G. Hummer. 2004. Diffusion and electrophoretic mobility of single-stranded RNA from molecular dynamics simulations. *Biophys. J.* **86:**681–689.

Yin, J., Z. Kuang, U. Mahankali, and T. L. Beck. 2004. Ion transit pathways and gating in ClC chloride channels. *Proteins* **57:**414–421.

Yool, A. J., and A. M. Weinstein. 2002. New roles for old holes: ion channel function in aquaporin-1. *News Physiol. Sci.* **17:**68–72.

Zhou, F., and K. Schulten. 1995. Molecular dynamics study of a membrane-water interface. *J. Phys. Chem.* **99:**2194–2208.

Zhou, F., and K. Schulten. 1996. Molecular dynamics study of phospholipase A_2 on a membrane surface. *Proteins* **25:**12–27.

Zhu, F., and K. Schulten. 2003. Water and proton conduction through carbon nanotubes as models for biological channels. *Biophys. J.* **85:**236–244.

Zhu, F., E. Tajkhorshid, and K. Schulten. 2001. Molecular dynamics study of aquaporin-1 water channel in a lipid bilayer. *FEBS Lett.* **504:**212–218.

Zhu, F., E. Tajkhorshid, and K. Schulten. 2002. Pressure-induced water transport in membrane channels studied by molecular dynamics. *Biophys. J.* **83:**154–160.

Zhu, F., E. Tajkhorshid, and K. Schulten. 2004a. Collective diffusion model for water permeation through microscopic channels. *Phys. Rev. Lett.* **93:**224501.

Zhu, F., E. Tajkhorshid, and K. Schulten. 2004b. Theory and simulation of water permeation in aquaporin-1. *Biophys. J.* **86:**50–57.

Bacterial Ion Channels and Their Eukaryotic Homologs
Edited by A. Kubalski and B. Martinac

Chapter 10

Bacterial Na Channels: Progenitors, Progeny, or Parallel Evolution?

Evgeny Pavlov, Christopher Bladen, Catherine Diao, and Robert J. French

A FAMILY OF VOLTAGE-GATED BACTERIAL SODIUM CHANNELS

In 1998, a tidal wave of excitement rushed over the beachfront of ion channel studies with the report of the crystal structure of the bacterial potassium channel, KcsA (Doyle et al., 1998; MacKinnon et al., 1998). The frenetic activity, which involved determination of additional high-resolution structures along with various complementary studies by spectroscopic and electrical techniques, was founded on the discovery of a number of bacterial homologs of eukaryotic ion channels (MacKinnon et al., 1998). The identification and recognition of these proteins have facilitated high-level expression and purification and have thus enabled crystallization of an unprecedented variety of ion channels. A spectacular array of high-resolution structures has appeared in print, including open and closed pore domains of bacterial potassium channels (Doyle et al., 1998; Jiang et al., 2002b; Morals-Cabral et al., 2001; Zhou et al., 2001), cytoplasmic domains involved in assembly and the sensing of redox state and calcium levels (Jiang et al., 2001; Jiang et al., 2002a; Nishida and MacKinnon, 2002), and high-resolution images of bacterial homologs of eukaryotic voltage-sensitive chloride channels (Aggarwal and MacKinnon, 1996; Dutzler et al., 2002; Dutzler et al., 2003). Further insight into channel structural biology has come in parallel from studies on aquaporins (Walz et al., 1997) and bacterial mechanosensitive channels (Chang et al., 1998).

In this flurry of work, sodium and calcium channels faded briefly from the structure-function limelight for lack of a candidate for formation of high-quality crystals. Thus, it was cause for considerable excitement when, in early 2001, a putative bacterial homolog of voltage-gated calcium channels was recognized (Durell and Guy, 2001). The tale took an unexpected twist later that year with a report from the Clapham laboratory that the putative calcium channel, on expression in mammalian cells, turned out to be highly selective for sodium. The channel was thus dubbed NaChBac (Ren et al., 2001).

Evgeny Pavlov, Christopher Bladen, Catherine Diao, and Robert J. French • Department of Physiology and Biophysics, University of Calgary, Calgary, Alberta T2N 4N1, Canada.

The description of the gene and the functional expression of NaChBac raise a number of interesting questions and possibilities. Eukaryotic voltage-sensitive sodium and calcium channels have a major structural subunit that consists of four linked, homologous domains, which contain six putative transmembrane (TM) segments for an overall total of 24. In contrast, the NaChBac gene sequence appears to encode a single, 6-TM subunit (Fig. 1). Presumably, this assembles with three identical partners to form a homotetramer as in the case of many voltage-dependent, eukaryotic potassium channels. The S4 segment in the sequence shows a series of arginine residues at every third position, a motif that is the instantly recognizable fingerprint of the voltage-sensitive ion channel superfamily and is thought to be the primary voltage sensor (Noda et al., 1984; Stühmer et al., 1989), which controls voltage-dependent gating (Fig. 2).

In the putative pore-lining S5-S6 linker, a glutamate residue appears in the position homologous to that acknowledged as the selectivity filter in eukaryotic sodium and calcium channels. In a ring made up of one residue from each of four contributing subunits, these

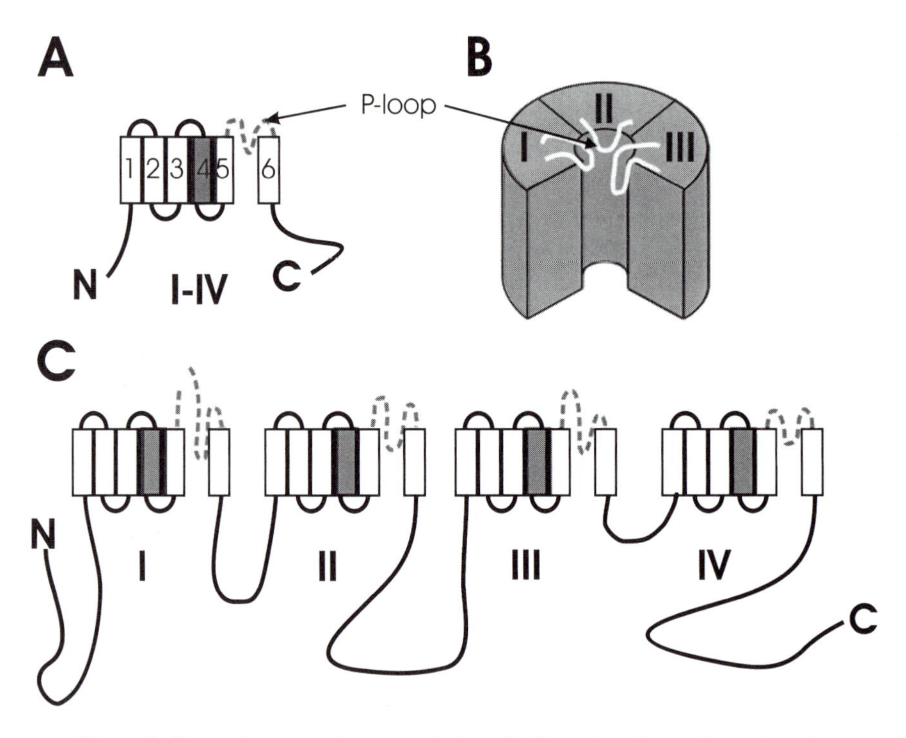

Figure 1. Proposed transmembrane topologies of voltage-dependent sodium channels from prokaryotes and eukaryotes. (A) A 6-TM segment structure of NaChBac. (B) Four identical 6-TM domains of the NaChBac protein presumably coassemble to form the functional channel. (C) In contrast to NaChBac, the α subunit of eukaryotic voltage-gated sodium channels is formed from four homologous but nonidentical 6-TM domains linked into a single polypeptide chain. The shaded S4 segments contain several positively charged residues (see Fig. 4), which play a primary role in voltage sensing and carry most or all of the measurable gating charge movement. In general, the cytoplasmic loops of eukaryotic Na channels are longer than those of NaChBac, presumably offering more opportunity for modulation of channel activity via various signaling pathways.

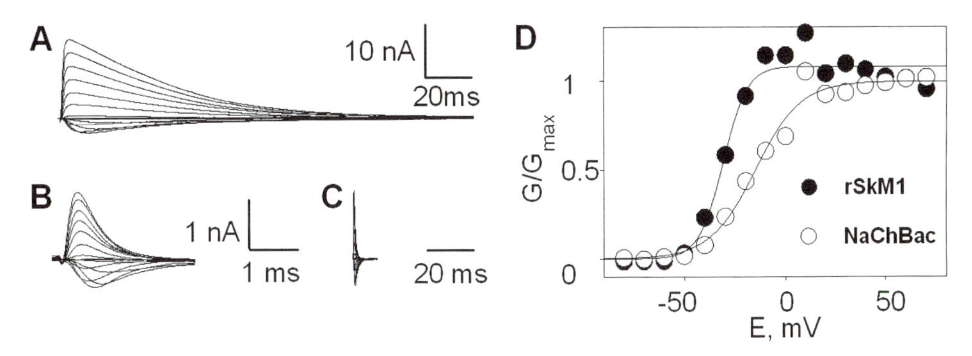

Figure 2. Comparison of voltage-dependent activation kinetics of Na_V and NaChBac channels. Kinetics of NaChBac activation and inactivation is significantly slower than that of rSkM1 channels; overall voltage dependence of these processes is similar, but here activation in NaChBac appears as a less steep function of voltage than it is for rSkM1 ($rNa_V1.4$). (A) Whole-cell patch-clamp recording of current through the NaChBac channel. Currents were induced by membrane depolarization from the holding potential of −140 mV to the range of voltages from −90 to 70 mV. The external (bathing) solution contained 140 mM NaCl, 5 mM KCl, 1 mM $CaCl_2$, 1 mM $MgCl_2$, and 10 mM HEPES (pH 7.4); internal (pipette) solution contained 130 mM NaF, 10 mM EGTA, 10 mM NaCl, and 10 mM HEPES (pH 7.4). Records were collected at room temperature. (B) Current recordings of rSkM1 channel done under experimental conditions identical to those of panel A. (C) Same current trace as on panel B shown on the timescale of panel A. Note the dramatic difference in kinetics between the two channels. (D) G-V relationship of the NaChBac and rSkM1 channels under identical conditions.

glutamates would make a filter possessing four negative charges at physiological pH, a feature that in eukaryotes distinguishes the voltage-sensitive calcium channels from their sodium-selective cousins (Sather et al., 1994; Sather and McCleskey, 2003; Yang et al., 1993). Members of the sodium channel family typically have a selectivity ring consisting of the aspartate, glutamate, lysine, and alanine residues contributed from domains I through IV, respectively. For this reason, Durell and Guy (2001) originally suggested that the gene encoded a prokaryotic voltage-gated calcium channel. After the surprising demonstration that NaChBac was selective for sodium over both potassium and calcium, point mutagenesis studies from the Clapham laboratory showed that the selectivity of NaChBac could simply be converted to one typical of a calcium channel by the addition of negative charge(s) in the pore-lining segment (Yue et al., 2002) (Fig. 3). These results echoed those of earlier work on eukaryotic channels by Heinemann, Terlau, and others, in which insertion of additional negative charges in the pore of the four-domain channels produced calcium-selective channels with a sodium channel backbone (Heinemann et al., 1992).

The discovery of NaChBac and, more recently, a number of additional bacterial homologs (Koishi et al., 2004) raises further questions. How have these proteins evolved, and what is their relationship to other membrane proteins in both prokaryotes and eukaryotes? What function do they serve in bacteria? What are the molecular requirements for sodium selectivity? What are the essential elements of the voltage sensor, and how is sensor motion coupled to channel opening? It is likely that NaChBac and its relatives will provide answers to certain aspects of each of these questions, and more.

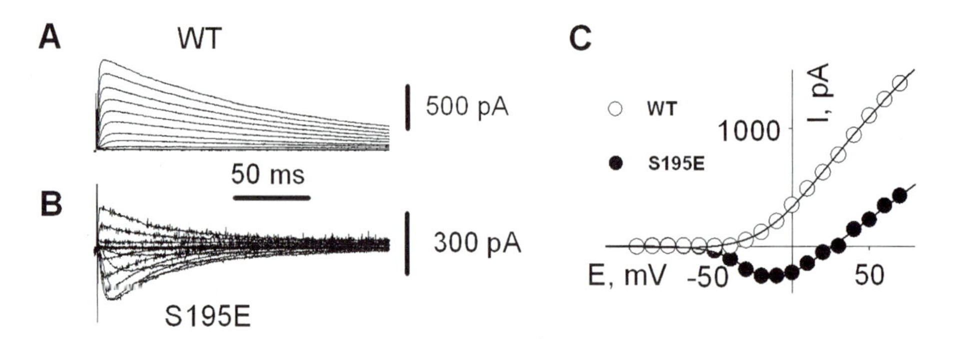

Figure 3. Selectivity changes are induced by insertion of an extra negative charge(s) in the pore of NaChBac. Currents conducted by NaChBac and the mutant S195E, recorded when external sodium has been replaced by calcium. (A) Wild-type NaChBac shows no detectable inward current, indicating that it is not permeable for calcium. (B) NaChBac S195E channels, with a point mutation in the pore region substituting neutral serine at position 195 with a negatively charged glutamate (S195E mutant), demonstrate significant inward currents, indicating that this mutated channel is permeable to calcium ions. In both experiments, the external bath solution consisted of 20 mM $CaCl_2$, 1 mM $MgCl_2$, 10 mM HEPES, 40 mM tetraethylammonium chloride, 10 mM glucose, and 65 mM CsCl (pH 7.2 with tetraethylammonium hydroxide); the internal pipette solution contained 130 mM NaF, 10 mM EGTA, 10 mM NaCl, and 10 mM HEPES (pH 7.4). (C) I-V relationship plotted from the current traces shown in panels A and B. Note the difference between reversal potentials for wild-type and mutant channels, consistent with a substantial permeability to Ca^{2+} ions induced by the mutation. Conversion of NaChBac into a functional calcium channel has been explored in some detail by Yue and colleagues (2002).

Evolution

Functionally, it is often convenient to consider ion channels as a distinct group of proteins from stoichiometric transporters, such as pumps, exchangers, and cotransporters, for which transport of a specific number of substrate molecules is linked, lock-step, to a specific sequence of conformational changes in the protein. Such stoichiometric transporters include members of the well-known major facilitator superfamily (MFS). Nonetheless, more than one line of evidence points to the idea that ion channels and other transport proteins are, in reality, closely related. Among ion channels, a recurring theme is a pore-lining connector suspended between pairs of membrane-spanning membrane helices in pore-forming subunits of 8 to 24 TM segments (MacKinnon, 1995; Moczydlowski, 1998).

Channels and Transporters

A series of studies by Saier and collaborators led to the conclusion that a primordial 2-TM pore-forming domain is a common ancestor of both voltage-gated channels and the MFS. It is suggested, however, that different evolutionary paths led, in one direction, to the assembly of the 6-TM units that form the monomers or homologous domains of the voltage-gated channels and, in another, to the two halves of the 12-TM MFS carrier proteins. In the case of voltage-gated channels, they suggest that a 4-TM element encoding a voltage sensor domain fused to the 2-TM channel domain. In contrast, regarding the evolution of the MFS, the conclusion is that the primordial 2-TM unit joined with an additional TM element to make a 3-TM structure, which by sequential duplications formed a 6-TM motif, and then the 12-TM structure now identified with the MFS proteins (Hvorup and Saier, 2002).

That channel and carrier molecules may sometimes be related even more closely than by the sharing of a distant common ancestor is dramatically illustrated by a recent article that shows that the bacterial homolog of eukaryotic ClC channels, in fact, functions not as a channel but as a stoichiometrically coupled H^+-Cl^- transporter (Accardi and Miller, 2004). This is the very molecule for which high-resolution structures were recently determined (Dutzler et al., 2002; Dutzler et al., 2003) and interpreted as a model for the eukaryotic chloride channels. For the latter, however, function as channels is unambiguously supported by the unitary current fluctuations seen in electrical recordings using patch-clamp and bilayer techniques (Ludewig et al., 1996; Middleton et al., 1996). Even for ClC channels showing distinct unitary fluctuations, evidence of mixed identity appears in the form of Cl-dependent, nonequilibrium gating transitions (Richard and Miller, 1990; Chen and Miller, 1996). Thus, there may be an influence of the permeant ion on gating, despite the absence of a fixed stoichiometric coupling between ion transport and the conformational transitions.

A Modular, Multipurpose Sensor

Blount and collaborators (Kumanovics et al., 2002) generalized the notion of a 4-TM voltage-sensing domain paired with a 2-TM pore-forming domain (Jiang et al., 2003a; Jiang et al., 2003b) and explored the possibility that a multipurpose sensor domain may have been further subdivided into additional modules as microbial channels evolved into eukaryotic forms. This fascinating focus led the authors to an analysis using the Cluster of Orthologous Groups database, which is available at http://www.ncbi.nlm.nih.gov/COG/ (Natale et al., 2000; Tatusov et al., 2003; Tatusov et al., 2001; Tatusov et al., 2000). In addition to the well-known 2-TM (S5-P-S6) and 6-TM channel motifs, Kumanovics and coworkers found genes encoding modules containing S1-S3, S1-S5, and a P-S6 fragment (without a complete selectivity filter pore). They then considered a motif that is highly conserved in the S3 segment of voltage-gated channels (N-h-h-D, where h represents a hydrophobic residue). Remarkably, this motif appears in the bacterial large mechanosensitive channel, MscL, and substitution of the conserved aspartate (D18C), followed by a viability analysis, suggested that the N-h-h-D motif was essential for normal channel function. On the basis of these studies, the authors proposed that a critical element in evolution of the voltage-sensitive channels was an ancestral S3-like domain, which, by early separating pathways, gave rise both to the 4-TM sensor domain of voltage-gated channels and to MscL (Kumanovics et al., 2002). With such a venerable lineage in mind, it perhaps is not surprising that among members of the 6-TM channel family a variety of stimuli may trigger gating transitions, including changes in temperature, osmolarity, voltage, and chemical environment. Furthermore, a given channel type may be biophysically capable of responding to a number stimuli, only one of which is important in the normal physiological setting.

A Biological Puzzle: What Do Voltage-Gated Sodium Channels Do in Bacteria?

The biological function of NaChBac and its relatives in bacteria, to our knowledge, remains a mystery, as do the functions of numerous other microbial channels (Kung and Blount, 2004). Bacteria have little need for the kind of rapid signaling for which voltage-gated

Na channels of eukaryotes are famous. That *Bacillus halodurans*, from which the NaChBac gene was isolated, is tolerant to salty, strongly alkaline environments raises the possibility that a sodium-selective channel may play some role in this salt tolerance. In the case of bacterial ClC, evidence suggests that it plays a role in the extreme acid response of *Escherichia coli*, enabling the bacterium to survive the low pH environment of the stomach and progress to inhabit the intestines (Iyer et al., 2002). The possibility of bacterial sodium channels being involved in rapid flagellar movement has been raised by Clapham and collaborators (Koishi et al., 2004). These authors note that flagellar movement is Na^+ dependent in a number of species, including *Vibrio cholerae*, the cholera bacterium, and that a mechanism driven by proton entry is unlikely at the high pH under which *B. halodurans* lives. Also, they report that NaChBac homologs are expressed by bacteria with Na^+-powered flagella. However, no means of coupling the energy dissipation of channel-mediated Na^+ entry to flagellar motion has been described to date. Furthermore, other studies on Na^+-dependent flagellar motors implicate other proteins as the Na-coupled motor and demonstrate that motility is inhibited by amiloride (Kojima et al., 1999a; Kojima et al., 1999b; Asai et al., 2003). Although amiloride is a somewhat fickle drug, this suggests that the current driving flagella may pass through a channel more akin to the epithelial sodium channel, ENaC, than a close relative of NaChBac or a member of the Na_V family. Overall, confirmation of the details and mechanisms of the physiological role of the NaChBac family members is yet to come. A recent article provides new evidence of multiple roles in motility, chemotaxis, and pH homeostasis for Na_VBP, a NaChBac relative from the alkaliphile *Bacillus pseudofirmus* (Ito et al., 2004).

NaChBac OR CaChBac? FROM SEQUENCE TO FUNCTION

Properties of a channel are defined by its selectivity, voltage-dependent gating, and regulation by different inhibitors and agonists. Although a number of bacterial sodium channels have now been identified, most of the experimental results now available have been obtained from study of NaChBac. Here, we briefly review the experimental data that reflect these functional properties of the bacterial voltage-gated sodium channels and compare key parts of the sequence with other voltage-gated channels.

Voltage-gated ion channels (VICs) are a subset of the larger P-loop ion channel family (MacKinnon, 1995; Moczydlowski, 1998). The motif that defines VIC channels is a 4-TM domain that appears in the sequence attached to the N-terminal end of the 2-TM pore-forming unit that defines the P-loop family. The simplest members of the P-loop family are typified by the 2-TM monomers that coassembled in fourfold symmetry to make up inward rectifier potassium channels. The sine qua non of the voltage-sensing domain of VICs is the S4 segment, which shows a characteristic set of arginine or lysine residues spaced at every third position across the putative membrane-spanning helix.

Voltage Sensor

The S4 segment (Fig. 4) was recognized immediately as a repeated motif in the first determined sequence of a voltage-gated sodium channel (from electric eel) when it was determined in the 1980s (Noda et al., 1984; Noda et al., 1986a; Noda et al., 1986b). The cluster of S4 charges was an obvious candidate for a voltage sensor. That proposition has held

		S4 segment		% identity
NaChBac	112	VLRILRVLRVLRAISV	127	100
Shaker	292	ILRVIRLVRVFRIFKL	307	37
KvAP	128	LFRLVRILRFLRILLI	143	37
rSkM1-DI	217	ALRTFRVLRALKTITV	232	56
rSkM1-DII	661	VLRSFRLLRVFKLAKS	676	43
rSkM1-DIII	1119	SLRTLRALRPLRAISR	1134	68
rSkM1-DIV	1442	VIRLARIGRVLRLIRG	1457	50
hCav1.3-DI	238	ALRAFRVLRPLRLVSG	253	50
hCav1.3-DII	617	VFRCVRLLRIFKVTKY	632	43
hCav1.3-DIII	982	ILRVLRVLRPLRAINR	997	68
hCav1.3-DIV	1317	FFRLFRVMRLVKLLSR	1332	25

Figure 4. Amino acid sequence alignments for the N-terminal part of S4 voltage sensor segments from 6-TM and 24-TM voltage-gated channels. Note the conserved pattern of basic residues repeated every three amino acids, provided exclusively by arginine in the first three positions for the channels shown, with lysine appearing in the fourth position for some domains in the eukaryotic Na_V and Ca_V channels. Other basic residues are seen when some of the sequences are extended in the C-terminal direction, but the residues shown here are thought to account for essentially all of the gating charge movement. National Center for Biotechnology Information (NCBI) GeneBank accession numbers for sequences used in the alignment are Q01668 (Human $hCa_V1.3$–α1D, an L-type Ca channel), CAA76659 (rSkM1 or $Na_V1.4$), Q9YDF8 (K_VAP), NP_037102 (Shaker), and AAR21291 (NaChBac). The program used for the alignments was Clustal W version 1.8, which defines "highly conserved" residues as showing ≥50% identity among the sequence set used for the search.

through the years, despite ongoing debate about the details of how S4 might move and what other components should be considered as part of the voltage sensor. With the hindsight of present knowledge, it is now accepted that the 4-TM module that precedes the 2-TM pore-forming domain is the unit that can be consistently associated with voltage-gating through the VIC family. Although a subset of the S4 charges appears to account for most of the measurable charged movement (gating current) that precedes channel opening, it is now recognized that there are also conserved negative charges that help stabilize the voltage-sensing domain (Papazian et al., 1995; Tiwari-Woodruff et al., 2000; Tiwari-Woodruff et al., 1997).

Recently, a high-resolution structure for the separately crystallized voltage-sensing domain of a bacterial potassium channel was determined, in conjunction with studies of the structure of the entire molecule. These studies from the MacKinnon laboratory led to the working hypothesis that the voltage sensor may move, paddle-like, across the lipid phase of the membrane (Jiang et al., 2003a; Jiang et al., 2003b). This suggestion has drawn a widespread response from experimentalists and triggered energetic debate regarding the mechanism by which measurable charge movement takes place during the conformational changes of channel activation (Blaustein and Miller, 2004; Ahern and Horn, 2004; Bezanilla and Perozo, 2003; Laine et al., 2004).

We illustrate, in Fig. 4, the highly conserved S4 segment of NaChBac aligned with homologous stretches of sequence from a selection of other voltage-gated channels. There is striking conservation of the S4 motif among channels ranging from prokaryotes to eukaryotes and including potassium, sodium, and calcium channels. In the broad-ranging examples shown, conservation of positive charge at the first four positions in the S4 segments is absolute. These charges are almost always provided by arginine, although in a few cases lysine takes its place at the fourth charged locus. Additional positive charges are frequently seen as one progresses further toward the C terminus of the protein; however, in general these are less strictly conserved as illustrated at the end of the segment shown in the figure. Oddly, among the examples shown, the highest degree of identity with NaChBac is exhibited not by another prokaryotic channel (K_VAP) but by the domain III S4 segments of rat skeletal muscle sodium channels and the domain III S4 of a human voltage-gated calcium channel. The similarity among the sequences is further underscored if one notes that at many positions where there is not identity of residues separating the charged arginines or lysines, the loci are nonetheless occupied by residues of a similar, hydrophobic nature.

A more extended sequence alignment provided in Fig. 8 of the article by Blunck and collaborators (Blunck et al., 2004) shows one very highly conserved negative residue in the C-terminal section of the S3 segment along with the common occurrence of one or more additional negative charges scattered in the succeeding 20 or so positions. Despite this additional similarity, there is large variability in the number of residues making up the connecting segment between S3 and S4, with NaChBac showing one of the shortest connectors, if not the shortest, among voltage-gated channels. Changes in this linker region in Shaker-IR have been shown to induce changes in gating kinetics but do not qualitatively disrupt the voltage-dependent gating process. They do, however, show notable periodicity in their effects on activation kinetics and the half-activation voltage, consistent with the idea that S4 helical structure may extend, in the N-terminal direction, several residues beyond the first charge-carrying arginine (Gonzalez et al., 2000, 2001; Sorensen et al., 2000). For NaChBac, the particularly short S3-S4 linker may well be a factor contributing to its distinctive, slow-gating kinetics and to a concerted movement of S3 and S4, as has been suggested for K_VAP, based on its structure. Fluorescence signals from the NaChBac S3-S4 loop are small, dependent on the position and identity of the labeling fluorophore, and consistent with the idea that the structure of the NaChBac voltage sensor is similar to that of K_VAP but different from that of Shaker (Blunck et al., 2004). Shaker, with a longer S3-S4 linker and quenching glutamate residues in the vicinity, gives much larger fluorescence signals associated with gating than does NaChBac.

Gates and Gating

If a channel is to open and close in response to changes in membrane voltage, it must possess not only a means to sense the electric field but also the structural requirements to couple that electric signal to the physical opening and closing of the conducting pathway. Most voltage-activated channels exhibit two competing responses to membrane depolarization, which initiates both an activation process that results in channel opening and an inactivation process that ultimately results in channel closing. Although the norm is that activation is a substantially faster process than inactivation, the kinetics of these processors varies widely, and in many channels there appear to be at least two distinct mechanisms of inactivation. A fairly recent discovery is that in the HERG family of potassium channels the "inactivation" process is substantially faster than activation and consequently the normal opening of the channel is triggered by the second half of a depolarization-repolarization cycle, a peculiar phenomenon that has powerful implications for the physiological role of the HERG channel in the repolarization of the membrane (Smith et al., 1996; Sanguinetti et al., 1996).

In the case of the bacterial sodium channels, detailed studies on the mechanism of gating and the coupling of the voltage sensor to gating are just beginning. Below we will briefly consider both activation and inactivation processes.

Activation

As expected, the NaChBac channel can be activated by applying a depolarizing pulse to the membrane. Voltage dependence of NaChBac activation is qualitatively similar to the voltage dependence of the activation of eukaryotic sodium channels. However, the speed of NaChBac activation is dramatically slower ($\tau_{act} = 12.9 + 0.4$ ms at −10 mV for NaChBac versus <2 ms for eukaryotic channels). An example of some data for NaChBac and rat $Na_V1.4$, recorded under identical ionic conditions, is given in Fig. 2. In this case, it also appears that activation of NaChBac is shifted in the depolarizing direction and depends less steeply on voltage than $Na_V1.4$.

Analyses of the crystal structures of two 2-TM bacterial potassium channels led to the conclusion that the structure of KcsA represented a closed state and that of the related calcium-gated channel, MthK, crystallized in the presence of bound calcium, represented an open state. In KcsA (Doyle et al., 1998), the S6 helices, which line the cytoplasmic segment of the pore, project more or less straight across thc membrane, forming a narrow bundle crossing near the cytoplasmic end of the pore (the "smoke hole" of the "inverted tepee"). The structure of MthK shows an outward bend of about 30° in the S6 segments near this position. This results in a conspicuous, open pathway through the pore structure. The outer section of the pore, incorporating the selectivity filter, is, however, similar in the KcsA and MthK structures. The recently determined structure of a 6-TM voltage-gated potassium channel (K_VAP) is similar to that of MthK.

The S6 segments of the 6-TM VIC family show a highly conserved "hinge" glycine residue, which could promote the outward bending of the α-helix to promote channel opening. A glycine is present in this position in the S6 segment of NaChBac (Fig. 5), and its role has been examined in a study that has just appeared from the Catterall laboratory (Zhao et al., 2004). This study examines the effects on activation gating of substitution of a rigid, kink-promoting proline residue for the hinge glycine (G219P). The G219P mutation at S6 caused a −51-mV shift in NaChBac activation. This negative shift in activation following the G219P substitution suggests that the kinked or bent conformation represents the open state. Although the analysis in that article concentrates on activation gating, it is worth noting that inactivation on the scale of a second or so was completely abolished by this substitution (at −60 mV, inactivation was slowed 1,200-fold). The authors found that the proline substitution strongly stabilized the open state and dramatically slowed channel closure by deactivation. Much less dramatic effects resulted from alanine substitution (G219A) and from the substitution of proline in other nearby positions. From these results, the authors proposed a gating model in which all four S6 segments undergo a concerted transition to the open state by bending the S6 residues at glycine 219, and further suggest that this gating mechanism is general for 2-TM and most 6-TM ion channels.

		S6 segment		% identity
NaChBac	207	WSWLY**F**VSFVLIGTFIIF**N**LFIGVIVNN-	234	100
Shaker	454	WGKIVGSLCVVAGVLTIAALPVPVIVSN-	481	32
KvAP	208	IGKVIGIAVMLTGISALTLLIGTVSNMN	235	17
rSkM1-DI	418	-YMIF**F**VVIIFLGSFYLI**N**LILAVVAMAY	445	25
rSkM1-DII	772	-CLTV**F**LMVMVIGNLVVL**N**LFLALLLSSF	799	31
rSkM1-DIII	1263	YMYLY**F**VIFIIFGSFFTL**N**LFIGVIIDN-	1290	53
rSkM1-DIV	1570	----F**F**CSYIIIGFLIVV**N**MYIAIILEN-	1593	37
hCav1-DI	394	YNFIY**F**ILLIIMGSFFMI**N**LCLVVIATQ-	421	28
hCav1-DII	993	---LY**F**VALMTFGNYVLF**N**LLVAILV---	1015	39
hCav1-DIII	1530	WMLLY**F**ISFLLIVSFFVL**N**MFVGVVVEN-	1557	53
hCav1-DIV	1836	-SPVY**F**VTFMLVAQFVLV**N**VVVAVLM---	1860	32

Figure 5. A putative gate alignment of S6 sequences from NaChBac and other voltage-gated channels shows a highly conserved glycine (shown in white on black), which is thought to provide a flexible link, or gating hinge, near the middle of S6 in most cases. NCBI GeneBank accession numbers for sequences used in the alignment are AAC67239 (Human $hCa_V3.2$–α1H, a T-type Ca channel), CAA76659 (rSkM1 or $Na_V1.4$), Q9YDF8 (K_VAP), NP_037102 (Shaker), and BAB05220 (NaChBac). Identical residues are highlighted in bold type. Search software used was as in the legend of Fig. 4.

Inactivation

A cursory examination of the sequence and proposed topology for NaChBac raises interesting questions with regard to the mechanism of inactivation. Among the voltage-dependent channel family an "N-type" inactivation has been ascribed to a ball-and-chain mechanism that rapidly blocks the channel after opening. Although the term "ball and chain" was coined much earlier based on the work of Armstrong and collaborators (Armstrong et al., 1973; Armstrong and Bezanilla, 1977), it was placed on a more concrete basis by an analysis of the Shaker channel performed in the Aldrich laboratory (Hoshi et al., 1990; Zagotta et al., 1990). Fast inactivation in eukaryotic sodium channels is now considered to take place by a "hinged-lid" mechanism (Catterall, 2000), which might be considered as the domain IV homolog of the Shaker ball-and-chain mechanism where the ball and chain have constrained at the two ends rather than one. The argument for this particular analogy might be further strengthened by considering that considerable evidence points to a particular role of domain IV of sodium channels in coupling the inactivation process to channel activation. In some potassium channels, it has also been shown that the role of the ball and chain is played, not by the N-terminal segment of the protein, but rather, part of an accessory subunit. In some cases, this may incorporate a redox sensor.

The inactivation of NaChBac in response to a sustained depolarization is, like activation kinetics, much slower than that exhibited by fast-inactivating eukaryotic potassium, sodium, or calcium channels ($\tau = 166$ ms at -10 mV, in the study of Ren et al. [2001]) than inactivation of eukaryotic channels ($\tau < 10$ ms). This difference is illustrated dramatically in Fig. 2. Coupled with the absence of any obvious candidate ball-and-chain or hinged-lid structure, this leaves the hypothesis of default that inactivation is likely to occur by a "C-type" or pore-based mechanism (Hoshi et al., 1991; Ogielska et al., 1995; Todt et al., 1999; Ong et al., 2000). A preliminary report from our own laboratory (Pavlov et al., 2004) supports this idea on two grounds: first, linking of the NaChBac gene to express either two domain 12-TM or four domain 24-TM units has little effect on NaChBac inactivation kinetics; and second, certain pore mutations do indeed substantially modify inactivation. NaChBac inactivation may thus involve rearrangement of the pore region, similar perhaps to ultraslow inactivation in eukaryotic sodium channels (Todt et al., 1999).

P-Region and Selectivity Filter: Prokaryotes Do It Their Way

The strongest connection initially identified (Durell and Guy, 2001) between the NaChBac sequence (then termed $Ca_{Bac1.1}$) and a eukaryotic channel was the conservation of a consensus sequence from eukaryotic voltage-dependent calcium channel pore regions as follows: FxxxTxExW. This motif appears exactly in the following sequence taken from the NaChBac S5-S6 linker: FQQVTLESWAS (Fig. 6). The last six residues in this fragment

		Pore loop		% identity
NaChBac	173	YFGNLQLSLLTLFQVVTLESWASGVMR	199	100
rSkM1-DI	382	SYDTFSWAFLALFRLMTQDYWENLFQL	408	18
rSkM1-DII	737	HMNDFFHSFLIVFRILTGEWIETMWDC	763	14
rSkM1-DIII	1119	NYDNVGLGYLSLLQVATFKGWMDIMYA	1145	29
rSkM1-DIV	1510	FNFETFGNSICLFEITTSAGWDGLLNP	1536	14
hCav1.2-DI	345	NFDNFAFAMLTVFQCITMEGWTDVLYW	371	33
hCav1.2-DII	688	TFDNFPQSLLTVFQILTGEDWNSVMYD	714	44
hCav1.2-DIII	1117	DFDNVLAAMMALFTVSTFEGWPELLYR	1143	33
hCav1.2-DIV	1446	NFQTFPQAVLLPLRCATGEAWQDIMLA	1469	22
Shaker	423	NSFFKSIPDAFWWAVVTMTTV*GYG*DMT	449	7
KvAP	177	NSSIKSVFDALWWAVVTATTV*GYG*DMT	203	18

Figure 6. Sequence alignment for the pore regions of different voltage-gated channels. Charged groups of the selectivity filter of NaChBac, rSkM1, and Ca_V channels are shown in white on black; the highly conserved GYG selectivity filter motif of the potassium channels is underlined and italicized. Note that selectivity filter alignment as well as aligned score suggests closer homology between NaChBac and Ca_V than between NaChBac and rSkM1 channels. The motif FxxxTxExW appears in NaChBac and in eukaryotic Na and Ca channels, but the E is present in all domains only in the Ca channel. The alignments for the K channels are ambiguous due to the low sequence identity and might be better portrayed shifted left three residues so that the TVGYG motif begins under the fully conserved T in the Na and Ca channel sequences. NCBI GeneBank accession numbers for sequences used in the alignment are Q13936 (HumanCa_V1.2–α1C, an L-type Ca channel), CAA76659 (rSkM1 or Na_V1.4), Q9YDF8 (K_VAP), NP_037102 (Shaker), and BAB05220 (NaChBac). Search software used was as in the legend of Fig. 4.

from the NaChBac sequence, LESWAS, were the focus of studies from the Clapham laboratory on the selectivity of NaChBac and are discussed further below. With hindsight, it is possible to suggest that focusing on a strictly conserved motif from the pore segment indicated above caused Durell and Guy to discount the possibility that NaChBac might be sodium selective, in favor of its apparent closer relation to voltage-dependent calcium channels based on symmetry. However, this viewpoint failed to give weight to the fact that, in voltage-dependent calcium channels, the initial glutamate that appears in all four domains and is thought to contribute directly to the narrow part of the selectivity filter (Sather and McCleskey, 2003) always appears in the context of additional negative charges (Yue et al., 2002). Highly conserved, although present only in domain II of voltage-dependent calcium channels, is an additional negative charge in the form of an aspartate, located immediately C-terminal to the fully conserved DII glutamate in the selectivity ring. In addition, usually at least two additional negative charges appear in the overall pore structure of eukaryotic voltage-gated calcium channels either three or four residues downstream from the conserved glutamate. This "outer ring" of charge is generally complete in voltage-dependent sodium channels, in the sense that there is one negative charge in each of the four domains at this level of the pore. In rSkM1, E758 and D1241 in this ring are the major contributors to the high-affinity binding of the GIII μ-conotoxins (Dudley et al., 1995; Xue et al., 2003). By comparison with voltage-dependent sodium and calcium channels, NaChBac is noticeable both for the presence of the conserved glutamate and the absence of outer ring charges, as compared to the other channels.

Yue and collaborators showed that the tempting similarity with voltage-gated calcium channels, although misleading in one sense, was well worth some attention (Yue et al., 2002). They demonstrated that NaChBac channels could be progressively converted from strong sodium selectivity to channels that allow sodium and calcium both to permeate freely, and then to strong calcium selectivity, by the successive addition of negative charges. In that study, additional negative charges were added to the pore by replacing one or two of the native residues with aspartate. However, we have shown that similar results can be obtained by adding additional glutamate residues (Fig. 3).

The more general biophysical principle that is suggested by these results is that, in contrast to the strategy that has evolved in eukaryotes, sodium selectivity does not require asymmetry among the four domains of the channel. The studies on NaChBac show unequivocally that a symmetric channel, with four glutamates in the critical selectivity positions, can be highly selective for sodium over calcium. Addition of one negative charge in the adjacent position (i.e., addition of four charges to the intact tetrameric channel) has a particularly telling effect in the conversion of the channel toward calcium selectivity, but at least one additional charge per subunit (making a total of eight negative charges more than for the wild-type channel) appears to be necessary to exclude sodium in the presence of low concentrations of calcium. When the pore sequence of NaChBac, LESWAS (net charge, −1/monomer), is changed to LDDWAD (net charge, −3/monomer), the resulting currents show the hallmark features of a eukaryotic calcium channel: at millimolar calcium concentrations, virtually ideal selectivity for calcium over sodium, and at micromolar calcium concentrations, block of sodium currents by calcium, behavior that reflects the middle range of the anomalous mode fraction effect—a non-monotonic dependence of calcium current and conductance on the mole fraction of calcium in the bathing solution.

Interactions with Drugs and Toxins

The pharmacology of NaChBac has not yet been studied in great detail. Preliminary data suggest that NaChBac is more sensitive to the blockers of voltage-gated calcium channels than to agents that target sodium channels. For example, NaChBac is insensitive to tetrodotoxin and saxitoxin, classic sodium channel blockers. No block by μ-conotoxin has been observed (E. Pavlov, unpublished data). Less selective sodium channel-blocking drugs, lidocaine and phenytoin (dilantin), block only with low affinity. The 50% inhibitory concentrations (IC_{50}s) for block of single pulse peak current (−10 mV from a holding potential of −140 mV) were 0.28 mM lidocaine and 0.12 mM phenytoin (Pavlov et al., 2003). More potent inhibition of NaChBac is produced by dihydropyridine blockers of L-type calcium channels (Ren et al., 2001): nifedipine (IC_{50} of 2.2 μM) and nimodipine (IC_{50} of 1 μM). In contrast, NaChBac is relatively insensitive to mibefradil (IC_{50} of 720 μM). Mibefradil is often used as a T-type calcium channel blocker, but it is not highly selective and also blocks high-voltage activated Ca channels and voltage-dependent Na channels (Eller et al., 2000; Huang et al., 2004). A detailed discussion of the possible interactions of small hydrophobic drugs with NaChBac is beyond the scope of this chapter. Briefly, however, for both eukaryotic sodium and calcium channels, it is likely in most cases that several residues, from the S5 and/or S6 segments of more than one domain, contribute to channel-drug interactions. Overviews of the voluminous literature can be found in numerous recent reviews (Doering and Zamponi, 2003; Ragsdale and Avoli, 1998).

NaChBac is not blocked by ω-conotoxin GVIA nor by ω-agatoxin IVA. NaChBac channel currents, similar to eukaryotic calcium channels and some sodium channels, are also reduced by Cd^{2+}, Co^{2+}, and La^{3+} ions. Overall, in pharmacological terms, NaChBac seems more closely related to L-type voltage-gated calcium channels than to other channel types.

PROSPECTS FOR THE FUTURE: A NEW BIOPHYSICAL MODEL

Crystallization and Detailed Structure-Function Studies

Sodium channels, once the experimental archetype for voltage-gated channels, have more recently felt hot competition for attention from Shaker and other members of the VIC family, such as HERG and hyperpolarization-activated (HCN) channels, which exhibit striking features that adapt them to unique physiological roles. The 6-TM channels such as Shaker offer easier genetic manipulation, including the possibility of changing four residues in a functional channel for the price of a single mutation. Also, Shaker mutants yield higher expression levels than their 24-TM cousins. HERG, HCN, and other members of the broader VIC family offer striking illustrations of the adaptability of the 6-TM scaffold: contrasting kinetic patterns for activation and inactivation, diametrically opposed voltage dependencies based on similar voltage sensors, and a variety of activating ligands, ranging from Ca^{2+} to cyclic nucleotides and vanilloid compounds. What do NaChBac and its immediate relatives have to contribute? Despite their similar selectivity, NaChBac may give us less insight into the behavior of eukaryotic Na_V channels than into the general principles of voltage-dependent gating mechanisms, and the roles to which these have been adapted in prokaryotes. Will NaChBac yield yet another exquisite high-resolution structure dotted with unsuspected details? If so, will it differ in important ways from KvAP, other than in the particulars of the conducting pathway? Will it offer insight into one or more forms of inactivation? Will it be used as a building block to reconstruct properties of multidomain channels and therefore lead to better understanding of the role of asymmetry in channel design? The future will tell.

Progenitors, Progeny, or Parallel Evolution?

Do the bacterial sodium channels represent progenitors, progeny, or parallel evolution with respect to eukaryotic voltage-gated channels? Perhaps the best answer is that all three descriptors apply. A strong argument has been made that they represent, or at least are closely related to, progenitors of eukaryotic channels. Among the 6-TM voltage-gated channels, they may well be progeny of an earlier evolved, voltage-gated potassium channel related, for example, to KvAP. Finally, however, in the emergence of a Na-selective channel, microbes have clearly adopted a parallel strategy to attain an impressive level of discrimination between the closely related alkali cations, Na^+ and K^+, and the size-mates, Na^+ and Ca^{2+}.

Acknowledgments. Our work is supported by grants from the Canadian Institutes of Health Research and the Heart and Stroke Foundation of Alberta, NWT and Nunavut. R.J.F. is a Medical Scientist of the Alberta Heritage Foundation for Medical Research.

We thank Gerald Zamponi for comments on a draft of the manuscript.

REFERENCES

Accardi, A., and C. Miller. 2004. Secondary active transport mediated by a prokaryotic homologue of ClC Cl-channels. *Nature* **427:**803–807.

Aggarwal, S. K., and R. MacKinnon. 1996. Contribution of the S4 segment to gating charge in the *Shaker* K^+ channel. *Neuron* **16:**1169–1177.

Ahern, C. A., and R. Horn. 2004. Stirring up controversy with a voltage sensor paddle. *Trends Neurosci.* **27:**303–307.

Armstrong, C. M., and F. Bezanilla. 1977. Inactivation of the sodium channel. II. Gating current experiments. *J. Gen. Physiol.* **70:**567–590.

Armstrong, C. M., F. Bezanilla, and E. Rojas. 1973. Destruction of sodium conductance inactivation in squid axons perfused with pronase. *J. Gen. Physiol.* **62:**375–391.

Asai, Y., T. Yakushi, I. Kawagishi, and M. Homma. 2003. Ion-coupling determinants of Na+-driven and H+-driven flagellar motors. *J. Mol. Biol.* **327:**453–463.

Bezanilla, F., and E. Perozo. 2003. The voltage sensor and the gate in ion channels. *Adv. Protein Chem.* **63:**211–241.

Blaustein, R. O., and C. Miller. 2004. Ion channels: shake, rattle or roll? *Nature* **427:**499–500.

Blunck, R., D. M. Starace, A. M. Correa, and F. Bezanilla. 2004. Detecting rearrangements of Shaker and NaChBac in real-time with fluorescence spectroscopy in patch-clamped mammalian cells. *Biophys. J.* **86:** 3966–3980.

Catterall, W. A. 2000. From ionic currents to molecular mechanisms: the structure and function of voltage-gated sodium channels. *Neuron* **26:**13–25.

Chang, G., R. H. Spencer, A. T. Lee, M. T. Barclay, and D. C. Rees. 1998. Structure of the MscL homolog from *Mycobacterium tuberculosis*: a gated mechanosensitive ion channel. *Science* **282:**2220–2226.

Chen, T.-Y., and C. Miller. 1996. Nonequilibrium gating and voltage dependence of the ClC-0 Cl^- channel. *J. Gen. Physiol.* **108:**237–250.

Doering, C. J., and G. W. Zamponi. 2003. Molecular pharmacology of high voltage-activated calcium channels. *J. Bioenerg. Biomembr.* **35:**491–505.

Doyle, D. A., J. M. Cabral, R. A. Pfuetzner, A. Kuo, J. M. Gulbis, S. L. Cohen, B. T. Chait, and R. MacKinnon. 1998. The structure of the potassium channel: molecular basis of K^+ conduction and selectivity. *Science* **280:**69–77.

Dudley, S. C., H. Todt, G. Lipkind, and H. A. Fozzard. 1995. A μ-conotoxin-insensitive Na^+ channel mutant: possible localization of a binding site at the outer vestibule. *Biophys. J.* **69:**1657–1665.

Durell, S. R., and H. R. Guy. 2001. A putative prokaryote voltage-gated Ca^{2+} channel with only one 6TM motif per subunit. *Biochem. Biophys. Res. Comm.* **281:**741–746.

Dutzler, R., E. B. Campbell, M. Cadene, B. T. Chait, and R. MacKinnon. 2002. X-ray structure of a ClC chloride channel at 3.0 Å reveals the molecular basis of anion selectivity. *Nature* **415:**287–294.

Dutzler, R., E. B. Campbell, and R. MacKinnon. 2003. Gating the selectivity filter in ClC chloride channels. *Science* **300:**108–112.

Eller, P., S. Berjukov, S. Wanner, I. Huber, S. Hering, H. G. Knaus, G. Toth, S. D. Kimball, and J. Striessnig. 2000. High affinity interaction of mibefradil with voltage-gated calcium and sodium channels. *Br. J. Pharmacol.* **130:**669–677.

Gonzalez, C., E. Rosenman, F. Bezanilla, O. Alvarez, and R. Latorre. 2000. Modulation of the Shaker K(+) channel gating kinetics by the S3-S4 linker. *J. Gen. Physiol.* **115:**193–208.

Gonzalez, C., E. Rosenman, F. Bezanilla, O. Alvarez, and R. Latorre. 2001. Periodic perturbations in Shaker K+ channel gating kinetics by deletions in the S3-S4 linker. *Proc. Natl. Acad. Sci. USA* **98:**9617–9623.

Heinemann, S. H., H. Terlau, W. Stühmer, K. Imoto, and S. Numa. 1992. Calcium channel characteristics conferred on the sodium channel by single mutations. *Nature* **356:**441–443.

Hoshi, T., W. N. Zagotta, and R. W. Aldrich. 1990. Biophysical and molecular mechanisms of *Shaker* potassium channel inactivation. *Science* **250:**533–538.

Hoshi, T., W. N. Zagotta, and R. W. Aldrich. 1991. Two types of inactivation in Shaker K+ channels: effects of alterations in the carboxy-terminal region. *Neuron* **7:**547–556.

Huang, L., B. M. Keyser, T. M. Tagmose, J. B. Hansen, J. T. Taylor, H. Zhuang, M. Zhang, D. S. Ragsdale, and M. Li. 2004. NNC 55-0396 [(1S,2S)-2-(2-(*N*-[(3-benzimidazol-2-yl)propyl]-*N*-methylamino)ethyl)-6-fluoro-1,2,3,4-tetrahydro-1-isopropyl-2-naphtylcyclopropanecarboxylate dihydrochloride]: a new selective inhibitor of T-type calcium channels. *J. Pharmacol. Exp. Ther.* **309:**193–199.

Hvorup, R. N., and M. H. Saier, Jr. 2002. Sequence similarity between the channel-forming domains of voltage-gated ion channel proteins and the C-terminal domains of secondary carriers of the major facilitator superfamily. *Microbiology* **148:**3760–3762.

Ito, M., H. Xu, A. A. Guffanti, Y. Wei, L. Zvi, D. E. Clapham, and T. A. Krulwich. 2004. The voltage-gated Na+ channel NaVBP has a role in motility, chemotaxis, and pH homeostasis of an alkaliphilic Bacillus. *Proc. Natl. Acad. Sci. USA.* **101:**10566–10571.

Iyer, R., T. M. Iverson, A. Accardi, and C. Miller. 2002. A biological role for prokaryotic ClC chloride channels. *Nature* **419:**715–718.

Jiang, Y., A. Lee, J. Chen, M. Cadene, B. T. Chait, and R. MacKinnon. 2002a. Crystal structure and mechanism of a calcium-gated potassium channel. *Nature* **417:**515–522.

Jiang, Y., A. Lee, J. Chen, M. Cadene, B. T. Chait, and R. MacKinnon. 2002b. The open pore conformation of potassium channels. *Nature* **417:**523–526.

Jiang, Y., A. Lee, J. Chen, V. Ruta, M. Cadene, B. T. Chait, and R. MacKinnon. 2003a. X-ray structure of a voltage-dependent K^+ channel. *Nature* **423:**33–41.

Jiang, Y., A. Pico, M. Cadene, B. T. Chait, and R. MacKinnon. 2001. Structure of the RCK domain from the *E. coli* K^+ channel and demonstration of its presence in the human BK channel. *Neuron* **29:**593–601.

Jiang, Y., V. Ruta, J. Chen, A. Lee, and R. MacKinnon. 2003b. The principle of gating charge movement in a voltage-dependent K^+ channel. *Nature* **423:**42–48.

Koishi, R., H. Xu, D. Ren, B. Navarro, B. W. Spiller, Q. Shi, and D. E. Clapham. 2004. A superfamily of voltage-gated sodium channels in bacteria. *J. Biol. Chem.* **279:**9532–9538.

Kojima, S., Y. Asai, T. Atsumi, I. Kawagishi, and M. Homma. 1999a. Na+-driven flagellar motor resistant to phenamil, an amiloride analog, caused by mutations in putative channel components. *J. Mol. Biol.* **285:** 1537–1547.

Kojima, S., K. Yamamoto, I. Kawagishi, and M. Homma. 1999b. The polar flagellar motor of *Vibrio cholerae* is driven by an Na+ motive force. *J. Bacteriol.* **181:**1927–1930.

Kumanovics, A., G. Levin, and P. Blount. 2002. Family ties of gated pores: evolution of the sensor module. *FASEB J.* **16:**1623–1629.

Kung, C., and P. Blount. 2004. Channels in microbes: so many holes to fill. *Mol. Microbiol.* **53:**373–380.

Laine, M., D. M. Papazian, and B. Roux. 2004. Critical assessment of a proposed model of Shaker. *FEBS Lett.* **564:**257–263.

Ludewig, U., M. Pusch, and T. J. Jentsch. 1996. Two physically distinct pores in the dimeric ClC-0 chloride channel. *Nature* **383:**340–343.

MacKinnon, R. 1995. Pore loops: an emerging theme in ion channel structure. *Neuron* **14:**889–892.

MacKinnon, R., S. L. Cohen, A. Kuo, A. Lee, and B. T. Chait. 1998. Structural conservation in prokaryotic and eukaryotic potassium channels. *Science* **280:**106–109.

Middleton, R. E., D. J. Pheasant, and C. Miller. 1996. Homodimeric architecture of a ClC-type chloride ion channel. *Nature* **383:**337–340.

Moczydlowski, E. 1998. Chemical basis for alkali cation selectivity in potassium-channel proteins. *Chem. Biol.* **5:**R291–R301.

Morals-Cabral, J. H., Y. Zhou, and R. MacKinnon. 2001. Energetic optimization of ion conduction rate by the K^+ selectivity filter. *Nature* **414:**37–42.

Natale, D. A., M. Y. Galperin, R. L. Tatusov, and E. V. Koonin. 2000. Using the COG database to improve gene recognition in complete genomes. *Genetica* **108:**9–17.

Nishida, M., and R. MacKinnon. 2002. Structural basis of inward rectification: cytoplasmic pore of the G protein-gated inward rectifier GIRK1 at 1.8 A resolution. *Cell* **111:**957–965.

Noda, M., T. Ikeda, T. Kayano, H. Suzuki, H. Takeshima, M. Kurasaki, H. Takahashi, and S. Numa. 1986a. Existence of distinct sodium channel messenger RNAs in rat brain. *Nature* **320:**188–191.

Noda, M., T. Ikeda, H. Suzuki, H. Takahashi, T. Takahashi, M. Kuno, and S. Numa. 1986b. Expression of functional sodium channels from cloned cDNA. *Nature* **322:**826–828.

Noda, M., H. Takahashi, T. Tanabe, M. Toyosato, Y. Furutani, T. Hirose, M. Asai, S. Inayama, T. Miyata, and S. Numa. 1984. Primary structure of *Electrophorus electricus* sodium channel deduced from cDNA sequence. *Nature* **312:**121–127.

Ogielska, E. M., W. N. Zagotta, T. Hoshi, S. H. Heinemann, J. Haab, and R. W. Aldrich. 1995. Cooperative subunit interactions in C-type inactivation of K channels. *Biophys. J.* **69:**2449–2457.

Ong, B. H., G. F. Tomaselli, and J. R. Balser. 2000. A structural rearrangement in the sodium channel pore linked to slow inactivation and use dependence. *J. Gen. Physiol.* **116:**653–662.

Papazian, D. M., X. M. Shao, S.-A. Seoh, A. F. Mock, Y. Huang, and D. H. Wainstock. 1995. Electrostatic interactions of S4 voltage sensor in Shaker K^+ channel. *Neuron* **14:**1293–1301.

Pavlov, E., C. Bladen, P. P. S. Dhaliwal, and R. J. French. 2003. Block of a prokaryotic voltage-gated sodium channel (NaChBac) by phenytoin and lidocaine. *Biophys. J.* **84:**66a.

Pavlov, E., C. Bladen, R. J. Winkfein, P. P. S. Dhaliwal, Q. Ma, and R. J. French. 2004. Inactivation in a bacterial voltage-gated sodium channel—a C-type mechanism? *Biophys. J.* **86:**167a.

Ragsdale, D. S., and M. Avoli. 1998. Sodium channels as molecular targets for antiepileptic drugs. *Brain Res. Rev.* **26:**16–28.

Ren, D., B. Navarro, H. Xu, L. Yue, Q. Shi, and D. E. Clapham. 2001. A prokaryotic voltage-gated sodium channel. *Science* **294:**2372–2375.

Richard, E. A., and C. Miller. 1990. Steady-state coupling of ion-channel conformations to a transmembrane ion gradient. *Science* **247:**1208–1210.

Sanguinetti, M. C., M. E. Curran, A. Zou, J. Shen, P. S. Spector, D. L. Atkinson, and M. T. Keating. 1996. Coassembly of K_vLQT1 and minK (IsK) proteins to form cardiac I_{Ks} potassium channel. *Nature* **384:**80–83.

Sather, W. A., and E. W. McCleskey. 2003. Permeation and selectivity in calcium channels. *Annu. Rev. Physiol.* **65:**133–159.

Sather, W. A., J. Yang, and R. W. Tsien. 1994. Structural basis of ion channel permeation and selectivity. *Curr. Opin. Neurobiol.* **4:**313–323.

Smith, P. L., T. Baukrowitz, and G. Yellen. 1996. The inward rectification mechanism of the HERG cardiac potassium channel. *Nature* **379:**833–836.

Sorensen, J. B., A. Cha, R. Latorre, E. Rosenman, and F. Bezanilla. 2000. Deletion of the S3-S4 linker in the Shaker potassium channel reveals two quenching groups near the outside of S4. *J. Gen. Physiol.* **115:** 209–222.

Stühmer, W., F. Conti, H. Suzuki, X. Wang, M. Noda, N. Yahagi, H. Kubo, and S. Numa. 1989. Structural parts involved in activation and inactivation of the sodium channel. *Nature* **339:**597–603.

Tatusov, R. L., N. D. Fedorova, J. D. Jackson, A. R. Jacobs, B. Kiryutin, E. V. Koonin, D. M. Krylov, R. Mazumder, S. L. Mekhedov, A. N. Nikolskaya, B. S. Rao, S. Smirnov, A. V. Sverdlov, S. Vasudevan, Y. I. Wolf, J. J. Yin, and D. A. Natale. 2003. The COG database: an updated version includes eukaryotes. *BMC Bioinformatics* **4:**41.

Tatusov, R. L., M. Y. Galperin, D. A. Natale, and E. V. Koonin. 2000. The COG database: a tool for genome-scale analysis of protein functions and evolution. *Nucleic Acids Res.* **28**:33–36.

Tatusov, R. L., D. A. Natale, I. V. Garkavtsev, T. A. Tatusova, U. T. Shankavaram, B. S. Rao, B. Kiryutin, M. Y. Galperin, N. D. Fedorova, and E. V. Koonin. 2001. The COG database: new developments in phylogenetic classification of proteins from complete genomes. *Nucleic Acids Res.* **29:**22–28.

Tiwari-Woodruff, S. K., M. A. Lin, C. T. Schulteis, and D. M. Papazian. 2000. Voltage-dependent structural interactions in the Shaker K(+) channel. *J. Gen. Physiol.* **115:**123–138.

Tiwari-Woodruff, S. K., C. T. Schulteis, A. F. Mock, and D. M. Papazian. 1997. Electrostatic interactions between transmembrane segments mediate folding of Shaker K+ channel subunits. *Biophys. J.* **72:**1489–1500.

Todt, H., S. C. Dudley, Jr., J. W. Kyle, R. J. French, and H. A. Fozzard. 1999. Ultra-slow inactivation in µ1 Na^+ channels is produced by a structural rearrangement of the outer vestibule. *Biophys. J.* **76:**1335–1345.

Walz, T., T. Hirai, K. Murata, J. B. Heymann, K. Mitsuoka, Y. Fujiyoshi, S. L. Smith, P. Agre, and A. Engel. 1997. The three-dimensional structure of aquaporin-1. *Nature* **387:**624–627.

Xue, T., I. L. Ennis, K. Sato, R. J. French, and R. A. Li. 2003. Novel interactions identified between µ-conotoxin and the Na^+ channel domain I P-loop: implications for toxin-pore binding geometry. *Biophys. J.* **85:**2299–2310.

Yang, J., P. T. Ellinor, W. A. Sather, J.-F. Zhang, and R. W. Tsien. 1993. Molecular determinants of Ca^{2+} selectivity and ion permeation in L-type Ca^{2+} channels. *Nature* **366:**158–161.

Yue, L., B. Navarro, D. Ren, A. Ramos, and D. E. Clapham. 2002. The cation selectivity filter of the bacterial sodium channel, NaChBac. *J. Gen. Physiol.* **120:**845–853.

Zagotta, W. N., T. Hoshi, and R. W. Aldrich. 1990. Restoration of inactivation in mutants of Shaker potassium channels by a peptide derived from ShB. *Science* **250:**568–571.

Zhao, Y., V. Yarov-Yarovoy, T. Scheuer, and W. A. Catterall. 2004. A gating hinge in Na+ channels; a molecular switch for electrical signaling. *Neuron* **41:**859–865.

Zhou, Y., J. H. Morals-Cabral, A. Kaufman, and R. MacKinnon. 2001. Chemistry of ion coordination and hydration revealed by a K^+ channel-Fab complex at 2.0 Å resolution. *Nature* **414:**43–48.

Bacterial Ion Channels and Their Eukaryotic Homologs
Edited by A. Kubalski and B. Martinac

Chapter 11

The CLC Family of Proteins: Chloride Transporters and Channels

Heather R. Rickard, Paul A. Bartley, Christopher J. Bagley, and Allan H. Bretag

Despite the ubiquity of chloride in both the extracellular and intracellular milieus of living organisms, there has long been a tendency to relegate Cl^- pores, channels, and transporters to the easily dismissed "passive leakage pathways" among the seemingly more important catalysts of cation transport (Bretag, 1987). It appears to have been with some reluctance that Cl^- permeation through cell membranes finally came to be accepted, at first in muscle cell membranes. There, sarcolemmal Cl^- permeability could no longer be ignored (Boyle and Conway, 1941), eventually being found to be several times K^+ permeability at rest, of the order of 100 times more than Na^+, and perhaps 1,000 times greater than Ca^{2+} permeability. An ultimate vindication for those who have researched the transmembrane movement of Cl^- has, however, now been realized with the award of the 2003 Nobel Prize for chemistry to Roderick MacKinnon of the Howard Hughes Medical Research Institute, Rockefeller University, New York, N.Y. Purification, crystallization, and the X-ray structural determination of a novel bacterial CLC protein (cf. Color Plates 22, 23, and 24) played a significant part in this achievement. *Salmonella enterica* serovar Typhimurium CLC protein, ClC-Setl (StClC), and its *Escherichia coli* homolog, ClC-Ec1 (EriC and YadQ), have become the conformational prototypes of the first large family of anion channels (the CLC family of voltage-gated Cl^- channels) to be analyzed in this way (Dutzler et al., 2002).

Medically, this family has been believed to be important since low membrane Cl^- conductance was associated with myotonic muscle disease in the mid-1960s (Lipicky and Bryant, 1966). Proof of this hypothesis, however, had to wait for the cloning of the skeletal muscle *CLCN1* gene and demonstration of mutations associated with myotonic disease. No one imagined the extent of the CLC family of proteins or of its association with the large number and variety of medical conditions now known.

Heather R. Rickard, Paul A. Bartley, and Allan H. Bretag • Centre for Advanced Biomedical Studies, University of South Australia, North Terrace, Adelaide, South Australia 5000, Australia. ***Christopher J. Bagley*** • Protein Laboratory, Hanson Institute, Institute of Medical and Veterinary Science, Frome Rd., Adelaide, South Australia 5000, and Department of Medicine, The University of Adelaide, Adelaide, South Australia 5005, Australia.

Table 1. CLC nomenclature

Suggested designation	Species identification	Former term
ClC-Tm1 or Tc1	*Torpedo marmorata* or *T. californica*	Torpedo ClC-0
ClC-Hs1, Hs2, etc.	*Homo sapiens*	Human ClC-1, 2, etc.
ClC-Rr1, Rr2, etc.	*Rattus rattus*	Rat ClC-1, 2, etc.
ClC-Rn1, Rn2, etc.	*Rattus norvegicus*	Rat ClC-1, 2, etc.
ClC-Mm1, Mm2, etc.	*Mus musculus*	Mouse ClC-1, 2, etc.
ClC-Cf1, Cf2, etc.	*Canis familiaris*	Dog ClC-1, 2, etc
ClC-Cp1, Cp2, etc.	*Cavia porcellus*	Guinea pig ClC-1, 2, etc.
ClC-Oc1, Oc2, etc.	*Oryctolagus cuniculus*	Rabbit ClC-1, 2, etc.
ClC-Dm1, Dm2, etc.	*Drosophila melanogaster*	Fruit fly ClC-1, 2, etc.
ClC-Ce1, Ce3b, etc.	*Caenorhabditis elegans*	Nematode clh-1, 3b, etc.
ClC-Ata, Atb, etc.	*Arabidopsis thaliana*	Arabidopsis AtClC-a, b, etc.
ClC-Sc1	*Saccharomyces cerevisiae*	Gef1
ClC-Ec1	*Escherichia coli*	EriC, YadQ
ClC-Set1	*Salmonella enterica* serovar Typhimurium	StClC
ClC-1, 2, etc.	Generic channel terminology	ClC-1, 2, etc.

To eliminate some of the confusion with respect to terminology and mutation sites relating to these channel proteins, we have utilized the scheme shown in Table 1, which includes an indication of species where appropriate. For generic usage, this is omitted. We have presumed that the presence of ClC-0 in torpedo skeletal muscle (Jentsch et al., 1993) implies that it is the ortholog of mammalian ClC-1. It could, however, be a paralog or splice variant of the major skeletal muscle Cl^- channel of elasmobranch fish.

Only after the first members of the CLC family had been cloned were proteins of similar amino acid sequence found to occur in prokaryotes, and it was proposed that these prokaryotic proteins might be amenable to structural analysis similar to that already achieved for K^+ channel proteins. As well, attention was given to whether the prokaryotic proteins might be functional orthologs of the eukaryotic Cl^- channels.

SOME GENERAL CHARACTERISTICS OF THIS FAMILY

Prokaryotic CLC Proteins

While there is no doubt from flux studies that the purified bacterial protein ClC-Ec1 confers Cl^- permeability on liposomes and black lipid bilayers, it now transpires that both its behavior and its structure are quite different from all preconceived notions of ion channels (Accardi and Miller, 2004). Potassium, Na^+, and Ca^{2+} channels are single pores, neatly formed at the center of a group of approximately parallel peptide helices, each tending to span the membrane and to be arranged perpendicular to the plane of the membrane lipid bilayer (Hille, 2001). In this way they mimic the classical gramicidin model of an ion channel (Hille, 2001). Typically, each "pore-forming" helix in the cation channels is a component of an independent protein subunit (composed of two to six similarly parallel helices). Usually four such subunits surround the channel pore or there is an equivalent number of concatemerized motifs mimicking the subunit arrangement. Even the less selective, ligand-gated channels follow a similar pattern, most commonly with five subunits surrounding the pore.

By contrast, the CLC protein monomer has 16 or 17 intramembrane helices, which are present at various oblique angles to the membrane (Color Plates 22 and 23 [see color insert]). Each monomer incorporates its own pore, but these are not tidily arranged with respect to helices and, as a result, are neither straight nor perpendicular to the plane of the membrane. Furthermore, the CLC protein occurs in membranes as a dimer (Middleton et al., 1996; Ludewig et al., 1996; Fahlke et al., 1997), creating a double-barreled structure with a separate pore in each of the monomeric subunits. This arrangement had long since been predicted from electrophysiological studies (Miller, 1983; Miller and White, 1984) and means that each pore must be lined entirely by sequences from within its own protein monomer (Mindell and Maduke, 2001).

Nevertheless, some similarities to cation channels do exist. In the CLC protein, as in the K^+ channel protein, the structure is arranged so as to stabilize the presence of the relevant ion in its pore (Color Plate 24 [see color insert]), and "hourglass-shaped" entranceways (not shown) tend to be guarded by residues of opposite charge to that of the permeating ion.

In ClC-Set1 there is also intriguing evidence of a simple gate, a glutamate residue that could swing in and out to close off the pore (Color Plate 25 [see color insert]) and that has been identified with the independent, fast gate of CLC channels (Dutzler, 2003). The X-ray crystallographic structure (Dutzler et al., 2002) of the prokaryotic CLC orthologs, ClC-Ec1 and ClC-Set1, confirmed earlier predictions that the amino and carboxyl termini both lie in the cytoplasm (Gründer et al., 1992) and showed the arrangement of the helices (designated B-R, with L of just three residues possibly not constituting a helix at all) with respect to the plane of the membrane lipid bilayer with an additional α-helical region (A) in the cytoplasmic amino-terminal sequence. The Cl^--binding site is stabilized within the pore by a radial arrangement of helix dipoles to which helices D, F, and N are the major contributors (Color Plate 24). It has since been shown, however, that ClC-Ec1 is not a channel (Accardi and Miller, 2004) and so the suggested location of a water-filled pore within the crystal structure of the prokaryotic CLC proteins is in some doubt.

Thus far, it appears as though we have learned a vast amount about the structure of a mediator of Cl^- transport in bacteria and, because of close similarity with eukaryotic CLC proteins, also about Cl^- transport in higher organisms, including humans. In reality, however, the CLC protein is an enigma. Prokaryotic CLC seems to be a proton transporter that incidentally exchanges protons for Cl^- (an H^+/Cl^- antiporter) (Accardi and Miller, 2004). By contrast, those eukaryotic members of the CLC family that have been amenable to study are obvious Cl^- channels. If they retain any of their evolutionary history, it is, first, that they still catalyze Cl^- flux and, second, that they retain an unusual ability to couple the conformational states involved in their gating to the transmembrane electrochemical gradient for Cl^- (Richard and Miller, 1990).

Eukaryotic CLC Proteins

Modulation by Voltage

Several members of the CLC family of proteins are voltage-gated, and this entire family is sometimes termed the voltage-gated family of Cl^- channels. Among these, the first to be described in electrophysiological terms was the Cl^- channel from the electric organ of the ray, *Torpedo californica* (White and Miller, 1979). The orthologous channel from *Torpedo marmorata* has subsequently been cloned and its cDNA sequence determined (ClC-Tm1, ClC-0) (Jentsch et al., 1990). Through homology screening with ClC-Tm1, a clone of another voltage-gated Cl^- channel from the skeletal muscle of the rat was obtained and its cDNA sequence determined. This channel is designated ClC-Rr1 (also ClC-Hs1 in humans) and is the major skeletal muscle Cl^- channel responsible for membrane potential stabilization and repolarization after an action potential (Steinmeyer et al., 1991b). Subsequently, a number of structurally related proteins have been described for other species and tissues, not all of which are strictly voltage-gated and some may not even be channels.

Modulation by Protons

The muscle-type members of the CLC family are sensitive to extracellular proton concentration, having distinctly different characteristics at either extreme of the physiological pH range (Hanke and Miller, 1983; Rychkov et al., 1996).

Modulation by Chloride

One of the most unusual features of these channels is that they are "self-gated." That is, the entry of Cl^- into the pore of the channel opens its gate in the manner of a person entering the sensor field of an automatic door causing it to open. This means that the channel opens intermittently in the presence of low concentrations of extracellular (external) Cl^- but will remain open continuously if the concentration is high, as does the automatic door. Conformational states of the channel associated with gating are thus coupled to the electrochemical gradient (Richard and Miller, 1990; Chen and Miller, 1996). As mentioned above, the Cl^--gating cycle is not microscopically reversible but demonstrates an unusual hysteresis. Single-channel patch clamping showed that the kinetic path from "one pore activated" to "both pores inactivated" to "both pores activated" was highly favored over the reverse path. While this might appear to be a violation of the second law of thermodynamics, it rather shows that the gating of these channels is normally occurring under conditions at a distance from thermodynamic equilibrium (Richard and Miller, 1990; Chen and Miller, 1996).

Being double-barreled, there is a gate in each channel, the "fast" gates that appear to have been identified as the glutamate residues in each channel pore and at least one further form of gating, formerly known as the "slow" gate from its characteristics in early studies of *T. californica* chloride channels (Miller, 1983; Miller and White, 1984) and later in expressed ClC-Tm1 (Pusch et al., 1997). This type of gating is now known as "common" gating because its rate of operation can be faster than the "fast" gate in some ClC-Hs1 mutants and because it operates simultaneously on both pores of a dimeric CLC channel (Bennetts et al., 2001).

Modulation According to Cellular Energy Status

Several recent papers have discussed the structure and possible function of the paired cystathionine β-synthase (CBS) domains now known to occur in 1,000 or so different proteins. Evidence is mounting for a role in sensing cellular energy levels (Adams et al., 2004; Scott et al., 2004). The paired CBS structures align themselves to create an adenosine-ligand-binding groove while mutations in the putative binding site modulate catalytic action in enzymes such as inosine monophosphate dehydrogenase and alter the K_d for ligand binding by several thousandfold. All eukaryotic CLC proteins have cytoplasmic carboxyl tails that include a pair of CBS domains (Color Plate 26 [see color insert]

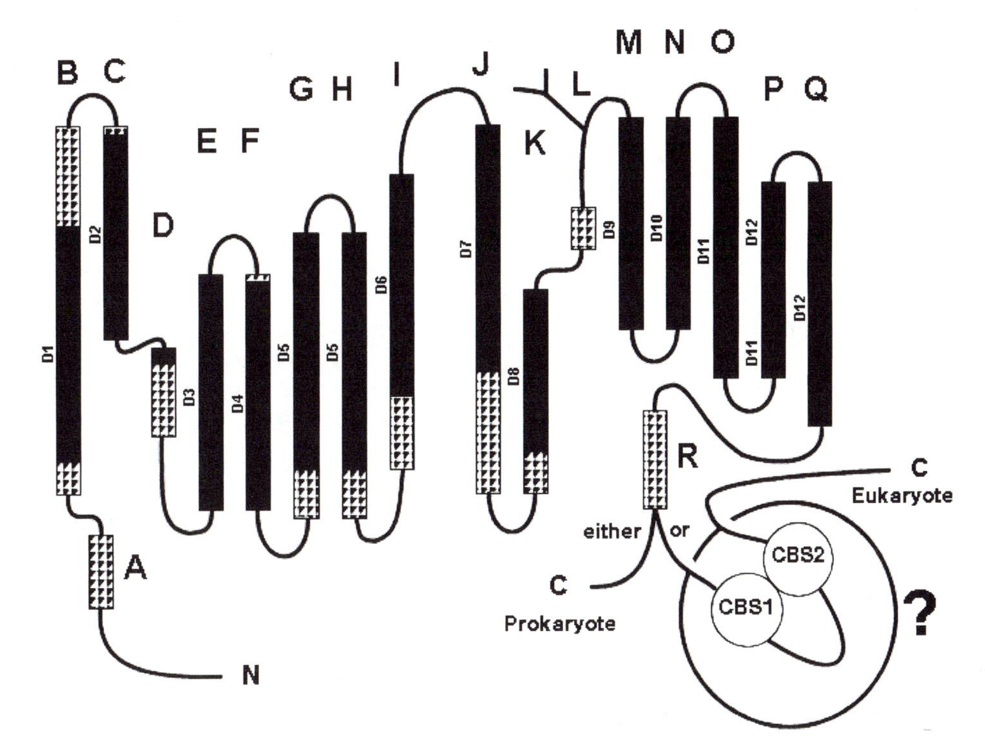

Figure 1. Dual topological diagram comparing the original 12 hydrophobic domains with the 18 helices (A to R) based on the X-ray structure. The lengthy eukaryotic C tail of largely indeterminate structure with its pair of CBS domains is also compared with the short prokaryotic C tail. Original hydrophobic domain D13 coincides roughly with CBS2.

and Fig. 1) and, hence, are also prime candidates for modulation according to cellular energy level.

Modulation by Ligands

Ligand-binding sites have been found for most ion channels that have been studied, not only those that are the classical, such as the nicotinic acetylcholine receptor, with its intrinsic nonselective cation channel for a pore, but also for several CLC family members. The most studied of these are ClC-1, where a ligand-binding site has been carefully characterized associated with the internal mouth of the conducting pore (Estévez et al., 2003), and ClC-RrK1, where another site is associated with the external mouth (Liantonio et al., 2004). Each of these is associated with channel block when their ligand partially occludes the pore. It would not be too surprising if further binding sites were to be found, which, when occupied, affected only channel gating or, by conformational modification, altered the characteristics of permeation or conduction.

Modulation by Volume or via the Cytoskeleton

Chloride channels are intimately involved in the regulation of cell volume (Hille, 2001). When a cell is exposed to hypotonic conditions, water enters because of the osmotic imbalance between the cytoplasm and the external fluid. Thus the cell swells. To control the amount of water entering, some cytoplasmic constituents must be lost, these being mostly K^+ and Cl^- ions, but also some organic osmolytes such as amino acids and sugars (Valverde et al., 1995). One Cl^- channel known to be regulated by changes in cell volume is ClC-Rn2 (Gründer et al., 1992), which, although definitely a member of the CLC family of voltage-gated channels, does not respond to voltage changes in the physiological range (Thiemann et al., 1992). Rather, it is activated, or opened, by cell swelling in response to the hypotonicity of the extracellular fluid (Gründer et al., 1992). Opening of the channels results in a loss of Cl^- ions to the external fluid, thus lowering the internal osmolarity of the cell to match the extracellular environment more closely. The ClC-Cp3 channel is another that has been proposed to be involved in volume-regulated Cl^- flux. This channel is, however, the subject of ongoing controversy. It is most closely related to other members of the CLC family that are known to be mainly intracellular and confined to endosomes (Fig. 2) and so the ability to monitor its function in the plasma membrane has been questioned. On the other hand, observations on both native cells and heterologous expression systems, and the use of what are believed to be specific blocking antibodies, point to volume-regulated Cl^- currents that can be directly attributed to the presence of ClC-3 in the plasma membrane (Jin et al., 2003; Wang et al., 2003). Finally, as will be discussed later, bioinformatics screening of CLC channels suggests the existence of regions

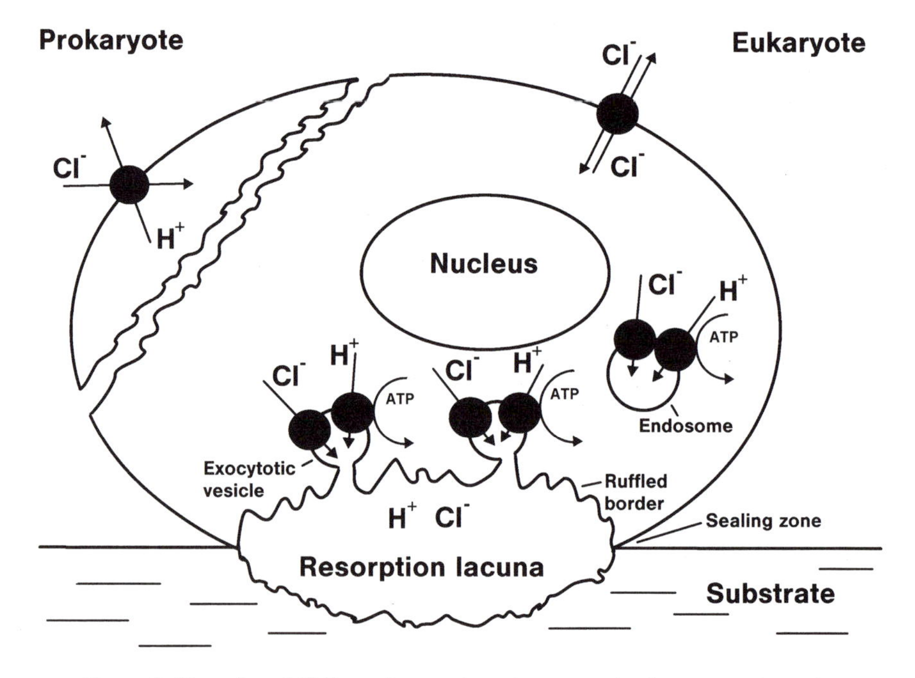

Figure 2. Illustration of CLC protein types in prokaryotes and eukaryotes. Prokaryotic ClC-Ec1 is shown as a Cl^-/H^+ exchanger. Eukaryotes have evolved a variety of CLC proteins. Some are confined to internal membranes, where they are channels associated with proton pumps and endosome acidification. Others are plasma membrane channels capable of accommodating the current flow necessary to stabilize muscle membrane potentials or to facilitate the stunning voltage discharges of electric rays.

in the cytoplasmic carboxyl tail of these proteins that have the propensity to bind actin and possibly other cytoskeletal proteins.

EVOLUTIONARY RELATIONSHIPS

Since the molecular characterization of the first member of the CLC family, ClC-Tm1 (ClC-0), nine mammalian homologs, many other vertebrate and invertebrate equivalents, a plant-derived group from *Arabidopsis thaliana* and other plants, one from the yeast *Saccharomyces cerevisiae*, and others from a variety of prokaryotes have been identified (Hechenberger et al., 1996; Mindell and Maduke, 2001). They display only some 20% amino acid sequence identity overall but result in proteins with similar structural properties. The first member of the group, ClC-Tm1 (ClC-Hs1), has been considered to be a muscle-type channel displaying strong voltage dependence. Other members may not, however, have voltage dependence as their primary gating mechanism.

All of the members of the CLC family are believed to have evolved from a common ancestral protein and can be found in individuals from all living kingdoms, including *Archaea*, implying these proteins are very ancient (Mindell and Maduke, 2001). Animal CLC proteins can be subdivided into three main subgroups, within which there is much higher sequence similarity than overall (Mindell and Maduke, 2001). These subgroups include the muscle type mentioned above, as well as ClC-2 and the kidney isoforms ClC-Ka and ClC-Kb. The second subgroup contains ClC-3, ClC-4, and ClC-5. The third subgroup includes ClC-6 and ClC-7. Of these, ClC-Hs1 and ClC-Hs2 are located in the plasma membrane of cells while ClC-Hs3, ClC-Hs4, ClC-Hs5, ClC-Hs6, and ClC-Hs7 and some of the orthologs from other species are mainly, if not completely, located intracellularly in organelle membranes (Pusch, 2002). Many of the intracellularly located forms are associated with acidification processes and possibly with H^+-ATPase activity (Fig. 2) (Devuyst et al., 1999; Kornak et al., 2001). Perhaps not surprisingly, the channels have frequently been found in association with transport processes where there is movement of cations and hence complementary movement of anions is necessary for electrostatic balance. As Cl^- is the most abundant extracellular anion, such an association is appropriate.

For relationships between the groups and other members of the CLC family, see Table 1 and Fig. 3. Prokaryotic proteins appear to suggest an ancestral form while those of the muscle type appear to be the most evolved (Mindell and Maduke, 2001). As mentioned above, the CLC family of proteins appears to represent an evolutionary continuum, demonstrating how a transporter may evolve into a channel.

ACTION POTENTIALS

Excitable cells (nerve, muscle, and some gland and plant cells) are conspicuous for their ability to undergo rapid membrane potential changes, impulses or action potential, for rapid cell-to-cell communication or as signals to initiate some physiological process. In a skeletal muscle fiber, the resting membrane potential is usually held at about −80 to −90 mV (inside negative relative to outside zero). If, for whatever reason, Na^+, K^+, or Ca^{2+} begins to enter (or Cl^- to leave) the cell, its membrane potential rises, opening Na^+ channels and allowing the influx of Na^+. When this occurs sufficiently rapidly so as to outweigh the accompanying increase in Na channel inactivation,

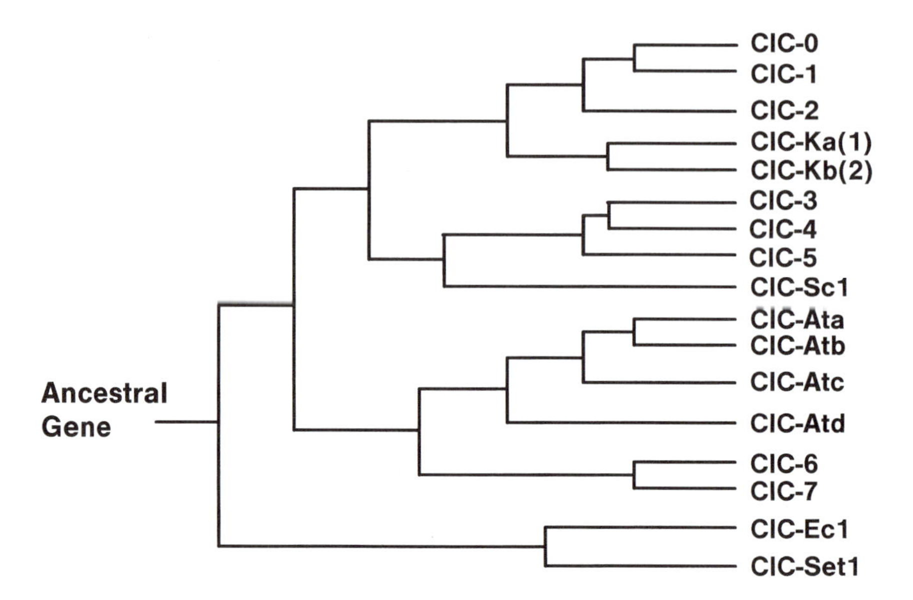

Figure 3. Diagram showing the relatedness of selected ClC sequences. Branch lengths are not proportional to evolutionary differences. Three main branches of the eukaryotic CLC family can be seen: that including ClC-0, ClC-1, ClC-2, ClC-Ka, and ClC-Kb; that including ClC-3, ClC-4, and ClC-5; and that including ClC-6 and ClC-7. Of the others, only those of the plant *A. thaliana*, the yeast *S. cerevisiae*, and the two prokaryotic CLCs, for which the crystal structures are known, have been included.

further Na channels open and increasing amounts of Na^+ enter the cell, allowing the membrane potential to reach threshold. Now, the chain reaction described above becomes self-sustaining (the regenerative component of the action potential). Within a few tenths of a millisecond, the membrane potential has risen to about +40 to +50 mV. In mammalian nerve fibers, voltage-coupled inactivation gates of the Na channels close in response to increasing depolarization, although these gates close rather more slowly than the activation gates open. This is sufficient to return the membrane potential to its resting value, the entire process constituting the typical action potential. Sodium pumps later restore ionic gradients to their pre-action potential states. In skeletal muscle cells, neurons, and amphibian and squid nerve fibers, K^+ channels begin to open slowly with depolarization and assist in ensuring the return of the membrane potential to its resting potential (repolarization) and to electrical stability at or near that potential. In skeletal muscle cells this process is complicated by the presence of transverse tubules invaginating from the sarcolemma, inside which K^+ accumulates during repetitive action potential discharges (tetanic muscle activation). Increasing transverse tubular (extracellular) K^+ depolarizes the muscle cells, thereby compromising membrane electrical stability. For this reason it appears that a high Cl^- conductance (via ClC-1) has evolved to assist in reducing intracellular positivity (by allowing Cl^- influx during and after the action potential) and bootstrapping the membrane potential to the Cl^- equilibrium potential near to the normal resting potential.

THE MUSCLE-TYPE CLC CHANNELS (ClC-0 AND ClC-1)

Cells (electrocytes) making up the electric organs (electroplax) of the rays of the genus *Torpedo* have evolved from fusiform muscle fibers into noncontractile flattened coin-shaped disks. These cells are innervated on their obverse surfaces and have such a high concentration of acetylcholine receptor channels that these form semicrystalline two-dimensional arrays. Membranes of their reverse surfaces are similarly rich in Cl^- channels. These arrangements enable serial electrocyte stacks to operate as high-voltage, low internal resistance batteries capable of producing powerful electric currents that the fish use to stun their prey. The unexpectedly high concentration of Cl^- channels in electrocyte membranes enabled extraction and incorporation of individual channels into black lipid bilayer membranes. This led to them being the first of the voltage-gated Cl^- channels to undergo biophysical characterization (White and Miller, 1979).

Although it has outward rectification and the comparatively large single channel conductance of 10 pS, ClC-0 (ClC-Tm1) is often considered representative of this unusual group of Cl^- channels. It has a selectivity sequence of $Cl^- > Br^- > I^-$, and I^- appears to produce a strong voltage-dependent channel block (Pusch and Jentsch, 1994). It was proposed quite early that gating of each subunit of the double-barreled dimer occurred independently on a millisecond timescale (Hanke and Miller, 1983). If both pores are open, conductance is approximately 20 pS, as was originally observed (White and Miller, 1979). Fast gating is dependent on external Cl^- concentration, with these ions binding to an anion-binding site accompanied by a conformational change (conferring voltage dependence) and opening the channel (Pusch et al., 1995; Pusch et al., 1997; Ludewig et al., 1997a, 1997b, 1997c). Gating of ClC-Tm1 shows the usual two components: fast and common. At depolarizing potentials, the faster gate opens, while at hyperpolarizing potentials, the common gate is activated. It has been postulated that common gating is a property of both pores operating simultaneously due to conformational interactions between each of the subunits of the dimer (Ludewig et al., 1996). A very high temperature coefficient for common gating suggests substantial conformational rearrangement during the process (Pusch et al., 1997).

The cDNA from *T. marmorata* predicts a protein of 805 amino acids with a molecular mass of 89 kDa (Jentsch et al., 1990). Although much of the earlier electrophysiological measurement had been carried out on channels from *T. californica*, it was certain to be highly relevant because the cDNA of this closely related species is only slightly different, predicting a protein with 810 amino acids and 97% identity (O'Neill et al., 1991). The ClC-Tm1 protein was, at first, proposed to comprise 13 hydrophobic domains, termed D1 to D13, with the N terminus being intracellular and some uncertainty about the location of the C terminus (Jentsch et al., 1990). Initially there appeared to be two possible glycosylation sites, one between D8 and D9 (helices K and M) (Fig. 1), and the other between D13 (CBS2) in the cytoplasmic C tail and the C terminus (Jentsch et al., 1990). Topology of all members of the CLC family has been extensively revised since the determination of the X-ray crystallographic structure of a prokaryotic channel (Dutzler et al., 2002). This study found the original D1-to-D12 domains to correspond only roughly with the 16 or 17 helical intramembrane domains, B to R, as discussed above (Fig. 1). That the D13 domain was not a membrane-spanning region but free in the cytoplasm along with the carboxyl terminus had been ascertained much earlier (Gründer et al., 1992). It has been suggested that the carboxyl tail, the region C-terminal to helix R (the old D12), is necessary for formation of functional channels and that a considerable part of the region between D12 and D13 and D13 itself are essential for common gating to occur (Fong et al., 1998; Maduke et al., 1998). This seems to imply that the D12-to-D13 region (helix R to CBS2) (Fig. 1) is essential for the intersubunit interaction necessary to achieve common gating. Fast gating also seems to depend on components of the C tail; in particular, a lysine residue at position 519 was earlier postulated to be part of the pore of the channel (Ludewig et al., 1997b). Iodide blocks wild-type channels (containing this K519), but mutated channels with no positively charged residue at this position show reduced block by I^- (Pusch et al., 1995). A model was proposed in which the gate is located at the cytoplasmic end of the pore with two anion-binding sites in the pore, one close to the gate and one on the external side of the pore (Pusch et al., 1995). This suggested that the region at the end of D12 forms part of the pore with a region that acts as the anion-binding site, possibly K519, which is conserved between ClC-Tm1, ClC-Hs1, ClC-Hs2, ClC-HsKa, and ClC-HsKb. The anion-binding site, however, may not be just a simplistic electrostatic interaction as mutations around K519 also affect gating (Ludewig et al., 1997b). This residue has now been shown to be part of helix R, the amino terminus of which is involved in the pore, and the carboxyl terminus of which (containing K519) protrudes into the cytoplasm (Dutzler et al., 2002). Hence K519 probably does not constitute a Cl^--binding site within the pore. It is, however, located in the intracellular mouth of the pore and influences the movement of the negatively charged chloride ions through the pore (Chen and Chen, 2003). This lysine also forms an integral part of an SH3-binding domain and may have important implications for cytoplasmic signaling on or by the channel.

Many mutational studies have been performed in ClC-Tm1 to identify residues important in gating and permeation. A region between the N-terminal end of helix D (the old D2-D3 linker), in particular a serine residue at position 123, has been shown to affect ion selectivity and gating, which is not surprising since this has now been shown to form an essential part of the central Cl^--binding site. Serine 123 forms part of a sequence, GSGIP, that is highly conserved in most forms of CLC (Ludewig et al., 1996). Several other regions have been investigated for their involvement in channel function. These include the area around D70 in helix B (the old D1) that corresponds to a conserved D136 in ClC-Hs1, mutations of which result in myotonic muscle stiffness in humans (Fahlke et al., 1995; Ludewig et al., 1997a). Others include several positions in the highly conserved P and Q helices (the old D11 and D12 domain) and a positively charged arginine residue between D3 and D4, all of which result in an inwardly rectifying phenotype similar to the D70 mutation. Although these mutations obviously change the channel gating, none of the mutated sites can be regarded as the voltage sensor in its own right, as all produce similar effects (Ludewig et al., 1997a). Unlike voltage-gated cation channels, in which membrane potential is detected by charged amino acids in the transmembrane domains, in ClC-Tm1 the gating charge is supplied by the permeating Cl^- ions and hence a voltage sensor, as such, does not exist in the ClC-Tm1 protein (Ludewig et al., 1997a; Dutzler et al., 2003).

ClC-1

Skeletal muscle has a high Cl^- conductance, in some species up to 85% of total conductance (Palade and Barchi, 1977; Bretag, 1987). This current is predominantly due to the presence in the skeletal muscle cell membrane of ClC-1, the second member of the CLC family to be cloned and sequenced (Steinmeyer et al., 1991b). This channel is expressed predominantly in skeletal muscle, and to a minor extent in cardiac and smooth muscle, liver, and kidney. Levels of mRNA for ClC-Rn1 increase rapidly in the first few weeks of life, mirroring a similar increase in Cl^- conductance in intact muscle over the same time period (Steinmeyer et al., 1991b). Expression of ClC-Mm1 and Rn1 appears to be related to the presence of an intact innervation, as mRNA is not detectable in denervated muscles or myoballs. The protein appears to have a very short half-life, of the order of less than a day, allowing tight up- or down-regulation of protein expression (Klocke et al., 1994).

Single channel conductance in ClC-1 is very low, of the order of 1 pS, and can be blocked by anthracene-9-carboxylate (A9C). Selectivity of the ClC-Hs1 pore follows the sequence $Cl^- > Br^- > I^-$, similar to ClC-Tm1, and it is also strongly blocked by extracellular I^-, which appears to affect both conductance and gating (Pusch et al., 1994; Pusch and Jentsch, 1994). Anomalous mole fraction effects on conductance and permeability in ClC-Rn1, Hs1, and Tm1 indicate that they are channels containing multiple binding sites for the permeating ion (Miller, 1983; Pusch et al., 1995; Rychkov et al., 1998). A fast gating component similar to that of ClC-Tm1 is seen in ClC-Hs1, but it appears to lack the common (slow) gating component with its inverse voltage dependence. On the other hand, a common gating process is present, but for wild-type ClC-Rn1 and Hs1 it is almost as fast as the fast gating process under physiological conditions (Aromataris et al., 2001) while it is also characterized by a high temperature coefficient, although not so significant as that for ClC-Tm1 (Bennetts et al., 2001).

Typical whole-cell patch-clamp currents through wild-type ClC-Hs1 channels and the means of assessing their apparent open probability are shown in Fig. 4A and B. Membrane depolarization results in very rapidly activating outward currents (positive/upward by convention and representing influx of Cl^- into the cell). Hyperpolarization initiates inward currents (negative/downward and representing Cl^- efflux). These currents deactivate (decay) with a double exponential time course that allows the open probability of the fast and common gates to be determined (e.g., for the mutant C278G) (Fig. 4C and D). Because the open probability for the channel increases at positive potentials, it will be opening at around the time the skeletal muscle voltage-gated Na^+ channels are closing near the peak of the action potential (Steinmeyer et al., 1991b; Pusch et al., 1994). Chloride ions will then be flowing into the cell as K^+ is flowing out, assisting repolarization. Functionally, ClC-1 is an inward rectifier, which means that maximum open probability for Cl^- influx is achieved and current is limited already at even small positive potentials (depolarizations) (Fig. 4B). On the other hand, increasingly large initial inward currents (Cl^- efflux) are obtained at increasingly negative potentials (hyperpolarization) (Fig. 4A).

At first, there was some disagreement over whether ClC-Hs1 has a fast gating mechanism that is similar to that of ClC-Tm1 and dependent on the permeating anion. The opposing suggestion was that an intrinsic part of the ClC-Hs1 protein molecule plays the part of a voltage sensor as in voltage-dependent cation channels, thus giving it a fast gating mechanism different from that of ClC-Tm1. A "ball-and-chain" model was first proposed for the gating mechanism, with voltage-dependent binding of the "ball" moiety to a site located at the cytoplasmic end of the channel pore, to form a cytoplasmic gate. Voltage dependence was suggested to be due to a pair of voltage sensors, located at or around D136,

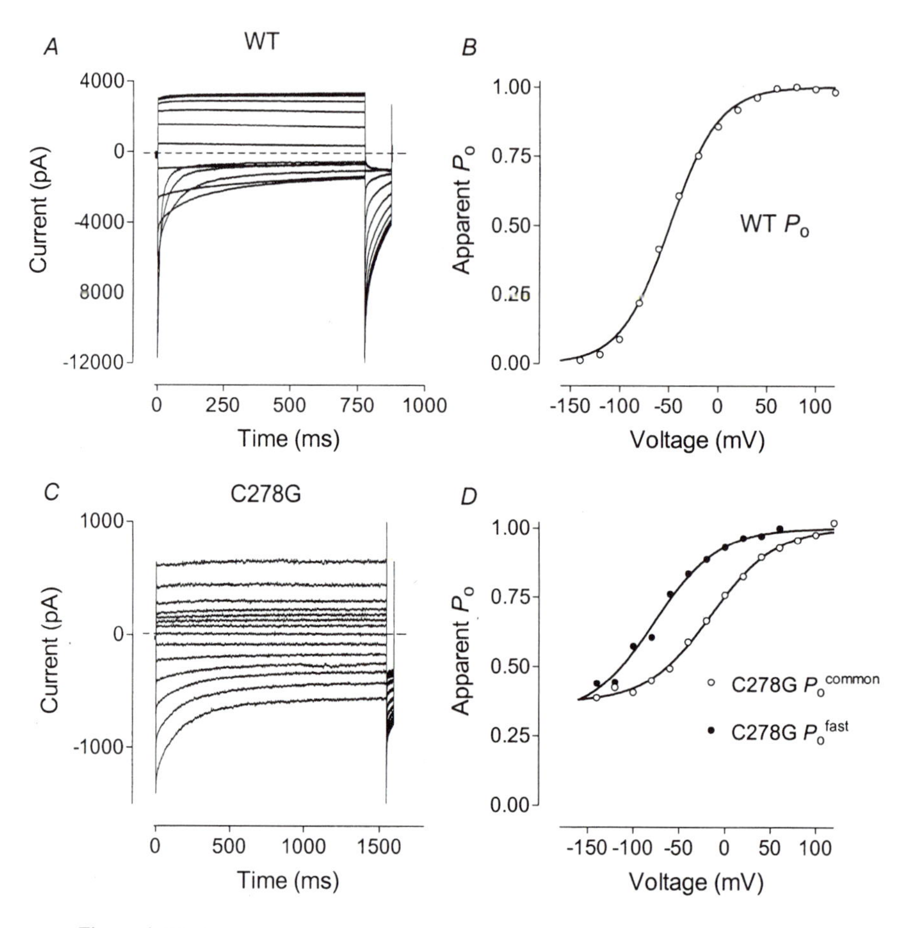

Figure 4. Whole-cell patch-clamp currents recorded from cultured HEK cells expressing the wild-type (WT) or mutant (C278G) human skeletal muscle chloride channel, ClC-Hs1. (A) A sequence of currents is shown (overlying each other) in response to conditioning voltage steps from −140 mV (inside negative compared to outside zero) to +100 mV. "Tail" currents are also shown in response to a constant test pulse of −100 mV. From the sizes of the tail currents, an apparent overall (for both gates) open probability, P_o, can be calculated for these channels at each conditioning voltage. (B) P_o for the WT channels is shown. Separated apparent open probabilities for the fast and common gates are shown in panel D for the C278G mutant whose currents are recorded in panel C.

which did not form an integral part of the pore. It was postulated that the function of the voltage sensor was to act as a transducer to couple changes in membrane potential to changes in the affinity of the binding site in the pore for the gate or ball moiety (Fahlke et al., 1996). Alternatively, it has been proposed that ClC-Hs1 has a gating mechanism similar to that of ClC-Tm1 and thus does not include an intrinsic voltage sensor. Gating is again dependent on the permeating anion binding to a site accessible only from the extracellular side of the membrane, with opening of the channel only occurring when the site is occupied (Rychkov et al., 1996). Gating is dependent on the extracellular but not intracellular Cl^- concentration, with Cl^- binding coupled to a conformational change (again conferring voltage dependence) and opening a gate to the cytoplasmic side (Rychkov et al., 1996). In accord with the crystal structure, it has been proposed that Cl^- ions activate (open) the channel by displacing a glutamate side chain that projects into the lumen of the pore (Dutzler et al., 2002; Dutzler et al., 2003).

It has long been recognized that pH affects Cl^- flux in intact skeletal muscle cells, low pH reducing Cl^- conductance (Hutter and Warner, 1967a; Warner, 1972; Palade and Barchi, 1977). For ClC-Rr1 expressed in Sf-9 insect cells, gating and conductance are also affected by pH (Rychkov et al., 1996). Lowering the external pH (increasing H^+ concentration) results in reduced peak currents and conductance, slowing of deactivation, and a shift in the deactivation curve to more hyperpolarized potentials. This effect of external pH could be titrated, with an apparent pK_a of 6, which is close to the pK_a of the imidazole side chains of histidine residues and, so, appeared to implicate these residues as putative protonation sites. The protonation site was again, like the Cl^--binding site, only accessible from the external side of the membrane. At low external pH, kinetics became independent of the Cl^- concentration, and it was postulated that protonation increased the affinity of Cl^- for its binding site and removed the voltage dependence displayed by Cl^- binding to its site at physiological pH (Rychkov et al., 1996). In light of the X-ray crystal structure of the CLC proteins, however, the site of protonation is almost certainly the glutamate gate, which is opened by protonation.

Lowering the internal pH had a different effect, slowing deactivation kinetics and shifting the open probability of the channel to more hyperpolarizing potentials. Concomitantly, raising the internal pH shifts the voltage dependence of gating to more depolarizing potentials, mimicking the effect of reducing extracellular Cl^- concentration. It was proposed that OH^- might bind to the inner side of the pore in a manner that reduces the affinity of Cl^- for its binding site, perhaps through a conformational change in the protein (Rychkov et al., 1996).

One of the effects of lowering external pH was to expose a previously unrecognized slow gate in ClC-Rr1, perhaps analogous to the common (slow) gate of ClC-Tm1, in that, at low external pH (5.5), a slowly activating current was produced at hyperpolarizing potentials. This gate might be closed at low pH and potentials near zero, only opening at high internal negativity. Under normal physiological conditions it might be constitutively open (Rychkov et al., 1996).

Just as for the effects of pH on skeletal muscle, it has been known for some time that certain transition metals are able to interfere with normal muscle function by blocking Cl^- currents (Hutter and Warner, 1967a, 1967b; Bretag, 1987). This effect has been examined using Zn^{2+} and Cd^{2+}, which exhibit a block of ClC-Rn1 from the extracellular side of the channel at neutral pH but show no effect at low pH. This is the opposite effect to that of A9C, which shows enhanced binding at low pH. Investigation of these blockers has shown a pK_a of 6.8 for Cd^{2+} block with the involvement of three residues in the binding site, and for A9C a pK_a of 6.4 involving the protonation of a single site. Both pK_a values suggest histidine as the residue most likely to be involved (although not excluding other residues such as cysteine or even glutamate or aspartate), but not the same residue in each case (Rychkov et al., 1997).

Extending the work on the effect of anions other than Cl^-, it has been shown that these foreign anions fall into three classes, those that do not permeate the channel and do not gate the channel, those that cannot permeate the channel but can gate it and block currents, and those that can permeate the channel and can gate it but block Cl^- passage through the pore (Rychkov et al., 1998). These authors proposed a model of the channel pore in which there are two sites important for channel function. One site is the Cl^- binding site that regulates opening and closing of the channel. The other is a selectivity filter that excludes some anions. The Cl^- binding site is proposed to be more external than the selectivity filter, with permeating ions binding first to the regulating site to open the gate and then passing through the selectivity filter. Due to the differing effects mentioned above, it is postulated that the regulatory site and the selectivity filter in ClC-Hs1 have different specificities and hence gating and permeation are not directly coupled. In ClC-Tm1, gating and permeation are coupled, so that the specificities of each site may be similar. From the size of the anions that were permeant and those that were not, the pore diameter was estimated as approximately 4.5 Å. It was also postulated that interactions within the pore involve more than simple electrostatic forces and that significant hydrophobic interactions with the wall of the pore also occur (Rychkov et al., 1998).

Rat skeletal muscle ClC-Rr1 was the first of the ClC-1 congeners to be cloned and sequenced, the cDNA encoding a protein of 994 amino acids and a molecular mass of 110 kDa, making ClC-Rr1 considerably larger than ClC-Tm1. Most of the difference lies in the termini, both the amino and carboxyl, and in an insertion between the predicted helix R and the CBS2 domain (old D12 and D13) (Steinmeyer et al., 1991b). Hydropathy profiles predict a topology similar to that of ClC-Tm1 (Gründer et al., 1992). The full sequence of ClC-Hs1 cDNA from human skeletal muscle has also been determined. The gene predicts a protein of 988 amino acids and similar topology to that of the rat protein, ClC-Rr1, with most of the membrane-spanning domains encoded by separate exons and with several putative binding sites for transcription regulation in the 5′ promoter region (Lorenz et al., 1994). In rats, during the first few weeks after birth, ClC-Rr1 is developmentally up-regulated while its mRNA decreases rapidly when a muscle is denervated. Hence muscle activity is essential for ClC-Rr1 expression (Klocke et al., 1994). It is assumed that ClC-Rr1, like ClC-Tm1, has a topology similar to that of the prokaryotic channels that have been characterized by X-ray crystallography (Dutzler et al., 2002), although that study could not address the structure of the carboxyl tail as it is absent in ClC-Ec1 and ClC-Set1.

Some cases of dominant myotonia congenita and dominant osteopetrosis are associated with mutations in the carboxyl tail of ClC-Hs1 and ClC-Hs7, respectively. This suggests that the tail region (Fig. 1) is also somehow important in common gating. A pair of CBS domains has been identified in the carboxyl tail of ClC-Hs1 (Ponting, 1997). These domains of known β-sheet, α-helix structure ($\beta_1\alpha_1\beta_2\beta_3\alpha_2$) occurred, one about 20 residues after the end of the R helix (the old D12) and the second some 150 residues further downstream, in the region roughly corresponding to the old D13 (Schmidt-Rose and Jentsch, 1997b). It has been shown that D13 and even the whole of the second CBS domain are not crucial for function, but that a stretch of 18 amino acids from lysine 877 to arginine 894 is essential (Hryciw et al., 1998). Until recently, therefore, no function had been assigned to these CBS domains although they occur frequently in a variety of other protein families. It now appears that they are involved as pairs in the formation of adenosine ligand-binding pockets, acting as detectors of cellular energy levels (Adams et al., 2004; Scott et al., 2004). Of possible relevance here is the arrangement of the CBS pairing (Color Plate 26). The immediate post-CBS2 sequence, believed to be essential for ClC-1 function, is juxtaposed with the sequence upstream from CBS1, the R helix and the cytoplasmic helix-helix linkers of the CLC protein. Such close alignment could provide an appropriate feedback path for regulation of channel function.

It might be hypothesized that it is the added carboxyl tail that turns a prokaryotic H^+/Cl^- antiporter into a eukaryotic Cl^- channel. In this regard, a good deal of the eukaryotic carboxyl tail can be truncated (although not back as far as CBS2) before in vitro function fails (Estévez et al., 2004). Furthermore, in concatenated monomer pairs, only one tail seems to be necessary for function to be maintained (Hebeisen et al., 2004). Lengthy sequences within the tail can be eliminated without detriment to function and reconstitution experiments have found that nonfunctional, truncated versions of ClC-Tm1, ClC-Hs1, and ClC-Hs5 (provided the truncation is after CBS1 and before CBS2 [Fig. 1]) can have their function restored by coexpression with their complementary tail portions (Maduke et al., 1998; Estévez et al., 2004; Mo et al., 2004). In some cases, just short coexpressed peptide sequences from the distal tail can restore function. While observations such as these may provide some scope for the future therapy of genetic diseases involving the carboxyl tails of CLC proteins, they may also be misleading. It would be surprising if the lengthy carboxyl tails of eukaryotic CLCs did not perform more than a single regulatory or accessory function and, therefore, if all tail functions could be restored by treatment with a short, unattached peptide.

Mutations that disrupt channel function in ClC-Hs1 are spread over almost the entire length of its gene, *CLCN1*. More than 70 disease-causing mutations are now known and at least 12 benign polymorphisms have been identified (Pusch, 2002; Simpson et al., 2004). Mutations range from Q68X in the amino terminus to P932L in the carboxyl tail. Only about 25% of the disease-causing mutations display a dominant or partial dominant inheritance, the remainder being recessive "loss-of-function" types of mutation, in which channel activity is partly or totally destroyed, leading to membrane destabilization. Dominant patterns of inheritance can occur in oligomeric proteins, where the mutation may affect subunit interaction, i.e., the mutant subunit exerts a dominant negative effect (Meyer-Kleine et al., 1995). Analysis of these dominant mutants could indicate which regions of the protein are important for subunit interaction. Initially it was believed that there was no concentration of dominant mutations in any particular region of the protein, although they did tend to occur more frequently in the membrane-spanning domains (Pusch, 2002). Most of these mutations shifted the P_o curve for the channels to more positive potentials (parallel but well to the right of that in Fig. 4B) such that their contribution to the maintenance of the resting membrane potential is reduced (Pusch, 2002) and hence the membrane potential is destabilized. In fact, this depolarizing shift is generally entirely due to an underlying rightward shift in the open probability of the common gate while fast gating is relatively unaffected. In turn, the effect on common gating is dependent on the dominant negative effect, probably transmitted conformationally from the defective subunit to its wild-type partner (Color Plates 22 and 23). That this is the case is supported by a preponderance of naturally occurring dominant myotonic mutants located on the dimer interface, e.g., helices H, I, P, and Q (Duffield et al., 2003), or in regions between the dimer interface and the postulated fast gate (E232) in ClC-Hs1, e.g., in helices D and E and in the E-F and M-N linkers (Color Plates 22 and 23). Site-directed mutations in this region (Color Plate 22) support this view. Here mutation G284S (just N terminal of the middle of helix H) appears to prevent the minimum open probability of the common gate from ever getting below 0.75, even at very negative potentials (Fig. 5A and B). By contrast, for mutation S289G, just C terminal from the middle of helix H, the open probability of gating is shifted by more than 100 mV to the right (Fig. 5C and D) whereas for C278G (at the C-terminal end of helix G, as it begins the G-H linker), both fast and common gates tended to be open at negative potentials (Fig. 4C and D).

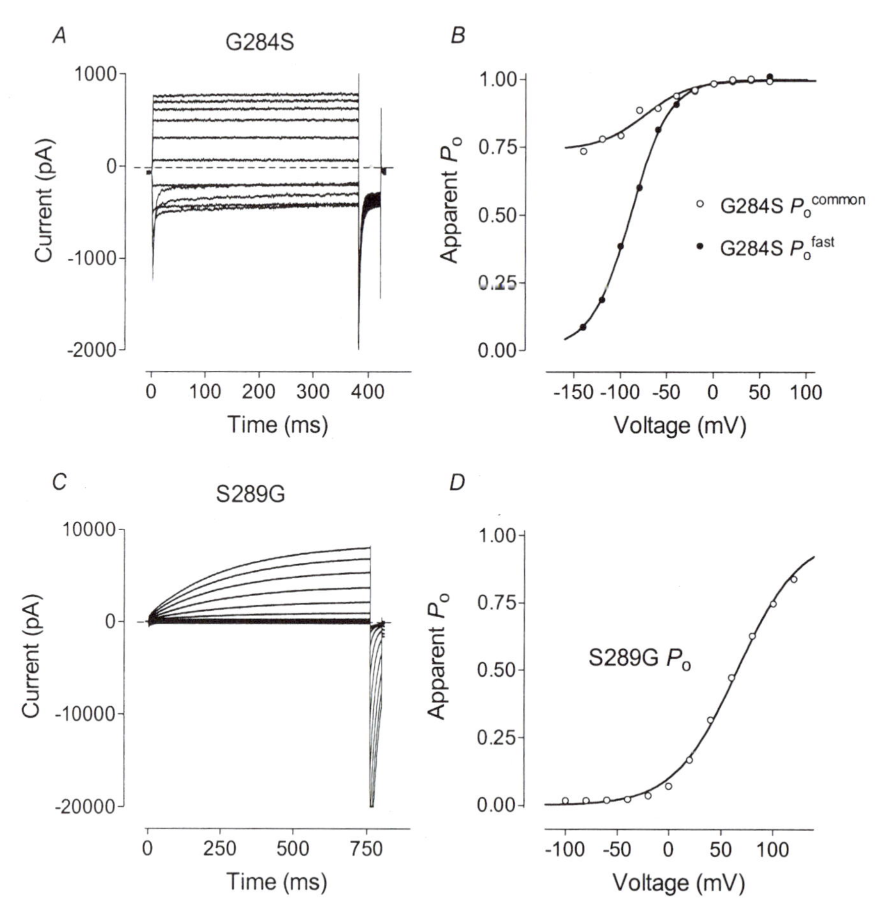

Figure 5. Whole-cell patch-clamp currents as in Fig 4. It is apparent (A) that currents from the G284S mutant display almost purely fast gating, and (B) this is reinforced by the separated P_o curves, where it can be seen that the P_o for the common gates is close to 1 even at very negative potentials. (C) In complete contrast, mutant S289G seems to show only common (slow) gating. This is because, as the overall P_o curve in panel D shows, the common gates do not begin to open in this mutant until around 0 mV, by which voltage the fast gates are fully open (cf. P_o^{fast} for G284S in panel B and P_o for WT in Fig. 4B).

Interestingly, some monocarboxylic aromatic acids (MCAA) affect common gating in a manner indistinguishable from that of the dominant genetic mutations, shifting the open probability curve for common gating far in the positive direction (Aromataris et al., 1999; Aromataris et al., 2001). A variety of analogs of these MCAA modifiers of channel function have been tested to discover those most effective in producing a depolarizing shift in the voltage dependence of gating, that is, those that would disallow common gate opening within the normal membrane potential range. Inverse agonists (MCAA analogs that increase open probability at normal membrane potentials) have also been sought that might be functional as small molecule therapeutic agents for dominant myotonia.

There have been a number of reports suggesting that protein kinase C (PKC) regulates macroscopic skeletal muscle membrane Cl^- conductance and even ClC-Hs1 function itself (Brinkmeier and Jockusch, 1987; Bryant and Conte-Camerino, 1991; De Luca et al., 1997; Rosenbohm et al., 1999). In support of these observations, the channel protein contains a number of possible phosphorylation sites although there are no reports of confirmatory studies that have used site-directed mutagenesis to determine the relevant phosphorylation sites. One group investigated the effects of growth hormone and insulin-like growth factor I on the Cl^- conductance of aged rat muscle and concluded that aging affects the phosphorylation state of the Cl^- channels, in that PKC maintains the channels in a phosphorylated, low-conducting state (De Luca et al., 1997). This effect can be countered by application of insulin-like growth factor I, which may act through a serine-threonine phosphatase to dephosphorylate the channel (De Luca et al., 1997). Another group has concluded that the phosphorylation site lies within the cytoplasmic vestibule of the channel (Rosenbohm et al., 1999).

PEPTIDE MOTIFS IN ClC-Hs1

Phosphorylation Sites

In theory, many potential phosphorylation sites exist in CLC proteins, but most of them occur in physiologically irrelevant positions. Phosphorylation has been postulated to be important for regulation of ClC-3 and has been linked to its S51 residue in the amino terminus (Duan et al., 1999). Other CLC proteins including ClC-1 have analogous residues in the area. Another possible phosphorylation site at a relevant position, accessible from the cytoplasm, occurs at the amino-terminal end of helix E. Many potential phosphorylation sites occur in the carboxyl tail of the eukaryotic forms of the proteins, but these sites are not conserved and their significance is unknown.

Myristoylation Sites

Potential myristoylation sites within a protein, although never myristoylated, may represent regions involved in hydrophobic interactions. Several such sites are found among the transmembrane helices where they may be important for helix packing, especially a number of overlapping sites in the helix G–helix H region, which suggests that it is particularly important for hydrophobic interactions.

Glycosylation Sites

Only one possible site for glycosylation is present in CLC proteins, and it is located in the helix L to M extracellular loop (old D8-D9 linker). This site has been shown to be glycosylated in ClC-Tm1 (ClC-0) (Schmidt-Rose and Jentsch, 1997a), while most of the other forms contain an analogous site in the same loop.

GxxxG Helix Interaction Motifs

One of the features of CLC proteins is the large number of conserved glycine residues, many of which occur in the recurring motif, GxxxG, found particularly in the membrane portion of the protein. This motif was originally described for glycophorin A in which it mediates dimer formation. Surrounding residues are also important and are preferably aliphatic residues such as leucine, isoleucine, or valine. Serine may substitute for the second glycine (Lemmon et al., 1994; Russ and Engelman, 2000). In ClC-1, core sequences with appropriate flanking residues can be found in helices G and H, while other core sequences are found in helices B, E, J, and N. Since functional CLC proteins are believed to be homodimers (Dutzler et al., 2002), these motifs may contribute to dimer formation, especially in helix H, which is located at the dimer interface (Dutzler et al., 2002). Motifs in transmembrane helices not located at the dimer interface presumably have another function. Being the smallest of the amino acids, glycines result in indentations or "notches" in helices and tend to occur in positions where helices cross over each other or interact (Javadpour et al., 1999). The X-ray crystal structure of ClC-Set1 revealed a complex arrangement of membrane helices where most of the helices interact with at least one other (Dutzler et al., 2002), and hence these motifs may be involved in helix packing within the CLC monomer. Similar motifs performing a similar function are found in transmembrane domains of other proteins with multiple helices, e.g., the phenylalanine/tyrosine transporter protein, PheP, from *E. coli* (Dogovski et al., 2003). A definitive topology for PheP has not yet been determined; hydropathy profiles predict a protein with 12 hydrophobic membrane domains and cytoplasmic amino and carboxyl tails (Dogovski et al., 2003). Many other bacterial transporters are believed to have this structure. It is intriguingly similar to that predicted for CLC proteins before their crystal structure was determined (Steinmeyer et al., 1991b). There is, however, minimal amino acid similarity between the two proteins, and their evolutionary relationship is unknown.

GxP(xG) Molecular Hinges

Proline residues are usually considered to be "helix breakers," causing bends or kinks in peptide chains. When they occur in association with a glycine, either two residues N terminal or C terminal from the proline, they may form a molecular hinge, allowing conformational changes to occur within the protein (Sansom and Weinstein, 2000). Several of these hinge motifs occur in CLC proteins, notably in the linkers between helices C and D, G and H, and M and N. Some occur within helices, such as in helix O. These hinge regions may allow the flexibility of movement required for gating (both fast and common) to occur (Tieleman et al., 2001).

Proline-Rich Motifs

Proteins often interact with their counterparts in regulatory proteins via proline-rich motifs. These regions can interact with regulatory or signal-transducing proteins carrying SH3, WW, or EVH-1 domains. A characteristic of many forms of CLC with long carboxyl tails is the presence of numerous proline residues. Some of these could be involved in regulatory functions.

Helix R is the final membrane-associated helix of CLC proteins and participates in Cl^- binding via a well-conserved tyrosine residue at the amino terminus of the helix. Most of the carboxyl terminus of the helix protrudes into the cytoplasm (Dutzler et al., 2002), providing the potential for a direct link between regulatory mechanisms operating in the cytoplasm and the conduction pore and selectivity filter of the channel. Located within the carboxyl terminus of helix R in ClC-Tm1, ClC-Hs1, and ClC-Hs2 is an SH3-binding motif

(KKLPYLP), providing a site for regulatory proteins carrying SH3 domains to act on the channel and, potentially, to influence events at the selectivity filter. Other ion channels from diverse families also possess SH3-binding motifs in their amino or carboxyl tails, e.g., Shaker family voltage-gated K^+ channels.

A distinct subgroup within the CLC family, ClC-Tm1, ClC-Hs1, and ClC-Hs2 appear to be the most evolved, i.e., furthest from the common ancestral protein of the family (Mindell and Maduke, 2001). This group also appears to contain the proteins most responsive to membrane potential, i.e., the channels that are strongly voltage-gated (ClC-Tm1 and ClC-Hs1). Prokaryotic forms of CLC, which have now been shown to be transporters and not channels (Accardi and Miller, 2004), do not have this motif in helix R. The core sequence for an SH3-binding motif is PxxP, usually flanked at either side by a positively charged residue; however, not all SH3 ligands display the charged residues (Kay et al., 2000). In ClC-Hs1, there are a number of core sequences found in the carboxyl tail of the protein, but none carries adjacent positively charged residues.

Recognition sequences for interaction with regulatory proteins carrying WW domains occur in ClC-5 (Schwake et al., 2001) but do not occur in any of the other members of the CLC family. None of the members of the CLC family carries recognition sequences for SH2 domain proteins.

Proteins of the Homer/Vesl family carry an EVH-1 domain that recognizes the sequence PPxxF(R) in target proteins (Kay et al., 2000). Just after the second CBS domain in ClC-1, there is a sequence similar to an EVH-1-binding motif (PPLASFR). Members of the Homer family of proteins are associated with actin rearrangements in the cytoplasmic processes of neurons in response to variations in synaptic activity (Brakeman et al., 1997; Kato et al., 1997). If the sequence in ClC-1 is indeed an EVH-1-binding motif, then it may indicate that this Cl^- channel interacts with the cytoskeleton.

The WH2 Domain

Further evidence for interaction of ClC-Hs1 with the cytoskeleton occurs at the tip of the carboxyl tail where there is a Wiskott-Aldrich syndrome protein homology domain type 2 (WH2 domain) located adjacent to the highly acidic carboxyl terminus. It has been shown for several classes of proteins that WH2 domains bind to monomeric actin and mediate actin polymerization, a process aided by nearby acidic regions binding to the Arp2/3 complex (Paunola et al., 2002; Weaver et al., 2003). Comparison of the ClC-Hs1 WH2 domain with other WH2 proteins such as human Wave 1, 2, and 3 shows that residues essential for actin binding are conserved (Paunola et al., 2002). These proteins also have a similar domain organization to ClC-Hs1 in the carboxyl tail with a proline-rich region, a single WH2 domain, and an acidic region in the terminus. Another member of the CLC family, ClC-Hs5, has been shown to bind to cofilin, a protein involved in actin filament depolymerization (Hryciw et al, 2003).

Proline-rich sequences tend to occur in proteins involved in situations requiring rapid associations or disassociations and turnover, such as signal transduction and cytoskeletal rearrangements (Kay et al., 2000). It appears that ClC-1 and possibly other members of the CLC family could be involved in actin rearrangements via a WH2 domain, a possible EVH-1-type domain, and/or a cofilin-binding site. This interaction, in the case of ClC-Hs1, could provide a direct link between electrical action potential activity (or recovery from this) and cytoskeletal rearrangements to accommodate the mechanical events of muscle contraction (or relaxation). Some individuals with recessive generalized myotonia experience muscle weakness, and some naturally occurring mutations (e.g., P932L) result in muscle wasting (Nagamitsu et al., 2000), phenomena that may be accounted for by a disruption of interaction with the cytoskeleton if the disease-causing mutations interfere with the actin-binding domains described. The mutation P932L is located in a proline-rich region upstream from the WH2 domain and thus may be influencing the behavior of this motif.

Cytoplasmic regulation of a predominantly voltage-gated ion channel, particularly within an excitable tissue such as skeletal muscle, may seem redundant as channel function appears to be controlled by changes in membrane potential. However, it is conceivable that there may be situations where it would be advantageous to the cell to override the effects of the action potential through cytoplasmic regulatory mechanisms. An example may be during fatigue, when cytoplasmic signals may alter the phosphorylation state of the channel, possibly via an interaction with SH3 regulatory proteins, and cause the Cl^- channels to open. This would push the cell toward hyperpolarization and thus further from threshold, preventing further contractions. Alternatively, it may be that exposure of the regulatory domains in ClC-Hs1 during an action potential is influencing events in the cytoplasm, especially in relation to cytoskeletal rearrangements. Thus ClC-Hs1 may be acting as a signal-transducing protein itself, coupling the action potential to cytoskeletal rearrangements. There is evidence that other proteins involved in excitation-contraction-coupling in skeletal muscle also carry actin-binding domains, e.g., the ryanodine receptor has a Homer-type EVH-1 domain (Sheng and Kim, 2000). These also may be acting as electromechanical signal transducers.

ClC-2

Homology screening with sequences derived from ClC-Tm1 and ClC-Rr1 was employed to identify the next member of the family, ClC-Rr2. This close relative of the muscle-type channels seems to be ubiquitously expressed, being found in heart, brain, pancreas, lung, kidney, gastrointestinal tract (GIT), and liver. It has approximately 50% identity with ClC-Tm1 and ClC-Rr1, its cDNA predicting a protein of 907 amino acids and a molecular mass of 99 kDa. The very wide range of tissues in which this channel is found suggests that it has a "housekeeping" function. It has been shown to be regulated by changes in cell volume associated with hypotonicity (Gründer et al., 1992) and thus probably contributes to regulatory volume decrease in cells subject to osmotically induced swelling. There appears to be a region in the amino terminus of the protein that is essential for voltage and volume regulation and that surprisingly can be transplanted to a distal part of the protein, the interdomain segment between CBS1 and CBS2 (between the old D12 and 13), while still remaining functional. Both the amino terminus and the interdomain segment are believed to be cytoplasmic, as in the other members of the CLC family (Gründer et al., 1992). This suggested a ball-and-chain type of mechanism for gating, with the essential region forming the ball. Hyperpolarization or cell swelling would decrease the affinity of the ball motif for a binding site in the cytoplasmic pore mouth and would thus open the channel. Binding of the ball motif to the opening of the pore could then be influenced by internal Cl^- concentration, which is sufficiently high in many of the tissues that express ClC-Hs2 to be of physiological relevance (Foskett, 1998).

When expressed in *Xenopus* oocytes, the channel shows slow activation at hyperpolarizing potentials but deactivates rapidly on return to more positive potentials. Thus the channel is not open under physiological voltages. It is slightly inwardly rectifying and has a selectivity sequence similar to ClC-Tm1 and ClC-Hs1 but appears to be permeant to I^-. It has a single channel conductance of 3 to 5 pS (Jentsch et al., 1995) but appears to lack the fast activation gate of ClC-Tm1 and ClC-Hs1 even though it shares considerable sequence identity within the regions believed to be involved in gating (Thiemann et al., 1992; Pusch and Jentsch, 1994). Its slow activation at hyperpolarizing potentials is similar to that of the ClC-Rr1 channels exposed to low extracellular pH (Rychkov et al., 1996). It has also been observed that ClC-Hs2 currents are increased at low pH by shifting the voltage dependence of gating (Pusch and Jentsch, 1994; Jordt and Jentsch, 1997). Mutations in the helix J-helix K cytoplasmic linker (the old D7-D8 linker) resulted in a constitutively open channel. It has been proposed that this region is part of the "receptor" for the ball motif, but that it does not form part of the pore since the mutations do not affect pore properties (Jordt and Jentsch, 1997). In contrast, a later study using mammalian cells (HEK 293 cells) for patch clamping has shown that a deletion mutant that removes the putative ball moiety from the amino terminus does not lead to a loss of voltage- or pH-dependent gating, arguing against a ball-and-chain mechanism of gating (Varela et al., 2002).

A splice variant of ClC-Hs2 in rabbit gastric parietal cells that is believed to be involved in acid secretion from these cells has been described (Malinowska et al., 1995). These channels are also activated by low pH, and a region believed to be the pH sensor has been identified in the extracellular helix K–helix M loop (between the old D8 and D9), a β-pleated sheet region that includes residues 416 to 419 (EELE), a highly negatively charged region (Stroffekova et al., 1998). Although E419 is not conserved in ClC-Hs2 where it has been replaced by glycine (E419G), it appears to be the most important residue. Interestingly, these channels appear to be regulated by protein kinase A (PKA). Compare the suggestion, above, that ClC-Hs1 is regulated by PKC, as ClC-Hs3 seems to be (Rosenbohm et al., 1999; Kawasaki et al., 1994). The putative binding sites for PKA are also potential binding sites for PKC (Malinowska et al., 1995). Several recent studies have elucidated some of the regulatory mechanisms operating on ClC-Hs2. It seems to be regulated by the cell cycle M-phase-specific cyclin-dependent kinase (p34/cyclin B complex) that phosphorylates the channel protein and attenuates channel activity (Furukawa et al., 2002). These authors suggest that the most likely residue for phosphorylation is S632, as determined by mutation analysis. They also suggest that ClC-Hs2 is counterregulated by protein phosphatase 1. Decreases in ClC-Hs2 expression levels after M-phase appear to be due to phosphorylation-dependent ubiquitination and degradation of the protein (Zheng et al., 2002). In another study utilizing an ortholog of ClC-Hs2 in the nematode *Caenorhabditis elegans*, ClC-Ce3 (clh-3), it was shown that the channel is inactivated by phosphorylation, as inhibition of kinase function activated the channel (Rutledge et al., 2002). In addition, it was shown that the type 1 protein phosphatase, CeGLC-7alpha and beta, mediates activation of the channel. Overall, it seems that ClC-Hs2 is able to be activated by hypotonic shock and cell swelling, low pH, and dephosphorylation (Kirk, 2000). Expression of ClC-Rn2 in the rat appears to be more abundant in some tissues (brain rather than lung, kidney, and intestine) in the embryo and is down-regulated after birth (Murray et al., 1996). This is the opposite of the situation for ClC-Rr1, which is upregulated after birth (Steinmeyer et al., 1991b). Expression of ClC-Oc2 may appear to be higher in the embryo due to its association with the M-phase of the cell cycle (Zheng et al., 2002).

Disruption of the ubiquitously expressed ClC-2 might be expected to result in severe consequences for the ClC-2-deficient animals. A knockout mouse model for ClC-Mm2 has been developed (Bösl et al., 2001); the mice unexpectedly displayed few overt abnormalities beyond severe retinal and testicular degeneration, resulting in blindness and male infertility. In these tissues, functional cells (germ cells and photoreceptors) are in close contact with epithelial supporting cells (Sertoli cells and retinal pigment epithelial cells, respectively). It was hypothesized that ClC-Mm2 regulates the ionic environment of these cells and thus plays a crucial role in interactions between the functional cells and their supporting cells. In particular, functional ClC-2 seems to be important during the establishment of a blood-organ barrier in these tissues (Bösl et al., 2001). Expression of ClC-Mm2 in murine small intestine epithelial cells has been shown to be associated with tight junction complexes linking adjacent epithelial cells (Gyömörey et al., 2000). More recently, it has been shown that ClC-Rn2 interacts with the retrograde motor dynein complex, such that cell surface expression of ClC-Rn2 is increased when dynein is inactivated (Dhani et al., 2003). Mutations in ClC-Hs2 resulting in both loss and gain of function have been implicated in certain forms of idiopathic generalized epilepsy (Haug et al., 2003). These authors proposed that altering the function of ClC-Hs2 destabilizes the Cl^- equilibrium potential such that gamma-aminobutyric acid-ergic inhibition cannot occur, resulting in depolarization and hyperexcitability in the neurons.

ClC-K1(a) AND ClC-K2(b)

The first member of the CLC family to be identified and cloned that is predominantly renally expressed was ClC-RnK1, a 688-amino-acid protein with about 40% identity to ClC-1. It is outwardly rectifying and activates on both depolarization and hyperpolarization. It is expressed mainly in the thin ascending limb of the loop of Henle and is up-regulated in response to dehydration, suggesting a role in the urinary concentration mechanisms (Uchida et al., 1993). When the gene for this channel was disrupted in mice, the animals could no longer produce concentrated urine, indicating that ClC-MmK1 is essential for urinary concentration and water retention (Matsumura et al., 1999).

Homology-cloning strategies were used to clone another kidney form of CLC localized to the thick ascending limb of the loop of Henle and collecting ducts. This 687-amino-acid protein with about 80% identity to ClC-RrK1 was called ClC-K2 (ClC-RrK2), and two isoforms were identified, one of which had a deletion of the second transmembrane domain (Adachi et al., 1994). The corresponding human genes, *CLCNKA* and *CLCNKB*, were also cloned, but their predicted proteins, ClC-HsKa and ClC-HsKb, could not be expressed in *Xenopus* oocytes. No colocalization of ClC-MmK1 and ClC-MmK2 could be demonstrated in the kidney (Uchida, 2000). Rather, ClC-MmK1 is located mainly in the thin ascending limb of the loop of Henle, in both apical and basolateral plasma membranes, whereas ClC-MmK2 shows a much wider distribution pattern, being found in distal tubules, collecting tubules and ducts, and the medullary thick ascending limb of the loop of Henle. It appears as though ClC-MmK2 is present only at the basolateral cell membrane along with Na^+-dependent Cl^- transporters, suggesting a role for the channels in Cl^- reabsorption (Uchida, 2000). Naturally occurring mutations in the orthologous human gene, *CLCNKB*, for ClC-HsKb have been shown to be associated with the salt-wasting disease, Bartter's syndrome (Simon et al., 1997). In addition, it has been shown that, in one form of Bartter's syndrome displaying congenital deafness as well as renal failure, an integral membrane protein, now known as barttin, associates with the ClC-HsKb channels in the form of a β-subunit and is itself essential for normal channel function (Estévez et al., 2001). Identification of a β-subunit accompanying this CLC-type channel may have implications for the other forms of CLC protein in terms of regulatory protein interactions.

ClC-3

While the ClC-0, ClC-1, ClC-2, and ClC-Ka and -Kb branch of the CLC family are generally believed to be plasma membrane channels, the location of the others is controversial, being totally or partially confined to intracellular membranes under normal circumstances (Fig. 2).

Through consensus sequences derived from ClC-Tm1, ClC-Hs1, and ClC-Hs2, the fourth member of the family was identified from a rat kidney cDNA library and was designated ClC-3 (ClC-Rn3) (Kawasaki et al., 1994). Its cDNA predicts a protein of 760 amino acids and a molecular mass of 85 kDa, hence being smaller than the previously discovered members of the family. It displays only about 24% amino acid identity with the others, but the hydropathy profile is similar in predicting the same number of putative transmembrane domains. Amino acid similarity is greatest in these regions. Two consensus sites for phosphorylation by PKC were found at serines 51 and 655 (Kawasaki et al., 1994).

The channel produces an outwardly rectifying current that is dependent on the presence of extracellular Cl^- and it appears to be open at physiological voltages. Thus this channel may be active at the normal resting membrane potential and contributing to the resting conductance. When PKC is activated, the channel closes, reducing resting conductance. This may result in an "after depolarization" leading to excitation of the membrane. Regions of the brain believed to be involved in short-term memory express ClC-3 in abundance, and the channel may therefore have some unforeseen role in establishing short-term memory or in its conversion to long-term memory. It is also expressed in lung, kidney, and adrenal gland (Kawasaki et al., 1994). Diversity in the function and in the widespread tissue expression of ClC-3 has led to the conclusion that the protein performs housekeeping functions in the manner of ClC-2. Dispute remains as to the exact role and location of ClC-3. Several groups have produced mice lacking the gene for ClC-Mm3, and all have remarked on degeneration of the retina and hippocampus with associated neurological deficits (Stobrawa et al., 2001; Yoshikawa et al., 2002; Dickerson et al., 2002). In addition, the Yoshikawa group noted a phenotype very similar to that of neuronal ceroid lipofuscinosis, with progressive degeneration of neurons in the hippocampus, cells lining the mucosa of the ileum, and retinal cells. They also saw an increased endosomal pH and hypothesized that the defect causing the cellular degeneration was in cytoplasmic protein degradation and turnover mechanisms. This work suggests a role for ClC-3 in acidification of endosomes, in a manner similar to the role of ClC-5, a structurally related member of the CLC family. The more recent study of ClC-Mm3 knockout mice (Dickerson et al., 2002) reported that the animals tended to exhibit spontaneous seizures and that this was possibly due to loss of gamma-aminobutyric acid-synthesizing cells in the dentate gyrus and subsequent reduced inhibitory signals. Again, ClC-3 may have a role in the acidification of synaptic vesicles or endosomes, but its role in ileal epithelial cells has not been established (Thevenod, 2002).

It has been known for some time that cardiac muscle displays Cl^- currents with similar characteristics to those produced by CLC-type channels (Hill et al., 1989). Following its initial discovery in the rat kidney as ClC-Rr3, the guinea pig heart ortholog of this channel, ClC-Cp3, was also cloned and shown to have a high identity with the kidney clone (Duan et al., 1997). When expressed in NIH 3T3 cells, it produced large, outwardly rectifying currents that inactivated at positive potentials and reversed at the estimated equilibrium potential for Cl^-. Thus these channels would be expected to be basally active or constitutively open. In hypotonic conditions, the cells swelled and large increases in membrane conductance were observed; hence the channels can be further activated by cell swelling. Activation of PKC inhibited both basal activity and swelling-induced activity. Channel activity increased with increasing Cl^- concentration, and the selectivity sequence showed that iodide was more permeant than Cl^-. Mutation of N579 at the end of the putative transmembrane domains resulted in a reversal of rectification and of selectivity, leading Duan et al. (1997) to propose that this region might be involved in the channel pore. From the crystal structure of related prokaryotic proteins (Dutzler et al., 2002), however, N579 lies within the R helix and projects into the cytoplasm, not being part of the pore but being likely to influence events at the cytoplasmic vestibule of the channel. Duan and colleagues have further shown that under isotonic conditions, PKC phosphorylates S51 in the amino terminus of the protein, and when the cell swells under hypotonic conditions, PKC is inhibited and S51 is dephosphorylated, activating the channel (Duan et al., 1999).

It has been demonstrated that currents similar to those mentioned above are produced by smooth muscle cells in the GIT and in vascular walls and that ClC-3 is expressed in both of these tissues (ClC-Cf3 and ClC-Mm3) (Yamazaki et al., 1998; Dick et al., 1998). There was also some suggestion that currents in the GIT smooth muscle cells are not only activated by osmotic cell swelling but also by simple physical distension as well, e.g., stretching of the GIT wall (Dick et al., 1998). It has since been shown that there is a link between ClC-Rn3 expression and proliferation of vascular smooth muscle cells induced by exposure to endothelin-1 (Wang et al., 2002). This group used an antisense strategy to inhibit expression of ClC-Rn3 in cultured vascular smooth muscle cells to show that if ClC-Rn3 is not expressed, proliferation is inhibited.

It has now been shown that there are at least two splice variants of ClC-3: ClC-Rn3A and ClC-Hs3B. These differ in that ClC-Hs3B has a PDZ-binding domain at its carboxyl terminus that appears to be important for trafficking of the protein (Ogura et al., 2002). The ClC-Hs3B variant appears to be expressed mainly in the Golgi compartment whereas the ClC-Hs3A form is mostly found in late endosomes (Gentzsch et al., 2003). When expressed together, they can form heterodimers.

ClC-4

The gene for ClC-Hs4, *CLCN4*, was discovered by mapping the Xp22.3 region of the human X chromosome (van Slegtenhorst et al., 1994). It encodes a protein of 760 amino acids with putative transmembrane domains similar to those of the other CLC proteins. It is naturally expressed in skeletal muscle, brain, heart, vascular smooth muscle, and endothelium (van Slegtenhorst et al., 1994; Lamb et al., 1999) and has been shown to produce an outwardly rectifying current when expressed in *Xenopus* sp. oocytes. It showed a reversed selectivity sequence to that of ClC-Hs3, Cl > I, and currents were reduced at low external pH (Friedrich et al., 1999). On the other hand, when expressed in Chinese hamster ovary cells, the channels showed activation at low external pH and displayed I > Cl selectivity (Kawasaki et al., 1999). A definitive study of ClC-Hs4 expressed in three different cell lines from three different species concluded that this channel has a small conductance (2 to 3 pS), is outwardly rectifying, shows Cl > I selectivity, requires ATP, and is inhibited by low external pH (Vanoye and George, 2002). The normal function of ClC-4, however, is still unclear. While the closely related channels, ClC-3 and ClC-5, appear to be present in intracellular compartments (endosomes, Golgi), ClC-4 appears to be expressed in the apical brush border of intestinal epithelial cells as well as in endosomes (Mohammed-Panah et al., 2002). Antibody-labeling studies suggest that the channels colocalize with the cystic fibrosis transmembrane conductance regulator protein and mediate Cl^- flux across the apical membrane (Mohammed-Panah et al., 2002).

ClC-5

Dent's disease is a renal tubular disorder characterized by low molecular weight proteinuria, hypercalciuria, kidney stones, calcification of renal tubules, and eventual renal failure. It is an X-linked syndrome, and genetic analysis identified a region on the X chromosome that was subsequently sequenced. Comparison of this sequence with sequences from other CLC family members revealed that the new gene had significant identity with ClC-MmK2 (Fisher et al., 1994). Many of the symptoms of Dent's disease indicate proximal tubule dysfunction, however, and ClC-RrK2 is predominantly expressed in the loop of Henle and collecting ducts (Adachi et al., 1994). Hence the gene responsible for Dent's disease was assumed to be yet another kidney form of CLC. Uptake of much of the material into proximal tubule cells is via endocytosis, and this process is dependent on acidification of the endosomal compartment, which in turn is dependent on voltage-gated Cl^- channels (Fisher et al., 1994). This suggested that the new form of CLC was likely to be intracellularly located in association with the endosomal membrane.

Eventually, the new kidney form of CLC, called ClC-5 (ClC-Rn5), was cloned and produced a predicted protein of 83 kDa having 80% identity with ClC-3 and ClC-4, with which it forms a distinct subgroup (Steinmeyer et al., 1995). According to these authors the channel is outwardly rectifying with a selectivity sequence similar to that of other CLCs. Another group, however, found the channel to be also outwardly rectifying but with a greater I^- permeability, similar to that in some of the ClC-Mm3 studies (Sakamoto et al., 1996).

There are a number of conflicting reports as to the intrarenal location of ClC-5. Using a stringent immunohistochemistry approach, one group has localized ClC-Rn5 protein to the cytoplasm of epithelial cells in the S3 segment of the proximal tubule and the thick ascending limb of the loop of Henle (Luyckx et al., 1998). These authors also determined that ClC-Rn5 is located in endosomal membranes and is involved in acidification and trafficking of endosomes in these cells. There is also some suggestion that ClC-Rn5 is expressed in intercalated cells in the collecting ducts (Obermüller et al., 1998; Devuyst et al., 1999). It has been shown that mutations in ClC-Hs5 not only result in Dent's disease but also in X-linked recessive nephrolithiasis and X-linked recessive hypophosphatemic rickets with low molecular weight proteinuria (Uchida, 2000). All of these conditions can result from impaired endocytosis by cells lining the proximal tubules.

ClC-6

ClC-Hs6 was discovered by first searching databases for unassigned sequences with homology to other CLCs and then cloning any that matched. One sequence was found that represented a partial clone of ClC-Hs6, and then a full clone was produced from a human brain cDNA library (Brandt and Jentsch, 1995). The predicted protein has 869 amino acids with a molecular mass of 97 kDa. It has only a 29% identity to the ClC-Hs3 group and about 23% identity to ClC-Hs1, with most of the similarity occurring in putative transmembrane segments. Subsequent to the elucidation of ClC-Hs6, another member, ClC-Hs7, was discovered that has approximately 45% identity to ClC-Hs6. This protein has 803 amino acids and a molecular mass of 89 kDa, due to a shorter interdomain segment between CBS1 and CBS2 (the old D12 to D13 stretch). Both of these channels have been claimed to be broadly expressed at an RNA level in tissues including skeletal muscle, kidney, and brain (Brandt and Jentsch, 1995), although later studies suggest that protein expression is not so broad as previously imagined (Schaller et al., 2004). It has subsequently been shown that when expressed in COS cells, ClC-Hs6 colocalizes with an endoplasmic reticulum marker and hence probably has an intracellular location in vivo (Buyse et al., 1998).

ClC-7

Similar to ClC-6, ClC-Hs7 appears to be expressed predominantly in late endosomal and lysosomal compartments, but relatively little is known about the *CLCN7* gene and its protein product, ClC-Hs7. Immunohistochemistry indicates that it is mainly localized in the membranes of intracellular organelles (Brandt and Jentsch, 1995). Although obviously a close relative of the known voltage-gated Cl^- channels, the biological function of this protein was not known until recently. Knockout of the *CLCN7* gene in mice results in a lethal osteopetrotic phenotype, coupled with severe retinal and central nervous system degeneration (Kornak et al., 2001). Osteoclasts are, however, normal in number and in their attachment to bone, but they form defective ruffled border membranes and resorption lacunae are absent. Resorption is completely suppressed in these osteoclasts in culture on artificial ivory substrates. These studies have demonstrated that the deficiency of ClC-Hs7 prevents acid secretion by osteoclasts, because without the accompanying efflux of Cl^-, proton extrusion fails when the electrical potential builds up for lack of a counter ion flux. Both the hydrochloric acid production and the osmotic water transport that normally follow it to form resorption lacunae (Fig. 2) are thereby abolished or greatly diminished. Any suppression of the relevant proton pump (above) in the presence of normal CLC-Hs7 would, naturally, have similar effects. In addition to its importance in bone homeostasis and, perhaps, in brain development, ClC-Hs7 has been suggested to play a role in chronic pain (Diewald et al. 2002).

Most malignant forms of osteopetrosis have been shown to be caused by mutations in the *OC116* gene, which encodes a subunit of an intracellular H^+-ATPase, the vacuolar proton pump (Frattini et al., 2000). Mutations in the *CLCN7* gene, however, have also been shown to be responsible for the most common, but less severe, form of osteopetrosis (autosomal dominant osteopetrosis type II) in humans (Cleiren et al., 2001; Letizia et al., 2004; Frattini et al., 2003; Henriksen et al., 2004), autosomal recessive osteopetrosis (Kornak et al., 2001; Frattini et al., 2003), and an intermediate autosomal recessive osteopetrosis as described by Campos-Xavier et al. (2003).

Today, the main focus of ClC-Hs7 research relates to its potential for involvement in the therapy of the far more common crippling disease osteoporosis. It has been postulated that pharmacological manipulation of ClC-Hs7 could provide a possible treatment for this socially and economically costly disorder. It is not known whether gain-of-function mutations or failure of feedback control of *CLCN7* expression is responsible for excessive bone resorption and osteoporosis, but ClC-Hs7 stands out as an obvious target for intervention. To highlight the validity of this approach to the treatment of osteoporosis, a Danish pharmaceutical company, Nordic Biosciences, recently published data showing that a Cl^- channel-blocking compound could prevent bone resorption in ovariectomized rats (Schaller et al., 2004). This represents an encouraging beginning, but much more work is required in this field.

Although significant characterization of the promoter region of the *CLCN2* gene has been performed (Holmes et al., 2003) and the three kidney-specific Cl^- channel genes, *CLCN5* (Hayama et al., 2000), *CLCNK1*, and *CLCNK2* (Rai et al., 1999; Uchida et al., 2000), have been analyzed, very little by way of promoter study has been performed for *CLCN7*. In both mice and humans, the *CLCN7* promoter region has a high G + C content and lacks any typical TATA or CAAT boxes. With respect to transcriptional regulation, Kornak et al. (1999) have predicted that two E-box sites found within the first 233 bp of the mouse promoter are important transcription-factor-binding sites. No confirmatory experimental work has been conducted, however, to show whether this prediction is accurate.

Our recent analysis of the human *CLCN7* promoter has identified several interesting consensus transcription-factor-binding sites. By mutation, we have demonstrated their importance in the transcriptional regulation of this gene (P. Bartley and A. Bretag, unpublished observations). Transcription factor binding has been demonstrated, and the identification of these factors is under way.

NONVERTEBRATE CLC PROTEINS

Apart from ClC-Tm1, which was isolated from an elasmobranch fish, all of the other CLC family members discussed above were first isolated from mammals. Indeed, even ClC-Tm1 is probably just the torpedo ray ortholog of the mammalian ClC-1 (Jentsch et al., 1993). Once the sequence for the early CLC proteins had been determined, orthologous sequences were found in other species, ranging through other vertebrates, invertebrates, plants, fungi, eubacteria, and archaea. The locations of the various prokaryotic and eukaryotic CLCs in their surface and intracellular membranes are shown in Fig. 2.

The Yeast CLC Protein, Gef1 (ClC-Sc1)

One of the first homologs to be identified was the *GEF1* gene from the yeast *S. cerevisiae*. Mutations in this gene result in very small colonies due to a slow growth rate, which could be improved on addition of iron to the medium. Lack of iron in the organisms resulted in reduced function in heme-containing proteins in the aerobic respiratory system of the yeast mitochondria. Comparison of sequences showed that the protein ClC-Sc1 (Gef1) displays homology with other CLC proteins, particularly in the putative transmembrane domains. The protein is 779 amino acids long, with a molecular mass of 87 kDa, and has a similar arrangement of hydrophobic domains coinciding with the original D1 to D13 (helix B to CBS2 in current terminology) of ClC-Tm1 and ClC-Hs1 (Greene et al., 1993). It was suggested that the channel might be present in the membrane of vacuoles (endosomes) in the cytoplasm of the yeast cells, a site for the intracellular storage of Fe^{3+}. It was also suggested, however, that this protein might be located in mitochondrial membranes. Thus this was the first account proposing that a CLC protein might be intracellular rather than incorporated within the plasma membrane.

It has since been shown that the ClC-Sc1 (Gef1) is involved in iron (and copper) uptake into post-Golgi vesicles, being essential for the correct acidification of the interior of these structures (Gaxiola et al., 1998; Schwappach et al., 1998).

Prokaryotic CLCs

During a project aimed at sequencing the entire genome for *E. coli*, a bacterial homolog for CLC was identified (YadQ and EriC) (Fujita et al., 1994) and is now known as ClC-ec1 (ClC-Ec1). Since that time, many CLC homologs have been identified via genome sequencing projects in a wide range of prokaryotic organisms. Not all, however, have been shown to be channels; it has been hypothesized that they may contribute to bacterial survival at low pH, consistent with the association of ClC-5 (and possibly ClC-3) with H^+-ATPase activity in eukaryotic cells (Iyer et al., 2002). Homologs from *E. coli* (ClC-Ec1) and *S. enterica* serovar Typhimurium (ClC-Set1) were used to determine the crystal structure of the protein (Dutzler et al., 2002).

Plant CLCs

Plant homologs of the animal CLCs have been identified in *A. thaliana* (Hechenberger et al., 1996), tobacco (Lurin et al., 2000), and several other plant species. In *Arabidopsis*, a CLC homolog appears to play a role in regulating the intracellular nitrate status (Geelen et al., 2000), whereas in tobacco, the channels are located in the inner mitochondrial membrane and may correspond to the inner membrane anion channel of plant and animal mitochondria (Lurin et al., 2000). This study is the first report confirming a CLC-type channel associated with mitochondria.

Drosophila CLCs

Three homologs of vertebrate CLCs are known from *Drosophila melanogaster*, but, unexpectedly, no mutants have been described and their function in invertebrates remains unknown. One of the homologs is most closely related to the ClC-1 and ClC-2 group, one to the ClC-3, ClC-4, and ClC-5 group, and the last to the ClC-6 and ClC-7 group. More than one laboratory is working on these, but there are, as yet, no published results.

Caenorhabditis CLCs

Six putative Cl^- channel genes of the CLC family are known from the genome of the nematode *C. elegans* (*CeClC-1* to *CeClC-6* or *clh-1* to *clh-6*) as well as a number of splice variants (*clh-3a, clh-3b*, etc.). As for *Drosophila*, these fit into the three main families of vertebrate CLC channels. It is by no means certain, however, whether corresponding isoforms have orthologous functions or whether these functions are performed by the various splice variants. Petalcorin et al. (1999) cloned *clh-1* and noted its expression, mainly in the hypodermis. Soon afterward, heterologous expression in Sf9 insect cells produced currents from clh-1 and clh-3b reminiscent of ClC-2 along with the suggestion that clh-3b and ClC-2 are orthologs (Rutledge et al., 2001). By contrast, these researchers were unable to express functional versions of clh-2b, clh-4b, or clh-5 (Nehrke et al., 2000).

Wild-type oocytes of *C. elegans* displayed a Cl^- current that was biophysically indistinguishable from that of clh-3b expressed in HEK293 cells (Denton et al., 2004), implying that clh-3b is the normal anion current carrier in oocytes. This channel appears to be counter-regulated by phosphorylation (Rutledge et al., 2002). By contrast, clh-3a currents differed significantly from clh-3b in voltage dependence and in extracellular Cl^- and pH sensitivity, suggesting a role in neuron and muscle excitability, in some ways like ClC-1 (Denton et al., 2004). Unlike the broad expression of mammalian ClC-6 and ClC-7, expression of their likely *C. elegans* counterpart, clh-6, is restricted to two gamma-aminobutyric acid-ergic neurons (Bianchi et al., 2001). This may provide a good opportunity to analyze the function of this understudied group of channels, although if they are endosomal, there are the obvious attendant difficulties associated with attempting to patch-clamp internal membranes.

ROLE OF CLC CHLORIDE CHANNELS IN DISEASE

Mutations in CLC channels have now been associated with a number of diseases in both humans and other species. Mutations of ClC-1 result in the muscle disease myotonia, of ClC-Hs5 in Dent's disease and related nephropathies, of ClC-HsKb in Bartter's syndrome, of ClC-Hs7 in osteopetrosis, and most recently, there has been a suggestion that mutations in ClC-Hs2 may produce alterations in function that result in epilepsy via their interaction with neuronal inhibitory processes (Haug et al., 2003). As the functions of this large family of proteins are elucidated, other disease associations may become evident.

Myotonia, the best understood of the CLC diseases, is characterized by a peculiar muscle stiffness that is normally painless, an inability of the muscle to relax after a voluntary contraction. This is purely a muscle phenomenon and does not involve nerve dysfunction. It is sometimes accompanied by weakness, and the stiffness may improve after exercise (the "warm-up" phenomenon). There is a spectrum of severity in patients, some exhibiting severe muscle cramping, others only very mild symptoms. Electrophysiologically, myotonia is distinguished from other myopathies by myotonic runs of action potentials, firing repetitively (often called "dive-bombers" due to their characteristic sound when played through an audio amplifier), with no additional neural input beyond the initial stimulus. This implies an instability of the membrane potential near the resting potential due to the reduced damping ability of Cl^- flow (decreased bootstrapping) or to enhanced Na^+ flow through Na^+ channels displaying abnormal gain of function. Sodium channel myotonia will only be discussed insofar as its therapy could involve increasing Cl^- current around the resting potential to offset the abnormal Na^+ current.

The CLC diseases are hereditary with no known toxic or autoimmune associations and display both a dominant form, myotonia congenita or Thomsen's disease, and a recessive form, recessive generalized myotonia or Becker's disease. These conditions are caused by mutations in ClC-Hs1 (Koch et al., 1992). In the recessive form, either no protein (truncation mutations) or defective protein would be produced and hence there are no functional channels. In the dominant form in a heterozygote, one normal allele of the gene produces normal subunits and the abnormal allele produces abnormal subunits, which, when forming a wild-type mutant heteromer, then interfere with the normal function of the wild-type subunits, a dominant negative effect (Meyer-Kleine et al., 1995). Of the resultant dimers, one-fourth can be expected to be wild-type homomers, one-fourth mutant homomers, and one-half heteromers. Some of the dominant forms of myotonia, however, display a reduced penetrance, and for some mutations there may be dominant expression in some families and recessive in others (Kubisch et al., 1998). This indicates that factors in addition to ClC-Hs1 may well be involved in the etiology of Cl^- channel myotonia.

It is generally not possible to predict whether a given mutation will lead to dominant or recessive myotonia, except for severe truncation mutations, which would be expected to disrupt protein structure and function and so would be anticipated to be recessive (Kubisch et al., 1998).

There are a number of models in animals for myotonia, including the myotonic goat (Bryant and Morales-Aguilera, 1971) and the arrested development of righting response (ADR) and the myotonic (MTO) mice (Mehrke et al., 1988). All of these have now been shown to be associated with mutations in the corresponding ClC-1 genes in these animals: a single missense mutation in the goat (Beck et al., 1996), a transposon insertion in the ADR mouse (Steinmeyer et al., 1991a), an early termination in the MTO mouse, and a mutation in the myotonic dog (equivalent to T268M in the rat) (Rhodes et al., 1999). Because of the association between ClC-1 function and myotonia, other conditions that produce myotonic symptoms may well be affecting the CLC channel in some way. For example, the hypocholesterolemic agent, 20,25-diazacholesterol, produces symptoms of myotonia in some patients and myotonic-type electrical discharges in muscle. It seems to do this by blocking Cl^- channels and reducing Cl^- conductance (Bretag, 1987) although the effect is not acute and probably results from metabolic factors of unknown origin.

THERAPY AND CLC CHANNELS

Genetic or pharmacological manipulation of the relevant Cl^- channels could treat CLC diseases or, conversely, mimicking some aspect of these diseases could point the way to therapies for other diseases. For example, muscle weakness might be able to be treated by inducing mild myotonia with an appropriate ClC-Hs1 blocker or osteoporosis might be able to be countered by inactivation of ClC-Hs7.

To date, treatment for myotonia has initially been empirical and latterly via mild depression of Na^+ channel activity (Bretag, 1983) rather than enhancement of Cl^- channel activity. Agents with some antimyotonic action and known to depress Na channel activity include mexiletine (an antiarrhythmic drug), phenytoin (an antiseizure drug), and procainamide (a local anaesthetic derivative). Some compounds that interact with the ClC-Hs1 pore are proving useful starting points for agents that might increase either the open probability of Cl^- channels or increase their unitary conductance (Aromataris et al., 1999; Aromataris et al., 2001). As well, the promoter region of ClC-Hs1 is being studied with respect to the possibility of regulating its expression. In the case of some dominant myotonic mutants, a doubling of expression, or perhaps even less (based on the absence of symptoms in heterozygotes carrying a recessive mutation), should be sufficient to eliminate afterdischarges and restore normal muscle function.

Acknowledgments. Aspects of this work received funding support from the Muscular Dystrophy Association of South Australia, the Australian Research Council, and the University of South Australia.

We are grateful to Jennie Cederholm and Grigori Rychkov for help with the manuscript and figures.

REFERENCES

Accardi, A., and C. Miller. 2004. Secondary active transport mediated by a prokaryotic homologue of ClC Cl^- channels. *Nature* **427:**803–807.

Adachi, S., S. Uchida, H. Ito, M. Hata, M. Hiroe, F. Marumo, and S. Sasaki. 1994. Two isoforms of a chloride channel predominantly expressed in thick ascending limb of Henle's loop and collecting ducts of rat kidney. *J. Biol. Chem.* **269:**17677–17683.

Adams, J., Z. P. Chen, B. J. Van Denderen, C. J. Morton, M. W. Parker, L. A. Witters, D. Stapleton, and B. E. Kemp. 2004. Intrasteric control of AMPK via the gamma1 subunit AMP allosteric regulatory site. *Protein Sci.* **13:**155–165.

Aromataris, E. C., D. S. Astill, G. Y. Rychkov, S. H. Bryant, A. H. Bretag, and M. L. Roberts. 1999. Modulation of the gating of ClC-1 by S-(-) 2-(4-chlorophenoxy) propionic acid. *Br. J. Pharmacol.* **126:**1375–1382.

Aromataris, E. C., G. Y. Rychkov, B. Bennetts, B. P. Hughes, A. H. Bretag, and M. L. Roberts. 2001. Fast and slow gating of CLC-1: differential effects of 2-(4-chlorophenoxy) propionic acid and dominant negative mutations. *Mol. Pharmacol.* **60:**200–208.

Beck, C. L., C. Fahlke, and A. L. George, Jr. 1996. Molecular basis for decreased muscle chloride conductance in the myotonic goat. *Proc. Natl. Acad. Sci. USA* **93:**11248–11252.

Bennetts, B., M. L. Roberts, A. H. Bretag, and G. Y. Rychkov. 2001. Temperature dependence of human muscle ClC-1 chloride channel. *J. Physiol.* **535:**83–93.

Bianchi, L., D. M. Miller III, and A. L. George, Jr. 2001. Expression of a ClC chloride channel in *Caenorhabditis elegans* gamma-aminobutyric acid-ergic neurons. *Neurosci. Lett.* **299:**177–180.

Bösl, M. R., V. Stein, C. Hübner, A. A. Zdebik, S. E. Jordt, A. K. Mukhopadhyay, M. S. Davidoff, A. F. Holstein, and T. J. Jentsch. 2001. Male germ cells and photoreceptors, both dependent on close cell-cell interactions, degenerate upon ClC-2 Cl^- channel disruption. *EMBO J.* **20:**1289–1299.

Boyle, P. J., and E. J. Conway. 1941. Potassium accumulation in muscle and associated changes. *J. Physiol. Lond.* **100:**1–63.

Brakeman, P. R., A. A. Lanahan, R. O'Brien, K. Roche, C. A. Barnes, R. L. Huganir, and P. F. Worley. 1997. Homer: a protein that selectively binds metabotropic glutamate receptors. *Nature* **386:**284–288.

Brandt, S., and T. J. Jentsch. 1995. ClC-6 and ClC-7 are two novel broadly expressed members of the CLC chloride channel family. *FEBS Lett.* **377:**15–20.

Bretag, A. H. 1983. Antimyotonic agents and myotonia. *Proc. Aust. Physiol. Pharmacol. Soc.* **14:**170–191.

Bretag, A. H. 1987. Muscle chloride channels. *Physiol. Rev.* **67:**618–724.

Brinkmeier, H., and H. Jockusch. 1987. Activators of protein kinase C induce myotonia by lowering chloride conductance in muscle. *Biochem. Biophys. Res. Commun.* **148:**1383–1389.

Bryant, S. H., and D. Conte-Camerino. 1991. Chloride channel regulation in the skeletal muscle of normal and myotonic goats. *Pflugers Arch.* **417:**605–610.

Bryant, S. H., and A. Morales-Aguilera. 1971. Chloride conductance in normal and myotonic muscle fibres and the action of monocarboxylic aromatic acids. *J. Physiol.* **219:**367–383.

Buyse, G., D. Trouet, T. Voets, L. Missiaen, G. Droogmans, B. Nilius, and J. Eggermont. 1998. Evidence for the intracellular location of chloride channel (ClC)-type proteins: co-localization of ClC-6a and ClC-6c with the sarco/endoplasmic-reticulum Ca^{2+} pump SERCA2b. *Biochem. J.* **330:**1015–1021.

Campos-Xavier, A. B., J. M. Saraiva, L. M. Ribeiro, A. Munnich, and V. Cormier-Daire. 2003. Chloride channel 7 (CLCN7) gene mutations in intermediate autosomal recessive osteopetrosis. *Hum. Genet.* **112:**186–189.

Chen, M. F., and T. Y. Chen. 2003. Side-chain charge effects and conductance determinants in the pore of ClC-0 chloride channels. *J. Gen. Physiol.* **122:**133–145.

Chen, T. Y., and C. Miller. 1996. Nonequilibrium gating and voltage dependence of the ClC-0 Cl^- channel. *J. Gen. Physiol.* **108:**237–250.

Cleiren, E., O. Bénichou, E. Van Hul, J. Gram, J. Bollerslev, F. R. Singer, K. Beaverson, A. Aledo, M. P. Whyte, T. Yoneyama, M. C. deVernejoul, and W. Van Hul. 2001. Albers-Schönberg disease (autosomal dominant osteopetrosis, type II) results from mutations in the ClCN7 chloride channel gene. *Hum. Mol. Genet.* **10:** 2861–2867.

De Luca, A., S. Pierno, D. Cocchi, and D. Conte Camerino. 1997. Effects of chronic growth hormone treatment in aged rats on the biophysical and pharmacological properties of skeletal muscle chloride channels. *Br. J. Pharmacol.* **121:**369–374.

Denton, J., K. Nehrke, E. Rutledge, R. Morrison, and K. Strange. 2004. Alternative splicing of N and C termini of a *C. elegans* ClC channel alters gating and sensitivity to external Cl^- and H^+. *J. Physiol.* **555:**97–114.

Devuyst, O., P. T. Christie, P. J. Courtoy, R. Beauwens, and R. V. Thakker. 1999. Intra-renal and subcellular distribution of the human chloride channel, CLC-5, reveals a pathophysiological basis for Dent's disease. *Hum. Mol. Genet.* **8:**247–257.

Dhani, S. U., R. Mohammad-Panah, N. Ahmed, C. Ackerley, M. Ramjeesingh, and C. E. Bear. 2003. Evidence for a functional interaction between the ClC-2 chloride channel and the retrograde motor dynein complex. *J. Biol. Chem.* **278:**16262–16270.

Dick, G. M., K. K. Bradley, B. Horowitz, J. R. Hume, and K. M. Sanders. 1998. Functional and molecular identification of a novel chloride conductance in canine colonic smooth muscle. *Am. J. Physiol.* **275:** C940–C950.

Dickerson, L. W., D. J. Bonthius, B. C. Schutte, B. Yang, T. J. Barna, M. C. Bailey, K. Nehrke, R. A. Williamson, and F. S. Lamb. 2002. Altered GABAergic function accompanies hippocampal degeneration in mice lacking ClC-3 voltage-gated chloride channels. *Brain Res.* **958:**227–250.

Diewald, L., J. Rupp, M. Dreger, F. Hucho, C. Gillen, and H. Nawrath. 2002. Activation by acidic pH of CLC-7 expressed in oocytes from *Xenopus laevis*. *Biochem. Biophys. Res. Commun.* **291:**421–424.

Dogovski, C., J. Pi, and A. J. Pittard. 2003. Putative interhelical interactions within the PheP protein revealed by second-site suppressor analysis. *J. Bacteriol.* **185:**6225–6232.

Duan, D., S. Cowley, B. Horowitz, and J. R. Hume. 1999. A serine residue in ClC-3 links phosphorylation-dephosphorylation to chloride channel regulation by cell volume. *J. Gen. Physiol.* **113:**57–70.

Duan, D., C. Winter, S. Cowley, J. R. Hume, and B. Horowitz. 1997. Molecular identification of a volume-regulated chloride channel. *Nature* **390:**417–421.

Duffield, M., G. Rychkov, A. Bretag, and M. Roberts. 2003. Involvement of helices at the dimer interface in ClC-1 common gating. *J. Gen. Physiol.* **121:**149–161.

Dutzler, R., E. B. Campbell, M. Cadene, B. T. Chait, and R. MacKinnon. 2002. X-ray structure of a ClC chloride channel at 3.0 Å reveals the molecular basis of anion selectivity. *Nature* **415:**287–294.

Dutzler, R., E. B. Campbell, and R. MacKinnon. 2003. Gating the selectivity filter in ClC chloride channels. *Science* **300:**108–112.

Estévez, R., T. Boettger, V. Stein, R. Birkenhäger, E. Otto, F. Hildebrandt, and T. J. Jentsch. 2001. Barttin is a Cl^- channel beta-subunit crucial for renal Cl^- reabsorption and inner ear K^+ secretion. *Nature* **414:**558–561.

Estévez, R., M. Pusch, C. Ferrer-Costa, M. Orozco, and T. J. Jentsch. 2004. Functional and structural conservation of CBS domains from CLC chloride channels. *J. Physiol.* **557:**363–378.

Estévez, R., B. C. Schroeder, A. Accardi, T. J. Jentsch, and M. Pusch. 2003. Conservation of chloride channel structure revealed by an inhibitor binding site in ClC-1. *Neuron* **38:**47–59.

Fahlke, C., T. Knittle, C. A. Gurnett, K. P. Campbell, and A. L. George, Jr. 1997. Subunit stoichiometry of human muscle chloride channels. *J. Gen. Physiol.* **109:**93–104.

Fahlke, C., A. Rosenbohm, N. Mitrovic, A. L. George, Jr., and R. Rüdel. 1996. Mechanism of voltage-dependent gating in skeletal muscle chloride channels. *Biophys. J.* **71:**695–706.

Fahlke, C., R. Rüdel, N. Mitrovic, M. Zhou, and A. L. George, Jr. 1995. An aspartic acid residue important for voltage-dependent gating of human muscle chloride channels. *Neuron* **15:**463–472.

Fisher, S. E., G. C. Black, S. E. Lloyd, E. Hatchwell, O. Wrong, R. V. Thakker, and I. W. Craig. 1994. Isolation and partial characterization of a chloride channel gene which is expressed in kidney and is a candidate for Dent's disease (an X-linked hereditary nephrolithiasis). *Hum. Mol. Genet.* **3:**2053–2059.

Fong, P., A. Rehfeldt, and T. J. Jentsch. 1998. Determinants of slow gating in ClC-0, the voltage-gated chloride channel of *Torpedo marmorata*. *Am. J. Physiol.* **274:**C966–C973.

Foskett, J. K. 1998. ClC and CFTR chloride channel gating. *Annu. Rev. Physiol.* **60:**689–717.

Frattini, A., P. J. Orchard, C. Sobacchi, S. Giliani, M. Abinun, J. P. Mattsson, D. J. Keeling, A. K. Andersson, P. Wallbrandt, L. Zecca, L. D. Notarangelo, P. Vezzoni, and A. Villa. 2000. Defects in TCIRG1 subunit of the vacuolar proton pump are responsible for a subset of human autosomal recessive osteopetrosis. *Nat. Genet.* **25:**343–346.

Frattini, A., A. Pangrazio, L. Susani, C. Sobacchi, M. Mirolo, M. Abinun, M. Andolina, A. Flanagan, E. M. Horwitz, E. Mihci, L. D. Notarangelo, U. Ramenghi, A. Teti, J. Van Hove, D. Vujic, T. Young, A. Albertini, P. J. Orchard, P. Vezzoni, and A. Villa. 2003. Chloride channel ClCN7 mutations are responsible for severe recessive, dominant, and intermediate osteopetrosis. *J. Bone Miner. Res.* **18:**1740–1747.

Friedrich, T., T. Breiderhoff, and T. J. Jentsch. 1999. Mutational analysis demonstrates that ClC-4 and ClC-5 directly mediate plasma membrane currents. *J. Biol. Chem.* **274:**896–902.

Fujita, N., H. Mori, T. Yura, and A. Ishihama. 1994. Systematic sequencing of the *Escherichia coli* genome: analysis of the 2.4–4.1 min (110,917-193,643 bp) region. *Nucleic Acids Res.* **22:**1637–1639.

Furukawa, T., T. Ogura, Y. J. Zheng, H. Tsuchiya, H. Nakaya, Y. Katayama, and N. Inagaki. 2002. Phosphorylation and functional regulation of ClC-2 chloride channels expressed in Xenopus oocytes by M cyclin-dependent protein kinase. *J. Physiol.* **540:**883–893.

Gaxiola, R. A., D. S. Yuan, R. D. Klausner, and G. R. Fink. 1998. The yeast CLC chloride channel functions in cation homeostasis. *Proc. Natl. Acad. Sci. USA* **95:**4046–4050.

Geelen, D., C. Lurin, D. Bouchez, J. M. Frachisse, F. Leliévre, B. Courtial, H. Barbier-Brygoo, and C. Maurel. 2000. Disruption of putative anion channel gene AtCLC-a in Arabidopsis suggests a role in the regulation of nitrate content. *Plant J.* **21:**259–267.

Gentzsch, M., L. Cui, A. Mengos, X. B. Chang, J. H. Chen, and J. R. Riordan. 2003. The PDZ-binding chloride channel ClC-3B localizes to the Golgi and associates with cystic fibrosis transmembrane conductance regulator-interacting PDZ proteins. *J. Biol. Chem.* **278:**6440–6449.

Greene, J. R., N. H. Brown, B. J. DiDomenico, J. Kaplan, and D. J. Eide. 1993. The GEF1 gene of *Saccharomyces cerevisiae* encodes an integral membrane protein; mutations in which have effects on respiration and iron-limited growth. *Mol. Gen. Genet.* **241:**542–553.

Gründer, S., A. Thiemann, M. Pusch, and T. J. Jentsch. 1992. Regions involved in the opening of ClC-2 chloride channel by voltage and cell volume. *Nature* **360:**759–762.

Gyömörey, K., H. Yeger, C. Ackerley, E. Garami, and C. E. Bear. 2000. Expression of the chloride channel ClC-2 in the murine small intestine epithelium. *Am. J. Physiol. Cell. Physiol.* **279:**C1787–C1794.

Hanke, W., and C. Miller. 1983. Single chloride channels from *Torpedo electroplax*. Activation by protons. *J. Gen. Physiol.* **82:**25–45.

Haug, K., M. Warnstedt, A. K. Alekov, T. Sander, A. Ramirez, B. Poser, S. Maljevic, S. Hebeisen, C. Kubisch, J. Rebstock, S. Horvath, K. Hallmann, J. S. Dullinger, B. Rau, F. Haverkamp, S. Beyenburg, H. Schulz, D. Janz, B. Giese, G. Müller-Newen, P. Propping, C. E. Elger, C. Fahlke, H. Lerche, and A. Heils. 2003. Mutations in CLCN2 encoding a voltage-gated chloride channel are associated with idiopathic generalized epilepsies. *Nat. Genet.* **33:**527–532.

Hayama, A., S. Uchida, S. Sasaki, and F. Marumo. 2000. Isolation and characterization of the human *CLC-5* chloride channel gene promoter. *Gene* **261:**355–364.

Hebeisen, S., A. Biela, B. Giese, G. Müller-Newen, P. Hidalgo, and C. Fahlke. 2004. The role of the carboxyl terminus in ClC chloride channel function. *J. Biol. Chem.* **279:**13140–13147.

Hechenberger, M., B. Schwappach, W. N. Fischer, W. B. Frommer, T. J. Jentsch, and K. Steinmeyer. 1996. A family of putative chloride channels from Arabidopsis and functional complementation of a yeast strain with a CLC gene disruption. *J. Biol. Chem.* **271:**33632–33638.

Henriksen, K., J. Gram, S. Schaller, B. H. Dahl, M. H. Dziegiel, J. Bollerslev, and M. A. Karsdal. 2004. Characterization of osteoclasts from patients harboring a G215R mutation in ClC-7 causing autosomal dominant osteopetrosis type II. *Am. J. Pathol.* **164:**1537–1545.

Hill, J. A., Jr., R. Coronado, and H. C. Strauss. 1989. Reconstitution of ionic channels from human heart. *J. Mol. Cell. Cardiol.* **21:**315–322.

Hille, B. 2001. *Ion Channels of Excitable Membranes*, 3rd ed. Sinauer Associates, Sunderland, Mass.

Holmes, K. W., R. Hales, S. Chu, M. J. Maxwell, P. J. Mogayzel, Jr., and P. L. Zeitlin. 2003. Modulation of Sp1 and Sp3 in lung epithelial cells regulates ClC-2 chloride channel expression. *Am. J. Respir. Cell. Mol. Biol.* **29:**499–505.

Hryciw, D. H., G. Y. Rychkov, B. P. Hughes, and A. H. Bretag. 1998. Relevance of the D13 region to the function of the skeletal muscle chloride channel, ClC-1. *J. Biol. Chem.* **273:**4304–4307.

Hryciw, D. H., Y. Wang, O. Devuyst, C. A. Pollock, P. Poronnik, and W. B. Guggino. 2003. Cofilin interacts with ClC-5 and regulates albumin uptake in proximal tubule cell lines. *J. Biol. Chem.* **278:**40169–40176.

Hutter, O. F., and A. E. Warner. 1967a. The pH sensitivity of the chloride conductance of frog skeletal muscle. *J. Physiol.* **189:**403–425.

Hutter, O. F., and A. E. Warner. 1967b. Action of some foreign cations and anions on the chloride permeability of frog muscle. *J. Physiol.* **189:**445–460.

Iyer, R., T. M. Iverson, A. Accardi, and C. Miller. 2002. A biological role for prokaryotic ClC chloride channels. *Nature* **419:**715–718.

Javadpour, M. M., M. Eilers, M. Groesbeek, and S. O. Smith. 1999. Helix packing in polytopic membrane proteins: role of glycine in transmembrane helix association. *Biophys. J.* **77:**1609–1618.

Jentsch, T. J., W. Günther, M. Pusch, and B. Schwappach. 1995. Properties of voltage-gated chloride channels of the ClC gene family. *J. Physiol.* **482:**19S–25S.

Jentsch, T. J., M. Pusch, A. Rehfeldt, and K. Steinmeyer. 1993. The ClC family of voltage-gated chloride channels: structure and function. *Ann. N. Y. Acad. Sci.* **707:**285–293.

Jentsch, T. J., K. Steinmeyer, and G. Schwarz. 1990. Primary structure of *Torpedo marmorata* chloride channel isolated by expression cloning in Xenopus oocytes. *Nature* **348:**510–514.

Jin, N. G., J. K. Kim, D. K. Yang, S. J. Cho, J. M. Kim, E. J. Koh, H. C. Jung, I. So, and K. W. Kim. 2003. Fundamental role of ClC-3 in volume-sensitive Cl– channel function and cell volume regulation in AGS cells. *Am. J. Physiol. Gastrointest. Liver Physiol.* **285:**G938–G948.

Jordt, S. E., and T. J. Jentsch. 1997. Molecular dissection of gating in the ClC-2 chloride channel. *EMBO J.* **16:**1582–1592.

Kato, A., F. Ozawa, Y. Saitoh, K. Hirai, and K. Inokuchi. 1997. *vesl*, a gene encoding VASP/Ena family related protein, is upregulated during seizure, long-term potentiation and synaptogenesis. *FEBS Lett.* **412:** 183–189.

Kawasaki, M., T. Fukuma, K. Yamauchi, H. Sakamoto, F. Marumo, and S. Sasaki. 1999. Identification of an acid-activated Cl^- channel from human skeletal muscles. *Am. J. Physiol.* **277:**C948–C954.

Kawasaki, M., S. Uchida, T. Monkawa, A. Miyawaki, K. Mikoshiba, F. Marumo, and S. Sasaki. 1994. Cloning and expression of a protein kinase C-regulated chloride channel abundantly expressed in rat brain neuronal cells. *Neuron* **12:**597–604.

Kay, B. K., M. P. Williamson, and M. Sudol. 2000. The importance of being proline: the interaction of proline-rich motifs in signaling proteins with their cognate domains. *FASEB J.* **14:**231–241.

Kirk, K. L. 2000. Chloride channels and tight junctions. Focus on "Expression of the chloride channel ClC-2 in the murine small intestine epithelium." *Am. J. Physiol. Cell Physiol.* **279:**C1675–C1676.

Klocke, R., K. Steinmeyer, T. J. Jentsch, and H. Jockusch. 1994. Role of innervation, excitability, and myogenic factors in the expression of the muscular chloride channel ClC-1. A study on normal and myotonic muscle. *J. Biol. Chem.* **269:**27635–27639.

Koch, M. C., K. Steinmeyer, C. Lorenz, K. Ricker, F. Wolf, M. Otto, B. Zoll, F. Lehmann-Horn, K. H. Grzeschik, and T. J. Jentsch. 1992. The skeletal muscle chloride channel in dominant and recessive human myotonia. *Science* **257:**797–800.

Kornak, U., M. R. Bösl, and C. Kubisch. 1999. Complete genomic structure of the CLCN6 and CLCN7 putative chloride channel genes(1). *Biochim. Biophys. Acta* **1447:**100–106.

Kornak, U., D. Kasper, M. R. Bösl, E. Kaiser, M. Schweizer, A. Schulz, W. Friedrich, G. Delling, and T. J. Jentsch. 2001. Loss of the ClC-7 chloride channel leads to osteopetrosis in mice and man. *Cell* **104:**205–215.

Kubisch, C., T. Schmidt-Rose, B. Fontaine, A. H. Bretag, and T. J. Jentsch. 1998. ClC-1 chloride channel mutations in myotonia congenita: variable penetrance of mutations shifting the voltage dependence. *Hum. Mol. Genet.* **7:**1753–1760.

Lamb, F. S., G. H. Clayton, B. X. Liu, R. L. Smith, T. J. Barna, and B. C. Schutte. 1999. Expression of CLCN voltage-gated chloride channel genes in human blood vessels. *J. Mol. Cell. Cardiol.* **31:**657–666.

Lemmon, M. A., H. R. Treutlein, P. D. Adams, A. T. Brünger, and D. M. Engelman. 1994. A dimerization motif for transmembrane alpha-helices. *Nat. Struct. Biol.* **1:**157–163.

Letizia, C., A. Taranta, S. Migliaccio, C. Caliumi, D. Diacinti, E. Delfini, E. D'Erasmo, M. Iacobini, M. Roggini, O. M. Albagha, S. H. Ralston, and A. Teti. 2004. Type II benign osteopetrosis (Albers-Schönberg disease) caused by a novel mutation in CLCN7 presenting with unusual clinical manifestations. *Calcif. Tissue Int.* **74:**42–46.

Liantonio, A., M. Pusch, A. Picollo, P. Guida, A. De Luca, S. Pierno, G. Fracchiolla, F. Loiodice, P. Tortorella, and D. Conte Camerino. 2004. Investigations of pharmacologic properties of the renal CLC-K1 chloride channel co-expressed with barttin by the use of 2-(p-chlorophenoxy)propionic acid derivatives and other structurally unrelated chloride channels blockers. *J. Am. Soc. Nephrol.* **15:**13–20.

Lipicky, R. J., and S. H. Bryant. 1966. Sodium, potassium, and chloride fluxes in intercostal muscle from normal goats and goats with hereditary myotonia. *J. Gen. Physiol.* **50:**89–111.

Lorenz, C., C. Meyer-Kleine, K. Steinmeyer, M. C. Koch, and T. J. Jentsch. 1994. Genomic organization of the human muscle chloride channel ClC-1 and analysis of novel mutations leading to Becker-type myotonia. *Hum. Mol. Genet.* **3:**941–946.

Ludewig, U., T. J. Jentsch, and M. Pusch. 1997a. Inward rectification in ClC-0 chloride channels caused by mutations in several protein regions. *J. Gen. Physiol.* **110:**165–171.

Ludewig, U., T. J. Jentsch, and M. Pusch. 1997b. Analysis of a protein region involved in permeation and gating of the voltage-gated Torpedo chloride channel ClC-0. *J. Physiol.* **498:**691–702.

Ludewig, U., M. Pusch, and T. J. Jentsch. 1997c. Independent gating of single pores in CLC-0 chloride channels. *Biophys. J.* **73:**789–797.

Ludewig, U., M. Pusch, and T. J. Jentsch. 1996. Two physically distinct pores in the dimeric ClC-0 chloride channel. *Nature* **383:**340–343.

Lurin, C., J. Güclü, C. Cheniclet, J. P. Carde, H. Barbier-Brygoo, and C. Maurel. 2000. CLC-Nt1, a putative chloride channel protein of tobacco, co-localizes with mitochondrial membrane markers. *Biochem. J.* **348** (Part 2): 291–295.

Luyckx, V. A., F. O. Goda, D. B. Mount, T. Nishio, A. Hall, S. C. Hebert, T. G. Hammond, and A. S. Yu. 1998. Intrarenal and subcellular localization of rat CLC5. *Am. J. Physiol.* **275:**F761–F769.

Maduke, M., C. Williams, and C. Miller. 1998. Formation of CLC-0 chloride channels from separated transmembrane and cytoplasmic domains. *Biochemistry* **37:**1315–1321.

Malinowska, D. H., E. Y. Kupert, A. Bahinski, A. M. Sherry, and J. Cuppoletti. 1995. Cloning, functional expression, and characterization of a PKA-activated gastric Cl^- channel. *Am. J. Physiol.* **268:** C191–C200.

Matsumura, Y., S. Uchida, Y. Kondo, H. Miyazaki, S. B. Ko, A. Hayama, T. Morimoto, W. Liu, M. Arisawa, S. Sasaki, and F. Marumo. 1999. Overt nephrogenic diabetes insipidus in mice lacking the CLC-K1 chloride channel. *Nat. Genet.* **21:**95–98.

Mehrke, G., H. Brinkmeier, and H. Jockusch. 1988. The myotonic mouse mutant ADR: electrophysiology of the muscle fiber. *Muscle Nerve* **11:**440–446.

Meyer-Kleine, C., K. Steinmeyer, K. Ricker, T. J. Jentsch, and M. C. Koch. 1995. Spectrum of mutations in the major human skeletal muscle chloride channel gene (CLCN1) leading to myotonia. *Am. J. Hum. Genet.* **57:**1325–1334.

Middleton, R. E., D. J. Pheasant, and C. Miller. 1996. Homodimeric architecture of a ClC-type chloride ion channel. *Nature* **383:**337–340.

Miller, C. 1983. Integral membrane channels: studies in model membranes. *Physiol. Rev.* **63:**1209–1242.

Miller, C., and M. M. White. 1984. Dimeric structure of single chloride channels from *Torpedo electroplax*. *Proc. Natl. Acad. Sci. USA* **81:**2772–2775.

Mindell, J. A., and M. Maduke. 2001. ClC chloride channels. *Genome Biol.* **2:**REVIEWS 3003.1-3003.6.

Mo, L., W. Xiong, T. Qian, H. Sun, and N. K. Wills. 2004. Coexpression of complementary fragments of ClC-5 and restoration of chloride channel function in a Dent's disease mutation. *Am. J. Physiol. Cell Physiol.* **286:**C79–C89.

Mohammad-Panah, R., C. Ackerley, J. Rommens, M. Choudhury, Y. Wang, and C. E. Bear. 2002. The chloride channel ClC-4 co-localizes with cystic fibrosis transmembrane conductance regulator and may mediate chloride flux across the apical membrane of intestinal epithelia. *J. Biol. Chem.* **277:**566–574.

Murray, C. B., S. Chu, and P. L. Zeitlin. 1996. Gestational and tissue-specific regulation of C1C-2 chloride channel expression. *Am. J. Physiol.* **271:**L829–L837.

Nagamitsu, S., T. Matsuura, M. Khajavi, R. Armstrong, C. Gooch, Y. Harati, and T. Ashizawa. 2000. A "dystrophic" variant of autosomal recessive myotonia congenita caused by novel mutations in the CLCN1 gene. *Neurology* **55:**1697–1703.

Nehrke, K., T. Begenisich, J. Pilato, and J. E. Melvin. 2000. Into ion channel and transporter function. *Caenorhabditis elegans* ClC-type chloride channels: novel variants and functional expression. *Am. J. Physiol. Cell Physiol.* **279:**C2052–C2066.

Obermüller, N., N. Gretz, W. Kriz, R. F. Reilly, and R. Witzgall. 1998. The swelling-activated chloride channel ClC-2, the chloride channel ClC-3, and ClC-5, a chloride channel mutated in kidney stone disease, are expressed in distinct subpopulations of renal epithelial cells. *J. Clin. Investig.* **101:**635–642.

Ogura, T., T. Furukawa, T. Toyozaki, K. Yamada, Y. J. Zheng, Y. Katayama, H. Nakaya, and N. Inagaki. 2002. ClC-3B, a novel ClC-3 splicing variant that interacts with EBP50 and facilitates expression of CFTR-regulated ORCC. *FASEB J.* **16:**863–865.

O'Neill, G. P., R. Grygorczyk, M. Adam, and A. W. Ford-Hutchinson. 1991. The nucleotide sequence of a voltage-gated chloride channel from the electric organ of *Torpedo californica. Biochim. Biophys. Acta* **1129:**131–134.

Palade, P. T., and R. L. Barchi. 1977. Characteristics of the chloride conductance in muscle fibers of the rat diaphragm. *J. Gen. Physiol.* **69:**325–342.

Paunola, E., P. K. Mattila, and P. Lappalainen. 2002. WH2 domain: a small, versatile adapter for actin monomers. *FEBS Lett.* **513:**92–97.

Petalcorin, M. I., T. Oka, M. Koga, K. Ogura, Y. Wada, Y. Ohshima, and M. Futai. 1999. Disruption of clh-1, a chloride channel gene, results in a wider body of *Caenorhabditis elegans. J. Mol. Biol.* **294:** 347–355.

Ponting, C. P. 1997. CBS domains in ClC chloride channels implicated in myotonia and nephrolithiasis (kidney stones). *J. Mol. Med.* **75:**160–163.

Pusch, M. 2002. Myotonia caused by mutations in the muscle chloride channel gene CLCN1. *Hum. Mutat.* **19:**423–434.

Pusch, M., and T. J. Jentsch. 1994. Molecular physiology of voltage-gated chloride channels. *Physiol. Rev.* **74:**813–827.

Pusch, M., U. Ludewig, and T. J. Jentsch. 1997. Temperature dependence of fast and slow gating relaxations of ClC-0 chloride channels. *J. Gen. Physiol.* **109:**105–116.

Pusch, M., U. Ludewig, A. Rehfeldt, and T. J. Jentsch. 1995. Gating of the voltage-dependent chloride channel ClC-0 by the permeant anion. *Nature* **373:**527–531.

Pusch, M., K. Steinmeyer, and T. J. Jentsch. 1994. Low single channel conductance of the major skeletal muscle chloride channel, ClC-1. *Biophys. J.* **66:**149–152.

Rai, T., S. Uchida, S. Sasaki, and F. Marumo. 1999. Isolation and characterization of kidney-specific *CLC-K2* chloride channel gene promoter. *Biochem. Biophys. Res. Commun.* **261:**432–438.

Rhodes, T. H., C. H. Vite, U. Giger, D. F. Patterson, C. Fahlke, and A. L. George, Jr. 1999. A missense mutation in canine C1C-1 causes recessive myotonia congenita in the dog. *FEBS Lett.* **456:**54–58.

Richard, E. A., and C. Miller. 1990. Steady-state coupling of ion-channel conformations to a transmembrane ion gradient. *Science* **247:**1208–1210.

Rosenbohm, A., R. Rüdel, and C. Fahlke. 1999. Regulation of the human skeletal muscle chloride channel hClC-1 by protein kinase C. *J. Physiol.* **514:**677–685.

Russ, W. P., and D. M. Engelman. 2000. The GxxxG motif: a framework for transmembrane helix-helix association. *J. Mol. Biol.* **296:**911–919.

Rutledge, E., L. Bianchi, M. Christensen, C. Boehmer, R. Morrison, A. Broslat, A. M. Beld, A. L. George, D. Greenstein, and K. Strange. 2001. CLH-3, a ClC-2 anion channel ortholog activated during meiotic maturation in *C. elegans* oocytes. *Curr. Biol.* **11:**161–170.

Rutledge, E., J. Denton, and K. Strange. 2002. Cell cycle- and swelling-induced activation of a *Caenorhabditis elegans* ClC channel is mediated by CeGLC-7alpha/beta phosphatases. *J. Cell Biol.* **158:**435–444.

Rychkov, G. Y., D. S. Astill, B. Bennetts, B. P. Hughes, A. H. Bretag, and M. L. Roberts. 1997. pH-dependent interactions of Cd2+ and a carboxylate blocker with the rat C1C-1 chloride channel and its R304E mutant in the Sf-9 insect cell line. *J. Physiol.* **501:**355–362.

Rychkov, G. Y., M. Pusch, D. S. Astill, M. L. Roberts, T. J. Jentsch, and A. H. Bretag. 1996. Concentration and pH dependence of skeletal muscle chloride channel ClC-1. *J. Physiol.* **497:**423–435.

Rychkov, G. Y., M. Pusch, M. L. Roberts, T. J. Jentsch, and A. H. Bretag. 1998. Permeation and block of the skeletal muscle chloride channel, ClC-1, by foreign anions. *J. Gen. Physiol.* **111:**653–665.

Sakamoto, H., M. Kawasaki, S. Uchida, S. Sasaki, and F. Marumo. 1996. Identification of a new outwardly rectifying Cl^- channel that belongs to a subfamily of the ClC Cl^- channels. *J. Biol. Chem.* **271:** 10210–10216.

Sansom, M. S., and H. Weinstein. 2000. Hinges, swivels and switches: the role of prolines in signalling via transmembrane alpha-helices. *Trends Pharmacol. Sci.* **21:**445–451.

Schaller, S., K. Henriksen, C. Sveigaard, A. M. Heegaard, N. Hélix, M. Stahlhut, M. C. Ovejero, J. V. Johansen, H. Solberg, T. L. Andersen, D. Hougaard, M. Berryman, C. B. Shiødt, B. H. Sørensen, J. Lichtenberg, P. Christophersen, N. T. Foged, J. M. Delaissé, M. T. Engsig, and M. A. Karsdal. 2004. The chloride channel inhibitor n53736 prevents bone resorption in ovariectomized rats without changing bone formation. *J. Bone Miner. Res.* **19:**1144–1153.

Schmidt-Rose, T., and T. J. Jentsch. 1997a. Transmembrane topology of a CLC chloride channel. *Proc. Natl. Acad. Sci. USA* **94:**7633–7638.

Schmidt-Rose, T., and T. J. Jentsch. 1997b. Reconstitution of functional voltage-gated chloride channels from complementary fragments of CLC-1. *J. Biol. Chem.* **272:**20515–20521.

Schwake, M., T. Friedrich, and T. J. Jentsch. 2001. An internalization signal in ClC-5, an endosomal Cl– channel mutated in Dent's disease. *J. Biol. Chem.* **276:**12049–12054.

Schwappach, B., S. Stobrawa, M. Hechenberger, K. Steinmeyer, and T. J. Jentsch. 1998. Golgi localization and functionally important domains in the NH2 and COOH terminus of the yeast CLC putative chloride channel Gef1p. *J. Biol. Chem.* **273:**15110–15118.

Scott, J. W., S. A. Hawley, K. A. Green, M. Anis, G. Stewart, G. A. Scullion, D. G. Norman, and D. G. Hardie. 2004. CBS domains form energy-sensing modules whose binding of adenosine ligands is disrupted by disease mutations. *J. Clin. Investig.* **113:**274–284.

Sheng, M., and E. Kim. 2000. The Shank family of scaffold proteins. *J. Cell Sci.* **113**(Part 11)**:**1851–1856.

Simon, D. B., R. S. Bindra, T. A. Mansfield, C. Nelson-Williams, E. Mendonca, R. Stone, S. Schurman, A. Nayir, H. Alpay, A. Bakkaloglu, J. Rodriguez-Soriano, J. M. Morales, S. A. Sanjad, C. M. Taylor, D. Pilz, A. Brem, H. Trachtman, W. Griswold, G. A. Richard, E. John, and R. P. Lifton. 1997. Mutations in the chloride channel gene, *CLCNKB*, cause Bartter's syndrome type III. *Nat. Genet.* **17:**171–178.

Simpson, B. J., T. A. Height, G. Y. Rychkov, K. J. Nowak, N. G. Laing, B. P. Hughes, and A. H. Bretag. 2004. Characterization of three myotonia-associated mutations of the CLCN1 chloride channel gene via heterologous expression. *Hum. Mutat. MIB* **24:**185.

Steinmeyer, K., R. Klocke, C. Ortland, M. Gronemeier, H. Jockusch, S. Gründer, and T. J. Jentsch. 1991a. Inactivation of muscle chloride channel by transposon insertion in myotonic mice. *Nature* **354:** 304–308.

Steinmeyer, K., C. Ortland, and T. J. Jentsch. 1991b. Primary structure and functional expression of a developmentally regulated skeletal muscle chloride channel. *Nature* **354:**301–304.

Steinmeyer, K., B. Schwappach, M. Bens, A. Vandewalle, and T. J. Jentsch. 1995. Cloning and functional expression of rat CLC-5, a chloride channel related to kidney disease. *J. Biol. Chem.* **270:**31172–31177.

Stobrawa, S. M., T. Breiderhoff, S. Takamori, D. Engel, M. Schweizer, A. A. Zdebik, M. R. Bösl, K. Ruether, H. Jahn, A. Draguhn, R. Jahn, and T. J. Jentsch. 2001. Disruption of ClC-3, a chloride channel expressed on synaptic vesicles, leads to a loss of the hippocampus. *Neuron* **29:**185–196.

Stroffekova, K., E. Y. Kupert, D. H. Malinowska, and J. Cuppoletti. 1998. Identification of the pH sensor and activation by chemical modification of the ClC-2G Cl^- channel. *Am. J. Physiol.* **275:**C1113–C1123.

Thevenod, F. 2002. Ion channels in secretory granules of the pancreas and their role in exocytosis and release of secretory proteins. *Am. J. Physiol. Cell Physiol.* **283:**C651–C672.

Thiemann, A., S. Gründer, M. Pusch, and T. J. Jentsch. 1992. A chloride channel widely expressed in epithelial and non-epithelial cells. *Nature* **356:**57–60.

Tieleman, D. P., I. H. Shrivastava, M. R. Ulmschneider, and M. S. Sansom. 2001. Proline-induced hinges in transmembrane helices: possible roles in ion channel gating. *Proteins* **44:**63–72.

Uchida, S. 2000. In vivo role of CLC chloride channels in the kidney. *Am. J. Physiol. Renal Physiol.* **279:**F802–F808.

Uchida, S., S. Sasaki, T. Furukawa, M. Hiraoka, T. Imai, Y. Hirata, and F. Marumo. 1993. Molecular cloning of a chloride channel that is regulated by dehydration and expressed predominantly in kidney medulla. *J. Biol. Chem.* **268:**3821–3824.

Uchida, S., Y. Tanaka, H. Ito, F. Saitoh-Ohara, J. Inazawa, K. K. Yokoyama, S. Sasaki, and F. Marumo. 2000. Transcriptional regulation of the CLC-K1 promoter by myc-associated zinc finger protein and kidney-enriched Krüppel-like factor, a novel zinc finger repressor. *Mol. Cell Biol.* **20:**7319–7331.

Valverde, M. A., S. P. Hardy, and F. V. Sepulveda. 1995. Chloride channels: a state of flux. *FASEB J.* **9:** 509–515.

Vanoye, C. G., and A. L. George, Jr. 2002. Functional characterization of recombinant human ClC-4 chloride channels in cultured mammalian cells. *J. Physiol.* **539:**373–383.

van Slegtenhorst, M. A., M. T. Bassi, G. Borsani, M. C. Wapenaar, G. B. Ferrero, L. de Conciliis, E. I. Rugarli, A. Grillo, B. Franco, H. Y. Zoghbi, et al. 1994. A gene from the Xp22.3 region shares homology with voltage-gated chloride channels. *Hum. Mol. Genet.* **3:**547–552.

Varela, D., M. I. Niemeyer, L. P. Cid, and F. V. Sepulveda. 2002. Effect of an N-terminus deletion on voltage-dependent gating of the ClC-2 chloride channel. *J. Physiol.* **544:**363–372.

Wang, G. L., X. R. Wang, M. J. Lin, H. He, X. J. Lan, and Y. Y. Guan. 2002. Deficiency in ClC-3 chloride channels prevents rat aortic smooth muscle cell proliferation. *Circ. Res.* **91:**E28–E32.

Wang, G. X., W. J. Hatton, G. L. Wang, J. Zhong, I. Yamboliev, D. Duan, and J. R. Hume. 2003. Functional effects of novel anti-ClC-3 antibodies on native volume-sensitive osmolyte and anion channels in cardiac and smooth muscle cells. *Am. J. Physiol. Heart Circ. Physiol.* **285:**H1453–H1463.

Warner, A. E. 1972. Kinetic properties of the chloride conductance of frog muscle. *J. Physiol.* **227:**291–312.

Weaver, A. M., M. E. Young, W. L. Lee, and J. A. Cooper. 2003. Integration of signals to the Arp2/3 complex. *Curr. Opin. Cell Biol.* **15:**23–30.

White, M. M., and C. Miller. 1979. A voltage-gated anion channel from the electric organ of *Torpedo californica. J. Biol. Chem.* **254:**10161–10166.

Yamazaki, J., D. Duan, R. Janiak, K. Kuenzli, B. Horowitz, and J. R. Hume. 1998. Functional and molecular expression of volume-regulated chloride channels in canine vascular smooth muscle cells. *J. Physiol.* **507:** 729–736.

Yoshikawa, M., S. Uchida, J. Ezaki, T. Rai, A. Hayama, K. Kobayashi, Y. Kida, M. Noda, M. Koike, Y. Uchiyama, F. Marumo, E. Kominami, and S. Sasaki. 2002. CLC-3 deficiency leads to phenotypes similar to human neuronal ceroid lipofuscinosis. *Genes Cells* **7:**597–605.

Zheng, Y. J., T. Furukawa, T. Ogura, K. Tajimi, and N. Inagaki. 2002. M phase-specific expression and phosphorylation-dependent ubiquitination of the ClC-2 channel. *J. Biol. Chem.* **277:**32268–32273.

Bacterial Ion Channels and Their Eukaryotic Homologs
Edited by A. Kubalski and B. Martinac

Chapter 12

The Bacterial Mechanosensitive Channel MscS and Its Extended Family

Paul Blount, Irene Iscla, Yuezhou Li, and Paul C. Moe

THE DISCOVERY OF MscS

In the late 1980s, Ching Kung and his colleagues, working with giant *Escherichia coli* cells, embarked on the electrophysiological characterization of bacterial channels. When such native bacterial membranes were first investigated under patch clamp, it was not voltage- or ligand-gated channels that were observed, but a channel activity that was seen only when a vacuum or suction was placed in the pipette electrode (Martinac et al., 1987). The authors speculated, we now know correctly, that this activity could play a role in osmotic regulation, allowing the cell to adjust rapidly to acute hypoosmotic environmental shifts (osmotic downshock), a process that could explain previous studies that demonstrated solute fluxes under such conditions (Britten and McClure, 1962). Another work showed that, while in the whole-cell patch configuration, the channels could indeed be gated by an osmotic gradient (Cui et al., 1995).

Currently, three clearly distinguishable mechanosensitive (MS) channel activities in native *E. coli* membranes are recognized: MscL, a mechanosensitive channel of large conductance; MscS, with a smaller conductance (Sukharev et al., 1993); and MscM, with miniconductance (Berrier et al., 1996). The names of the last two channels can be misleading; the conductance of even the "mini" activity is large compared to that of eukaryotic channels. The channel activities range in conductance from 3.6 nS for MscL to 300 pS for MscM. Hence, although smaller than MscL, the MscS channel has a conductance that is still quite large at 1 nS, or approximately 20- to 50-fold larger than that of the average eukaryotic channel. With the exception of MscM, whose molecular identity remains elusive, these channels have served as models for determining how proteins sense and respond to membrane tension. Although many of the channel characteristics of MscL (see chapter 13) and MscS are similar, suggesting conserved mechanisms, the molecules are distinct and appear to reflect two independent families with no obvious homology (Pivetti et al., 2003; but see Martinac, 2004). By comparing and contrasting the structures and proposed dynamics

---

Paul Blount, Irene Iscla, Yuezhou Li, and Paul C. Moe • Department of Physiology, University of Texas Southwestern Medical Center, Dallas, TX 75390-9040.

of the protein complexes, researchers are now gaining insight into unique as well as conserved molecular mechanisms underlying sensors that detect membrane stretch.

CHARACTERIZING MscS ACTIVITY

Early studies of channels within *E. coli* native membranes defined many of the fundamental properties of bacterial MS channel activities. The observed activity was reported to be modulated, but not gated, by voltage (Martinac et al., 1987), and a subsequent study demonstrated that amphipaths that intercalate into the membrane asymmetrically can modulate the sensitivity of the channel (Martinac et al., 1990). In these early publications that described *E. coli* MS channel activities, only one activity was reported. However, as mentioned above, subsequent reports have described three distinct channel activities that are distinguished by conductance: MscL, MscS, and MscM (Berrier et al., 1996). From the conductance of the channels described, it appears that these first published reports, describing the modulation of bacterial MS channels by voltage and amphipaths, centered on the MscS activity. This notion is bolstered by subsequent studies in which MscM was seen only infrequently and MscL was shown to require more membrane tension than MscS to be observed.

The MscS and MscL activities have many similarities. They can both be purified and functionally reconstituted (Häse et al., 1995; Blount et al., 1996; Okada et al., 2002; Sukharev, 2002), demonstrating that no auxiliary proteins are necessary for the MS channel activity observed, both appear to directly sense tension within the membrane, as opposed to the pressure across it (Sukharev et al., 1999; Sukharev, 2002), and both are sensitive to amphipaths (Martinac et al., 1990; Perozo et al., 2002b), suggesting that they directly sense biophysical changes in the membrane. On the other hand, the two activities also show several fundamental differences in addition to conductance. Once open, the MscS channel stays open much longer than MscL; the open-dwell times can reach hundreds of milliseconds or more (Y. Li and P. Blount, unpublished data), whereas MscL dwells in the open state in the tens of milliseconds (Ou et al., 1998). Also in contrast with MscL, current-voltage plots have demonstrated that MscS activity shows a slight anionic preference, and a slight rectification is noted at higher positive voltages (Martinac et al., 1987; Li et al., 2002). Finally, MscS inactivates, or desensitizes, upon several seconds of stimulation. This inactivation does not appear to be voltage dependent (Koprowski and Kubalski, 1998) but has been observed to be pH dependent, occurring more readily at lower pH (6.0 versus 7.5) (Blount et al., 1999). The inactivation can be reversed by relaxing the stretched membrane for a few seconds. A typical trace record of MscS activity is presented in Fig. 1.

UNEARTHING THE *mscS* GENE

The *mscS* gene family was discovered by classical genetics. Ian Booth and his colleagues had isolated an *E. coli* strain, generated by UV mutagenesis, that showed impaired growth in the presence of both high K^+ and betaine or proline (both are compatible solutes that are transported into the cell to maintain turgor in a high-osmotic environment) (McLaggan et al., 2002). The lesion was identified as a missense mutation within a gene called *kefA* (also called *aefA*). Further characterization of KefA and its homologs led to another gene, *yggB*, that correlated with MscS activity (Levina et al., 1999). Similar to MscL (Häse et al., 1995; Blount et al., 1996), subsequent reconstitution of the *yggB* gene product demonstrated that

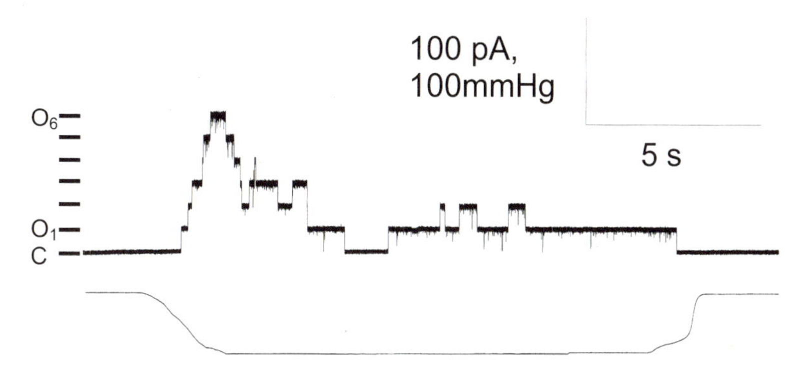

Figure 1. A typical trace obtained from an excised patch from native *E. coli* membranes. Giant spheroplasts were generated as previously described (Blount et al., 1999). The top trace is the channel activity, while the bottom shows the pressure (negative relative to atmospheric). The strain used is derived from MJF431 (Levina et al., 1999), which is null for *mscS* and *mscK* activities. MscS activity was replaced in *trans* using a pB10 expression vector (Blount et al., 1999). Note the long open-dwell times and the inactivation of the activity with time. Recording conditions were as described (Blount et al., 1999), pH 7.0 and −20 mV in the pipette channel openings are shown as upward for convenience; the closed (C) and open (O_1 to O_6) states are labeled to the left.

only the reconstituted protein and a lipid bilayer are necessary for channel activity; no other proteins are required (Okada et al., 2002; Sukharev, 2002). Hence, MscS responds directly to mechanical force, presumably transduced through the lipid bilayer. Investigation of the channel encoded by the wild-type *kef*A led to the discovery that the activity was only observed in high K^+ buffer, which suggested its current name, MscK (for K^+ regulated) (Li et al., 2002); YggB was renamed MscS. Throughout the rest of the chapter, this nomenclature will be maintained. MscS and MscK activities are nearly identical in conductance and are essentially impossible to distinguish (Levina et al., 1999). So, does this draw into question the earlier studies where conclusions were assumed to be drawn from a homogeneous MscS population? Perhaps not. Although all of the early studies were performed in high-K^+ buffers in which MscK would also be active, we now know that the vast majority of MscS-like activity truly is MscS. MscK activity is detected in only one out of five patches, and even then represents only a handful of the channel activities observed (or roughly 5% of the MscS-like activity) (Levina et al., 1999; Li et al., 2002). On the other hand, MscS inactivates more readily than MscK (Levina et al., 1999); so, if many of the MscS activities were inactivated in the patches reported, the contaminating MscK activity could have led to misleading interpretations. Now that null mutants exist, expressing only MscS or MscK activity, these studies can and perhaps will be repeated, leading to firmer conclusions.

REVEALING STRUCTURAL FEATURES OF THE MscS PROTEIN

Early structural studies of MscS, utilizing classic bacterial genetic and biochemical approaches, suggested that MscS subunits contained three transmembrane (TM) domains with a relatively large cytoplasmic region (Miller et al., 2003a). A similar study with MscK suggested a comparable profile at its C-terminal segments, where homologous domains are found. However, the MscK protein is considerably larger (1,120 amino acids versus 286 for MscS), including a large periplasmic domain at its N-terminal region, followed by eight additional TM domains, all preceding the conserved MscS-like region (Fig. 2) (McLaggan et al., 2002).

The structure of *E. coli* MscS was solved to 3.9-Å resolution by X-ray crystallography (Bass et al., 2002); residues 27 to 280 were resolved. As expected, three helical TM domains were found at the N-terminal region of the protein. The channel, however, appeared to be a homoheptamer (Bass et al., 2002), rather than the hexamer predicted from cross-linking experiments (Sukharev, 2002). The third TM domain, which is rich in glycine and alanine, forms the pore. A glycine at position 113 appears to induce a turn within the α-helical domain, realigning the helix along the presumed membrane-cytoplasmic interface. Distal to this structure is a region that is high in β-sheet character. The cytoplasmic region forms a cage with seven pores, or portals, at the subunit interface, all terminating in a short β-barrel-like "crown" at the extreme C-terminal end (Color Plate 27 [see color insert]). This cytoplasmic cage-like structure may function as a molecular sieve, effectively determining the selectivity of the pore. A visual inspection of the crystal structure strongly suggests that the channel was captured in an open conformation. The TM pore appears to reveal an 11-Å opening, the portals at the cytoplasmic subunit interfaces appear to be open to about 14 Å, and the crown forms an 8-Å hole. These pore sizes could account for the 1-nS conductance observed for the MscS activity. A more recent study, however, implied that water and ions cannot pass through the TM pore when the channel is in this conformation, leading the authors to speculate that the structure reflects an inactive or desensitized state rather than a truly open one (Anishkin and Sukharev, 2004). The hypothesis may not be unreasonable; outside the constraints of lateral pressures of the membrane, one might suppose that the inactivated state is of lower energy than the closed state. On the other hand, this study used only molecular simulation to model the permeation of the pore, an approach that inherently entertains many assumptions. In addition, the authors modeled MscS mutations that are known to lead to decreased-growth gain-of-function (GOF) phenotypes but have yet to

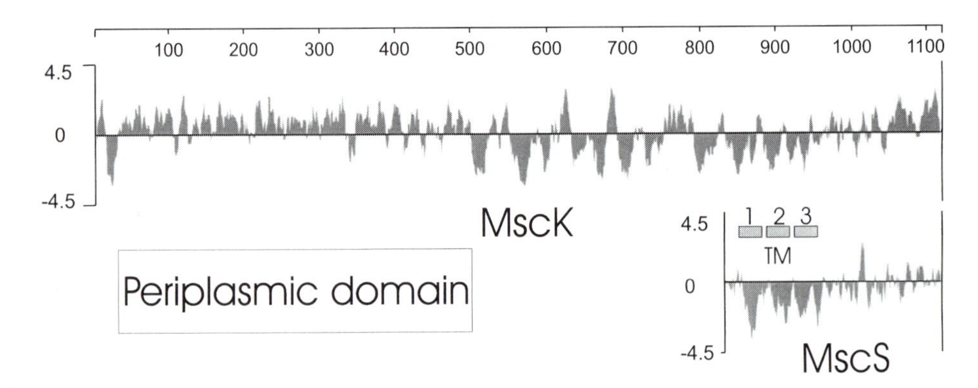

Figure 2. Hydrophilicity plots of MscK (top) and MscS (bottom). The scale bar is in amino acids. Note that MscK is much larger than MscS and that it is the C terminus of MscK that shows homology. The three TM domains of MscS (and last three of MscK) are indicated by gray boxes between the plots. Proximal to the homologous region, MscK contains an N-terminal periplasmic domain (which is labeled) and eight additional TM domains.

be fully characterized physiologically (Miller et al., 2003a). Hence, further experimental investigation will be necessary to determine "how open" is the MscS model derived from crystallographic analysis.

MODELS FOR GATING AND MODULATING MscS ACTIVITY

As described in chapter 13, random and site-directed mutagenesis studies of *mscL* identified the first TM domain (TM1) as an important functional region of the channel long before the crystallization of the channel. Many of the mutations isolated were substitutions to more hydrophilic residues. When the crystal structure of MscL was solved, this region was found to line the pore of the channel, thus leading to a relatively comprehensive hypothesis that this constitutes the channel gate and that the separation of these relatively hydrophobic domains (and their transient exposure to an aqueous environment) establishes the primary energy boundary between open and closed states of the channel (Blount and Moe, 1999; Yoshimura et al., 1999; Chiang et al., 2004), gating model for MscL that remains viable. In contrast, little mutagenesis had been performed on the MscS channel by the time the structure was solved. One missense mutation, valine to aspartate at position 40, within TM1 had been fully characterized (Okada et al., 2002). Although expression of this mutated channel led to a severe slowed-growth GOF phenotype and inappropriate ion flux, no channel activity could be recorded by patch clamp under normal conditions. In addition, at approximately the same time as publication of the crystal structure, mutations were generated in TM3, now known to be near the pore, and were reported to effect GOF phenotypes (T93R, L109S, and A102P), but full electrophysiological characterization had not been presented (Miller et al., 2003a). Hence, because little mutagenesis data existed at the time of crystallization, much of what we know and speculate for the molecular mechanisms of gating and modulation of the MscS channel is derived from the solved structure. Only recently have experimental studies shed additional light on the structural changes that occur upon gating.

Although the crystal structure gives us a wealth of information about one state of the channel, it reveals little of its dynamics. Therefore, some of the fundamental questions in the field concern the structural modifications necessary for the channel to acquire a closed state. The model derived from X-ray crystallography shows TM1 and TM2 to be distant from the central pore (Color Plate 27). Given the predicted lateral pressures within the membrane, this seems an unlikely arrangement when within a lipid bilayer. The authors noted this and predicted that, in the closed structure, the extension of this wing-like structure may be more proximal to the pore-forming TM3 (Bass et al., 2002). Perhaps it is the extension of this domain, when the lateral pressures of the membrane change upon tension, that effects the opening of the channel.

An interesting property of the structure is the placement of positively charged (arginine) residues in TM domains 1 and 2, found at positions 46 and 74, respectively. Additional positively charged residues including arginines at positions 54, 59, and 88 and a lysine (58) are found near the membrane-water interface. In the seminal paper on crystallographic structure, the authors speculated that these charged residues, especially those suspected to lie within the membrane, might provide the basis for voltage modulation of the channel; membrane potential would influence the probability of movement of the TM1/TM2 wing-like sensor domain (Bass et al., 2002). Such a mechanism is reminiscent of one of the models describing the structural changes that occur upon the gating of voltage-gated channels (Jiang et al., 2003). This model for MscS gating is also consistent with data derived from the V40D missense mutation described above; given the location of this residue within the crystal structure, combined with the in vivo and channel phenotypes, we can now speculate that this lesion leads to a channel in which the tension sensor is malfunctioning, thus forming a channel that slowly "leaks." Presumably, positive charges near the pore also underlie the anionic preference observed for the MscS activity. An additional charge, conferred by arginine 88, appears to be oriented into the pore and has been spotlighted as a primary candidate (Bass et al., 2002); however, at this time one cannot rule out the possibility that the numerous charges near the cytoplasmic portal impose the observed anionic preference. Hence, although the structure of MscS has inspired hypotheses for mechanisms of voltage modulation and anion preference of the channel activity, these theories have yet to be tested experimentally.

The constriction defining the pore of the MscS channel is composed of a region of TM3 in which glycine and alanine residues appear to be tightly packed, with the closest association between glycine 108 and alanine 106 (Fig. 3 and Color Plate 27). A model for the closure of MscS has been derived from mutagenesis experiments in which each of the glycines in this region has been substituted with serine (Edwards et al., 2005). This model proposes that the TM domains become more vertical, or normal to the membrane plane; to accomplish this, the

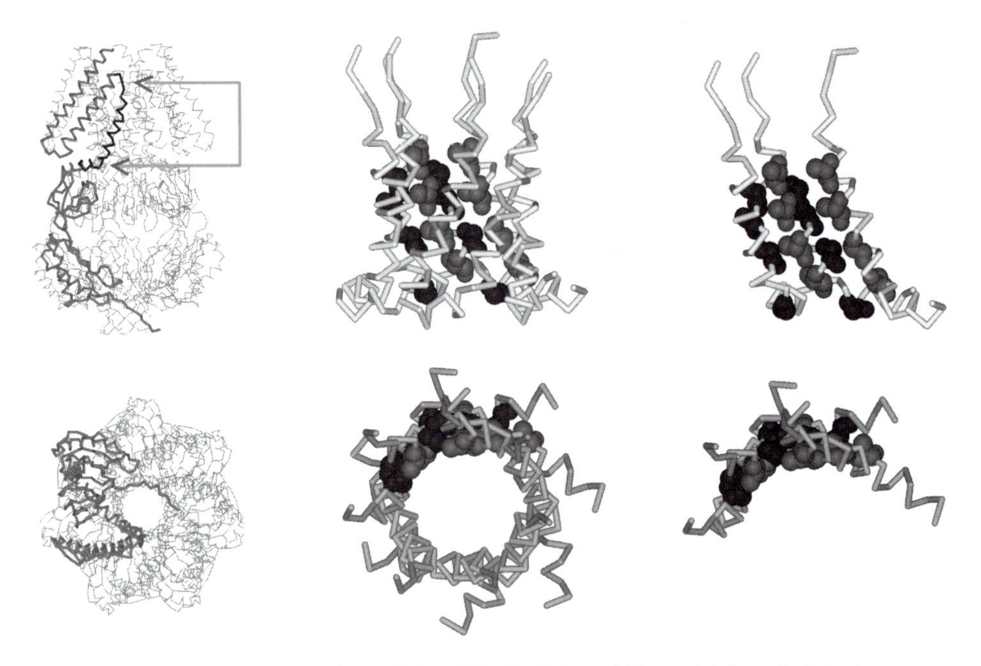

Figure 3. The structure and pore design of the MscS channel. The model shown is derived from X-ray crystallographic data (Bass et al., 2002). The side (top panels) and top (bottom panels) views are shown. One subunit of the homoheptamer is shown in darker gray for distinction; the pore-forming third TM domain of this subunit is in black (the bracketed arrows point to this TM region). The heptameric (center) and simple trimeric (right; the obstructing subunit structures have been removed) structures are shown. Note the tight packing of the glycines (black CPK residues) and alanines (gray CPK residues) of the three subunits in the right panel.

TM domains must slide along each other and rotate. This results in a change in the packing structure of the small amino acids, glycine and alanine, thus leading to a channel with a smaller pore. In the closed structure, leucines at positions 105 and 109 would come into closer proximity, thus forming a tighter constriction point and a hydrophobic barrier leading to a nonconducting conformation. The data from these mutations suggest that closure of the pore may involve a slight clockwise rotation of the TM domains (as observed from the periplasm). Note that this model is analogous to a current model for MscL gating: opening involves greater tilt of the TM domain relative to the bilayer, small amino acids are found within the pore presumably to give greater packing and flexibility in the region, and the pore-lining TM domain rotates upon the gating transition. However, these changes within MscS are of smaller magnitude: the domains tilt only about 25°, compared with the >65° tilt predicted for the open structure of MscL (Sukharev et al., 2001), and the rotation is of only a few degrees, compared with the 110° rotation of the domain predicted for MscL (Perozo et al., 2002a; Bartlett et al., 2004; also see chapter 13). Presumably, larger structural changes are required to achieve the larger pore and conductance of the MscL channel.

Structural changes are also predicted to occur in the cytoplasmic cage-like structure of MscS. In one study, Ni^{2+} ions were shown to inhibit gating of MscS bearing a C-terminal polyhistidine tag (Koprowski and Kubalski, 2003). The inhibition was primarily noted when the metal was bound to the channel in the closed rather than open state. The authors proposed that the C-terminal region of the protein separates significantly upon gating. Thus, intersubunit metal binding to the histidine residues prevents the structural arrangements required for gating, whereas the intrasubunit binding within individual poly-histidine tags, which occurs in the open state, does not. Although the large separation predicted to occur between the C termini of the complex is not apparent in the crystal structure, this study suggests that structural rearrangements can and do occur in the C-terminal bundle and that preventing this dilation will inhibit gating. A more recent study found that deletions at the C-terminal end of the protein are poorly tolerated, often yielding proteins that are not expressed well in the membrane (Schumann et al., 2004). Deletion of only the β-barrel-like crown (residues 266 through 286) was, however, tolerated, although the channel was less stable. Specifically, the Δ266-286 channel gates and inactivates, but does not recover from inactivation. Interestingly, addition of a hexahistidine tag at the C-terminal end of this deletion-mutated channel leads to an activity that can recover from desensitization, suggesting that the extreme C-terminal region of the protein is indeed a dynamic region where length is important to allow the structural transitions that occur from the inactive to the closed state (Schumann et al., 2004). A more quantitative determination of the proximity of specific residues within the C-terminal region of the closed MscS channel has been obtained by disulfide trapping and cross-linking studies (Miller et al., 2003b). We now know that this corresponds to the cytoplasmic cage-like structure. Here, the data imply that, when closed, this region of the protein assumes a much more compact structure than that observed in the solved crystal structure. Hence, the TM pore may not be the only permeation barrier in the closed state of the channel; the portal and β-barrel-like crown may also collapse into impermeable or micropermeable conformations. This hypothesis has been referred to as the "Chinese-lantern" model by analogy with such lanterns, whose intensity is adjusted by either collapsing or expanding the lamp (Edwards et al., 2004).

In sum, the data suggest large rearrangements of the MscS protein upon gating, as depicted in Fig. 4. The first two TM domains may constitute the primary sensors of membrane

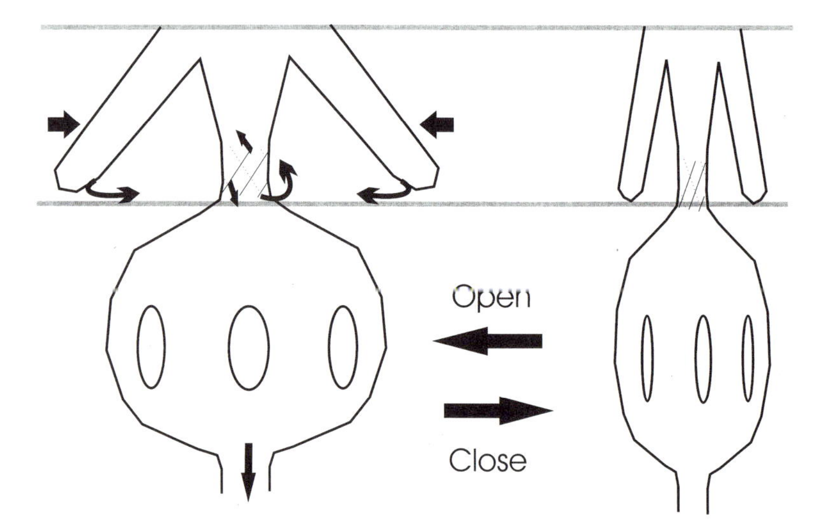

Figure 4. A cartoon depicting the current model for the structural rearrangements thought to occur upon closure of the MscS channel. Note that it is thought that it is the open (or inactivated) state of the channel whose structure was solved in the crystallographic study (Bass et al., 2002). The three TM domains are within the bilayer depicted by gray horizontal lines. Note that there are nine pores: one in the bilayer formed by TM3, seven at the interface of the subunits within the cytoplasmic cage, and one at the extreme cytoplasmic tip. Evidence suggests that each of these may constrict upon closure of the channel (Koprowski and Kubalski, 2003; Miller et al., 2003b; Edwards et al., 2005).

tension and modulation by voltage potential, flapping out like wings upon stimulation. Similar to MscL, the orientation of the TM domains forming the pore may adjust, becoming more tilted, relative to the normal, in the open structure. Note, however, that this reorientation may be at least partially shielded from the membrane, so one of the modulators of MscL activity, thinning of the membrane, may play little role in MscS gating; although this hypothesis has been previously proposed (Blount, 2003), it has yet to be explored experimentally. Also in contrast with MscL, the structural rearrangements in MscS gating may not be limited to the TM domains. The potential collapse of the cytoplasmic cage domain may define, at least in part, the potential permeation barriers in the closed structure.

THE MscS EXTENDED FAMILY: SPECULATION OF FUNCTION AND FORM

One function for MscL and MscS channels is clear; they play a vital role in osmotic regulation. When an *E. coli* cell grows in a high-osmotic environment, it accumulates compatible solutes, such as potassium, proline, betaine, and glutamate, by activating pumps or accumulating synthesized products (Dinnbier et al., 1988; Poolman et al., 2002). When the environment suddenly shifts to lower osmolarity, water accumulates within the cell, the cell swells, and MscS and MscL are activated, thus relieving the pressure. Hence, these channels serve as biological "emergency release valves." Consistent with a vital role in osmoregulation, *mscS* has been identified in a microarray study as a gene whose expression is modulated by osmotic conditions and potentially regulated by *rpoS*, which encodes the stationary phase/stress-activated sigma factor, σ^s (Cheung et al., 2003). The latter hypothesis has been further, and more directly, substantiated in a study demonstrating *mscS* and *mscL* regulation by *rpoS* (Stokes et al., 2003). In their role as emergency release valves, MscS and MscL appear to serve a redundant function; an osmotic-sensitive phenotype is observed only in cells that express neither channel (Levina et al., 1999).

The *mscS*-like gene family is more extended and diverse than that of *mscL*. Even within *E. coli*, several *mscS* homologs exist. As discussed above, the identity of MscS was first revealed through a homolog, MscK. This latter channel has the fundamental properties of MscS: gated by membrane tension, long open-dwell times, and a 1-nS conductance. However, MscK has fundamental differences. The channel has a decreased propensity for inactivation and is only observed in high concentrations of K^+, NH_4^+, or similar ions. The latter observation has fostered speculation that MscK protects the cell from osmotic stress only in very specific environments (Li et al., 2002). However, to date, no study has reported a loss-of-function phenotype for the MscK-null strain, nor has an amplification of the MscL/MscS double-null phenotype been observed when all three channels (MscL/MscS/MscK) are deleted. Another puzzling finding is that MscK is not seen in all patches, even though the amount of membrane within a single patch derived from a giant bacterial spheroplast is greater than the total area from a normal bacterial cell. When MscK activity is observed, it is often more than one channel per patch (Li et al., 2002), which could be explained by channel clustering. However, perhaps a more likely interpretation is that patches differ in the amount of cytoplasmic proteins, residual cell wall, and outer membrane that they contain. Since MscK is the only MS channel activity that has not been reconstituted into synthetic membranes, it could be that this residual material is required for its activation. There may be precedent: one of the models for gating of eukaryotic MS channels requires a linkage of the channel to cytoskeletal and extracellular proteins (Tavernarakis and Driscoll, 1997). Could it be that such interactions are also required for MscK gating? If so, MscK could represent a channel that senses membrane tension but, similar to eukaryotic sensors, requires additional interactions for activation. Hence, the study of regulation and modulation of the MscK channel holds promise for defining the mechanisms underlying the complexities that are often associated with more sophisticated systems.

A survey of the *E. coli* genome has revealed additional potential *mscS*-like genes (Levina et al., 1999). A hypothetical gene, F343 (D90771), is predicted to encode a protein with some homology to MscS. In addition, both *yjeP* and F786 (AE000183) are predicted to encode proteins that, like MscK, contain not only a C-terminal domain similar to that of MscS but also a large periplasmic N terminus and approximately eight additional TM domains. While it is possible that some of these homologs may encode the elusive MscM activity, it is unlikely that they all do. Why do these putative channels elude patch clamp? One potential explanation is that they are not truly channels but serve some other function. Indeed, an ortholog of YjeP from *Erwinia chrysanthemi*, BspA, has been associated with the retention of glycine betaine at high osmolarity; the null mutant specifically jettisons this osmoprotectant at high osmolarity (Touze et al., 2001). If true, perhaps even MscS is a multifunctional protein. Alternatively, and perhaps more likely, the activity of these channels may elude electrophysiological detection because the conditions required for their activity have not yet been determined. If Na^+, rather than K^+, had been the major conducting ion in the buffers used to patch bacterial membranes, then MscK may also have eluded detection. There are indications from whole-cell physiological data that some channels may function in vivo yet avoid detection by patch clamp. As an example, the flux of glycine betaine in *Corynebacterium glutamicum*, independent of MscL and MscS, upon osmotic downshock has been reported; although the estimated transport capacity is predicted to be similar to bacterial MS channels, no additional activity has yet been observed by patch clamp (Nottebrock et al., 2003). Hence, these additional putative channels may have requirements, perhaps even auxiliary proteins, that are not present in the standard patch-clamp setting.

As implied by some of the studies cited above, the MscS extended family is not confined to *E. coli*. Essentially all bacteria, as well as many of the archaea examined (Kloda and Martinac, 2002), contain at least one, and often multiple, copies of *mscS*- and/or *mscK*-like genes (Levina et al., 1999; Pivetti et al., 2003). A survey of some of the MscS family members from archaea has shown researchers just how diverse the activities encoded by family members can be in conductance, sensitivity, and ionic preference; some homologs even demonstrate cationic rather than anionic preferences (Kloda and Martinac, 2001d; Kloda and Martinac, 2001c; Kloda and Martinac, 2001b; Kloda and Martinac, 2001a; Kloda and Martinac, 2002). Even the fission yeast *Schizosaccharomyces pombe* appears to have multiple forms of *mscS*-like genes (Pivetti et al., 2003), although it is as yet uncertain if these encode the MS channel activity characterized in this species (Zhou and Kung, 1992). In contrast to the MscL family of proteins, the MscS family extends beyond the microbial world and is found in *Arabidopsis thaliana* (Pivetti et al., 2003) as well as several other plant species (A. Kumánovics and P. Blount, unpublished data). While one may speculate that in plants MscS-like channels play a role in solute flux in the root or in seed germination, the function of these putative channels remains unclear.

CONCLUDING REMARKS

The MscS-like family of channels is extremely large and diverse. In its most streamlined form, e.g., MscS in *E. coli*, it has many similarities with its counterpart, MscL. It appears to directly sense membrane tension, and it appears to utilize modification of TM domain tilt and rotation in its gating sequence. Because the MscS and MscL families are so far removed from each other, these preserved features may reflect conserved mechanisms found in many MS channel families. In addition, however, MscS has functions, including inactivation and voltage modulation, that are normally associated with more complex channels. As we investigate more distant homologs, the regulation and modulation may become more complex. The complex properties of MscK indicate dependence for K^+ but not Na^+ ions and the prospect of modulation by accessory proteins. Microbial physiology, combined with genomic analysis, hints that the full picture will inevitably be more complicated: novel channels of which we understand too little to predict the correct stimulus. Hence, the study of the MscS family of channels promises to lay the groundwork for understanding the basic design of a mechanosensor as well as the modulation and regulation that are usually associated with eukaryotic organisms.

Acknowledgments. We acknowledge Ian R. Booth for stimulating discussion, fruitful collaboration, and critical reading of the manuscript.

We are supported by grants GM61028 and DK60818 from the National Institutes of Health, grant I-1420 from the Welch Foundation, and grant F49620-01-1-0503 from the Air Force Office of Scientific Review.

REFERENCES

Anishkin, A., and S. Sukharev. 2004. Water dynamics and dewetting transitions in the small mechanosensitive channel MscS. *Biophys. J.* **86:**2883–2895.

Bartlett, J. L., Levin, G., and Blount, P. 2004. An in vivo assay identifies changes in residue accessibility on mechanosensitive channel gating. *Proc. Natl. Acad. Sci. USA* **101:**10161–10165.

Bass, R. B., P. Strop, M. Barclay, and D. C. Rees. 2002. Crystal structure of *Escherichia coli* MscS, a voltage-modulated and mechanosensitive channel. *Science* **298:**1582–1587.

Berrier, C., M. Besnard, B. Ajouz, A. Coulombe, and A. Ghazi. 1996. Multiple mechanosensitive ion channels from *Escherichia coli*, activated at different thresholds of applied pressure. *J. Membr. Biol.* **151:**175–187.

Blount, P. 2003. Molecular mechanisms of mechanosensation: big lessons from small cells. *Neuron* **37:**731–734.

Blount, P., and P. Moe. 1999. Bacterial mechanosensitive channels: integrating physiology, structure and function. *Trends Microbiol.* **7:**420–424.

Blount, P., S. I. Sukharev, P. C. Moe, B. Martinac, and C. Kung. 1999. Mechanosensitive channels of bacteria, p. 458–482. *In* P. M. Conn (ed.), *Methods in Enzymology*. Academic Press, San Diego, Calif.

Blount, P., S. I. Sukharev, P. C. Moe, M. J. Schroeder, H. R. Guy, and C. Kung. 1996. Membrane topology and multimeric structure of a mechanosensitive channel protein of *Escherichia coli. EMBO J.* **15:**4798–4805.

Britten, R. J., and F. T. McClure. 1962. The amino acid pool in *Escherichia coli. Bacteriol. Rev.* **26:**292–335.

Cheung, K. J., V. Badarinarayana, D. W. Selinger, D. Janse, and G. M. Church. 2003. A microarray-based antibiotic screen identifies a regulatory role for supercoiling in the osmotic stress response of *Escherichia coli. Genome Res.* **13:**206–215.

Chiang, C. S., A. Anishkin, and S. Sukharev. 2004. Gating of the large mechanosensitive channel in situ: estimation of the spatial scale of the transition from channel population responses. *Biophys. J.* **86:**2846–2861.

Cui, C., D. O. Smith, and J. Adler. 1995. Characterization of mechanosensitive channels in *Escherichia coli* cytoplasmic membrane by whole-cell patch clamp recording. *J. Membr. Biol.* **144:**31–42.

Dinnbier, U., E. Limpinsel, R. Schmid, and E. P. Bakker. 1988. Transient accumulation of potassium glutamate and its replacement by trehalose during adaptation of growing cells of *Escherichia coli* K-12 to elevated sodium chloride concentrations. *Arch. Microbiol.* **150:**348–357.

Edwards, M. D., I. R. Booth, and S. Miller. 2004. Gating the bacterial mechanosensitive channels: MscS a new paradigm? *Curr. Opin. Microbiol.* **7:**163–167.

Edwards, M. D., Y. Li, S. Kim, S. Miller, W. Bartlett, I. Iscla, P. Blount, J. U. Bowie, and I. R. Booth. 2005. Pivotal role of the glycine-rich TM3 helix in gating the MscS mechanosensitive channel. *Nat. Struct. Mol. Biol.* **12:**113–119.

Häse, C. C., A. C. Le Dain, and B. Martinac. 1995. Purification and functional reconstitution of the recombinant large mechanosensitive ion channel (MscL) of *Escherichia coli. J. Biol. Chem.* **270:**18329–18334.

Jiang, Y., A. Lee, J. Chen, V. Ruta, M. Cadene, B. T. Chait, and R. MacKinnon. 2003. X-ray structure of a voltage-dependent K+ channel. *Nature* **423:**33–41.

Kloda, A., and B. Martinac. 2001a. Mechanosensitive channel of Thermoplasma, the cell wall-less archaea: cloning and molecular characterization. *Cell Biochem. Biophys.* **34:**321–347.

Kloda, A., and B. Martinac. 2001b. Mechanosensitive channels in archaea. *Cell Biochem. Biophys.* **34:**349–381.

Kloda, A., and B. Martinac. 2001c. Molecular identification of a mechanosensitive channel in archaea. *Biophys. J.* **80:**229–240.

Kloda, A., and B. Martinac. 2001d. Structural and functional differences between two homologous mechanosensitive channels of *Methanococcus jannaschii. EMBO J.* **20:**1888–1896.

Kloda, A., and B. Martinac. 2002. Common evolutionary origins of mechanosensitive ion channels in archaea, bacteria and cell-walled eukarya. *Archaea* **1:**35–44.

Koprowski, P., and A. Kubalski. 1998. Voltage-independent adaptation of mechanosensitive channels in *Escherichia coli* protoplasts. *J. Membr. Biol.* **164:**253–262.

Koprowski, P., and A. Kubalski. 2003. C termini of the *Escherichia coli* mechanosensitive ion channel (MscS) move apart upon the channel opening. *J. Biol. Chem.* **278:**11237–11245.

Levina, N., S. Totemeyer, N. R. Stokes, P. Louis, M. A. Jones, and I. R. Booth. 1999. Protection of *Escherichia coli* cells against extreme turgor by activation of MscS and MscL mechanosensitive channels: identification of genes required for MscS activity. *EMBO J.* **18:**1730–1737.

Li, Y., P. C. Moe, S. Chandrasekaran, I. R. Booth, and P. Blount. 2002. Ionic regulation of MscK, a mechanosensitive channel from *Escherichia coli. EMBO J.* **21:**5323–5330.

Martinac, B. 2004. Mechanosensitive ion channels: molecules of mechanotransduction. *J Cell Sci.* **117:**2449–2460.

Martinac, B., J. Adler, and C. Kung. 1990. Mechanosensitive ion channels of *E. coli* activated by amphipaths. *Nature* **348:**261–263.

Martinac, B., M. Buechner, A. H. Delcour, J. Adler, and C. Kung. 1987. Pressure-sensitive ion channel in *Escherichia coli. Proc. Natl. Acad. Sci. USA* **84:**2297–2301.

McLaggan, D., M. A. Jones, G. Gouesbet, N. Levina, S. Lindey, W. Epstein, and I. R. Booth. 2002. Analysis of the kefA2 mutation suggests that KefA is a cation-specific channel involved in osmotic adaptation in *Escherichia coli. Mol. Microbiol.* **43:**521–536.

Miller, S., W. Bartlett, S. Chandrasekaran, S. Simpson, M. Edwards, and I. R. Booth. 2003a. Domain organization of the MscS mechanosensitive channel of *Escherichia coli. EMBO J.* **22:**36–46.

Miller, S., M. D. Edwards, C. Ozdemir, and I. R. Booth. 2003b. The closed structure of the MscS mechanosensitive channel. Cross-linking of single cysteine mutants. *J. Biol. Chem.* **278:**32246–32250.

Nottebrock, D., U. Meyer, R. Kramer, and S. Morbach. 2003. Molecular and biochemical characterization of mechanosensitive channels in *Corynebacterium glutamicum. FEMS Microbiol. Lett.* **218:**305–309.

Okada, K., P. C. Moe, and P. Blount. 2002. Functional design of bacterial mechanosensitive channels. Comparisons and contrasts illuminated by random mutagenesis. *J. Biol. Chem.* **277:**27682–27688.

Ou, X., P. Blount, R. J. Hoffman, and C. Kung. 1998. One face of a transmembrane helix is crucial in mechanosensitive channel gating. *Proc. Natl. Acad. Sci. USA* **95:**11471–11475.

Perozo, E., D. M. Cortes, P. Sompornpisut, A. Kloda, and B. Martinac. 2002a. Open channel structure of MscL and the gating mechanism of mechanosensitive channels. *Nature* **418:**942–948.

Perozo, E., A. Kloda, D. M. Cortes, and B. Martinac. 2002b. Physical principles underlying the transduction of bilayer deformation forces during mechanosensitive channel gating. *Nature Struct Biol.* **9:**696–703.

Pivetti, C. D., M. R. Yen, S. Miller, W. Busch, Y. H. Tseng, I. R. Booth, and M. H. Saier. 2003. Two families of mechanosensitive channel proteins. *Microbiol. Mol. Biol. Rev.* **67:**66–85.

Poolman, B., P. Blount, J. H. Folgering, R. H. Friesen, P. C. Moe, and T. van der Heide. 2002. How do membrane proteins sense water stress? *Mol. Microbiol.* **44:**889–902.

Schumann, U., M. D. Edwards, C. Li, and I. R. Booth. 2004. The conserved carboxy-terminus of the MscS mechanosensitive channel is not essential but increases stability and activity. *FEBS Lett.* **572:**233–237.

Stokes, N. R., H. D. Murray, C. Subramaniam, R. L. Gourse, P. Louis, W. Bartlett, S. Miller, and I. R. Booth. 2003. A role for mechanosensitive channels in survival of stationary phase: regulation of channel expression by RpoS. *Proc. Natl. Acad. Sci. USA* **100:**15959–15964.

Sukharev, S. 2002. Purification of the small mechanosensitive channel of *Escherichia coli* (MscS): the subunit structure, conduction, and gating characteristics in liposomes. *Biophys. J.* **83:**290–298.

Sukharev, S., S. Durell, and H. Guy. 2001. Structural models of the MscL gating mechanism. *Biophys. J.* **81:**917–936.

Sukharev, S. I., B. Martinac, V. Y. Arshavsky, and C. Kung. 1993. Two types of mechanosensitive channels in the *Escherichia coli* cell envelope: solubilization and functional reconstitution. *Biophys. J.* **65:**177–183.

Sukharev, S. I., W. J. Sigurdson, C. Kung, and F. Sachs. 1999. Energetic and spatial parameters for gating of the bacterial large conductance mechanosensitive channel, MscL. *J. Gen. Physiol.* **113:**525–540.

Tavernarakis, N., and M. Driscoll. 1997. Molecular modeling of mechanotransduction in the nematode *Caenorhabditis elegans. Annu. Rev. Physiol.* **59:**659–689.

Touze, T., G. Gouesbet, C. Bolangiu, M. Jebbar, S. Bonnassie, and C. Blanco. 2001. Glycine betaine loses its osmoprotective activity in a *bspA* strain of *Erwinia chrysanthemi. Mol. Microbiol.* **42:**87–99.

Yoshimura, K., A. Batiza, M. Schroeder, P. Blount, and C. Kung. 1999. Hydrophilicity of a single residue within MscL correlates with increased channel mechanosensitivity. *Biophys. J.* **77:**1960–1972.

Zhou, X., and C. Kung. 1992. A mechanosensitive ion channel in *Schizosaccharomyces pombe. EMBO J.* **11:**2869–2875.

Bacterial Ion Channels and Their Eukaryotic Homologs
Edited by A. Kubalski and B. Martinac

Chapter 13

MscL, a Bacterial Mechanosensitive Channel

Sergei Sukharev, Andriy Anishkin, Chien-Sung Chiang, Monica Betanzos, and H. Robert Guy

BACTERIAL MECHANOSENSITIVE CHANNELS AS MODEL SYSTEMS

The mechanosensitive (MS) channel of large conductance, MscL, is an integral protein present in the cytoplasmic membrane of essentially all free-living bacteria. Together with other MS channels (the small conductance MscS, the potassium-activated MscK, and the still elusive minichannel MscM), MscL constitutes a turgor-driven osmolyte efflux system that limits the pressure inside the cell and rescues bacteria from lysis in the event of strong osmotic downshock. The entire efflux system provides graded response to an osmotic imbalance, with the smallest MscM opening first, MscK and MscS activating later, and the large MscL acting as an emergency release valve that activates only at extreme pressures approaching the membrane lytic limit (Berrier et al., 1996; Levina et al., 1999). Under typical laboratory conditions, MscL and MscS are partially redundant as either of them rescues bacteria upon a strong osmotic downshift. The specific function of each individual component of the system in different environmental situations is yet to be clarified (Wood, 1999). On the other hand, bacterial MS channels MscL and MscS have emerged as convenient model systems for biophysical studies of tension-dependent channel gating.

Why have model systems for studies of primary mechanotransduction mechanisms come to the spotlight? Mechanosensation is a multifaceted phenomenon. While the slow processes of load-dependent tissue remodeling in vertebrates have been linked to integrin signaling pathways (Ingber, 2003), the majority of "fast" responses to sound, touch, tissue stretch, or osmotic imbalance are attributed to direct force-dependent gating of MS channels (Sachs and Morris, 1998; Hamill and Martinac, 2001; Sukharev and Corey, 2004). Analysis of available phenomenology and molecular information suggests that the entire group of mechanogated channels is diverse and includes representatives from unrelated protein families. Subunits of the epithelium sodium channel and related acid-sensing ion

Sergei Sukharev, Andriy Anishkin, Chien-Sung Chiang, and Monica Betanzos • Biology Department, University of Maryland, College Park, MD 20742. ***H. Robert Guy*** • Laboratory of Experimental and Computational Biology, CCR, National Cancer Institute, National Institutes of Health, Bldg. 12B, Bethesda, MD 20892-5567.

channels (Welsh et al., 2002), transient receptor potential channels, homologous polycystins (Corey, 2003), certain splice variants of the BK channel (Naruse et al., 2003), some inward rectifiers (Van Wagoner, 1993), and two-pore potassium channels (Patel and Honore, 2001) have been implicated in mechanosensation in different parts of the mammalian body. Multiple types of MS channels have been suggested by genetic studies of sensory transduction in nematodes (Tavernarakis and Driscoll, 1997; Colbert et al., 1997) and flies (Gillespie and Walker, 2001). Even the "simple" prokaryote *Escherichia coli* has at least two types of functionally similar MS channels, MscL and MscS, sharing no homology or structural similarity (Booth and Louis, 1999; Pivetti et al., 2003; Perozo and Rees, 2003). Perhaps MS channels have evolved many times in the course of evolution within different ancestral groups. As a result, somewhat similar functions have been implemented with dissimilar structural designs. This functional convergence, in our opinion, emphasizes the importance of exploration of the basic biophysical principles of MS channel function and leaves only a limited role to commonly used homology analysis. The attempts to understand these principles call for use of simple and convenient model systems.

Pioneering work by the Kung and Booth laboratories (Martinac et al., 1987; Sukharev et al., 1994a; Levina et al., 1999) has demonstrated numerous advantages offered by microbial MS channels. Small genes (*Eco mscL* of 408 bp, *Eco mscS* of 858 bp) enable exhaustive mutagenesis. Mutations often manifest themselves as either growth-suppressing gain-of-function (GOF) or osmotically fragile loss-of-function (LOF) bacterial phenotypes, permitting massive screens for unbiased identification of functionally important regions and domain interactions (Ou et al., 1998; Maurer and Dougherty, 2003; Yoshimura et al., 2004; Li et al., 2004). Exceptionally large conductances facilitate time-resolved patch-clamp recordings (Sukharev et al., 1999b; Shapovalov et al., 2003; Shapovalov and Lester, 2004), supplying information not only about kinetics of main transitions but also about gating intermediates. Now that several types of MS channels have been cloned from bacteria and archaea, a plethora of identified homologs (Spencer et al., 1999; Oakley et al., 1999; Maurer et al., 2000; Kloda and Martinac, 2001; Pivetti et al., 2003) provide a broad base for comparing natural sequence variations and functional traits and offer a variety of proteins for crystallization trials. Determination by Rees and coworkers of the crystal structures of *E. coli* MscS (Bass et al., 2002) and MscL homolog from *Mycobacterium tuberculosis* (Chang et al., 1998) now allows precise structure-based predictions. Although homology models and computer representations of different functional states derived from these structures are subjective, they are useful in formulating hypotheses about the gating mechanics (Sukharev et al., 2001b). Molecular dynamics and other computational approaches applied to crystal structures and models are becoming powerful predictive tools guiding the experiments (Gullingsrud and Schulten, 2003; Valadie et al., 2003). The experimental data, in turn, help in correcting the models (Anishkin et al., 2003). Last, but not least, reconstitutions of purified proteins have shown that these channels are gated directly by tension in the lipid bilayer (Hase et al., 1995; Blount et al., 1996a). Stability of protein complexes in detergents and foreign lipids allows application of magnetic resonance techniques (electron paramagnetic resonance [EPR]) and detailed studies of protein-lipid interactions that underlie the primary function of these channels (Perozo et al., 2002b).

In this chapter we will consider properties, structure, and the mechanism of gating of the large MS channel MscL, which is probably the best understood tension-gated channel to date. The progress has been rapid, and within 10 years of MscL cloning we have a reasonably supported structural model of gating. It is unlikely, however, that the unique structural design of these bacterial channels will permit direct replication of their exact mechanisms beyond their own orthologs. Instead of being prototypes for sequence- or structure-based generalizations, the simplicity of these channels allows us to comprehend basic principles that are likely to be common to many systems. Obtaining a self-consistent physical picture of channel gating by membrane stretch that includes the protein and surrounding lipids and solvent would be a stepping stone to understanding the mechanisms of the growing families of eukaryotic sensory channels modulated by lipids, hydrophobic or amphipathic substances, temperature, cell volume, or membrane stretch (Clapham, 2003; Patapoutian et al., 2003).

PREPARATIONS FOR BACTERIAL CHANNEL RECORDING

With the discovery of chemiosmotic coupling (Mitchell, 1979), it became clear that the electrical integrity of bacterial and mitochondrial membranes is critical for many processes including ATP synthesis, solute transport, flagellar motion, and osmoregulation. A strong interest in electrical events in coupling membranes arose, and preparations of bacterial and mitochondrial membranes amenable to direct electrical recordings were attempted. Generation of giant *E. coli* spheroplasts (Ruthe and Adler, 1985) overcame the small cell size limitation and permitted application of the standard patch-clamp technique (Hamill et al., 1981) to bacteria. The first publication by Martinac and coworkers described a complex behavior of a 1-nS suction-activated channel in *E. coli*, which was also sensitive to voltage and monovalent cation species present in the recording solution (Martinac et al., 1987). The abundance of this type of channel was demonstrated by whole-cell recordings in bacterial spheroplasts stimulated with positive pressure applied to the pipette (Cui et al., 1995). A variety of large MS channels was also found in gram-positive bacteria (Zoratti and Petronilli, 1988; Zoratti et al., 1990). Reconstitution of fragments of native membranes with multilamellar liposomes (Tank et al., 1982; Criado and Keller, 1987) allowed an alternative type of preparation suitable for observation of bacterial channels (Delcour et al., 1989; Berrier et al., 1989; Szabo et al., 1993; Sukharev et al., 1994b). Soon it was recognized that several different types of MS channel activities are constitutively present in bacterial membranes (Sukharev et al., 1993; Berrier et al., 1996). In *E. coli*, two types of channels, the smaller 1-nS channel (MscS type) with sustained openings and large channels (MscL type) of 3.2 nS with faster kinetics and subconducting states, were predominant. Solubilization of bacterial channels in mild detergents, column chromatography, and reconstitution of separate fractions with lipids yielded normal MscL and MscS activities in liposome patches (Sukharev et al., 1993). This was the key methodological step that led to isolation and characterization of the MscL protein.

IDENTIFICATION OF MscL

The intact MscL activities in liposomes reconstituted with crude membrane extracts permitted Sukharev et al. to use patch-clamp sampling as an assay technique for the presence of active channels in chromatographic separations (Sukharev et al., 1994a; Sukharev et al., 1994b). After each step of chromatography, individual fractions were reconstituted with soybean phospholipids, dialyzed from the detergent, and examined with patch clamp. Active fractions were pooled and applied to the next column, which further simplified the protein pattern. The number of active channels per patch was correlated with the presence of specific protein bands in particular fractions. Two separate series of chromatographic enrichment steps have been carried out. The most active fractions at the end of each series had relatively simple protein patterns with only one common band of apparent molecular mass of ~17 kDa. Microsequencing revealed 37 N-terminal residues of an orphan protein encoded 3′ of *trkA* on the 72nd minute of the *E. coli* chromosome. The open reading frame predicted a 15-kDa membrane protein. Insertional disruption of the identified gene removed the characteristic 17-kDa band and abolished MscL activities in spheroplasts and reconstituted chromatographic fractions. Reexpression of the identified gene in bacteria, yeast, or reticulocyte lysates produced active MscL (Sukharev et al., 1994a).

MscL PROTEIN

E. coli MscL is a 136-amino-acid (15-kDa) protein with two transmembrane domains (M1 and M2) predicted from the hydrophobicity profile (Sukharev et al., 1994a) residing in the inner (cytoplasmic) membrane of the bacterium (Blount et al., 1996a). The membrane topology determined with the PhoA fusion approach indicated that the short N-terminal (S1, ~15 residues) and the larger C-terminal (S3, ~40 residues) segments are cytoplasmic, whereas the loop connecting M1 and M2 segments (S2, ~25 residues) resides on the extracellular side (periplasm) (Blount et al., 1996a). Spectroscopic characterization of MscL predicted about 80% of the protein structured as alpha-helical elements with a net transbilayer orientation. The secondary structure was found highly thermostable (Arkin et al., 1998). A considerably larger size of MscL particle (~70 kDa) determined by size-exclusion chromatography indicated multimerization, and cross-linking patterns and electron microscopy suggested that the active complex likely consists of five or six identical subunits (Blount et al., 1996a; Saint et al., 1998; Sukharev et al., 1999a). Purification of the tagged protein and reconstitution with exogenous lipids confirmed that MscL is gated directly by tension in the lipid bilayer and no auxiliary elements are required for ion conduction or mechanosensitivity (Hase et al., 1995; Blount et al., 1996a).

HOMOLOGS OF MscL

Extensive sequencing projects targeting microbial genomes have identified *mscL*-like genes in essentially all free-living and pathogenic bacteria, with a few exceptions of utterly simplified obligatory parasites (mycoplasmas and rickettsiae) and possibly some marine microorganisms not subjected to environmental osmotic variations in their native habitats (Pivetti et al., 2003). All bacterial MscL homologs are 128- to 157-amino-acid proteins in which the two transmembrane segments are preserved. The highest degree of conservation occurs within the N-terminal and the first transmembrane domains (Spencer et al., 1999; Maurer et al., 2000; Sukharev et al., 2001b; Pivetti et al., 2003). Functional characterization of MscLs from nine gram-positive and gram-negative bacteria confirmed that they are all pressure-gated channels with variable kinetic properties, but a more consistent 3- to 4-nS unitary conductance (Moe et al., 1998). Among the functionally tested MscL homologs, channels from a *Synechocystis* sp. and the mycobacterial (*M. tuberculosis*) MscL required the highest pressure for activation when expressed in *E. coli*, possibly due to a specific lipid requirement (Moe et al., 2000). MscLs are also found outside the bacterial kingdom. One archaeal *mscL* homolog found in *Methanosarcina acetivorans* is shorter

(101 residues) than any of the bacterial homologs (Pivetti et al., 2003). The archaeal channel MJ0170 from *Methanococcus jannashii*, although not a direct homolog of MscL, exhibits clear sequence resemblance to the transmembrane domains of MscL (Kloda and Martinac, 2001). A longer *mscL* homolog (373 residues) was found in the eukaryote *Neurospora crassa* (Koprowski and Kubalski, 2001), but its properties and function have not been tested yet. The availability of multiple natural versions of MscL allowed not only identification of the most conserved segments and comparison of some functional traits but also provided a broad variety of material for protein crystallization trials, the advantage successfully utilized by the Rees group.

STRUCTURE OF TbMscL

After screening and refining 24,000 crystallization conditions for nine MscL homologs in the presence of 20 different detergents, Rees and collaborators (Chang et al., 1998) found that crystals of the MscL homolog from *M. tuberculosis* (TbMscL, 151 amino acids) exhibited the best diffraction quality. The structure of TbMscL was solved at a 3.5-Å resolution, revealing a pentameric assembly of identical subunits (Fig. 1). The transmembrane helices are tilted at ~28° relative to the axis of the pore, forming a fairly tightly packed right-handed bundle. The centrally positioned M1 helices line most of the funnel-like pore, and peripheral M2 helices face the lipid. The peculiar trait of the barrel organization is that interhelical contacts between M1 and M2 of the same subunit are relatively limited; instead, M1 of one subunit has a large buried contact with M2 of the neighboring subunit. The pore radius varies from 18 Å in the outer chamber to about 2 Å in a constriction near the cytoplasmic surface. The outer chamber is polar and lined with multiple threonines of M1 helices, whereas the constriction is lined by tightly packed girdles of aliphatic residues. The hydrophobic nature of the constriction makes it impermeant to ions. The S2 loops connecting the M1 and M2 helices are in extended conformation, flapping into the pore. Although predominantly polar, loops form hydrophobic contacts with aliphatic residues on M1, thus stabilizing the conformation of the flaps. The periphery of the barrel formed mainly by M2 helices anchors the channel into the lipid bilayer with aliphatic residues and numerous phenylalanines, with only one tyrosine per subunit located on the cytoplasmic side. In addition to the transmembrane barrel, the structure revealed the cytoplasmic domain formed by the bundle of C-terminal (S3) helices, coaxial with the pore. The linkers connecting M2 to the S3 helix contain a characteristic cluster of charged residues that may represent a binding site for a specific ligand or lipid. The initial hypothesis postulated that the channel opening is a result of an outward motion of M1 helices and their insertion between M2 helices, with the conducting pathway forming through a cylindrical barrel-stave assembly of 10 almost parallel helices (Chang et al., 1998).

It should be mentioned that the construct used in crystallization trials contained a 23-amino-acid N-terminal extension including a 10-His tag. Perhaps as a result of this modification, the native N-terminal segment (S1) was not fully resolved in the crystal structure. Reconstitution of TbMscL from the crystals has shown that the modified channel remained functional throughout the crystallization procedure and gated in liposomes in a similar fashion as freshly isolated His-tagged TbMscL or wild-type TbMscL expressed in *E. coli* spheroplasts (Moe et al., 2000). The latter preparation, however, exhibited "stiff" channels activated at pressures two times higher than EcoMscL in the same setting.

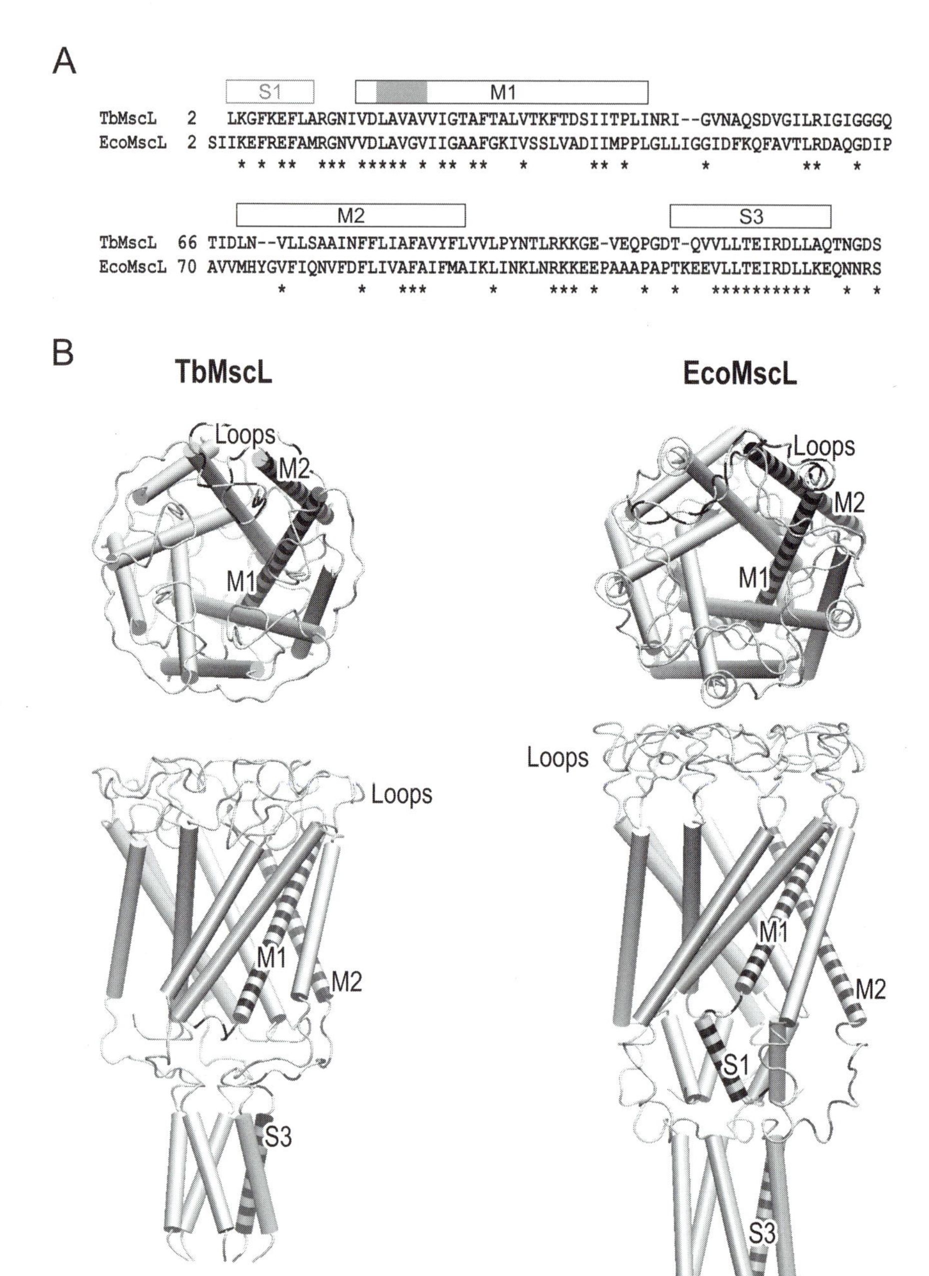

Figure 1. The *M. tuberculosis* (TbMscL) and *E. coli* (EcoMscL) MscL homologs. (A) Alignment of the two sequences indicates 37% identity (asterisks). Boxes denote the short N-terminal domain (S1), the first (M1) and second (M2) transmembrane domains, and the cytoplasmic helical domain (S3). The region of M1 forming the main gate is shaded in gray. The alignment is truncated at the end of EcoMscL; thus, 23 extra residues at the C-terminal segment of TbMscL are not shown. (B) The TbMscL crystal structure and the model of EcoMscL built by homology. Upper panels represent the views from the top (periplasm) and the bottom panels are the side views. One subunit in each structure is represented as striped rods to show the topology of the polypeptide comprising S1, periplasmic loop

The TbMscL structure became a cornerstone for the field as it gave the first structural basis for detailed exploration of channel gating driven by membrane stretch. TbMscL shares the same domain organization and 37% identity with its *E. coli* counterpart (Fig. 1A). Although mycobacterial MscLs form a subfamily somewhat distant from the group of closest homologs clustering near EcoMscL (Maurer et al., 2000), the degree of similarity clearly permits use of the crystal structure to interpret the broad phenomenology obtained primarily in the *E. coli* version of the channel. It would be a fair assumption that a large-scale transition leading to exactly the same conductance in both homologs is achieved via similar motions of key structural elements. Before we consider alternative structural models of MscL gating, we have to describe the major experimental findings that guide the quest for the open conformation and transition pathway.

MscL ACTIVITIES IN NATIVE MEMBRANES AND LIPOSOMES

Activating Tension

MscL is activated directly by tension in the lipid bilayer in which the protein is embedded. Upon a strong osmotic downshift, hydrostatic pressure building up inside the cell causes a distension of the elastic cell wall and eventually stresses the inner membrane. MscL opens at tensions just below the lytic threshold, under shocks of 300 to 800 mosM in magnitude, depending on the strain and presence of other channels. The conformational transition occurring under such conditions has been demonstrated by a drastic increase in accessibility of residues buried at rest (Batiza et al., 2002), and MscL function as a release valve has been supported by potassium efflux measurements in whole-cell experiments (Blount et al., 1997). Peculiarly, when expressed in *Vibrio alginolyticus*, EcoMscL confers considerably higher osmotic stability to this marine bacterium, which is naturally prone to osmotic lysis under downshifts beyond 500 mosM (Nakamaru et al., 1999).

In a typical patch-clamp experiment, channels are stimulated by suction applied to the pipette. Membrane tension (γ) generated in this manner is related to the pressure gradient across the patch (Δp) and the radius of curvature (r) by the law of Laplace ($\gamma = \Delta p \times r/2$). Imaging of large liposome patches while recording currents from reconstituted MscL showed that the midpoint of channel activation (pressure at which half of channels are open, $\gamma_{1/2}$) corresponds to 11.8 dynes/cm. It should be noted that this tension exceeds the mechanical strength of many lipid bilayers (Olbrich et al., 2000). Independent measurements of water permeability rates through MscL in large unilamellar phosphatidylcholine vesicles with the pipette aspiration technique suggested the beginning of activation at tensions of about 5 to 6 dynes/cm, with the midpoint near 9 dynes/cm (E. Evans, personal communication). A comparison of MscL activation pressures in spheroplasts with those of MscS, previously shown to be half-activated at tensions of about 5.5 dynes/cm (Cui and Adler, 1996; Sukharev, 2002), suggested $\gamma_{1/2}$ for MscL in its native environment to be around

---

(S2), cytoplasmic domains (S3), and M2-S3 connecting linkers. As seen from the alignment, the loops as well as S2-M3 linkers are divergent in the two species, and the correspondence between the model and the crystal structure is best in the more conserved transmembrane part. The N-terminal domain and the C-terminal end of TbMscL were not fully resolved; thus the S1 of EcoMscL was modeled anew as a short bundle of amphipathic helices.

10 dynes/cm (Chiang et al., 2004a). Gauging MscL against MscS, usually present in the same spheroplast patch, has been a simple and practical procedure for comparing activation points in different mutants and homologs (Blount et al., 1996b).

Conductive Properties and Permeability

MscL does not discriminate between anions and cations and its unitary conductance increases linearly with the specific conductivity of the bath electrolyte. Such a nonsaturable behavior suggests a wide aqueous pore, where ion movement is similar to that in the bulk. On the assumption that the channel pore is a 50-Å-long cylinder, the conductance measured in different buffers predicts a pore of 30 to 35 Å in diameter (Cruickshank et al., 1997; Sukharev et al., 1999b). Measurements of MscL single-channel conductance in the presence of polylysines of different sizes revealed a strong interference by molecules with an average diameter larger than 37 Å, suggesting that this might be close to the size of the pore or pore vestibule (Cruickshank et al., 1997). Having such a big pore, MscL was proposed to mediate the release of small proteins such as 12-kDa thioredoxin but conspicuously was found to be practically impermeable to glutamate, trehalose, or glycine-betaine as no difference in osmotic release of these substances from wild-type and *mscL*-null bacteria was detected (Ajouz et al., 1998). The report of thioredoxin permeation could not be confirmed (Vazquez-Laslop et al., 2001). It appeared that ATP is not freely permeable through native MscL either, but its permeability increases upon removal of the cytoplasmic domain (Anishkin et al., 2003).

Gating Patterns

The primary kinetic analysis of closed-to-open transitions in spheroplast patches at a low probability of being open, Po, indicated that the open dwell time distribution can be fit by three exponents with characteristic times of about 37, 6, and <1 ms (Blount et al., 1996b). This distribution points to the possibility that MscL has several kinetically distinct open states, but not necessarily because these responses were recorded from multichannel populations, usually nonhomogeneous as will be discussed below. True single-channel patches recorded in liposomes displayed distributions of both open and closed dwell times close to monoexponential and were fit satisfactorily in the simple two-state ($C \leftrightarrow O$) approximation. Both rate constants for opening (k_{on}) and closing (k_{off}) were tension dependent, with k_{on} growing with tension more steeply than k_{off} declining (Sukharev et al., 1999b).

Time-resolved recordings detected a number of short-lived subconducting states, with three of them visited most frequently. These sublevels were interpreted as gating intermediates, and a linear five-state gating scheme, $C1 \leftrightarrow S2 \leftrightarrow S3 \leftrightarrow S4 \leftrightarrow O5$, was initially proposed (Sukharev et al., 1999b). The more extended amplitude histogram analysis (Chiang et al., 2004a) suggested that the number of short-lived substates is likely to be more than three. The detectable peaks on all-point histograms suggested the presence of sublevels with amplitudes of 0.13, 0.22, 0.45, 0.70, and 0.93, relative to the fully open conductance (Fig. 2). These peaks represent short-lived subconducting states visited en route up to the open state or back down to the closed state, as well as characteristic brief downward deflections from the fully open state. In addition to these short substates, long-lived substates with relative amplitudes of 0.78 and 0.64 were observed, more prominent at high hyperpolarizing voltages. Their occupancy increased with tension

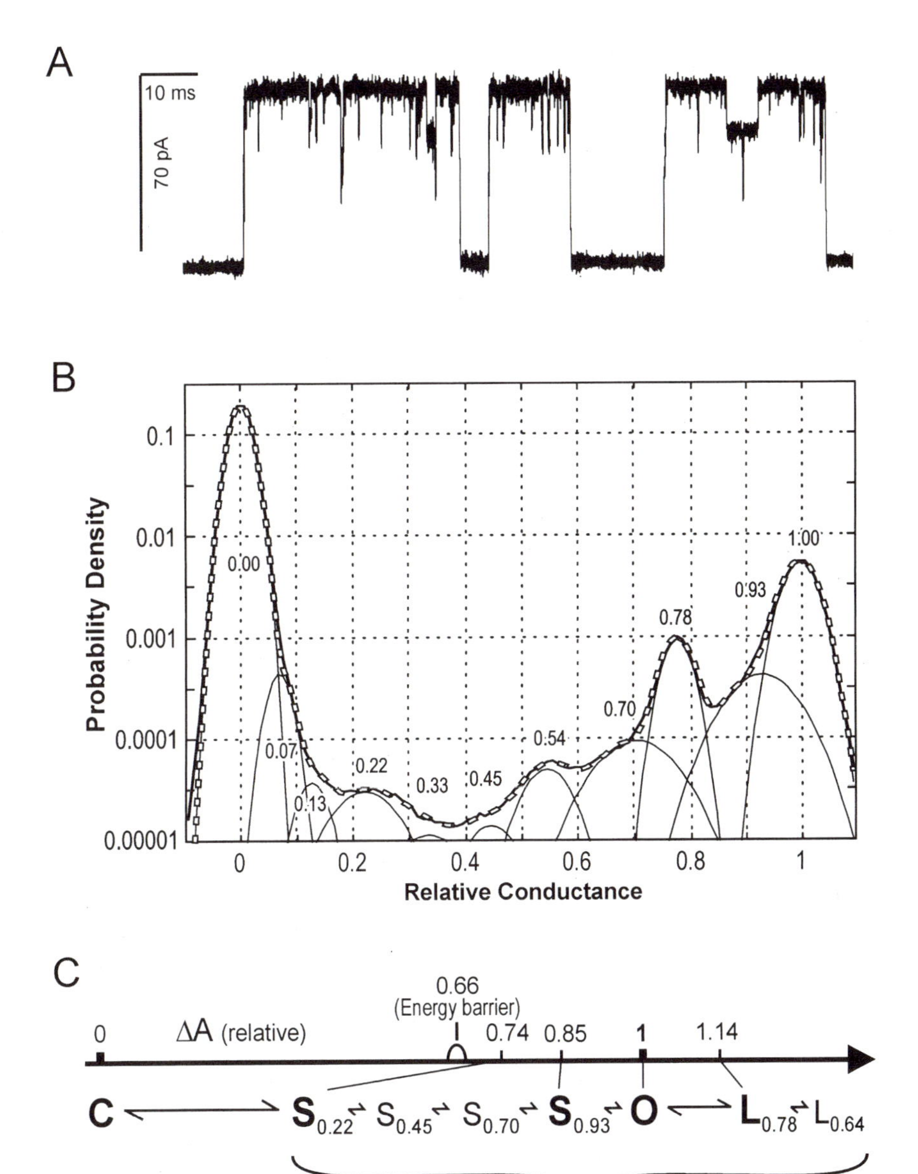

Figure 2. MscL activities and conduction states. (A) A portion of a typical current trace recorded at –20 mV with openings shown as upward transitions. Downward deflections from the fully open state represent short and long subconducting states. (B) Amplitude histogram based on a 5-min recording containing about 10^3 opening events. The histogram was fit with 11 gaussian peaks defining the amplitudes of the closed, fully open states, and nine intermediate levels. (C) The linear kinetic scheme aligned with the protein area scale chosen as a reaction coordinate. The positions of each conducting state and the main energy barrier were concluded from analyses of state occupancies and subtransition rates as a function of tension.

more steeply than the occupancy of the fully open state, suggesting that the long substates represent even more expanded conformations, which are not intermediates between the closed and open states in the gating scheme (Fig. 2C), but rather alternative open states (Chiang et al., 2004a).

The attempts to resolve the kinetics of elementary transitions in the complex multistate behavior of MscL prompted Shapovalov and Lester to design a special high-bandwidth recording system with a response time approaching 3 μs. They demonstrated that the time course of conductance changes between the substates usually occurs with characteristic times longer than 3 μs. The larger "jumps" of conductance correlated with longer transition times, which were between 10 and 30 μs for the rarely observed straight $C \rightarrow O$ transitions without discernible sublevels. In most instances, however, the opening transitions were composite and a multitude of sublevels were detected, confirming the need to account for at least nine intermediate levels (Shapovalov and Lester, 2004).

An important finding regarding the subconducting states was that their combined occupancy at low voltages was always about 30% of the total open (nonclosed) time. Because their amplitudes are smaller than that of the fully open state, their presence inevitably reduces the integral current, but this reduction (about 6% of the maximal current of an ideal channel without substates) is nearly independent of tension or open probability. This finding validated the use of integral current as an accurate measure of open probability in a wide range of tensions and enabled simplification of the multistate kinetic scheme to two effective states ($C \leftrightarrow O_{eff}$ [Fig. 2C]) for the purpose of gross thermodynamic description of the gating (Chiang et al., 2004a).

The Energetic and Spatial Scales of the Gating Transition

The simplest physical model of a tension-gated MS channel implies that the protein embedded in the lipid bilayer undergoes a $C \rightarrow O$ transition accompanied by in-plane expansion, ΔA. In the absence of tension, the ratio of open and closed state occupancies can be related to the intrinsic free-energy difference between the states (ΔE) with the Boltzmann law. When the protein complex expands under tension γ, the energy difference between the states is biased by the work of external forces, γΔA, and the ratio of probabilities can be written as:

$$Po/Pc = \exp(-[\Delta E - \gamma \Delta A]/kT) \qquad (1)$$

This equation describes the dose-response relationship for a generic two-state MS channel (Howard et al., 1988; Sachs, 1992). By measuring Po at different tensions one can extract both the energy change, ΔE, and the spatial parameter, ΔA, defining the slope of open probability on γ. Fitting of MscL dose-response curves recorded in multichannel liposome patches using the assumption that all channels are equal produced a ΔE of 19 kT and a ΔA of 6.5 nm^2 (Sukharev et al., 1999b). The latter value appeared to be too small to accommodate the opening of a pore 10 nm^2 in cross-section. A revisit of this problem by measuring dose-response curves in spheroplasts (Chiang et al., 2004a) revealed that even a slight nonhomogeneity of channels in patches may pose a substantial problem for accurate determination of the slope of MscL activation curves. Statistical simulations indicated that a small scatter of either ΔE or ΔA (standard deviation ≈ 5% of the mean) is enough to reduce the midpoint slope of the combined population response curve twice, leading to a severe underestimation of both ΔA and ΔE. The simulations also predicted that only the initial slope of left-most parts of such dose-response curves is close to the intrinsic slope for an individual channel, whereas the slope near the midpoint can be substantially lower and should not be trusted. Fitting the initial slopes of inflected experimental dose-response curves produced ΔA of 20 nm^2 and ΔE of 51 kT as gross parameters for the complex $C \rightarrow O$ transition (Chiang et al., 2004a). With this approach, the estimated ΔA is highly consistent with the area change (23 nm^2) predicted by molecular models of the two end states described below. A similar value for channel expansion (ΔA of 20 nm^2) was deduced from the slope of water permeability on tension in pipette aspiration experiments with MscL-containing liposomes (E. Evans, personal communication).

Substate Occupancy and Kinetics of Subtransitions Reveal Expanded Low-Conducting Intermediate States

Analysis of occupancies of substates and rates of subtransitions as functions of tension provided valuable information about the positions of intermediate states and major barriers on the reaction coordinate. Note that the Boltzmann-type relationship (equation 1) can be applied to any pair of states and the spatial and energy parameters ΔA and ΔE for each subtransition can be deduced. The combined data of detailed kinetic (Sukharev et al., 1999b) and equilibrium occupancy (Chiang et al., 2004a) analyses indicated that the entire $C \rightarrow O$ transition in wild-type MscL can be kinetically separated into two distinct steps. The first subtransition from the closed to a low-conducting substate ($C \rightarrow S$; S here denotes $S_{0.13}$ or $S_{0.22}$) was steeply tension-dependent and always rate-limiting. This subtransition was followed by a chain of fast transitions from the sub to the fully open state ($S_{0.13} \rightarrow \cdots \rightarrow O$), collectively characterized by high and virtually tension-independent rates. The slopes of major rate constants and occupancies on tension strongly suggested that the $C \rightarrow S$ subtransition is characterized by the major energy and area gain. A total expansion of 75% occurring at this step yields only about 20% of conductance, whereas the following $S \rightarrow O$ transition gains more than 80% conductance and is associated with only a minor expansion. The disparity between the area and conductance changes indicated that the dramatic increase in the outer dimensions of the channel protein occurs before the major opening of the conductive pathway. The low-conducting expanded intermediate state must involve a structural element that partially occludes the pore when the barrel is greatly expanded. This structural element can be provisionally called a second gate.

PROBING THE STRUCTURE

Early Mutagenesis, Deletions, Proteolysis, and Coreconstitution from Separate Parts

The primary deletion mutagenesis (Blount et al., 1996b; Hase et al., 1997) has shown that the N-terminal domain of MscL is a sensitive part of the protein possibly involved in gating and/or assembly. Small (three-residue) deletions were tolerated; extensive modifications within the eight-amino-acid stretch remarkably changed the gating phenotype, whereas the 11-residue deletion completely abolished function. As discussed below, the

N-terminal (S1) domain is mutable (Maurer and Dougherty, 2003), but at the same it is one of the most conserved parts of the protein. The C-terminal domain also contains a conserved stretch of about 12 residues, but it appeared to be somehow dispensable for the main channel function. The Δ110-136 deletion mutant formed functional channels, with a more pronounced variability of gating kinetics and higher occupancy of long-lived alternative open states of amplitude lower than the fully open state (Hase et al., 1997; Anishkin et al., 2003). A larger deletion (Δ104-136) removing the characteristic cluster of charges (R_{104}KKEEP) completely abolished activity.

Deletion of 25 residues comprising the poorly conserved M1-M2 loop produced a nonfunctional channel (Hase et al., 1997). A less radical probing of the loop by specific proteases has revealed that cleavage at a number of sites considerably reduces the pressure threshold, shifting the activation curve to the left (Ajouz et al., 2000). A coreconstitution of the separately expressed N-terminal and C-terminal halves of the protein containing the M1 and M2 helices, respectively, resulted in functional channels lacking continuous loops, which were also characterized by a reduced activation threshold (Park et al., 2004).

Probing different regions with point mutations revealed that perturbing the loops with aromatic substitutions (at position Gln56, in particular) dramatically increases the open dwell time, whereas deletion of this residue results in a very "stiff" channel with a right-shifted activation curve. Charge reversal in the first transmembrane domain (K31E/D) shifted the pressure threshold to the left, but at the same time increased the rate of transitions by two orders of magnitude, rendering a fast-flickering phenotype (Blount et al., 1996b; Blount et al., 1997). Thus, several complementary approaches agreed that the loops influence the activation tension, probably acting as springs countering the channel barrel expansion. The results of primary mutational analysis led to the general conclusion that essentially all domains of this small protein, except for the C-terminal tail, are somehow involved in defining the tension sensitivity and kinetics of gating.

Random Mutagenesis Points to the Gate Region and Areas Receiving Tension from the Lipid

The ingenious idea that perturbations of the gate region must result in "leaky" channels with an adverse effect on cell growth was successfully implemented by Ou et al. (1998). Error-prone PCR mutagenesis of the *mscL* gene was combined with screens for growth-suppressing phenotypes. Mutant-harboring colonies were selected that grew under noninducing conditions but were partially or completely retarded after induction of protein synthesis. The pattern of substitutions revealed that toxic effects had hydrophilic substitutions in the cytoplasmic (N-terminal) half of M1, rendering the predicted pore constriction more polar. The most severe growth phenotypes occurred for substitutions at equally spaced glycines Gly22, Gly26, and Gly30, which formed a characteristic "glabrous" strip on M1 helices. Electrophysiological survey showed that the majority of these mutations produced an easy-to-open GOF phenotype; thus the leaky mutants were in fact hyperactive channels with an inappropriately low pressure threshold. The following site-directed mutagenesis of residue 22 has demonstrated that the pressure threshold for channel activation increases in proportion with the hydrophobicity of the substitute at this position, consistent with the idea that the primary gate is formed by cytoplasmic halves of M1 helices and gating involves a "hydrophobic lock" mechanism (Yoshimura et al., 1999).

It should be mentioned that the partial redundancy of MscL and MscS in rescuing bacteria under acute osmotic shock initially precluded observation of the *mscL*-null phenotype on an *mscS*$^+$ background. Cloning of MscS and preparation of the double knockout (*mscL*$^-$ *mscS*$^-$) strain by Booth and coworkers (Levina et al., 1999) demonstrated that such bacteria are indeed osmotically fragile and reexpression of either of the channels alleviates the phenotype. This permitted development of screens not only for GOF but also LOF phenotypes, i.e., for alleles that produce assembled protein but do not rescue the osmotically challenged cells. The high-throughput screens for both GOF and LOF mutations were developed and implemented by the Dougherty laboratory (Maurer and Dougherty, 2001). The result of this extensive work was generation and primary characterization of a large library of mutants with multiple substitutions at essentially every residue (Maurer and Dougherty, 2003). While some of the isolated GOF mutations were the same as reported previously (Ou et al., 1998), multiple LOF mutations were found near the extra- and intracellular ends of transmembrane helices and in the loop region. The patterns of GOF and LOF mutations clearly indicated the importance of protein-lipid interactions, through which tension is conveyed to the protein and eventually to the gate. Interpretations of GOF and LOF phenotypes in the framework of current gating models will be discussed below.

MODELS OF THE GATING

Homology Models of EcoMscL

After crystallization of TbMscL, it was soon discovered that this ortholog is not particularly convenient for single-channel analysis in spheroplasts, as it activates only at extremely high tension (Moe et al., 2000). Instead of trying to rediscover all the gating phenomenology in TbMscL, a homology model of EcoMscL was built (Sukharev et al., 2001b), which was feasible because the two proteins share the same architecture and 37% identity. The model (Fig. 1B) retraces the prototype particularly well in the transmembrane region, forming a characteristic hydrophobic constriction (M1 gate) lined by side chains of Leu19 and Val23 homologous to Leu17 and Val21 in TbMscL. The extensive buried contact between M1 and M2 helices (belonging to adjacent subunits) was preserved and additionally stabilized by the identified Lys31-Asp84 salt bridge absent in TbMscL. The periplasmic loop, which is dissimilar in the two orthologs, was modeled only to satisfy general modeling criteria (Sukharev et al., 2001b). The very N-terminal end, unresolved in the original TbMscL structure, was represented as a bundle of short helices (S1) proposed to act as the second gate. The necessity of a second gate was previously suggested by the presence of highly expanded low-conducting intermediates (Sukharev et al., 1999b), which could be hardly explained by a partial opening of the primary M1 gate (constriction). The cytoplasmic C-terminal (S3) helices are connected to the M2 helices with extended and flexible linkers as in TbMscL. The S3 helices have conserved stretches forming characteristic amphipathic helices. While forming a fivefold helical bundle as in the Tb prototype, the S3 domains in EcoMscL were reoriented and put together in a structure similar to that of the cartilage oligomeric matrix protein (Anishkin et al., 2003), thus better satisfying the hydrophobicity criterion. It should be mentioned that another homology model of EcoMscL has been recently created (Valadie et al., 2003) using essentially the same alignment and the Modeler algorithm (Sali et al., 1995). The new model, having shorter M2 helices, shows a closer spatial correspondence to the crystal structure of TbMscL; however, the "normal mode" analysis suggested similar flexibility parameters and preferential directions of distortions for both the previous and the new model (Valadie et al., 2003).

Models of the Transition

The exceptionally large conductance of MscL implies that the channel complex must undergo a large-scale conformational change, but the nature of the change is not evident from the compact closed-state model. The originally proposed barrel-stave (10-helix pore) model of open MscL pore (Chang et al., 1998; Spencer et al., 1999; Batiza et al., 1999) is shown in Fig. 3 (left). Although simple and vivid, this model did not satisfy several key criteria; e.g., the predicted protein expansion and pore conductance were too small and M2 helices were too hydrophobic, especially in their cytoplasmic halves, to line an aqueous pore (Sukharev et al., 2001b). In an alternative iris-like (five-tilted-helix pore) model (Fig. 3, right), the opening process was presented as a gradual channel expansion associated with tilting and outward movement of pairs of transmembrane helices (Sukharev et al., 2001a). By introducing small (~1-Å) displacements and changes in helical tilts, 11 sequential conformations were generated representing the continuous opening process (Sukharev et al., 2001b). Three of these models of the "relaxed" closed (2nd), intermediate (9th), and fully open (11th) states are illustrated in Color Plate 28 (see color insert). The entire opening process proceeds in two main stages, the expansion of the barrel (opening of the hydrophobic M1 gate) followed by separation of the N-terminal (S1) gate. Up to the ninth conformation, in which the transmembrane helices are substantially tilted and separated (M1 gate already open), the S1 helices remain associated, largely occluding the pore. This state of the barrel models a low-conducting expanded state. As the barrel expands further, the stress building up in S1-M1 linkers partially separates and then completely disrupts the S1 bundle, leading to the fully open conformation. The S1 domains freed from the bundle swing and attach to the hydrophobic surfaces of highly tilted M1 and M2 helices; their "docking" was proposed to additionally stabilize the open conformation of the barrel. In

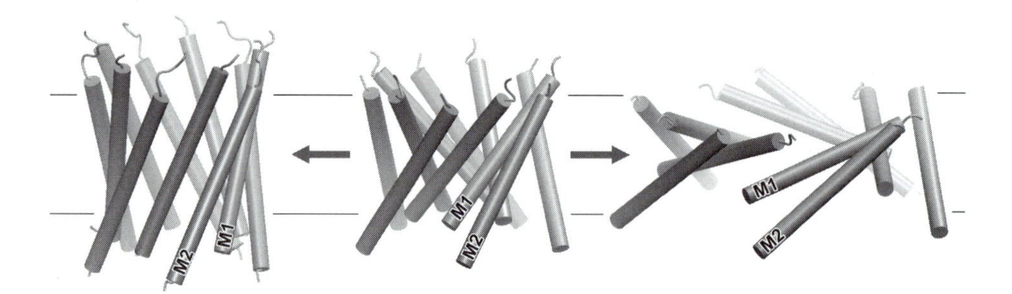

Figure 3. Transmembrane helices of MscL in the crystal-like closed conformation (center) and two possible pathways for channel expansion. The 10-helix barrel-stave model with 10 almost parallel helices all participating in pore lining is shown on the left. The alternative five-helix tilted model is shown on the right. The highly tilted arrangement permits a larger pore lined primarily by M1 helices.

the initial models, the cytoplasmic S3 domains were also presented as separating (Sukharev et al., 2001a); however, more recent studies (Anishkin et al., 2003) proved that cytoplasmic domains remain in a stable bundle-like structure in all conformations, forming a prefilter at the pore entrance. In some of the literature, the iris-like model of MscL opening has been called the SG (Sukharev-Guy) model. The fully open state (conformation 11) predicts an aqueous conductive pathway of nearly 10 nm^2 in cross-section, satisfying the channel conductive properties; the overall in-plane area change of the protein associated with the $C \rightarrow O$ transition is 23 nm^2, produced by the increase of helical tilt from about 32 to 71° relative to the membrane normal and a 13-Å radial displacement away from the pore axis (Sukharev et al., 2001b). A detailed analysis of spatial organization and tilting of helices associated with channel gating has been presented by Spencer and Rees (2002).

Computational Studies of the Transition

Equilibrium molecular dynamics (MD) simulations of the crystal-like TbMscL embedded in phosphatidylethanolamine (PE) bilayer (Elmore and Dougherty, 2001) illustrated the overall stability of the transmembrane region and a pronounced time-dependent evolution in the C-terminal and loop regions at neutral pH. Indeed, TbMscL was crystallized at pH 3.6, where the protonated state of acidic residues stabilized the particular bundle-like arrangement of C-terminal helices. These domains tended to separate after acquiring charges. Introduction of the V21A GOF mutation in simulations led to a slight widening of the pore radius, whereas the Q51E GOF mutation changed the character of intersubunit interactions in the loop region. An interesting effect of water orientation in the channel has been noticed. An extensive hydrogen bonding of both extracellular loops and C-terminal regions to the surrounding PE molecules suggested a specific character of protein-lipid interactions. Further studies showed that TbMscL forms many fewer hydrogen bonds with a phosphatidylcholine (PC) bilayer (Elmore and Dougherty, 2003). An independent MD study (Gullingsrud and Schulten, 2004) has shown that the difference between PE and PC bilayers is not only in the extent of headgroup to protein hydrogen bonding, but also substantially different lateral pressure profiles, as the two lipid species are characterized with different spontaneous curvatures.

In an effort to envision the transition pathway between the two modeled end states (fully open and fully closed), Kong et al. performed a targeted molecular dynamic simulation and confirmed that the barrel expansion should precede the separation of N-terminal domains if their predicted conformations and interactions are correct (Kong et al., 2002). The increase of the inner pore diameter during the simulated channel expansion lagged behind the increase of the outer dimension of the barrel supporting the presence of expanded low-conducting intermediate. Further attempts to implement less biased approaches by applying anisotropic pressure to the system mimicking membrane tension (Gullingsrud et al., 2001; Bilston and Mylvaganam, 2002; Colombo et al., 2003), or a steering force, led to expanded conformations generally associated with increase in helical tilt. Because the amount of tension applied in such simulations was 1 to 2 orders of magnitude larger than used in real experiments, the end conformations were often distorted with a substantially perturbed secondary structure. The detailed "normal modes" analysis of concerted harmonic oscillations of much smaller amplitudes also concluded that helix tilting around the pivotal point near residue 23 (EcoMscL) would be the most probable pathway for channel expansion (Valadie et al., 2003).

A series of studies by Gullingsrud and Schulten utilized steered MD to observe the transition from the initial closed to expanded open-like state with the steering force applied only to specific groups of atoms on the barrel periphery (Gullingsrud and Schulten, 2003). Importantly, the protein regions for application of tension were chosen in accordance with the lateral pressure distribution in the lipid bilayer precomputed in separate trials (Gullingsrud and Schulten, 2004). The studies concluded that the pathway for the opening transition is sensitive to the point of force application and the barrel expansion must be a result of tilting and outward movement of transmembrane helices driven by tension acting at the boundary regions between the hydrocarbon and polar group layers of the bilayer. An interesting observation was that in the closed state the constriction region of the pore was completely dehydrated and the partial opening coincided with pore wetting.

EXPERIMENTS VERSUS MODELS

Open Conformation of the Barrel Tested by Disulfide Trapping Experiments

The experimental testing of the iris-like transition (SG) model was approached by the analysis of distances between pairs of residues in the resting and open conformations. Single or double cysteine substitutions at several positions were made to test how probable the hypothetical conformations in disulfide trapping experiments really are. Color Plate 28 summarizes the cross-links confirmed for the resting and open states of EcoMscL, mapped on models for the two conformations. The intersubunit cross-links were identified by Western blotting, and several of them were tested for their functional impacts on the gating in patch-clamp experiments (Sukharev et al., 2001a; Betanzos et al., 2002). The tight packing of M1 helices in the resting state predicts that they are touching each other at residues I24 and G26. Cysteines in these positions readily cross-linked the entire pentamer with mild oxidation, and this abolished channel gating. When the barrel expands, M1 helices slide one along the other and the contacts between adjacent helices separate by four helical turns, bringing A20 on one helix close to L36 on the neighbor. The highly specific A20C-L36C cross-link was trapped in whole-cell osmotic shock experiments under stronger oxidizing conditions (Betanzos et al., 2002). The amount of cross-linked pentamers rose with the strength of the downshock. Patch-clamping of this double mutant under reducing conditions revealed nearly normal single-channel currents. Application of 0.08 mM iodine in the pipette led to either channel inactivation (presumably due to coupling of A20C to A20C in the closed state) or locking in a conducting state, consistent with the highly expanded tilted conformation satisfying the A20C-L36C bond.

Unlike the A20C-L36C bond, which is strictly state-specific, I32C and N81C cross-link in both open and closed states. This bond covalently links M1 and M2 helices belonging to adjacent subunits, and complete pentamers were observed upon mild oxidation. In the model, the adjacent M1 and M2 helices are predicted to remain in contact throughout the transition. Although this double mutant initially showed faster kinetics than wild-type MscL under ambient or reducing conditions, formation of cross-links in the presence of an oxidant did not abolish gating, indicating that M1 and M2 helices do not change their relative position substantially during gating and likely tilt together (Betanzos et al., 2002). This result does not exclude the possibility that normally the helices experience some limited rotations and scissoring motions as well (Yoshimura et al., 1999; Perozo et al., 2002a; Gullingsrud and Schulten, 2003).

While the opening pathway presented above was modeled as a symmetrical barrel expansion, the picture of the opening may be more complex. It was shown that with the V15C substitution, TbMscL becomes considerably "softer" (acquires GOF phenotype) and permits extended recordings in spheroplasts (Shapovalov et al., 2003). Although these cysteines on different subunits, according to the crystal structure, must be 12 Å apart, activation of the channel under ambient oxidizing conditions led to formation of cross-links manifested as stable or flickery subconducting levels. This behavior illustrated that during the gating process a multitude of asymmetric conformations may occur, thus permitting encounters of distant residues.

It must be noted that the biochemically detected cross-links, although suggestive of the proximity for certain residues, still may trap rare and nonnative conformations due to the irreversible character of reaction under oxidizing conditions. With this respect, patch-clamp recordings on corresponding mutants performed under oxidizing, ambient, or reducing conditions become equally important and in certain instances more informative as they directly report on the channel behavior before and after the cross-link formation (Sukharev et al., 2001a; Betanzos et al., 2002; Shapovalov et al., 2003; Levin and Blount, 2004).

Site-Directed Spin Labeling and EPR Experiments: Lipid-Borne Effects on MscL Conformation

MscL remains stable and functional in liposomes made of exogenous lipids. Initial characterization of MscL using scanning cysteine mutagenesis, site-specific spin labeling, and EPR spectroscopy demonstrated that the transmembrane region of EcoMscL has essentially the same organization as TbMscL, validating the correctness of the homology-based alignment of the EcoMscL model. An unexpected immobility of the carboxy terminus of the M2 helix pointed to specific intramolecular or protein-lipid interactions in this region (Perozo et al., 2001). It was clear that finding a chemical factor that would shift the equilibrium toward the open state would open a precious opportunity to compare the EPR spectra of two different conformations. Having previous experience with MscS-type channels, which were opened in patch-clamp experiments by asymmetrically added amphipathic substances (Martinac et al., 1990), Perozo and Martinac found that MscL can also be "pried" open by asymmetric application of lysolipids. Patches of PC liposomes with reconstituted MscL demonstrated substantial spontaneous activity upon exposure to 3 μM lysophosphatidylcholine (LPC) in the bath (Perozo et al., 2002b). Under similar conditions, LPC externally added to liposomes dramatically increased the mobility of the probe attached to residues in the pore constriction and decreased the spin-spin coupling between probes attached to the same residue on M1 on different subunits, indicating that the pore has drastically expanded. Spontaneous curvature asymmetrically introduced in one leaflet was considered to be the main physical factor that biased the equilibrium toward the open state. A partial compensation of curvature stress by PE led to a retardation of LPC effects on the mean width of the pore as reported by EPR spectral parameters (Perozo et al., 2002b). The spectra obtained with probes at every residue position in the transmembrane region under resting and curvature-inducing conditions resulted in a complete set of water and lipid accessibilities for each residue. A model of the open state was developed to satisfy these data. The model predicted a highly expanded and at the same time flattened structure, with a pore of at least 25 Å in diameter lined predominantly with the M1 helix. Most of the M2 helix, except for the very ends, remained in a buried or lipid-exposed state. The positions of M1 and M2 in this model resemble those in the SG model, except that M1 rotates clockwise by 110° about its axis during activation (Perozo et al., 2002a), whereas in the SG model M1 rotates about 27° counterclockwise (when observed from the periplasm).

The predicted strong flattening of the transmembrane part was highly consistent with a lower pressure threshold for channel activation in thinner bilayers made of short-chain lipids (PC14) and a higher threshold in thicker (PC20) bilayers relative to that in a bilayer of regular thickness (PC18). This strongly suggested that the distortion of boundary lipids surrounding the flattened channel brings an extra energetic cost to the transition. In addition, EPR measurements suggested that MscL reconstituted in a thick bilayer is packed tighter, being driven deeper to the closed state by the lipid environment. Peculiarly, the channel reconstituted into the thinner (PC14) bilayer was also packed tighter at the extracellular end, but helices were slightly more separated at the cytoplasmic end of the barrel. This conformation was interpreted as an intermediate state between the closed and open conformations (Perozo et al., 2002a). A recent theoretical treatment of hydrophobic mismatch explained such a constricting effect by line tension in the perturbed lipid layer near the protein boundary (Wiggins and Phillips, 2004). MD simulations have shown that thinning of the surrounding bilayer may lead to a kinking of helices and overall distortions of the barrel as a possible mode of protein reaction to a strong hydrophobic mismatch (Elmore and Dougherty, 2003). More studies are necessary to find the correspondence between the effects of short-chain lipids and lateral tension on the conformation of the flexible MscL barrel. From the above experiments it was evident that thickness alone is unable to drive the opening transition in MscL, whereas asymmetric area and curvature stresses induced by intercalated LPC can produce opening.

Other Support for the Tilting Model

Aromatic residues (tryptophans or tyrosines) that anchor the ends of transmembrane helices to membrane interfaces in many integral proteins (White and Wimley, 1999) are not common in MscLs. More common lipid-facing residues in the MscL family are phenylalanines, which reside with about equal frequency within the hydrocarbon layer or at its boundaries. This characteristic absence of "anchors" on both sides may be vital as strong helical reorientation may be necessary to open the channel. In EcoMscL, a single tyrosine (Y75) per subunit resides at the extracellular side of M2. It is absent in TbMscL, which instead has a single tyrosine (Y87) on its cytoplasmic side. Capping the M2 helix of EcoMscL on both sides (F93Y/W) increased the pressure threshold for activation, which correlated with a partial LOF in osmotic shock survival assays (Chiang et al., 2004b). The data show that restraining helical positions in MscL by introducing specific protein-lipid interactions at membrane interfaces compromises MscL function, which is more consistent with the tilting model of MscL expansion rather than the barrel-stave model.

A direct visualization of expanded and open-like conformations of MscL has been obtained using atomic force microscopy (Ornatska et al., 2003). The MscL homolog from *Salmonella enterica* serovar Typhimurium was reconstituted in supported alkylsilane monolayers in which lateral pressure was adjusted by varied density in Langmuir-Blodgett deposition. In compressed monolayers the protein looked more compact, whereas in more

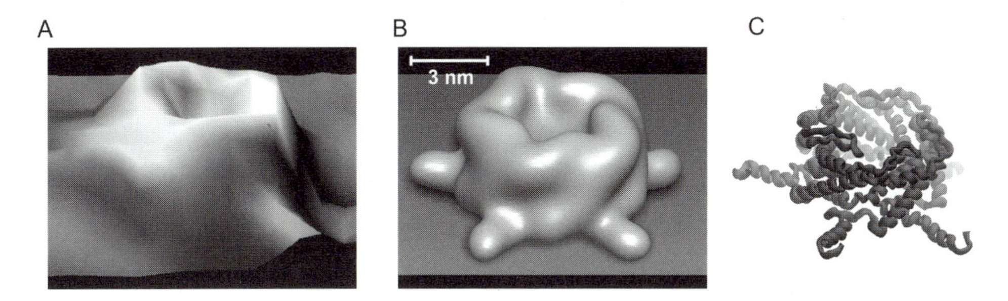

Figure 4. The AFM image of the expanded conformation of MscL (A) next to the space-filled "tilted" model of the open MscL (B). (C) Arrangement of helices with S3 domains separated. The AFM image is taken from Ornatska et al. (2003), where the MscL homolog from *S. enterica* serovar Typhimurium was studied in supported alkylsilane monolayers. Courtesy of Vladimir Tsukruk.

spread films channels were flat and looked more like an open doughnut. An atomic force microscopy (AFM) image of MscL obtained at low lateral pressure mimicking high tension is shown next to the space-filled model of the open conformation in Fig. 4. The image resembles the tilted model and seems to be inconsistent with the barrel-stave arrangement, in which the helices had to be staggered normally to the surface, forming tall narrow cylinders.

Residues Accessible through the Pore

Probing residue accessibilities using cysteine substitutions and methanethiosulfonate (MTS) reagents is a well-established technique in studies of ion channels (Karlin and Akabas, 1998). Yoshimura et al. (2001) were first to apply this approach to MscL. They found that residue 22, predicted to be deep in the pore at the interface between the adjacent M1 helices, is in fact accessible from the periplasmic side in the resting state. From the cytoplasmic side this locus could be reached only when the channel is open. Consistent with the previous mutagenesis results (Yoshimura et al., 1999), charged or hydrophilic MTS adducts to this cysteine decreased the activation threshold, whereas hydrophobic MTS adducts tended to increase it. The following study demonstrated that cysteines in position 19 (L19C) are inaccessible in the closed state but become accessible through the open channel from either side (Batiza et al., 2002). By shocking cells osmotically with simultaneous application of the reagent, the authors showed that the reaction of C19 with the positively charged MTS-ethyltrimethylammonium (MTSET) created a severe GOF phenotype, making this channel highly toxic for the cell. This was a direct indication that MscL indeed can be opened in vivo by an osmotic downshift of about 400 mosM in an *mscS*-null strain and at twice as strong shocks when MscS is present. The buried state of L19 is consistent with the position of homologous L17 in the TbMscL closed-like structure, whereas reactivity within the open channel suggests an orientation of the M1 helix in the open state that exposes this site to reagents. Employing a similar strategy combined with the scanning cysteine mutagenesis of both transmembrane helices, Blount and coworkers (Bartlett et al., 2004) have recently shown that cysteines substituting not only for L19 but also for V24 become accessible through the open pore under osmotic shock. Partial exposure of G22, V23, A27, and V37 sites observed at rest was augmented by osmotic shock, whereas G26, G30, S34, and A38 sites were accessible even in the resting state. The groups of residues accessible to MTS in different states thus suggested the facets of the M1 helix exposed to the pore lumen in the open and closed states. Mapping the fully exposed residues at rest (26, 30, 34, and 38) and residues more exposed under shock (19, 22, 23, 24, and 27) on the M1 helix revealed that the two groups are organized as two spiral strips on the same facet of the helix. The spiral arrangement of accessible sites led to a proposition that glycines G22, G26, and G30 line the closed pore at rest and that M1 helix should undergo a clockwise rotation during the opening transition (Bartlett et al., 2004).

Another attempt to probe the pore interior involved histidine substitutions in that region (Iscla et al., 2004). In the homopentameric structure of MscL, a single substitution in the M1 helix brings five identical histidines close, thus creating a high-affinity binding site for divalent ions. Coordination of an ion by multiple histidines is expected to stabilize the conformation that favors binding. Channels with histidines at positions 13, 22, 23, 24, and 26 showed increased activation pressure thresholds in the presence of 5 mM Ni^{2+} or Cd^{2+}. The effect of metals on activation thresholds was considerably stronger in the G22H and G26H mutants. At the same time, the effect of ions on gating of V23H MscL was relatively weak. The authors suggested that G26 forms the real constriction, not the previously proposed V23. Note that the highly conserved V23 occluding the pore in the EcoMscL model (Sukharev et al., 2001b) is in the position homologous to V21 forming the pore constriction in the TbMscL crystal structure (Chang et al., 1998). Although interesting and provocative, the data seem to be insufficient to question major predictions of the crystal structure. A proper reconciliation between pore geometry, helical dynamics, hydration, and the stereochemistry of metal coordination in particular mutants requires more data and structural analysis.

The orientations of helices in the SG model of the EcoMscL closed and open conformations generally explain the accessibility data (Fig. 5). Residues L19 and V23 line the dehydrated constriction in the closed state; correspondingly, the L19 site deep in the pore is completely buried from polar MTSET, whereas V23, also located in the dehydrated zone, is only partially exposed. Residue V24, pointing away from the pore, must also be inaccessible in the closed state. Residue A27 facing the lumen is partially accessible as it resides near the boundary of the dehydrated zone. From the data presented by Bartlett et al., one may conclude that the pore-exposed residues K31 and S35 are also accessible to MTSET, but the modifications at these positions evoked only mild changes in growth rate of host bacteria, the only parameter used to detect the accessibility (Bartlett et al., 2004). The right panel of the figure illustrates the orientation of the pore-lining helices in the open state. Because the wide pore is lined mostly by M1, the neighbors surround these helices less and therefore larger surfaces are exposed. The positions shown to be accessible to MTSET under osmotic shock (L19, V23, V24, G26 A27, S34, V37, and A38) are all seen from the lumen. The complete transition between the two states shown in Fig. 5 is associated with a net 27° rotation of the M1 helix about its main axis (counterclockwise when looking from the periplasmic side). This amount of rotation is considerably less than that concluded from spin-labeling data (Perozo et al., 2002a). The EPR-derived model predicts that the M1 helix turns clockwise by more than 110°, exposing V17, V21, V25, A28, and I32, which were not among residues found accessible to MTSET (Bartlett et al., 2004). A large rotation of M1 about its axis during gating is not consistent with the A20C-L36C disulfide bridge, and the observation that linking I32C on M1 to N81C on M2 of an adjacent subunit

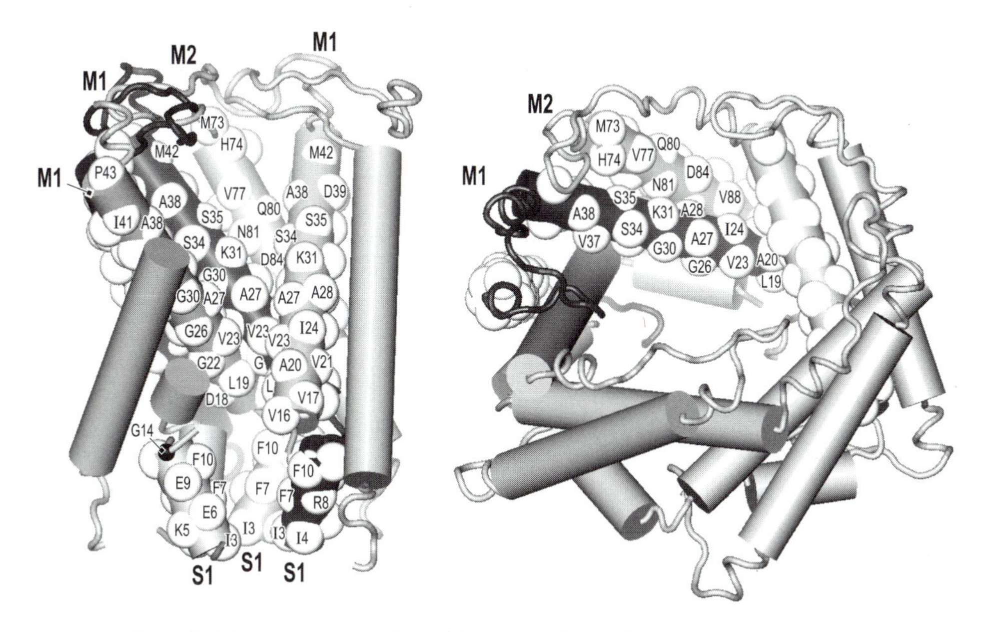

Figure 5. Schematic representations of the closed- (left) and open-state models (right) of the transmembrane barrel of EcoMscL. The helices are depicted as cylinders and positions of alpha-carbons of the pore-lining residues are shown as spheres. The closed-to-open transition in the SG model is accompanied by a relatively small (20 to 30°) counterclockwise rotation of M1, which does not change substantially the pore-exposed face of the helix.

has no detectable effect on channel gating (Betanzos et al., 2002). MD simulations of the pore expansion process detected only small and disconcerted rotations of M1 helices (Gullingsrud and Schulten, 2003).

What Is a GOF Phenotype?

The genetic term "gain of function" originally meant that a mutant receptor or enzyme is intrinsically hyperactive compared to the wild type. Conversely, the term "loss of function" denotes an impairment of the function by a mutation. Peculiarly, the term gain of function, when applied to bacterial MS channels, has the primary meaning of growth suppression upon expression of the mutant channel. This effect is presumed to be a result of channel hyperactivity causing solute leaks at resting tensions. The implied correlation between the toxicity of the mutation and the leftward shift of the activation set point generally holds for most of the GOF mutations isolated in the random screens (Ou et al., 1998) and engineered at specific sites (Yoshimura et al., 1999). Although GOF mutations are scattered in different domains, the most severe ones are hydrophilic or charged substitutions in the cytoplasmic half of M1 forming the constriction. Strong growth-suppressing effects were caused by bulky and hydrophilic substitutions for glycines forming the glabrous strip of M1. G22N, G26S, and G30R exemplify very severe GOF substitutions, and this clearly points to the importance of a close steric match (knob-into-hole) between helical surfaces for the stability of the closed state. Normally, residues G22, G26, and G30, forming the most polar (bare) face of M1, are buried in the M1-M1 junction. Ostensibly, when a bulky polar side chain is inserted in any of these positions, this promotes exposure of the entire polar strip to the pore, thus making the pore effectively more hydrophilic. Qualitatively similar GOF effects were produced by charged or hydrophilic mutations at other sites, such as V23D, that cause no obvious effects on the helical packing but simply make the constriction more hydrophilic. The apparent decrease of energy separating the closed and open states, manifested as a leftward shift of activation curves, was always accompanied by a much faster kinetics signifying a lower transition barrier (Ou et al., 1998; Yoshimura et al., 1999). Since hydrophilic substitutions at position 22 increased the rate of transitions, whereas more hydrophobic ones slowed them down, it was hypothesized that M1 helices experience rotations in the process of transition such that the G22 site is buried in the resting state, then becomes exposed to the aqueous phase in the transition state, and then turns away from the lumen again (Yoshimura et al., 1999). Similar trends were observed when G22C was modified with hydrophilic or hydrophobic MTS reagents (Yoshimura et al., 2001), but a sequence of hidden-exposed-hidden states would be harder to explain by limited helical rotations for longer and bulkier cysteine-attached probes.

An interesting feature observed in severe GOF mutants G22N and G22K was a high and stable occupancy of a low-conducting substate at low pipette pressures before achieving full openings at higher pressures (Yoshimura et al., 1999). Similar patterns of gating, nicknamed "spikes riding on plateaus," were recorded from G22C channels modified with charged MTS probes (Yoshimura et al., 2001). It has been concluded that the G22 site, buried in the closed state, becomes water-exposed in the low-conducting substate (S). Having in mind that the constriction of wild-type MscL in the resting state is dehydrated (Perozo et al., 2001; Gullingsrud and Schulten, 2003), this finding leads to the notion that the transition from the fully closed state to the low substate is associated with pore hydration. Note that the previous kinetic analysis of wild-type MscL showed that the major area and energy changes occur specifically during the $C \rightarrow S$ stage, whereas the following $S \rightarrow O$ stage brings the main gain in conductance associated with only a minor area change (Sukharev et al., 1999b; Chiang et al., 2004a). Thus, the question of whether the first $C \rightarrow S$ stage involving major protein expansion and passing the rate-limiting barrier is directly associated with pore hydration becomes mechanistically important.

Recent thermodynamic analysis of three GOF mutants, including mild V23T, severe V23D, and the very toxic G22N, concluded that polar substitutions in the constriction dramatically decrease the energy gap specifically between the C and S states, commensurably with the severity of the mutation (Anishkin et al., 2005). The increase of the energetic cost of the second $S \rightarrow O$ transition observed at the same time explains the stabilization of the intermediate S state. The analysis also suggested that the resting conformation of severe GOF mutants is substantially preexpanded relative to the closed conformation of wild type, consistent with a larger internal volume of prehydrated pore. Reconstruction of energetic landscapes illustrated that the barrel expansion coupled to the critical hydration event is energetically very costly in the wild-type channel (~50 kT) but is drastically facilitated by hydrophilic substitutions in the M1 gate, reducing the energy to 2 to 15 kT. This stage approximates the M1 gate opening (Color Plate 28) in the SG model, which results in the intermediate water-filled conformation occluded only by the "leaky" S1 gate. This state may correspond to the stable low substate (plateau), producing spikes when the S1 gate transiently opens.

It should be mentioned that although the facilitated hydration of the pore exerts the most drastic reductions in activating pressure, other parts of the protein clearly influence this parameter as well. Mutations perturbing hydrogen bonding between R45 and Q51 in the loops of TbMscL result in a measurable GOF effect (Maurer et al., 2000). An E104Q mutation in the cytoplasmic M2-S3 linker of TbMscL was also shown to reduce the activation threshold; however, the effect was not as strong as exerted by the V15C substitution (Shapovalov et al., 2003), analogous to a very mild GOF mutant V17C in EcoMscL (Levin and Blount, 2004).

LOF Mutations and Intragenic Second-Site Suppressors

While leftward shifts of activation point cause toxic GOF phenotypes, rightward shifts decrease the ability of the mutant channels to rescue a sensitive bacterial strain from lysis under strong osmotic downshock, i.e., cause LOF. It should be mentioned that the two categories are not mutually exclusive. A mutation that substantially perturbs the structure may cause a leak in the gate (GOF) and at the same time impair the normal opening process, leading to a LOF effect. For instance, the V23D mutation in EcoMscL previously characterized as a strong GOF (Ou et al., 1998) was later classified as LOF as well (Maurer and Dougherty, 2003). LOF mutations may potentially represent functional but unstable proteins with a short lifetime in the membrane. Extremely low levels of protein expression shown for many GOF mutants may be another reason for functional insufficiency manifested as LOF. The large mutational library generated for EcoMscL (Maurer and Dougherty, 2003) was not tested in this respect, but nevertheless it revealed a striking pattern of LOF sites primarily at the protein-lipid interface and loop regions. When mapped on the closed-state model of EcoMscL, most of the mutations concentrated at the level of lipid headgroups. This location of LOF mutations was interpreted as perturbing hydrogen bonding between the protein and phospholipids. It seems even more important that these mutations specifically affect the boundaries between the lipid hydrocarbon layer and polar groups, where the maximum of tension in the bilayer was computed in MD simulations (Gullingsrud and Schulten, 2004). The abundance of LOF mutations also in the loops was consistent with the view that during the transition the loops act as springs connecting rigid pairs of helices. Because none of the GOF or LOF mutations were found in N-terminal (S1) and C-terminal (S3) segments, it was concluded that these parts are not critical for the primary channel function (Maurer and Dougherty, 2003).

An attempt to randomly generate second-site suppressors for the G22D GOF mutation resulted in a small set of mutations that were mapped primarily along the extracellular rim of the barrel, in the same area where many LOF sites were located previously (Yoshimura et al., 2004). To explore protein-lipid interactions further, the authors engineered 21 asparagine substitutions in that region. Seven of these mutations were especially strong in reducing the tension sensitivity of the channel, effectively causing LOF effect. These were groups of apolar residues on the periplasmic ends of M1 and M2 apparently anchoring the barrel to the lipid bilayer. This work independently indicated the tension-receiving surface of MscL, but in addition to the role of hydrogen bonding proposed by Maurer and Dougherty, it emphasized the importance of hydrophobic protein-lipid interactions. The recent observation that the survival rate of F78N-expressing *mscL⁻ mscS⁻* cells upon osmotic shock is lower than that of empty-vector control illustrates that MscL with an inappropriately hydrophilic lipid-exposed surface may also compromise the mechanical strength of the membrane (T. Nomura and K. Yoshimura, unpublished data).

Another study based on the same strategy was designed to find intragenic suppressor mutations that would alleviate toxic effects of the V23A and G26S substitutions (Li et al., 2004). Nineteen and seventeen suppressors residing in both transmembrane domains, loops, and M2-S3 linkers were independently isolated for the two GOF mutations, respectively. Twelve suppressors were common for each of the primary GOF mutants and two more were different substitutions at the same sites. One GOF suppressing mutation was found in the S1 domain, providing the very first genetic hint about possible involvement of the N terminus in gating. Interestingly, while many mutations provided some offset for the low activation thresholds without changing their flickery kinetics, one mutation (V23I) slowed the kinetics down, apparently by making the lumen of the pore narrower and more hydrophobic.

Do S1 Domains Form the Second Gate?

The sequence of the N-terminal segment is highly conserved across the entire family of MscL and is predicted to fold as an amphipathic helix by most of the structure-predicting protocols. The two 100% conserved phenylalanines (F7 and F10 in EcoMscL) and preceding aliphatic residues (I3 and I4) form the putative hydrophobic facet that may mediate bundling of five identical S1 helices. The polar face includes charged Lys5, Glu6, Arg8, and Glu9, which are also very conserved. Early deletion analysis indicated that a removal of three residues (Ser2-Ile4) was functionally tolerated, whereas removal of 11 residues (Ser2-Met12) from the N terminus did not produce functional channels (Blount et al., 1996b; Hase et al., 1997). A replacement of the N-terminal sequence from Ser2 through Arg8 by plasmid-derived linkers, mostly polar in nature, resulted in "stiff" channels activated at pressures more than two times the pressure activating wild type. Simultaneous deletions of 8 residues from the N terminus and 25 residues from the C terminus produced extremely stiff channels characterized by unusually long openings (Hase et al., 1997). The data clearly indicated the involvement of the N terminus in the gating and suggested interactions between N- and C-terminal domains. That cysteines in the 19th position (L19C) in the closed state are completely inaccessible to MTS reagents from the cytoplasmic side (Batiza et al., 2002), although indirectly, points to the presence of a second gate.

The hypothetical S1 bundle was proposed to act as the second gate because a pore-occluding element was needed to explain the postulated expanded low-conducting substate (Sukharev et al., 2001a). This expanded intermediate was detected kinetically in wild-type MscL (Sukharev et al., 1999b) and found to be stable in certain GOF mutants (Yoshimura et al., 1999; Anishkin et al., 2005). According to the model, the S1 gate remains closed when the central M1 gate opens and the barrel expands, and then separates as the channel enters the fully open state (Color Plate 28). Cysteine substitutions supported the proximity of conserved phenylalanines as intersubunit cross-links were easily formed in F7C or F10C mutants under ambient conditions. These channels displayed only incomplete openings unless the bonds were thoroughly reduced by ME, indicating that this assembly should separate during the opening. Consistent with the effects of cysteines, F7H and F10H substitutions apparently create a binding site for divalent ions, and in the presence of trace Cd^{2+} such channels were stiff and gated predominantly between the closed and low-conducting substate (Anishkin et al., 2005). In the fully open conformation, the predicted "docked" state of S1 domains on the perimeter or the pore was tested by I3C-I96C cross-links (Ile96 is in M2 near its C terminus). When this bond formed in patch-clamp experiments, the channel gated between the fully open state and one of the low-conducting substates, practically without complete closing events (Sukharev et al., 2001a). The experiment illustrated the feasibility of the docked open conformation, although, based on the patterns of subconducting states, the actual behavior of the S1 helices is expected to be very dynamic.

While the normal channel function/assembly required most of the S1 domain (Blount et al., 1996b), the analysis of the random library of mutants indicated that the N terminus of MscL can be altered at many sites without a change of the channels' osmotic rescuing function in vivo (Maurer and Dougherty, 2003). Mutations of phenylalanines 29, 85, and 93 in the putative docking site presumed to be occupied by S1 in the open state (Sukharev et al., 2001a) were tolerated as well. The authors rightfully noted, "Although one could argue that mutational tolerance at any given site might not make a compelling case, the range of conserved residues that can, in fact, be mutated without phenotypical consequences suggests that the S1 helix developed for the SG model may not play a crucial role in MscL function." Out of the 26 mutations randomly introduced in S1, however, only two were drastically nonconserved changes for Phe7 (F7S and F7C) and, being introduced one at a time, none of the mutations changed the amphipathic character of the helix substantially. At the same time, mutations that make S1 considerably more hydrophilic drastically change the threshold for activation (Hase et al., 1997). In this regard, it would be very important to determine whether the strong amphipathicity (hydrophobic moment) is the only requirement for the normal function of S1 domains or whether more specific interactions within the S1 bundle must be satisfied.

While analyzing the random library, Maurer and Dougherty also proposed that the blockage in the preexpanded state of the channel, which is necessary to explain the electrophysiology of MscL, may be due to some parts of the protein other than S1. Patch-clamp data, nevertheless, point to S1 and the S1-M1 connecting linker. According to the SG model, short S1-M1 linkers containing absolutely conserved glycine 14 serve as force transmitting elements between the expanding barrel and the S1 domains. Extension of these linkers by one extra glycine (GG14 mutation) led to a higher occupancy of low substates indicative of a lower rate for the $S \rightarrow O$ transition (Sukharev et al., 2001a). An obvious decrease of driving force separating the bundle is consistent with a small "slack" in extended linkers. Extension of linkers on a GOF background (V23D/GG14) dramatically reduces the occupancy of the fully open state and leaves the channel lingering in intermediate conducting states apparently due to inability of the S1 bundle to separate completely (Anishkin et al., 2005). Deletion of Gly14, on the other hand, produced channels that often switched to a low subconducting state and remained there in a tension-independent manner (Liu et al., 1999).

It should be pointed out that separation of S1 domains does not appear to be an energetically costly process as most of the transition energy is consumed by the separation of the main hydrophobic (M1) gate. Estimations based on occupancies of the closed (C), open (O), and low-conducting (S) substate suggested that the energy gap between C and S states in wild-type MscL is about 45 to 48 kT, whereas between S and O it is only 3 to 5 kT. This is consistent with most of the mutagenesis data (Ou et al., 1998; Yoshimura et al., 1999; Yoshimura et al., 2001; Batiza et al., 2002; Li et al., 2004), indicating that the integrity of the hydrophobic constriction is the most crucial factor determining the setpoint for activation by tension, and M1 gate separation is indeed the rate-limiting step preceding the opening of the S1 gate. But if S1 domains play essentially no role in the normal channel energetics, why do we still call them the "second gate"? The reason is that separation of S1 domains appears to be the last step in the gating transition, which yields 80% of channel conductance (Anishkin et al., 2005).

The combined data illustrate that mutations in M1 gates, on one hand, and in S1 domains and linkers, on the other, affect two separate stages of the opening process. Substitutions in the constriction specifically facilitate the C→S transition ascribed to the opening of the main M1 gate (see above), whereas mutations in the S1 domain and the S1-M1 linker affect primarily the second S→O subtransition associated with the function of the hypothetical second gate. The locations and the character of the latter mutations suggest that the S1 domains indeed form the second gate. More work is required to transform this attractive hypothesis into a proven statement. The recently achieved complete chemical synthesis of MscL (Clayton et al., 2004) opens rich opportunities for unconventional modifications that may facilitate mechanistic studies of the channel and clarify the role of S1 domains.

Cytoplasmic S3 Domains Are Stably Associated in all Conformations, Forming a Prefilter

In the crystal structure of TbMscL, C-terminal (S3) domains are in an unusual orientation with charged residues buried inside the bundle and hydrophobic side chains on the periphery. One should remember that TbMscL was crystallized at pH 3.6 in the presence of detergent and heavy metals (Chang et al., 1998); thus some of the observed interactions may be nonnative. An alternative configuration of the S3 bundle was modeled by analogy to the cartilage oligomeric matrix protein, whose sequence is similar to the amphipathic S3 helix, and the crystal structure revealed a fivefold coiled-coil arrangement (Malashkevich et al., 1996). In the SG model, S3 helices are held together by apolar interactions of centrally positioned L121, L122, V125, L128, and L129 and by two salt bridges per subunit on the periphery of the bundle. Although early models depicted an open channel with separated S3 helices (Sukharev et al., 2001b), MD simulations suggested high stability of the bundle and predicted distances for disulfide cross-linking. The easily formed L121C-L122C and L128C-L129C cross-links (Color Plate 28) gave support to the modeled helical orientation but had no effect on tension sensitivity, indicating that the bundle remains assembled in all conformations (Anishkin et al., 2003). The data also suggested that the S3 bundle is physically uncoupled from the motions of the transmembrane barrel. ATP, used as a molecular probe, did not permeate significantly through wild-type MscL in in vivo osmotic shock experiments; however, bacteria expressing the truncated mutant lacking S3 (Δ110-136) leached more ATP. This suggests that the S3 bundle and connecting linkers act as a prefilter at the cytoplasmic entrance to the pore, preventing pore occlusion by large molecules and loss of essential metabolites. Independent EPR study gave similar predictions to the orientations of the S3 helices and their stable association (Martinac et al., 2004).

CURRENT TRENDS AND PERSPECTIVES

The above sections described the phenomenology of MscL presented in the context of the crystal structure of TbMscL and the SG gating model. The model of MscL transition is nothing more than a hypothesis about the gating mechanism and obviously a lot of work is still ahead. Although incomplete, the model seems to be plausible. Developing, changing, and refining the model in accordance with new data have constituted a dynamic process. Results obtained by many groups, both confirming and disproving certain aspects of the model, have been incorporated, thus making current models a result of a collective effort. Skepticism and rightful critique help us present MscL gating more rigorously and consistently.

There is now a consensus that pore opening is a result of tilting of both transmembrane helices such that M1 lines most of the pore. Probing of residues lining the pore in different conformations has just begun, and the amount of helical rotation accompanying barrel expansion is a matter of debate. It is likely that further phenotypical assays and direct chemical techniques such as MTS labeling combined with mass spectrometry (Kelly and Gross, 2003) will soon clarify the general positions of all accessible residues. However, the dynamic nature of the open conformation may make identification of a single precise open structure difficult. Transition conformations are likely to be even less stable and less defined, and some conformational states may be quite asymmetric.

There is little doubt that periplasmic loops act as springs countering barrel expansion. However, why and how mutations at many sites drastically change the tension sensitivity and kinetics of the channel remain unclear. It appears that modern computational methods of structure prediction and molecular simulations may help in identifying intramolecular and protein-lipid interactions that affect the energetics of the loops.

It is becoming evident that wetting of the hydrophobic pore constriction is the critical event defining activating tension and the rates for the opening and closing transitions. Yet, some observations remain puzzling. Why, for instance, does the V23A mutation, which apparently does not perturb helical packing and makes the pore surface only slightly more hydrophilic, exert a GOF phenotype as strong as V23D? Detailed analysis of GOF mutants and computational studies of pore hydration are currently under way.

The question of what structural element forms the second gate in MscL has not been answered unequivocally. Structure prediction methods and simulations in conjunction with detailed experimental analysis of N-terminal mutants may give further support to the hypothesis of S1 domains acting as the second gate. If during channel opening these N-terminal domains separate last, do they prime the closure?

It has been strongly suggested by molecular dynamics and multiple LOF mutations that tension developed in the lipid bilayer is transmitted to MscL through narrow boundary regions between membrane hydrocarbon and polar headgroups. MscL gating has been shown to be strongly modulated by membrane-active and amphipathic substances. The effects of lysolipids were especially strong. How can we separate effects of area stress brought by amphipath intercalation from the effects of curvature stress? How does the lateral pressure profile in the bilayer influence the state equilibrium in MscL? Answering all these questions should help produce a consistent picture of MscL gating. A detailed biophysical description of this phenomenon will help us understand other channels.

Acknowledgments. The work in the Sukharev laboratory was supported by research grants from the National Institutes of Health and the National Aeronautics and Space Administration.

REFERENCES

Ajouz, B., C. Berrier, M. Besnard, B. Martinac, and A. Ghazi. 2000. Contributions of the different extramembranous domains of the mechanosensitive ion channel MscL to its response to membrane tension. *J. Biol. Chem.* **275:**1015–1022.

Ajouz, B., C. Berrier, A. Garrigues, M. Besnard, and A. Ghazi. 1998. Release of thioredoxin via the mechanosensitive channel MscL during osmotic downshock of *Escherichia coli* cells. *J. Biol. Chem.* **273:**26670–26674.

Anishkin, A., C.-S. Chiang, and S. Sukharev. 2005. Gain-of-function mutations reveal expanded intermediate states and a sequential action of two gates in MscL. *J. Gen. Physiol.* **125:**155–170.

Anishkin, A., V. Gendel, N. A. Sharifi, C. S. Chiang, L. Shirinian, H. R. Guy, and S. Sukharev. 2003. On the conformation of the COOH-terminal domain of the large mechanosensitive channel MscL. *J. Gen. Physiol.* **121:**227–244.

Arkin, I. T., S. I. Sukharev, P. Blount, C. Kung, and A. T. Brunger. 1998. Helicity, membrane incorporation, orientation and thermal stability of the large conductance mechanosensitive ion channel from *E. coli*. *Biochim. Biophys. Acta* **1369:**131–140.

Bartlett, J. L., G. Levin, and P. Blount. 2004. An in vivo assay identifies changes in residue accessibility on mechanosensitive channel gating. *Proc. Natl. Acad. Sci. USA* **101:**10161–10165.

Bass, R. B., P. Strop, M. Barclay, and D. C. Rees. 2002. Crystal structure of *Escherichia coli* MscS, a voltage-modulated and mechanosensitive channel. *Science* **298:**1582–1587.

Batiza, A. F., M. M. Kuo, K. Yoshimura, and C. Kung. 2002. Gating the bacterial mechanosensitive channel MscL in vivo. *Proc. Natl. Acad. Sci. USA* **99:**5643–5648.

Batiza, A. F., I. Rayment, and C. Kung. 1999. Channel gate! Tension, leak and disclosure. *Structure Fold. Des.* **7:**R99–R103.

Berrier, C., M. Besnard, B. Ajouz, A. Coulombe, and A. Ghazi. 1996. Multiple mechanosensitive ion channels from *Escherichia coli*, activated at different thresholds of applied pressure. *J. Membr. Biol.* **151:**175–187.

Berrier, C., A. Coulombe, C. Houssin, and A. Ghazi. 1989. A patch-clamp study of ion channels of inner and outer membranes and of contact zones of *E. coli*, fused into giant liposomes. Pressure-activated channels are localized in the inner membrane. *FEBS Lett.* **259:**27–32.

Betanzos, M., C. S. Chiang, H. R. Guy, and S. Sukharev. 2002. A large iris-like expansion of a mechanosensitive channel protein induced by membrane tension. *Nat. Struct. Biol.* **9:**704–710.

Bilston, L. E. and K. Mylvaganam. 2002. Molecular simulations of the large conductance mechanosensitive (MscL) channel under mechanical loading. *FEBS Lett.* **512:**185–190.

Blount, P., M. J. Schroeder, and C. Kung. 1997. Mutations in a bacterial mechanosensitive channel change the cellular response to osmotic stress. *J. Biol. Chem.* **272:**32150–32157.

Blount, P., S. I. Sukharev, P. C. Moe, M. J. Schroeder, H. R. Guy, and C. Kung. 1996a. Membrane topology and multimeric structure of a mechanosensitive channel protein of *Escherichia coli*. *EMBO J.* **15:**4798–4805.

Blount, P., S. I. Sukharev, M. J. Schroeder, S. K. Nagle, and C. Kung. 1996b. Single residue substitutions that change the gating properties of a mechanosensitive channel in *Escherichia coli*. *Proc. Natl. Acad. Sci. USA* **93:**11652–11657.

Booth, I. R., and P. Louis. 1999. Managing hypoosmotic stress: aquaporins and mechanosensitive channels in *Escherichia coli*. *Curr. Opin. Microbiol.* **2:**166–169.

Chang, G., R. H. Spencer, A. T. Lee, M. T. Barclay, and D. C. Rees. 1998. Structure of the MscL homolog from *Mycobacterium tuberculosis*: a gated mechanosensitive ion channel. *Science* **282:**2220–2226.

Chiang, C. S., A. Anishkin, and S. Sukharev. 2004a. Gating of the large mechanosensitive channel in situ: estimation of the spatial scale of the transition from channel population responses. *Biophys. J.* **86:**2846–2861.

Chiang, C.-S., L. Shirinian, and S. Sukharev. 2004b. Capping transmembrane helices of the mechanosensitive channel MscL with aromatic residues changes its sensitivity to stretch. *Biophys. J.* **86:**546a.

Clapham, D. E. 2003. TRP channels as cellular sensors. *Nature* **426:**517–524.

Clayton, D., G. Shapovalov, J. A. Maurer, D. A. Dougherty, H. A. Lester, and G. G. Kochendoerfer. 2004. Total chemical synthesis and electrophysiological characterization of mechanosensitive channels from *Escherichia coli* and *Mycobacterium tuberculosis*. *Proc. Natl. Acad. Sci. USA* **101:**4764–4769.

Colbert, H. A., T. L. Smith, and C. I. Bargmann. 1997. OSM-9, a novel protein with structural similarity to channels, is required for olfaction, mechanosensation, and olfactory adaptation in *Caenorhabditis elegans*. *J. Neurosci.* **17:**8259–8269.

Colombo, G., S. J. Marrink, and A. E. Mark. 2003. Simulation of MscL gating in a bilayer under stress. *Biophys. J.* **84:**2331–2337.

Corey, D. P. 2003. New TRP channels in hearing and mechanosensation. *Neuron* **39:**585–588.

Criado, M., and B. U. Keller. 1987. A membrane fusion strategy for single-channel recordings of membranes usually non-accessible to patch-clamp pipette electrodes. *FEBS Lett.* **224:**172–176.

Cruickshank, C. C., R. F. Minchin, A. C. Le Dain, and B. Martinac. 1997. Estimation of the pore size of the large-conductance mechanosensitive ion channel of *Escherichia coli*. *Biophys. J.* **73:**1925–1931.

Cui, C., and J. Adler. 1996. Effect of mutation of potassium-efflux system, KefA, on mechanosensitive channels in the cytoplasmic membrane of *Escherichia coli*. *J. Membr. Biol.* **150:**143–152.

Cui, C., D. O. Smith, and J. Adler. 1995. Characterization of mechanosensitive channels in *Escherichia coli* cytoplasmic membrane by whole-cell patch clamp recording. *J. Membr. Biol.* **144:**31–42.

Delcour, A. H., B. Martinac, J. Adler, and C. Kung. 1989. Modified reconstitution method used in patch-clamp studies of *Escherichia coli* ion channels. *Biophys. J.* **56:**631–636.

Elmore, D. E., and D. A. Dougherty. 2001. Molecular dynamics simulations of wild-type and mutant forms of the *Mycobacterium tuberculosis* MscL channel. *Biophys. J.* **81:**1345–1359.

Elmore, D. E., and D. A. Dougherty. 2003. Investigating lipid composition effects on the mechanosensitive channel of large conductance (MscL) using molecular dynamics simulations. *Biophys. J.* **85:**1512–1524.

Gillespie, P. G., and R. G. Walker. 2001. Molecular basis of mechanosensory transduction. *Nature* **413:**194–202.

Gullingsrud, J., D. Kosztin, and K. Schulten. 2001. Structural determinants of MscL gating studied by molecular dynamics simulations. *Biophys. J.* **80:**2074–2081.

Gullingsrud, J., and K. Schulten. 2003. Gating of MscL studied by steered molecular dynamics. *Biophys. J.* **85:**2087–2099.

Gullingsrud, J., and K. Schulten. 2004. Lipid bilayer pressure profiles and mechanosensitive channel gating. *Biophys. J.* **86:**3496–3509.

Hamill, O. P., and B. Martinac. 2001. Molecular basis of mechanotransduction in living cells. *Physiol. Rev.* **81:**685–740.

Hamill, O. P., A. Marty, E. Neher, B. Sakmann, and F. J. Sigworth. 1981. Improved patch-clamp techniques for high-resolution current recording from cells and cell-free membrane patches. *Pflugers Arch.* **391:**85–100.

Hase, C. C., A. C. Le Dain, and B. Martinac. 1995. Purification and functional reconstitution of the recombinant large mechanosensitive ion channel (MscL) of *Escherichia coli*. *J. Biol. Chem.* **270:**18329–18334.

Hase, C. C., A. C. Le Dain, and B. Martinac. 1997. Molecular dissection of the large mechanosensitive ion channel (MscL) of *E. coli*: mutants with altered channel gating and pressure sensitivity. *J. Membr. Biol.* **157:**17–25.

Howard, J., W. M. Roberts, and A. J. Hudspeth. 1988. Mechanoelectrical transduction by hair cells. *Annu. Rev. Biophys. Biophys. Chem.* **17:**99–124.

Ingber, D. E. 2003. Tensegrity I. Cell structure and hierarchical systems biology. *J. Cell Sci.* **116:**1157–1173.

Iscla, I., G. Levin, R. Wray, R. Reynolds, and P. Blount. 2004. Defining the physical gate of a mechanosensitive channel, MscL, by engineering metal-binding sites. *Biophys. J.* **87:**3172–3180.

Karlin, A., and M. H. Akabas. 1998. Substituted-cysteine accessibility method. *Methods Enzymol.* **293:**123–145.

Kelly, B. L., and A. Gross. 2003. Potassium channel gating observed with site-directed mass tagging. *Nat. Struct. Biol.* **10:**280–284.

Kloda, A., and B. Martinac. 2001. Mechanosensitive channels in archaea. *Cell Biochem. Biophys.* **34:** 349–381.

Kong, Y., Y. Shen, T. E. Warth, and J. Ma. 2002. Conformational pathways in the gating of *Escherichia coli* mechanosensitive channel. *Proc. Natl. Acad. Sci. USA* **99:**5999–6004.

Koprowski, P., and A. Kubalski. 2001. Bacterial ion channels and their eukaryotic homologues. *Bioessays* **23:**1148–1158.

Levin, G., and P. Blount. 2004. Cysteine scanning of MscL transmembrane domains reveals residues critical for mechanosensitive channel gating. *Biophys. J.* **86:**2862–2870.

Levina, N., S. Totemeyer, N. R. Stokes, P. Louis, M. A. Jones, and I. R. Booth. 1999. Protection of *Escherichia coli* cells against extreme turgor by activation of MscS and MscL mechanosensitive channels: identification of genes required for MscS activity. *EMBO J.* **18:**1730–1737.

Li, Y., R. Wray, and P. Blount. 2004. Intragenic suppression of gain-of-function mutations in the *Escherichia coli* mechanosensitive channel, MscL. *Mol. Microbiol.* **53:**485–495.

Liu, W., W. Dietmer, and B. Martinac. 1999. Glycine G14, the amino acid essential for electromechanical coupling in gating the MscL of *E. coli* by mechanical force. *Biophys. J.* **76:**A203.

Malashkevich, V. N., R. A. Kammerer, V. P. Efimov, T. Schulthess, and J. Engel. 1996. The crystal structure of a five-stranded coiled coil in COMP: a prototype ion channel? *Science* **274:**761–765.

Martinac, B., J. Adler, and C. Kung. 1990. Mechanosensitive ion channels of *E. coli* activated by amphipaths. *Nature* **348:**261–263.

Martinac, B., M. Buechner, A. H. Delcour, J. Adler, and C. Kung. 1987. Pressure-sensitive ion channel in *Escherichia coli*. *Proc. Natl. Acad. Sci. USA* **84:**2297–2301.

Martinac, B., D. M. Cortes, and E. Perozo. 2004. Structural dynamics of MscL C-terminal domain. *Biophys. J.* **86:**547a.

Maurer, J. A. and D. A. Dougherty. 2001. A high-throughput screen for MscL channel activity and mutational phenotyping. *Biochim. Biophys. Acta* **1514:**165–169.

Maurer, J. A., and D. A. Dougherty. 2003. Generation and evaluation of a large mutational library from the *Escherichia coli* mechanosensitive channel of large conductance, MscL: implications for channel gating and evolutionary design. *J. Biol. Chem.* **278:**21076–21082.

Maurer, J. A., D. E. Elmore, H. A. Lester, and D. A. Dougherty. 2000. Comparing and contrasting *Escherichia coli* and *Mycobacterium tuberculosis* mechanosensitive channels (MscL). New gain of function mutations in the loop region. *J. Biol. Chem.* **275:**22238–22244.

Mitchell, P. 1979. Keilin's respiratory chain concept and its chemiosmotic consequences. *Science* **206:**1148–1159.

Moe, P. C., P. Blount, and C. Kung. 1998. Functional and structural conservation in the mechanosensitive channel MscL implicates elements crucial for mechanosensation. *Mol. Microbiol.* **28:**583–592.

Moe, P. C., G. Levin, and P. Blount. 2000. Correlating a protein structure with function of a bacterial mechanosensitive channel. *J. Biol. Chem.* **275:**31121–31127.

Nakamaru, Y., Y. Takahashi, T. Unemoto, and T. Nakamura. 1999. Mechanosensitive channel functions to alleviate the cell lysis of marine bacterium, *Vibrio alginolyticus*, by osmotic downshock. *FEBS Lett.* **444:** 170–172.

Naruse, K., Q. Tang, Q. Zhi, and M. Sokabe. 2003. Cloning and functional expression of a stretch-activated BK channel (SAKCa) from chick embryonic cardiomyocyte. *Biophys. J.* **84:**234a.

Oakley, A. J., B. Martinac, and M. C. Wilce. 1999. Structure and function of the bacterial mechanosensitive channel of large conductance. *Protein Sci.* **8:**1915–1921.

Olbrich, K., W. Rawicz, D. Needham, and E. Evans. 2000. Water permeability and mechanical strength of polyunsaturated lipid bilayers. *Biophys. J.* **79:**321–327.

Ornatska, M., S. E. Jones, R. R. Naik, M. O. Stone, and V. V. Tsukruk. 2003. Biomolecular stress-sensitive gauges: surface-mediated immobilization of mechanosensitive membrane protein. *J. Am. Chem. Soc.* **125:** 12722–12723.

Ou, X., P. Blount, R. J. Hoffman, and C. Kung. 1998. One face of a transmembrane helix is crucial in mechanosensitive channel gating. *Proc. Natl. Acad. Sci. USA* **95:**11471–11475.

Park, K. H., C. Berrier, B. Martinac, and A. Ghazi. 2004. Purification and functional reconstitution of N- and C-halves of the MscL channel. *Biophys. J.* **86:**2129–2136.

Patapoutian, A., A. M. Peier, G. M. Story, and V. Viswanath. 2003. ThermoTRP channels and beyond: mechanisms of temperature sensation. *Nat. Rev. Neurosci.* **4:**529–539.

Patel, A. J., and E. Honore. 2001. Properties and modulation of mammalian 2P domain K+ channels. *Trends Neurosci.* **24:**339–346.

Perozo, E., D. M. Cortes, P. Sompornpisut, A. Kloda, and B. Martinac. 2002a. Open channel structure of MscL and the gating mechanism of mechanosensitive channels. *Nature* **418:**942–948.

Perozo, E., A. Kloda, D. M. Cortes, and B. Martinac. 2001. Site-directed spin-labeling analysis of reconstituted MscL in the closed state. *J. Gen. Physiol.* **118:**193–206.

Perozo, E., A. Kloda, D. M. Cortes, and B. Martinac. 2002b. Physical principles underlying the transduction of bilayer deformation forces during mechanosensitive channel gating. *Nat. Struct. Biol.* **9:**696–703.

Perozo, E., and D. C. Rees. 2003. Structure and mechanism in prokaryotic mechanosensitive channels. *Curr. Opin. Struct. Biol.* **13:**432–442.

Pivetti, C. D., M. R. Yen, S. Miller, W. Busch, Y. H. Tseng, I. R. Booth, and M. H. Saier, Jr. 2003. Two families of mechanosensitive channel proteins. *Microbiol. Mol. Biol. Rev.* **67:**66–85.

Ruthe, H. J., and J. Adler. 1985. Fusion of bacterial spheroplasts by electric fields. *Biochim. Biophys. Acta* **819:**105–113.

Sachs, F. 1992. Stretch-sensitive ion channels: an update. *Soc. Gen. Physiol. Ser.* **47:**241–260.

Sachs, F., and C. E. Morris. 1998. Mechanosensitive ion channels in nonspecialized cells. *Rev. Physiol. Biochem. Pharmacol.* **132:**1–77.

Saint, N., J. J. Lacapere, L. Q. Gu, A. Ghazi, B. Martinac, and J. L. Rigaud. 1998. A hexameric transmembrane pore revealed by two-dimensional crystallization of the large mechanosensitive ion channel (MscL) of *Escherichia coli*. *J. Biol. Chem.* **273:**14667–14670.

Sali, A., L. Potterton, F. Yuan, H. van Vlijmen, and M. Karplus. 1995. Evaluation of comparative protein modeling by MODELLER. *Proteins* **23:**318–326.

Shapovalov, G., R. Bass, D. C. Rees, and H. A. Lester. 2003. Open-state disulfide crosslinking between *Mycobacterium tuberculosis* mechanosensitive channel subunits. *Biophys. J.* **84:**2357–2365.

Shapovalov, G., and H. A. Lester. 2004. Gating transitions in bacterial ion channels measured at 3 microns resolution. *J. Gen. Physiol.* **124:**151–161.

Spencer, R. H., G. Chang, and D. C. Rees. 1999. 'Feeling the pressure': structural insights into a gated mechanosensitive channel. *Curr. Opin. Struct. Biol.* **9:**448-454. (Erratum, **9:**650–651.)

Spencer, R. H., and D. C. Rees. 2002. The alpha-helix and the organization and gating of channels. *Annu. Rev. Biophys. Biomol. Struct.* **31:**207–233.

Sukharev, S. 2002. Purification of the small mechanosensitive channel of *Escherichia coli* (MscS): the subunit structure, conduction, and gating characteristics in liposomes. *Biophys. J.* **83:**290–298.

Sukharev, S., M. Betanzos, C. S. Chiang, and H. R. Guy. 2001a. The gating mechanism of the large mechanosensitive channel MscL. *Nature* **409:**720–724.

Sukharev, S., and D. P. Corey. 2004. Mechanosensitive channels: multiplicity of families and gating paradigms. *Sci. STKE* **2004:**re4.

Sukharev, S., S. R. Durell, and H. R. Guy. 2001b. Structural models of the MscL gating mechanism. *Biophys. J.* **81:**917–936.

Sukharev, S. I., P. Blount, B. Martinac, F. R. Blattner, and C. Kung. 1994a. A large-conductance mechanosensitive channel in *E. coli* encoded by MscL alone. *Nature* **368:**265–268.

Sukharev, S. I., B. Martinac, V. Y. Arshavsky, and C. Kung. 1993. Two types of mechanosensitive channels in the *Escherichia coli* cell envelope: solubilization and functional reconstitution. *Biophys. J.* **65:** 177–183.

Sukharev, S. I., B. Martinac, P. Blount, and C. Kung. 1994b. Functional reconstitution as an assay for biochemical isolation of channel proteins: application to the molecular identification of a bacterial mechanosensitive channel. *Methods Companion Methods Enzymol.* **6:**51–59.

Sukharev, S. I., M. J. Schroeder, and D. R. McCaslin. 1999a. Stoichiometry of the large conductance bacterial mechanosensitive channel of *E. coli*. A biochemical study. *J. Membr. Biol.* **171:**183–193.

Sukharev, S. I., W. J. Sigurdson, C. Kung, and F. Sachs. 1999b. Energetic and spatial parameters for gating of the bacterial large conductance mechanosensitive channel, MscL. *J. Gen. Physiol.* **113:**525–540.

Szabo, I., V. Petronilli, and M. Zoratti. 1993. A patch-clamp investigation of the *Streptococcus faecalis* cell membrane. *J. Membr. Biol.* **131:**203–218.

Tank, D. W., C. Miller, and W. W. Webb. 1982. Isolated-patch recording from liposomes containing functionally reconstituted chloride channels from Torpedo electroplax. *Proc. Natl. Acad. Sci. USA* **79:** 7749–7753.

Tavernarakis, N., and M. Driscoll. 1997. Molecular modeling of mechanotransduction in the nematode *Caenorhabditis elegans*. *Annu. Rev. Physiol.* **59:**659–689.

Valadie, H., J. J. Lacapcre, Y. H. Sanejouand, and C. Etchebest. 2003. Dynamical properties of the MscL of *Escherichia coli*: a normal mode analysis. *J. Mol. Biol.* **332:**657–674.

Van Wagoner, D. R. 1993. Mechanosensitive gating of atrial ATP-sensitive potassium channels. *Circ. Res.* **72:**973–983.

Vazquez-Laslop, N., H. Lee, R. Hu, and A. A. Neyfakh. 2001. Molecular sieve mechanism of selective release of cytoplasmic proteins by osmotically shocked *Escherichia coli*. *J. Bacteriol.* **183:**2399–2404.

Welsh, M. J., M. P. Price, and J. Xie. 2002. Biochemical basis of touch perception: mechanosensory function of degenerin/epithelial Na+ channels. *J. Biol. Chem.* **277:**2369–2372.

White, S. H., and W. C. Wimley. 1999. Membrane protein folding and stability: physical principles. *Annu. Rev. Biophys. Biomol. Struct.* **28:**319–365.

Wiggins, P., and R. Phillips. 2004. Analytic models for mechanotransduction: gating a mechanosensitive channel. *Proc. Natl. Acad. Sci. USA* **101:**4071–4076.

Wood, J. M. 1999. Osmosensing by bacteria: signals and membrane-based sensors. *Microbiol. Mol. Biol. Rev.* **63:**230–262.

Yoshimura, K., A. Batiza, and C. Kung. 2001. Chemically charging the pore constriction opens the mechanosensitive channel MscL. *Biophys. J.* **80:**2198–2206.

Yoshimura, K., A. Batiza, M. Schroeder, P. Blount, and C. Kung. 1999. Hydrophilicity of a single residue within MscL correlates with increased channel mechanosensitivity. *Biophys. J.* **77:**1960–1972.

Yoshimura, K., T. Nomura, and M. Sokabe. 2004. Loss-of-function mutations at the rim of the funnel of mechanosensitive channel MscL. *Biophys. J.* **86:**2113–2120.

Zoratti, M., and V. Petronilli. 1988. Ion-conducting channels in a gram-positive bacterium. *FEBS Lett.* **240:**105–109.

Zoratti, M., V. Petronilli, and I. Szabo. 1990. Stretch-activated composite ion channels in *Bacillus subtilis*. *Biochem. Biophys. Res. Commun.* **168:**443–450.

Bacterial Ion Channels and Their Eukaryotic Homologs
Edited by A. Kubalski and B. Martinac

Chapter 14

The Role of Bacterial Channels in Cell Physiology

Ian R. Booth, Michelle D. Edwards, Ewan Murray, and Samantha Miller

One of the most powerful contributions of genome analysis to the field of microbial physiology has been the recognition of widespread occurrence of ion channel homologs in bacteria and archaea (Table 1) (Booth, 2003; Booth et al., 2003). This in turn has fueled gentle speculation on the roles that these channels might have in cell physiology. In contrast to the fury of work on the structures of these channels, with the promise of insights into the mechanisms of ion permeation and gating, the analysis of their role in cell physiology has been slower to develop. In part this reflects the intrinsic difficulties associated with working with the more commonplace ions, K^+, Na^+, and Cl^-, that are more difficult to analyze than cations such as Cu^{2+}, Zn^{2+}, and Fe^{3+}, which are required by cells in small quantities but can be toxic at high concentrations. It is generally the case that cells accumulate K^+ and exclude Na^+ and Cl^-; both K^+ and Na^+ are implicated in pH homeostasis, K^+ in turgor generation (Epstein, 2003), and Cl^- in assisting acid tolerance (Iyer et al., 2002). Sophisticated functions are in short supply, and this means that strong selectable phenotypes are rare. The mechanogated MscL and MscS channels have provided one of the few examples of a strong selectable phenotype (Blount et al., 1996), but even here the interpretation of the actual observed growth or survival defect must be made with extreme caution (see chapters 12 and 13). In plate assays gain-of-function mutations can cause loss of growth potential (Blount et al., 1996), but in survival assays both gain- and loss-of-function mutants may exhibit identical phenotypes (Miller et al., 2003). It is against this background that any analysis of the role of ion channels in cell physiology must be made. Consequently, this chapter must be strong on theory and speculation, while considering the existing data for potential support.

The presence of ion channels in the bacterial inner (cytoplasmic) membrane has often been controversial. Many scientists, particularly those with a strong bioenergetic leaning, doubted the existence of bacterial ion channels, and the epiphenomena associated with their existence were frequently dismissed as poor experimental design. Certainly, the central

Ian R. Booth, Michelle D. Edwards, Ewan Murray, and Samantha Miller • Bacterial Physiology Group, School of Medical Sciences, University of Aberdeen, Institute of Medical Sciences, Foresterhill, Aberdeen, AB25 2ZD, United Kingdom.

Table 1. Types of bacterial and archaeal ion channels

Channel type	Ion	Structural protein	Physiological role
Ion specific	K^+	KcsA, MthK, KirBac, KvAp,	Membrane potential regeneration? Potassium uptake?
		Kch, MVP	Prevention of membrane hyperpolarization?
	Cl^-[a]	ClcA, ClcB, YneE[b]	Extreme acid resistance
	Na^+	NaChBac	Na^+-based pH regulation
Ligand-gated	K^+, (Na^+, H^+)	KefC	Protection against electrophiles
Mechanogated	Nonspecific	MscL, MscS, MscK, MscM	Protection against hypoosmotic shock
Glyceroaquaporin	H_2O	AqpZ, GlpF	Osmoregulation, glycerol permeation
Gated solute-pores	K^+, NH_3, NH_4^+	AmtB, YhaSTU, AmhM-AmhT	NH_3/NH_4^+ permeation, K^+ release, growth at high pH
Porin[c]	Variable specificity	OmpC, OmpF, PhoE, FhuA	Outer membrane permeability

[a]The ClcA and ClcB proteins are listed here despite the recent reclassification of ClcA as a $2Cl^-/H^+$ antiport.
[b]YneE is the *E. coli* bestrophin homolog (Goodstadt and Ponting, 2001) that has been predicted to be a chloride channel.
[c]Porins are included for completeness here, although not discussed in this chapter.

role that the cytoplasmic membrane plays in energy transduction means that ion channels must be tightly regulated to allow ATP synthesis and other work functions to proceed at a significant rate. As a rule, the majority of bacterial cells possess high cytoplasmic concentrations of K^+ (0.3 to 0.7 M), Mg^{2+} (~10 mM), and glutamate (0.1 to 0.3 M) and low concentrations of H^+ (10^{-7} to 10^{-8} M), Ca^{2+} (low μM), Na^+ (~100 μM to 1 mM), and Cl^- (probably low mM). Organisms that live at extremes of pH, temperature, and salinity may have adopted different cytoplasmic balances, but ion concentrations in the majority of organisms lie within these ranges. The proton motive (and sodium motive) force provides an interconvertible energy source that links the chemical energy of the cell, in the form of ATP and oxidizable organic compounds, and the work functions of transport and motility. The cytoplasmic pH is often regulated within narrow ranges (usually from pH 6 for acidophiles through pH 8.5 for some alkalophiles), and this means that the pH gradient varies as a function of the external pH (Booth, 1985; Padan et al., 1976). Consequently, the membrane potential component of the proton and sodium electrochemical gradient varies with the pH of the environment and with the metabolic processes that drive growth. The range of −120 to −200 mV is not uncommon within a single organism, with lower values associated with fermentative metabolism and low pH and higher values found in cells performing electron transport and/or growth at alkaline pH. Membrane potentials appear to be particularly high in photosynthetic bacteria.

With such high values for the membrane potential and for the pH gradient it is clear that cytoplasmic ion concentrations are not at equilibrium with those observed in the environment, and consequently, there are substantial driving forces for ion movements through channels. When not in use, a channel must remain tightly closed if it is not to cause perturbation of both ion gradients and energy transduction (Fig. 1). This is particularly true for the mechanosensitive (MS) channels that have large conductance values (0.1 to 3 nS) and lack the specificity observed with the K^+, Na^+, and Cl^- channels (Martinac et al., 1987). The limited capacitance of the cytoplasmic membrane means that only small numbers of ions are required to pass across the membrane to achieve a substantial change in the membrane potential (Booth, 1985). Such changes are transient, since once the channel has closed again, respiration, light-driven electron transport, or ATPase activity will rapidly reestablish the resting membrane potential. In contrast, the small number of ions moving through the channel will make little difference to the cytoplasmic ion pools (Fig. 1). An exception to this is the MS channels, which have evolved specifically to facilitate the rapid loss of cytoplasmic solutes, and consequently, even when open for short periods (~1 ms), they may equilibrate the internal and external ion pools (Fig. 1) (Berrier et al., 1992; Schleyer et al., 1993). It is worth noting here that despite the common architecture of the pore in K^+ channels, among the different families of K^+ channels, there are very large differences in the rates of K^+ ion movement associated with the open state (Nimigean et al., 2003). These differences appear to be largely the influence of fixed anionic charges in the vestibule that maximize the ion gradient by concentrating cations. This maximizes the rate of permeation of the cation once the channel achieves the open state. Thus, channels with similar selectivity filters may attain significantly different rates of K^+ movement.

POTENTIAL ROLES OF BACTERIAL ION CHANNELS

The understanding of the function of ion channels in bacterial cells lags significantly behind the advances that have been made in the analysis of their structures (Booth, 2003). Significant diversity is already recognized in the potential gating signals for the range of channel homologs (Booth et al., 2003) (Fig. 2). At the time of writing, the number of specific functions that have been proposed for K^+- or Na^+-specific channels is limited, whereas a function has been suggested for the ClcA and ClcB chloride channels of *Escherichia coli* (Booth, 2003; Iyer et al., 2002). Consequently, we can explore channel roles by analysis of their potential functions.

Signaling

The classical role ascribed to ion channels is in the transmission of signals (Albuquerque et al., 1995; Nicholls, 2003). The activation of Cl^- channels can hyperpolarize the membrane as Cl^- ions enter down their chemical gradient, whereas activation of Na^+ channels can depolarize the membrane. However, the major "signaling" class is the Ca^{2+} channels, which by altering the cytoplasmic Ca^{2+} concentrations can activate a range of kinases, phosphatases, and regulatory proteins. The potential role of Ca^{2+} channels in bacterial cells is poorly understood since the actual distribution of this channel class is poorly understood. The filter of the recently discovered bacterial NaChBac channel (Ren et al., 2001) is Na^+-specific and mutants were created that allowed passage of Ca^{2+}, but these

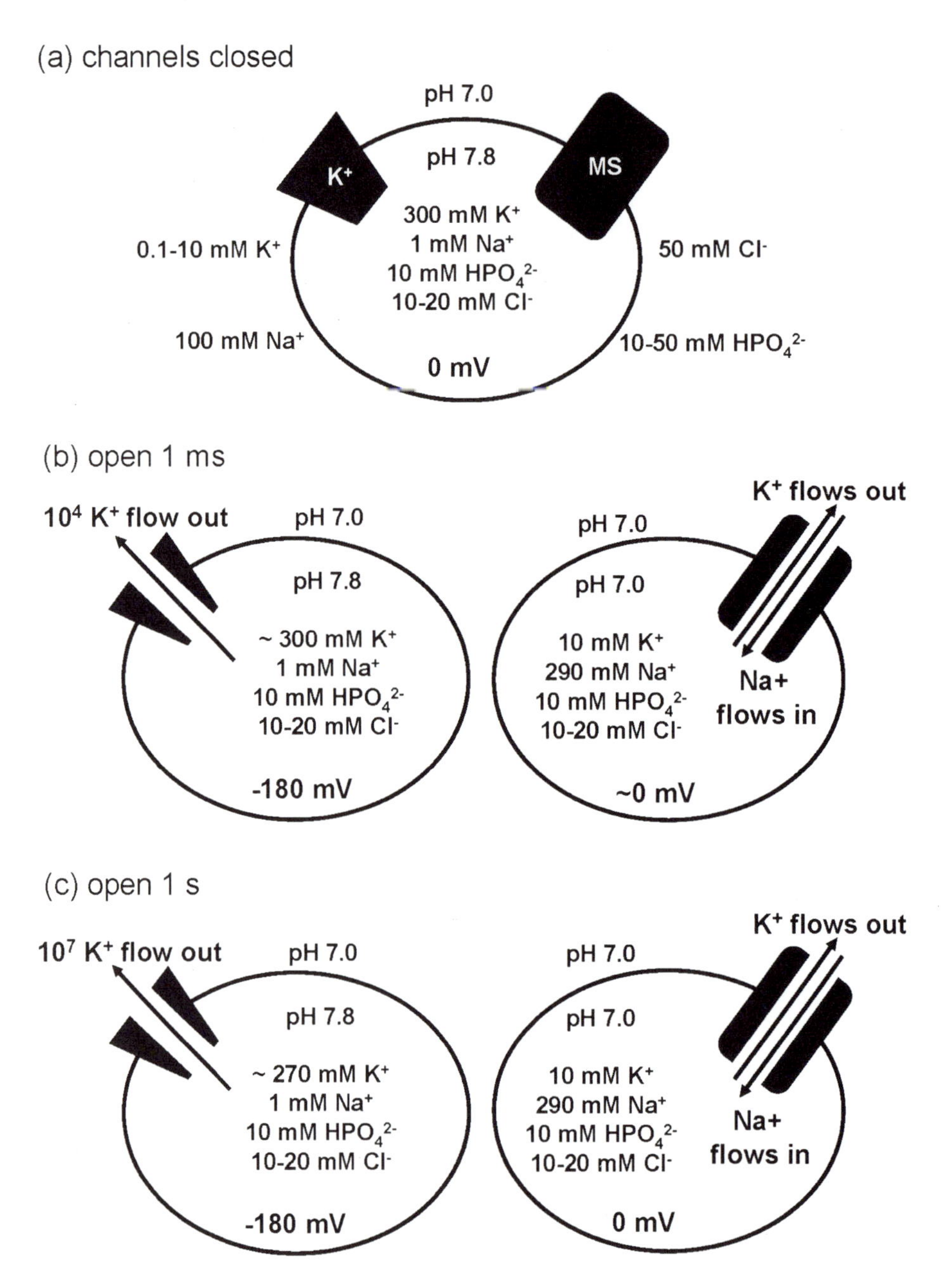

Figure 1. Consequences of channel gating for bacterial cells. The figure depicts the consequences of ion channel gating for a K^+-specific channel (left) and an MS channel (right). The internal and external ion concentrations depicted are based on published values (see text for detail), and the changes in membrane potential, ion concentrations, and cytoplasmic pH value are based on previously published calculations that use data for the capacitance of the cytoplasmic membrane (see text) (Booth, 1985). Initially the cell is depicted with the channels closed (a) with typical cell and environmental concentrations of inorganic ions (note that bacterial cells will also contain ~60 to 150 mM glutamate as the major osmotically active anion, plus lower concentrations of acetate and other organic anions [McLaggan et al., 1994; Roe et al., 1998]). Upon gating the channels for 1 ms (b) or 1 s (c), there are major changes in ion pools when MS channels fire, but much less so for K^+ channels, with the consequence that the latter, but not the former, can be used to generate a membrane potential. Note that unless there is a substantial inward leak current, a voltage-gated K^+ channel would be unlikely to sustain the open state for 1 s,

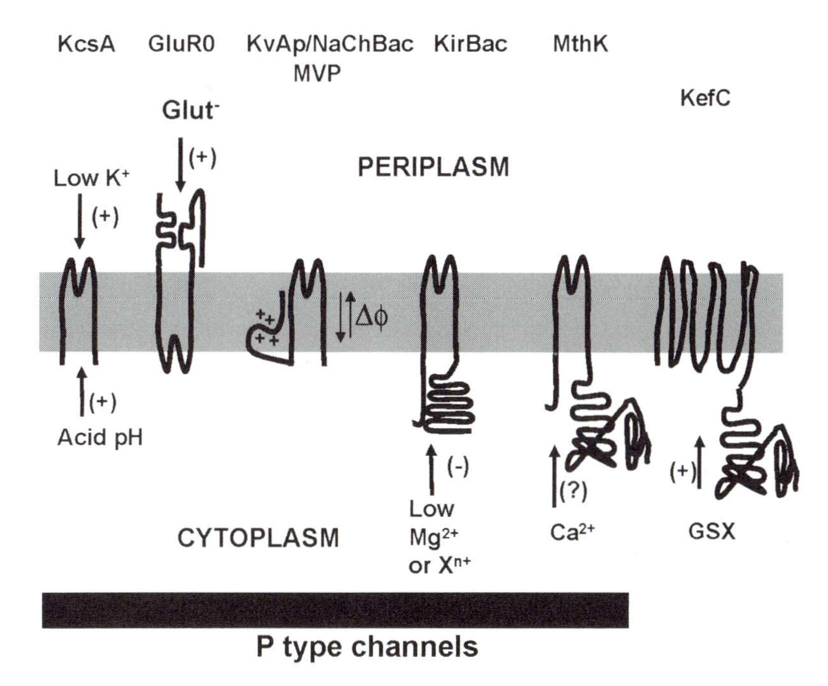

Figure 2. Ligand gating of bacterial channels. The figure illustrates the different types of K^+ and Na^+ channels (and pore) now known to be located in bacterial and archaeal cells. The majority of these possess the classical P-type structure, but there is another class, of which KefC is currently the best example, of gated systems. The nature of the ligand and the site of its action (periplasm, cytoplasm or membrane) are indicated. Further details on these channels are provided elsewhere (Booth, 2003; Booth et al., 2003; Ito et al., 2004; Mayer et al., 2001; Sesti et al., 2003). $\Delta\Phi$, membrane potential; GSX, glutathione adducts; X^{n+}, polyamines; $Glut^-$, glutamate.

were not studied in bacterial hosts (Yue et al., 2002). However, it was pointed out that at least two naturally occurring bacterial homologs (to these Ca^{2+}-selective mutants) exist and that their analysis might prove the presence of Ca^{2+} channels in bacteria and archaea. Glutamate-gated (GluR0) (Chen et al., 1999) channels have been analyzed and shown to have the potential to be sensors.

Ion Uptake

The most obvious ion that the cell requires to accumulate is K^+. Despite the failure to ascribe a function to the KcsA K^+ channel in *Streptomyces lividans*, its potential role in K^+

whereas a ligand-gated channel could remain open as long as the ligand remains bound. For an MS channel an open state of 1 s is frequently seen in patch-clamp recordings but is not considered physiologically relevant during hypoosmotic shock when channel activation would dissipate the pressure differential required for channel activation in milliseconds.

uptake has been demonstrated in *E. coli* mutants that lack K^+ uptake systems (Choi and Heginbotham, 2004; Irizarry et al., 2002). However, for this to take place, mutations were introduced into KcsA to increase the open probability. Similarly, it has proved possible to alter the gating of the Kch channel to allow K^+ uptake into a strain lacking the major K^+ uptake systems (Kuo et al., 2003). A native voltage-gated methanococcal K^+ channel, MVP, has been shown to have the ability to facilitate the accumulation of K^+ ions in both *E. coli* and yeast mutants devoid of other potassium transport systems (Sesti et al., 2003). This channel was activated by hyperpolarization, consistent with the channel opening when the membrane potential is high, a condition favorable to K^+ uptake. Conceivably, this channel might be used to set the upper limit for the membrane potential as well as facilitate K^+ uptake under conditions where other systems were nonfunctional. The widespread occurrence of recognized homologs of the *E. coli* K^+ transport systems in bacteria and archaea can be interpreted to indicate that there is little need for channels to play a role in K^+ uptake in these organisms. Even in the streptomycetes, from which the discovery of the KcsA channel was made, other K^+ transport systems are present that should fulfill the demand for K^+ uptake. Thus, one should be open-minded in seeking roles for these channels.

Channel-mediated Na^+ uptake has been proposed as a solution to some of the problems associated with Na^+-driven pH homeostasis in alkalophiles (Booth, 1985). Alkalophiles grow in the basic range of pH, by virtue of maintaining the cytoplasm more acidic than the environment (Krulwich et al., 1997). The well-documented mechanism for achieving this is the Na^+/H^+ antiport, which exchanges cytoplasmic Na^+ for external H^+ (Padan and Schuldiner, 1994). Cytoplasmic Na^+ poses a specific problem since, to achieve pH homeostasis, the cation must be available on demand, specifically to lower the pH inside the cell. Under normal conditions the antiport also functions to maintain Na^+ homeostasis, i.e., keeping the cytoplasmic Na^+ concentration low. Thus the same transporter achieves two essential functions and they are in conflict with each other (Booth, 1985). If the cell requires a bulk change in the pH of the cytoplasm, significant Na^+ must be available to be exchanged for H^+, and this in turn can only be the case if the cell is not achieving Na^+ homeostasis. An ion channel that responds solely to cytoplasmic pH (Booth, 1985) would provide an effective shortcut. A single bacterial Na^+ channel (NaChBac) has been described to date, and this has a structure predicted to be similar to the basic voltage-gated K^+ channel (i.e., KvAp) (Jiang et al., 2003; Ren et al., 2001; Ruta et al., 2003). However, other homologs are known from the database that retain the Na^+ selectivity filter but lack the sequence of positively charged residues in H3b/H4 that are associated with the voltage sensor paddle. It would be beneficial for a pH-sensing Na^+ channel to be insensitive to the membrane potential since cells that require a Na^+ pulse for pH homeostasis purposes often exhibit a high membrane potential and Na^+ influx will depolarize the membrane. Thus it seems plausible that channels with the right properties to serve in pH homeostasis may exist. Gating the channel would create a pulse of cytoplasmic Na^+ due to ion influx down its chemical gradient. Exchange of this accumulated Na^+ for H^+ would acidify the cytoplasm. A simple prediction of this model is that high external Na^+ should potentiate acidification of the cytoplasm, since opening the channel would be driven by the Na^+ gradient. This observation has been made for at least one alkalophile (McLaggan et al., 1984). Recently, a sodium channel in an obligate alkalophile has also been described, and its loss influences pH homeostasis (Ito et al., 2004).

Chloride ion function in bacterial cells has not been intensively studied. However, chloride ions are accumulated in some halophiles, and some organisms display complex and specific behavioral shifts in the presence of this ion (Muller and Oren, 2003). Chloride expulsion takes place during the bacterial metabolism of chlorinated molecules (Evans et al., 2000), and they have been proposed to enter acidophiles to allow the reversal of the membrane potential required to maintain a high cytoplasmic pH (McLaggan et al., 1990). More recently, chloride channels/transporters have been proposed to have a role in survival of acid pH in *E. coli*. These are discussed in detail in the following section.

Potassium Efflux

Over 30 years ago the bacterial K^+ gradient was first proposed as a residual energy source during times of starvation (Skulachev, 1992). With a cytoplasmic pool of >300 mM and an external concentration of 0.1 to 10 mM, a gradient of 1,000-fold could lead to the generation of a membrane potential of ~ −180 mV (Fig. 1). On the basis of the capacitance of a lipid bilayer and the volume of a bacterial cell, we have previously calculated that ~10^4 ions flowing through a K^+ channel would generate a potential of ~ −180 mV (Booth, 1985). A channel that was fully open for 1 to 10 ms would accomplish the regeneration of a membrane potential without significant loss of the K^+ pool. K^+ channels with voltage gates are found in bacterial and archaeal genomes and could play a role in the generation of membrane potential in such organisms when energy-generating nutrition was limited. Consequently, opening a K^+-specific channel could readily generate a membrane potential; for how long that potential could be sustained would depend on the rate of ion leakage through the membrane and the activity of various work functions.

KcsA was initially cloned and characterized in liposome-protoplasts and demonstrated to gate at physiological pH (Schrempf et al., 1995). However, subsequent reports suggested KcsA activity was only possible at an intracellular pH of 5 or less (Cuello et al., 1998). These studies differed from the original work in that they were conducted on purified KcsA protein reconstituted into planar lipid bilayers and the K^+ levels of the experimental solutions were of a symmetrical nature. An interesting recent study has tested the ability of this channel to open at pH 7.2 in the presence of asymmetrical K^+ and showed clearly that with high intracellular K^+ (200 mM) and a concentration gradient of greater than twofold ($[K^+_{in}]/[K^+_{out}]$) channels gate (Zakharian and Reusch, 2004). Further, current-voltage measurements depicted outward rectification. Thus KcsA appears to be influenced by membrane potential and the K^+ concentration gradient. KcsA activity could be involved in K^+ uptake, or in light of its rectification property, this channel may be activated when the membrane has been highly depolarized and functions to reestablish resting potential levels.

ESTABLISHED FUNCTIONS OF ION CHANNELS IN BACTERIAL PHYSIOLOGY

In higher organisms ion channels play their major roles in information processing and the transmission of signals over great distances to bring about the required actions. Even in single-celled eukaryotes, channels function to perceive external stimuli and bring about specific actions (Preston et al., 1992; Saimi et al., 1999). In these organisms one can perceive channels as central components of homeostasis with a major role in countering stress. As far as we can detect at this early stage in their analysis, the role that has evolved for bacterial channels is similar and thus one can speculate that this role arose early in evolution. Where we understand bacterial channel functions, it is their role in stress avoidance that is prominent.

MS CHANNELS

The proposed role of MS channels is the rapid and nonspecific release of solutes from the cell in response to the generation of excessive turgor pressure (Berrier et al., 1992). This model has now been tested and verified for *E. coli* (Levina et al., 1999), and supporting evidence has been presented for *Vibrio alginolyticus* (Nakamaru et al., 1999). In the gram-positive bacteria, mutants have been created that lack the MS channels, but the lysis phenomenon seen in *E. coli* is more variably observed (Nottebrock et al., 2003; Ruffert et al., 1999; E. Bremer, personal communication). This may reflect differences in the mechanical strength of the cell wall, but other possible reasons are explored below.

Gram-negative bacteria are surrounded by a thin layer of peptidoglycan (PTG), which defines the shape of the organism and constrains the expansion of the cell (Holtje, 1996; Holtje, 1998; Popham and Young, 2003). PTG structure consists of oligosaccharide chains cross-linked by peptides (Glauner et al., 1988). The mechanical strength of the wall is a critical determinant of cell integrity given the high pressure (≥4 atm) exerted on it during growth. PTG is a polymer of *N*-acetylglucosamine and *N*-acetylmuramic acid cross-bridged by peptides [L-Ala-D-Glu-*m*A2pm-D-Ala-D-Ala]. The average muropeptide chain is ~9 units (Glauner et al., 1988; Pisabarro et al., 1985). Given that the glycan strands run perpendicular to the axis of the cell, an average cell of circumference 2,500 nm requires 100 to 300 units to encircle the cell, and these would be held in position by their cross-bridges (Vollmer and Holtje, 2001). Overall, this produces an imperfect mesh that resists the outward pressure from the cytoplasm.

The cell wall undergoes significant remodeling. During each generation ~50% of the PTG is recycled by hydrolysis and recapture of the breakdown products via specific transport systems. Approximately 30 to 50% of peptides are cross-linked (depending on growth phase), and this has led to the proposal of helical strands of sugar chains, with each subsequent peptide oriented at 90° to the previous one. As the PTG is one layer thick, only every second peptide chain is oriented parallel to the cell surface such that cross-bridges can be formed. The degree of linkage is known to vary across the growth phase, with the highest frequency of cross-bridges observed in stationary-phase cells (Pisabarro et al., 1985). Entry into mammalian cells is also known to cause modification of PTG (Quintela et al., 1997). The PTG layer is surmounted by the outer membrane (OM), which contains lipopolysaccharide (LPS) in the outer leaflet. LPS molecules have a high negative charge and when purified are found to have ~5 Mg^{2+} or Ca^{2+} ions attached per molecule of LPS (Nikaido and Vaara, 1985; Nikaido, 1996). Chelation of Mg^{2+} has been shown to cause release of LPS and might be expected to compromise cell integrity. The PTG and OM are covalently linked through lipoproteins that are attached to the PTG and inserted into the OM (Braun and Wu, 1994). No complete analysis of the relative roles of these structures in determining wall strength has been completed (possibly due to a lack of assay methods), but many mutants that lack LPS, OM proteins, and/or lipoproteins have been reported to be osmotically sensitive, suggesting a defect in wall resilience.

The wall provides the essential resistance to the turgor pressure (TP) that is generated across the cell membrane by water influx in response to high cytoplasmic solute concentrations. The TP is the difference between the osmotic potential of the cytoplasm (OPi) and that of the environment (OPo), TP = OPi – OPo. Solute accumulation in the cytoplasm raises OPi and consequently water flows into the cell. In *E. coli* the TP has been estimated to be ~4 atm (≅160 mosM). The pressure required to activate an MS channel in an isolated patch is ~0.05 to 0.1 atm (≅5 to 6 mosM) (Martinac et al., 1987), which is obviously much lower than the estimated TP. Further, estimates of the change in the external salt concentration required to activate MS channels in whole cells are ~0.2 M (i.e., ~400 mosM) (Levina et al., 1999). Such a drop in the external salt concentration would lead to rapid inrush of water developing a pressure of up to 12 to 14 atm within seconds of the entry of cells into the more dilute medium. The difference between the pressure change required to activate MS channels in cells and in membrane patches represents the contribution of the cell wall to resisting the pressure. During steady-state growth, and upon imposition of a small osmotic downshock, the resistance of the wall balances the outwardly directed turgor, such that the net pressure on the cell membrane is below the limit needed to activate MS channels. The activation pressure for *E. coli* MS channels has evolved to be just below the cell lytic pressure of MS channel mutants (Levina et al., 1999).

The role for *E. coli* MS channels, MscL, MscS, and MscK, has been tested in a number of ways. First, the release of organic solutes from cells upon imposition of an osmotic downshock was shown to be sensitive to gadolinium ions, a known inhibitor of MS channels (Berrier et al., 1992). Second, mutants were created with different combinations of MscL, MscS, and MscK; such mutants display normal growth but double (MscL$^-$ MscS$^-$) and triple (MscL$^-$ MscS$^-$ MscK$^-$) mutants are very susceptible to osmotic downshock (Levina et al., 1999). For downshock greater than 0.2 M salt, greater than 95% loss of viability is observed, but single mutants display essentially no such sensitivity (Levina et al., 1999; Sukharev et al., 1994). Cell death is associated with lysis of the majority of the cells, resulting in a release of A_{260}-absorbing material into the medium and loss of light-scattering properties (Levina et al., 1999). Expression of either MscS or MscL from a plasmid is sufficient to protect cells against lysis, and this has been used as a simple assay for channel function. Relatively small numbers of channels are required for protection; the escape synthesis from a *trc* promoter was found to be sufficient to rescue 40 to 80% of the cells, depending on the assay conditions. Membrane patches derived from equivalent cells contain ~30 MscS channels per patch, a patch being equivalent to 2 to 3 cells, suggesting that ~10 to 15 channels per cell may be sufficient for complete protection against osmotic downshock.

This number of MS channels is probably close to the number normally found in cells when expression is driven from the chromosomal channel genes. Both MscL and MscS are expressed at a basal level during growth at low osmolarity and are then induced upon either entry into stationary phase or during adaptation to osmotic stress (Stokes et al., 2003). This change in expression is mediated, in part, through the stabilization of the alternative sigma factor RpoS by the changed growth conditions. RpoS mutants express much lower levels of channels than their isogenic RpoS$^+$ parents; however, during exponential growth, they do not display an increased sensitivity to osmotic downshock, suggesting that the basal level of expression is sufficient to provide protection. In sharp contrast, RpoS mutants that have entered stationary phase at high osmolarity become acutely sensitive to osmotic downshock, lysing to even greater extents than MscS$^-$ MscL$^-$ double mutants. *E. coli* cells remodel their cell wall during entry into stationary phase in an RpoS-dependent modulation of morphology, leading to the small spheroidal cells commonly seen in stationary-phase cultures (Popham and Young, 2003). It seems possible that cells require greater MS channel activity during this transition, although direct proof of this is still lacking. Similarly, preliminary data suggest that strict control of MS channel activity is required during outgrowth from the stationary phase, since again at this stage cell wall remodeling is required (L. Moir, W. Bartlett, and I. R. Booth, unpublished data).

Other evidence supports the view that while MS channels are essential under most conditions of osmotic downshock, there are other adaptations that can reduce the requirement for them. We and others (P. Blount, personal communication) have observed that cells that have reached close to the end of the exponential growth phase are somewhat resistant to osmotic lysis arising from downshock, suggesting that the wall might be more robust under these conditions. Certainly, stationary-phase cells of wild-type *E. coli* are more resilient to the osmotic shock protocol required to release periplasmic proteins than are exponential-phase cells (unpublished data). This would again fit the idea of a more tightly cross-linked cell wall in stationary phase. Finally, the nature of the solute accumulated affects the lysis phenotype of MS channel mutants. When *E. coli* cells are grown at high osmolarity in minimal medium, they accumulate K^+ glutamate and trehalose to maintain a positive TP. When cells are subjected to downshock, these solutes exit via the MS channels (Schleyer et al., 1993). If MS channel mutants are grown in the presence of betaine or proline, this solute is accumulated in preference to K^+ glutamate and trehalose. MS channel mutants accumulate betaine normally but fail to lose viability or lyse when subjected to the osmotic downshock (W. Bartlett, N. R. Stokes, and I. R. Booth, unpublished data). Since the rate of betaine efflux is largely unaffected by the presence or absence of the three major MS channels (MscL, MscS, and MscK), it may be that betaine efflux via a specific channel or carrier is sufficiently rapid that MS channels are rendered nonessential.

At first sight gram-positive bacteria pose a different problem since it might be expectecd that their much thicker PTG coat makes them more resistant to hypoosmotic shock. However, MS channels are present in these organisms, and when the MscL homologs have been expressed in *E. coli*, they gate at similar pressures to their *E. coli* equivalents (Moe et al., 1998). In contrast, the *Mycobacterium tuberculosis* MscL channel did not gate properly in *E. coli*, and this may reflect an adaptation to the lipid environment rather than structures placed on activity by the cell wall (Moe et al., 2000). Gram-positive bacteria MS channel mutants have been assayed in two major gram-positive organisms, *Bacillus subtilis* and *Corynebacterium glutamicum* (Ruffert et al., 1999; Szabo et al., 1992). Double $MscL^-$ $MscS^-$ mutants created in *B. subtilis* exhibited ~20% survival after an osmotic downshock and thus behaved similarly to *E. coli* (E. Bremer, personal communication). In *C. glutamicum* channel mutants showed only small changes in the rate of betaine efflux but were otherwise unaffected by the deletions (Nottebrock et al., 2003). In particular, the double $MscL^-$ $MscS^-$ mutant did not exhibit the lysis phenomenon seen with equivalent *E. coli* mutants. One of the characteristics of this organism is that it synthesizes proline when subjected to the osmotic stress that is a prerequisite for analyzing survival of osmotic downshock (Nottebrock et al., 2003). Above we indicated that even the *E. coli* double channel mutant survives downshock after growth with either proline or betaine. Together these data suggest either that proline and betaine efflux via other exit pathways is sufficiently rapid not to cause the same buildup of turgor or that the accumulation of these compatible solutes modifies the structure of the cell wall and membrane.

MS channels have been identified and studied in a number of archaea, including the halophile *Haloferax volcanii* and the thermophiles *Thermoplasma volcanium* and *Thermoplasma acidophilum* (Kloda and Martinac, 2001a, 2001b, 2001c, 2001d). Two MS channel homologs, MscMJ and MscMJLR, have also been cloned and characterized from *Methanococcus jannaschii* (Kloda and Martinac, 2001c). In contrast to the *E. coli* MS channels, which are non- or slightly anionic-selective, MscMJ and MscMJLR exhibit cation selectivity, a feature that may be related to the organism's environment and specific physiological role. A common characteristic of the archaeal and the bacterial MS channels is their ability to be activated after solubilization in detergent and reconstitution into liposomes, indicating that their mechanism of mechanosensitivity (involving the bilayer model) is closely related and thus that the likely physiological function of archaeal MS channels is based on cellular osmoregulation, as already demonstrated for *E. coli* homologs. No physiological functions have been demonstrated for these channels due to the absence of appropriate mutants.

In *E. coli* the MscL mechanogated channel, after the acquisition of a gating mutation, was shown to be able to utilize the membrane potential to drive K^+ uptake (Buurman et al., 2004). In this case the openings of the channel did not cause permanent depolarization of the membrane, as might have been expected (Fig. 1), probably because of the short duration and infrequency of the open state and the low numbers of the channel protein arising from expression from the chromosome. Although ion uptake is not a normal function for these channels, it is clear from this analysis that new functions can be acquired by single mutations.

CHLORIDE CHANNELS

In most bacterial cells Cl^- ions have not been predicted to have significant roles in cell physiology. In some halophilic bacteria Cl^- is accumulated to provide salt balance during growth in very high salt concentrations (Muller and Oren, 2003); here Cl^- ions are transported inward through the action of halorhodopsin, which utilizes the light-driven isomerization of retinal to produce the conformational cycle needed for chloride passage (Essen, 2002; Zimanyi and Lanyi, 1989), or via symport with Na^+ (Muller and Oren, 2003). In acidophiles it has been proposed that Cl^- entry is important to create the large pH gradient required for organisms growing at pH 1 to 2 with an internal pH close to 7 (McLaggan et al., 1990). Such cells display a low membrane potential, and in some cases this is reversed in polarity (i.e., positive inside) compared with bacteria growing at neutral and alkaline pH. In inverted membrane vesicles the presence of chloride stimulates the generation of a large pH gradient during respiration, possibly due to the collapse of the membrane potential (inside positive) by Cl^- influx. In *E. coli* (growing in the neutral pH range) the intracellular Cl^- concentration has been estimated to be two- to threefold lower than the external concentration (Schultz et al., 1962), suggesting that it has the ability to regulate its cytoplasmic Cl^- concentration. Furthermore, it has been shown that the production of cytoplasmic Cl^- by dehalogenase activity results in the rapid and quantitative loss of Cl^- from the cytoplasm (Evans et al., 2000), indicating either an active chloride pump or high passive membrane permeability to chloride that would allow chloride exit down the electrical gradient since the membrane potential is positive outside. The discovery of ClC homologs in bacteria (ClcA and ClcB in *E. coli*) has yet to explain any of these observed activities (Maduke et al., 1999). The situation is made even less clear by the discovery of a further Cl^- channel family, the bestrophins (Goodstadt and Ponting, 2001). An *E. coli* deletion mutant lacking the YneE bestrophin homolog has no readily observable phenotype (E. Murray, M. J. Pallen, L. Zhang, and I. R. Booth, unpublished data). Thus there remains considerable ignorance of the mechanisms of Cl^- transport in bacteria and the potential roles of such systems in physiology.

The clearest insight into the role of bacterial ClC channels has arisen from the analysis of acid tolerance in *E. coli* (Iyer et al., 2002). These Cl^- channels in *E. coli* appear functionally redundant since single mutants lacking either system are wild type in their acid survival. When *E. coli* cells are exposed to ~pH 2, cell death ensues (Foster and Moreno, 1999). However, a range of mechanisms, one of which has the ClcA/B as a component, are present to provide protection for cells placed in extreme acid. Amino acid decarboxylase systems provide protection against extreme acid stress (Bearson et al., 1997; Foster and Hall, 1990; Park et al., 1996). For the glutamate system the induced components include two functionally redundant glutamate decarboxylases and a glutamate-γ-aminobutyrate (GABA) antiport (Foster, 1999; Foster and Moreno, 1999). Provision of glutamate in the incubation medium at acid pH elicits enhanced survival, which requires transport of the amino acid across the membrane and decarboxylation in the cytoplasm to yield GABA, which exits the cell in exchange for another molecule of glutamate. Controversy still surrounds the precise mechanism by which acid protection is provided, since it critically depends on the mechanism of the antiport (Booth et al., 2002; Park et al., 1996). It has been reported that this protective system is significantly impaired in a mutant lacking both ClcA and ClcB (Iyer et al., 2002). GABA excretion in the mutant is reduced by 50%. Similar observations on survival at acid pH, though less striking in quantitative terms, have been made with arginine as the protective amino acid despite the observation that conversion of arginine to agmatine is strongly inhibited by deletion of ClcA and ClcB. Mutants lacking both Clc systems were restored close to normal when the ClcA channel was expressed from a low-copy-number plasmid. Consistent with this role in countering acid stress, the ClcA system has been shown to be activated by low pH (Iyer et al., 2002). Thus, the primary role envisaged for the protein is that of providing counter-ion movement.

However, there are problems with this model for the role of the Clc systems. First, deletion of ClcA and ClcB renders the cells significantly more resistant to acid stress induced by the acidification of medium with HCl (E. Murray and I. R. Booth, unpublished data). This phenomenon is quantitatively larger than the influence of loss of the Clc systems on the glutamate-protective effect. From this it can be inferred that possession of these systems at acid pH where HCl is an acidulant poses a threat to cellular homeostasis rather than an asset. In the original model for protection it was envisaged that passive HCl permeation was significant, with the Clc systems playing the role of efflux of cytoplasmic Cl^-, which itself was conceived to be a charge compensation for the entry of the glutamate anion. The recent development of the $H^+/2Cl^-$ antiport model for ClcA function (Accardi and Miller, 2004) does not fit so readily with the original model for integration with the glutamate and arginine protection system. The ClcA system displays a low turnover number for an ion channel (10^4 to 10^5 ions s^{-1}). Furthermore, the imposition of a pH gradient under symmetrical Cl^- concentrations causes a shift in the reversal potential, which cannot be due to Cl^- movement since this ion is symmetrically distributed. The transport reaction participated in by ClcA is now thought to follow the following pattern:

$$n\mathrm{Cl}^-_{\mathrm{out}} + m\mathrm{H}^+_{\mathrm{in}} \leftrightarrow n\mathrm{Cl}^-_{\mathrm{in}} + m\mathrm{H}^+_{\mathrm{out}}$$

where n and m are the stoichiometric coefficients of ions participating in a single cycle of the transport reaction and "out" and "in" refer to ions in the external and cytoplasmic locations, respectively (Accardi and Miller, 2004). This coupling of the H^+ and Cl^- movements changes the thermodynamics of the transport process and also has mechanistic implications for ClcA activity, since it is now carrying out secondary active transport. From thermodynamic considerations of the proposed mechanism of ClcA the reversal potential now becomes a function of both the Cl^- and H^+ gradients, a prediction that has been tested and verified. This analysis has determined the values of n and m to be 2 and 1, respectively (i.e., $2Cl^-/H^+$ antiport). This change in the proposed mechanism of ClcA does not immediately affect the understanding of the physiological role of the protein, which still eludes significant comprehension.

The crystal structure of the *E. coli* ClcA (EriC, or Clc-ec1) protein has been solved at 2.8 and 3.5 Å. The structure revealed a permeation pathway that is blocked by the negative side chain of residue E148 (Dutzler et al., 2002; Dutzler et al., 2003). ClcA is thought to gate in acidic conditions, where the side chain is more likely to be protonated and thus displaced from its partially positive "docking" site by a more preferential Cl^- ion, consequently unblocking the pore (Iyer et al., 2002). The crystal structure has proved useful in the analysis of the ion-conducting pathway and of gating of eukaryotic Clc0 channels (Dutzler et al., 2003). For example, gating by acidic pH is eliminated in both Clc0 and ClcA by equivalent mutations, E166A and E148A, respectively. However, an alternative interpretation of this mutant is emerging. ClcA has somewhat nonspecific anion selectivity, enabling the passage of several inorganic anions at significant rates. The selectivity profile $SCN^- > Br^- > Cl^- \sim NO_3^- > SO_4^{2-}$ has been determined (Accardi et al., 2004). The E148A mutation not only leads to activity at both neutral and acidic pH but also the selectivity profile is almost obliterated; the channel is still anion-selective but functions equally with SCN^- and SO_4^{2-}. Moreover, this mutation changes the reversal potential observed with ClcA such that it now displays almost perfect Nernstian behavior (the observed reversal potential is ~31 mV in the wild-type ClcA compared with ~41 mV for E148A and ~44 mV predicted from the experimental transmembrane Cl^- gradient). This observation suggests that E148 is essential for the conduction of H^+. However, the mutant protein can still conduct Cl^-, suggesting that Cl^- and H^+ translocations are independent of each other (Accardi et al., 2004). Recent progress in the analysis of the structure of the Lac and GlpT permeases (Abramson et al., 2003; Huang et al., 2003) has generated new insights into how protein conformational changes may achieve coupled transport that may help our understanding of how the complex ClcA antiport works.

GLUTATHIONE-GATED K^+ EFFLUX SYSTEMS

In *E. coli* and many gram-negative pathogens, there are KefC K^+ efflux systems (see chapter 2 for a more detailed description) (Douglas et al., 1991; Munro et al., 1991). KefC here is used simply to denote the observation that *E. coli* and its relatives have two closely related systems, KefB and KefC, that share a common mechanism of regulation (involving a Ktn domain) and the general structure of the protein components (MacLean et al., 1998; Miller et al., 1997; Miller et al., 2000; Ness and Booth, 1999). The differences are subtle; most importantly, each is activated by a different electrophile and each has a specific sequence that defines to which branch of the family it belongs. This is critical only in the sense that the KefB system, which in *E. coli* is strongly activated by methylglyoxal (MG)-generated glutathione adducts, is the most widespread of the systems in all bacterial genera examined to date (MacLean et al., 1998). In terms of physiology, KefB and KefC have clearcut roles. Each is maintained inactive by the binding of glutathione (GSH) (Meury and Kepes, 1982) and activation is achieved, under physiological conditions, by GSH adducts formed during the detoxification of electrophiles (Ferguson et al., 1993; MacLean et al., 1998; Meury et al., 1980). The first step in detoxification of electrophiles is their reaction with GSH to form an adduct (Fig. 3). This reaction can be spontaneous (MG, iodoacetate [IOA], and *N*-ethylmaleimide [NEM]) or may require a glutathione-*S*-transferase activity (e.g., chlorodinitrobenzene [CDNB]) (Vuilleumier, 1997; Vuilleumier and Pagni, 2002). The fate of each adduct is quite specific:

$$\text{MG} + \text{GSH} \rightarrow \text{HTA} \rightarrow \text{SLG} \rightarrow \text{D-lactate} + \text{GSH} \quad (1)$$
$$\text{NEM} + \text{GSH} \rightarrow \text{ESG} \rightarrow \text{NEMA} + \text{GSH} \quad (2)$$
$$\text{IOA} + \text{GSH} \rightarrow \text{CMG} \rightarrow \text{??} \quad (3)$$
$$\text{CDNB} + \text{GSH} \rightarrow \text{2,4DNP-GSH} \rightarrow \text{export} \rightarrow \text{degradation} \quad (4)$$

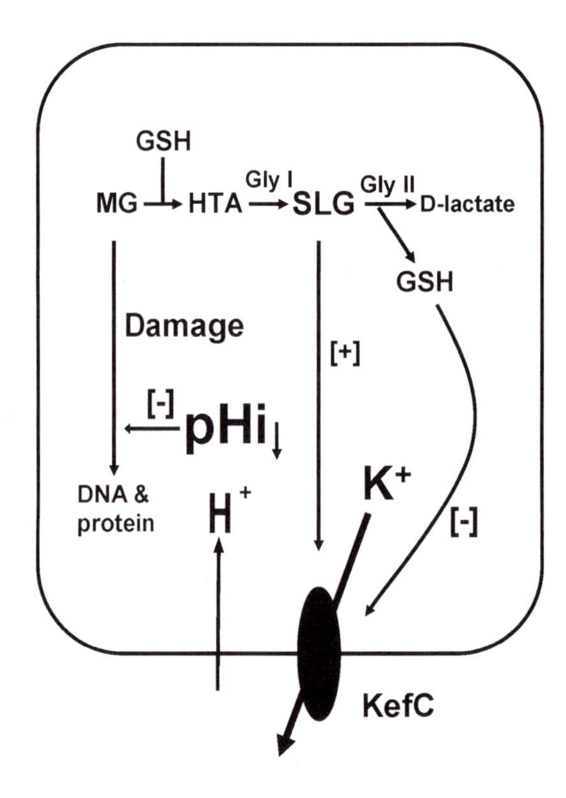

Figure 3. The consequences of electrophile activation of KefC. The formation of an adduct between MG and GSH provides the substrate, hemithiolacetal (HTA), for glyoxalase I (GlyI). The product of GlyI action is SLG, which is the primary activator of KefB and KefC. Further detoxification is undertaken by glyoxalase II (GlyII), regenerating GSH and forming D-lactate. GSH and SLG are negative and positive regulators, respectively, of KefB and KefC. The result is K^+ efflux from the cell, and this is accompanied by H^+ and Na^+ entry by a route not yet identified. It has been demonstrated that the acidification of the cytoplasm aids cell survival of *E. coli* cells (Ferguson et al., 1993; Ferguson et al., 1995; Ferguson et al., 1996; Ferguson and Booth, 1998), protecting the DNA against damage by MG (Ferguson et al., 2000).

where HTA is hemithiolacetal, SLG is *S*-lactoyl-GSH, ESG is *N*-ethylsuccinimido-*S*-GSH, NEMA is *N*-ethyl maleamic acid, CMG is carboxymethyl-GSH, and 2,4DNP-GSH is 2,4-dinitrophenyl-GSH. While the detoxification processes for HTA and ESG take place in the cytoplasm and regenerate cytoplasmic GSH, that for 2,4DNP-GSH takes place in the periplasm after export of the GSH adduct and results in net loss of GSH. After metabolism of 2,4DNP-GSH, probably by γ-glutamyltranspeptidase, the glycine and glutamate moieties may be accumulated by the bacterial cell (Zablotowicz et al., 1995). The fate of CMG is not known.

The KefB and KefC systems are protective to cells exposed to both MG and NEM but appear to play no role in protection against CDNB, and their activation can be a component of the inhibition experienced by cells exposed to IOA (Ferguson et al., 1995; Ferguson et al., 1997; Ferguson and Booth, 1998; Ness et al., 1997). However, all four electrophiles activate KefB and KefC, although that by IOA is the least strong but is apparently significant for cell physiology. The consequence of activation is that K^+ exits from the cell in an essentially first-order process and is partially replaced by Na^+ and H^+ (Bakker and Mangerich, 1982; Ferguson et al., 1995; Ferguson et al., 1997). The route of Na^+ and H^+ entry is not known, but it has been clearly established that it is the change in cytoplasmic pH consequent upon H^+ entry that is critical for protection against the toxicity of the electrophile (Ferguson et al., 1995; Ferguson et al., 1996; Ferguson et al., 1997; Ferguson and Booth, 1998; Ferguson et al., 2000). Mutants lacking the KefB and KefC systems die more rapidly when exposed to MG and NEM than do the parent cells, and either system is sufficient to provide protection if expressed from a multicopy plasmid (Ferguson et al., 1997; MacLean et al., 1998). No significant change in the cytoplasmic pH is seen when the mutant is exposed to the electrophile and these cells can be rescued by acidification of the cytoplasm by incubation with a weak organic acid. Similarly, expression of KefB from a plasmid results in enhancement of the rate of K^+ efflux and an extent of acidification of the cytoplasm leading to even greater protection than seen in the isogenic parent where expression of KefB is from the chromosomal copy of the gene (MacLean et al., 1998). The mechanism of protection is best understood for MG, but a similar type of mechanism may apply to NEM. It has been demonstrated that activation of KefB or acidification of the cytoplasm does not affect the rate of detoxification of MG although both enhanced survival. Cells that have been exposed to high concentrations of MG display significant amounts of DNA damage, and this can be ameliorated by acidification of the cytoplasm (Ferguson et al., 2000). MG is very reactive and can modify guanine bases in DNA, amino acids in proteins, and the headgroups of phospholipids. It is still not clear which of these targets is most critical to the mechanism of cell killing. However, protection does not require new protein synthesis, suggesting that the acidification of the cytoplasm affects existing processes and the most plausible mechanism would be lowering of the rate of modification of macromolecules, such that damage is limited.

SOLUTE-SPECIFIC (GATED) PORES

Aquaglyceroporins

The aquaglyceroporins are pores that allow the permeation of water and small linear polyhydric molecules through the bacterial membrane (Borgnia et al., 1999; Calamita, 2000; Fu et al., 2000). The glycerol transporter GlpF is now known to be a member of the aquaporin family. The protein conducts the transport of glycerol, urea, glycine, water, and DL-glyceraldehyde (Stroud et al., 2003). Crystal structures for the *E. coli* GlpF and AqpZ have been obtained at high resolution and their properties have been examined in detail by measurement of solute/water transport after expression of the pore gene in *Xenopus* oocytes (Borgnia et al., 1999; Fu et al., 2000; Stroud et al., 2003). Mutants lacking aquaporins have been created with both *E. coli* (Calamita et al., 1998) and *Brucella abortus* (Hernandez-Castro et al., 2003) and physiological studies conducted. The *E. coli* mutant displayed no obvious defect in growth and survival. However, in sophisticated analyses of plasmolysis the mutant was found to exhibit a slower rate of water flow (Delamarche et al., 1999). The *B. abortus* aquaporin mutant exhibited a significant loss of viability 50 h after incubation in hypoosmotic conditions. However, expression of the gene was correlated with growth in hyperosmotic conditions (Hernandez-Castro et al., 2003). In contrast, similar studies on the *E. coli aqpZ* gene indicated that the system was repressed during growth at high osmolarity (Calamita et al., 1998). There is, therefore, no doubt about the existence of the aquaporins in bacteria, but in laboratory conditions they do not play an essential role, which does not preclude the ultimate demonstration of an important function in more rigorous environments.

Ammonium Pores/Channels

The cell faces two types of problems that relate to ammonium ions. First, there is the need to scavenge for ammonium when the concentration in the environment is low, which is amply supported by the extensive literature on the regulation of glutamine synthetase and other assimilatory mechanisms (Kustu et al., 1985). Second, cells growing on broth may encounter a surfeit of cytoplasmic ammonium ions arising from the deamination of amino acids and may require to rapidly dispose of these ions to the medium. Recent work suggests that regulated pores may play roles in both processes.

AmtB-GlnK System for Regulated Ammonia Transport

In a wide range of bacterial genera the genes for AmtB and GlnK exist as an operon, and recent work suggests that the former is a pore that is used by cells to accelerate the passage of ammonium ions across the membrane and that GlnK is a specific component of the nitrogen-regulatory circuit that controls its activity (Javelle et al., 2004). The AmtB protein forms trimers in the membrane (Blakey et al., 2002) and is responsible for acquisition of ammonium ions when the concentration in the medium is very low (μM levels) (Coutts et al., 2002). The system is rapidly inactivated when the ammonium concentration in the environment rises. The reversible uridylation of GlnK has been shown to be critical to the affinity of this protein for AmtB (Javelle et al., 2004). Under conditions of high ammonium ions, uridyl groups are removed from GlnK and the protein moves from its cytoplasmic location to become associated with the AmtB protein in the membrane. Mutants that lack AmtB display very low rates of methylamine (a good analog of NH_3) transport, as do mutants of GlnK that cannot be uridylated. The physical association of GlnK and AmtB in the membrane has been demonstrated by pull-down assays, and this association has been shown to be blocked by ammonia limitation and stimulated by ammonia excess (Javelle et al., 2004). The mechanism of AmtB remains controversial (Soupene et al., 1998; Soupene et al., 2002).

The YhaU and AmhM Systems

This family has recently achieved prominence through the analysis of homologs from alkalophiles and *B. subtilis* (Fujisawa et al., 2004; Wei et al., 2003). The *yhaSTU* operon from *B. subtilis* encodes three proteins, YhaS, YhaT (a Ktn domain), and YhaU, two of which are homologs of elements of the YbaL-MagA-RosB proteins (see chapter 2) (Fujisawa et al., 2004). Expression of the YhaU protein introduced a K^+ leak into an *E. coli* strain that lacked K^+ uptake activities. The K^+ leak could be blocked by coexpression of YhaS and YhaT proteins with YhaU. Neither YhaS nor YhaT alone was sufficient to completely regulate the YhaU protein. In *B. subtilis* expression of the *yhaSTU* operon was induced at alkaline pH. Previous analysis of a homologous system AmhM-AmhT in *Bacillus pseudofirmus* strain OF4 led to the proposal that the system forms a K^+/NH_4^+ pore that is regulated by the Ktn domain and that is required for growth at alkaline pH (Wei et al., 2003). Clearly this is an interesting family of proteins. Not only do they create regulated pores, but as with the K^+ uptake systems, they are capable of coupling the separate Ktn domain to maintain the pore-forming domain in a closed state. One of the key questions remaining is what is their mechanism of activation?

CONCLUSIONS

Bacterial channels have been accepted as the molecular paradigm for understanding mechanisms of ion conduction and gating due to their ease of overexpression and genetic manipulation (Booth, 2003). The increased understanding of mechanistic aspects of the channel selectivity and gating has advanced rapidly, but an appreciation of the role of channels in cell physiology lags well behind. The sheer multiplicity of transport mechanisms for K^+ in most bacterial species, coupled with the less facile genetics in these organisms, means that for some K^+ channels good experimental support for their roles may take some time, due to the importance of this cation in maintaining cell turgor and in regulating cytoplasmic pH (Epstein, 2003). In contrast, one would expect that Cl^- and Na^+ channels will yield their physiological roles more quickly, and already there has been substantial progress (Ito et al., 2004; Iyer et al., 2002). In terms of microbial physiology, ion channels offer one of the most exciting challenges for the future. Allied to the ion channels are the new hybrid systems that seem to play important roles in cell growth by mechanisms that are poorly understood (Fujisawa et al., 2004; Wei et al., 2003). From a position of undeserved obscurity bacterial channels have emerged as one of the most exciting and challenging research fields.

Acknowledgments. We thank Tarmo Roosild, Senyon Choe, Wolf Epstein, Sergei Sukharev, Paul Blount, Boris Martinac, Ching Kung, Eduardo Perozo, Terry Krulwich, and Chris Miller.

Our work is supported by a program grant from The Wellcome Trust (040174).

REFERENCES

Abramson, J., I. Smirnova, V. Kasho, G. Verner, H. R. Kaback, and S. Iwata. 2003. Structure and mechanism of the lactose permease of *Escherichia coli. Science* **301:**610–615.

Accardi, A., L. Kolmakova-Partensky, C. Williams, and C. Miller. 2004. Ionic currents mediated by a prokaryotic homologue of CLC Cl^- channels. *J. Gen. Physiol.* **123:**109–119.

Accardi, A., and C. Miller. 2004. Secondary active transport mediated by a prokaryotic homologue of ClC Cl^- channels. *Nature* **427:**803–807.

Albuquerque, E. X., E. F. R. Pereira, N. G. Castro, and M. Alkondon. 1995. Neuronal nicotinic receptors: function, modulation and structure. *Semin. Neurosci.* **7:**91–101.

Bakker, E. P., and W. E. Mangerich. 1982. N-Ethylmaleimide induces K^+-H^+ antiport activity in *Escherichia coli* K-12. *FEBS Lett.* **140:**177–180.

Bearson, S., B. Bearson, and J. W. Foster. 1997. Acid stress responses in enterobacteria. *FEMS Microbiol. Lett.* **147:**173–180.

Berrier, C., A. Coulombe, I. Szabo, M. Zoratti, and A. Ghazi. 1992. Gadolinium ion inhibits loss of metabolites induced by osmotic shock and large stretch-activated channels in bacteria. *Eur. J. Biochem.* **206:** 559–565.

Blakey, D., A. Leech, G. H. Thomas, G. Coutts, K. Findlay, and M. Merrick. 2002. Purification of the *Escherichia coli* ammonium transporter AmtB reveals a trimeric stoichiometry. *Biochem. J.* **364:**527–535.

Blount, P., S. I. Sukharev, M. J. Schroeder, S. K. Nagle, and C. Kung. 1996. Single residue substitutions that change the gating properties of a mechanosensitive channel in *Escherichia coli. Proc. Natl. Acad. Sci. USA* **93:**11652–11657.

Booth, I. R. 1985. Regulation of cytoplasmic pH in bacteria. *Microbiol. Rev.* **49:**359–378.

Booth, I. R. 2003. Bacterial ion channels, p. 91–112. *In* J. K. Setlow (ed.), *Genetic Engineering—Principles and Methods*, vol. 25. Kluwer Academic/Plenum Publishers, New York, N.Y.

Booth, I. R., P. Cash, and C. O'Byrne. 2002. Sensing and adapting to acid stress. *Antonie Leeuwenhoek Int. J. Gen. Mol. Microbiol.* **81:**33–42.

Booth, I. R., M. D. Edwards, and S. Miller. 2003. Bacterial ion channels. *Biochemistry* **42:**10045–10053.

Borgnia, M., S. Nielsen, A. Engel, and P. Agre. 1999. Cellular and molecular biology of the aquaporin water channels. *Annu. Rev. Biochem.* **68:**425–458.

Braun, V., and H. C. Wu. 1994. Lipoproteins, structure, function, biosynthesis and model for protein export, p. 319–341. *In* J.-M. Ghuysen and R. Hakenbeck (ed.), *Bacterial Cell Wall.* Elsevier, Amsterdam, The Netherlands.

Buurman, E. T., D. McLaggan, J. Naprstek, and W. Epstein. 2004. Multiple paths for nonphysiological transport of K^+ in *Escherichia coli. J. Bacteriol.* **186:**4238–4245.

Calamita, G. 2000. The *Escherichia coli* aquaporin-Z water channel. *Mol. Microbiol.* **37:**254–262.

Calamita, G., B. Kempf, M. Bonhivers, W. Bishai, E. Bremer, and P. Agre. 1998. Regulation of the *Escherichia coli* water channel gene *aqpZ. Proc. Natl. Acad. Sci. USA* **95:**3627–3631.

Chen, G. Q., C. H. Cui, M. L. Mayer, and E. Gouaux. 1999. Functional characterization of a potassium-selective prokaryotic glutamate receptor. *Nature* **402:**817–821.

Choi, H., and L. Heginbotham. 2004. Functional influence of the pore helix glutamate in the KcsA K^+ channel. *Biophys. J.* **86:**2137–2144.

Coutts, G., G. Thomas, D. Blakey, and M. Merrick. 2002. Membrane sequestration of the signal transduction protein GlnK by the ammonium transporter AmtB. *EMBO J.* **21:**536–545.

Cuello, L. G., J. G. Romero, D. M. Cortes, and E. Perozo. 1998. pH-dependent gating in the *Streptomyces lividans* K^+ channel. *Biochemistry* **37:**3229–3236.

Delamarche, C., D. Thomas, J. P. Rolland, A. Froger, J. Gouranton, M. Svelto, P. Agre, and G. Calamita. 1999. Visualization of AqpZ-mediated water permeability in *Escherichia coli* by cryoelectron microscopy. *J. Bacteriol.* **181:**4193–4197.

Douglas, R. M., J. A. Roberts, A. W. Munro, G. Y. Ritchie, A. J. Lamb, and I. R. Booth. 1991. The distribution of homologues of the *Escherichia coli* KefC K^+ efflux system in other bacterial species. *J. Gen. Microbiol.* **137:**1999–2005.

Dutzler, R., E. B. Campbell, M. Cadene, B. T. Chait, and R. MacKinnon. 2002. X-ray structure of a ClC chloride channel at 3.0 angstrom reveals the molecular basis of anion selectivity. *Nature* **415:**287–294.

Dutzler, R., E. B. Campbell, and R. MacKinnon. 2003. Gating the selectivity filter in ClC chloride channels. *Science* **300:**108–112.

Epstein, W. 2003. The roles and regulation of potassium in bacteria. *Prog. Nucleic Acid Res. Mol. Biol.* **75:** 293–320.

Essen, L. O. 2002. Halorhodopsin: light-driven ion pumping made simple? *Curr. Opin. Struct. Biol.* **12:** 516–522.

Evans, G. J., G. P. Ferguson, I. R. Booth, and S. Vuilleumier. 2000. Growth inhibition of *Escherichia coli* by dichloromethane in cells expressing dichloromethane dehalogenase/glutathione *S*-transferase. *Microbiology* **146:**2967–2975.

Ferguson, G. P., A. W. Munro, R. M. Douglas, D. McLaggan, and I. R. Booth. 1993. Activation of potassium channels during metabolite detoxification in *Escherichia coli. Mol. Microbiol.* **9:**1297–1303.

Ferguson, G. P., D. McLaggan, and I. R. Booth. 1995. Potassium channel activation by glutathione-*S*-conjugates in *Escherichia coli*; protection against methylglyoxal is mediated by cytoplasmic acidification. *Mol. Microbiol.* **17:**1025–1033.

Ferguson, G. P., A. D. Chacko, C. Lee, and I. R. Booth. 1996. The activity of the high-affinity K^+ uptake system Kdp sensitizes cells of *Escherichia coli* to methylglyoxal. *J. Bacteriol.* **178:**3957–3961.

Ferguson, G. P., Y. Nikolaev, D. McLaggan, M. Maclean, and I. R. Booth. 1997. Survival during exposure to the electrophilic reagent *N*-ethylmaleimide in *Escherichia coli*: role of KefB and KefC potassium channels. *J. Bacteriol.* **179:**1007–1012.

Ferguson, G. P., and I. R. Booth. 1998. Importance of glutathione for growth and survival of *Escherichia coli* cells: detoxification of methylglyoxal and maintenance of intracellular K^+. *J. Bacteriol.* **180:**4314–4318.

Ferguson, G. P., J. R. Battista, A. T. Lee, and I. R. Booth. 2000. Protection of the DNA during the exposure of *Escherichia coli* cells to a toxic metabolite: the role of the KefB and KefC potassium channels. *Mol. Microbiol.* **35:**113–122.

Foster, J. W. 1999. When protons attack: microbial strategies of acid adaptation. *Curr. Opin. Microbiol.* **2:**170–174.

Foster, J. W., and H. K. Hall. 1990. Adaptive acidification tolerance response of *Salmonella typhimurium. J. Bacteriol.* **172:**771–778.

Foster, J. W., and M. Moreno. 1999. Inducible acid tolerance mechanisms in enteric bacteria. *Novartis Found. Symp.* **221:**55–74.

Fu, D. X., A. Libson, L. J. W. Miercke, C. Weitzman, P. Nollert, J. Krucinski, and R. M. Stroud. 2000. Structure of a glycerol-conducting channel and the basis for its selectivity. *Science* **290:**481–486.

Fujisawa, M., Y. Wada, and M. Ito. 2004. Modulation of the K^+ efflux activity of *Bacillus subtilis* YhaU by YhaT and the C-terminal region of YhaS. *FEMS Microbiol. Lett.* **231:**211–217.

Glauner, B., J. V. Holtje, and U. Schwartz. 1988. The composition of the murein of *Escherichia coli. J. Biol. Chem.* **263:**10088–10095.

Goodstadt, L., and C. R. Ponting. 2001. Sequence variation and disease in the wake of the draft human genome. *Hum. Mol. Gen.* **10:**2209–2214.

Hernandez-Castro, R., M. C. Rodriguez, A. Seoane, and J. M. G. Lobo. 2003. The aquaporin gene *aqpX* of *Brucella abortus* is induced in hyperosmotic conditions. *Microbiology* **149:**3185–3192.

Holtje, J. V. 1996. Molecular interplay of murein synthases and murein hydrolases in *Escherichia coli. Microb. Drug Resist.* **2:**99–103.

Holtje, J.-V. 1998. Growth of the stress-bearing and shape-maintaining murein sacculus of *Escherichia coli. Microbiol. Mol. Biol. Rev.* **62:**181–203.

Huang, Y. F., M. J. Lemieux, J. M. Song, M. Auer, and D. N. Wang. 2003. Structure and mechanism of the glycerol-3-phosphate transporter from *Escherichia coli. Science* **301:**616–620.

Irizarry, S. N., E. Kutluay, G. Drews, S. J. Hart, and L. Heginbotham. 2002. Opening the KcsA K^+ channel: tryptophan scanning and complementation analysis lead to mutants with altered gating. *Biochemistry* **41:** 13653–13662.

Ito, M., H. Xu, A. A. Guffanti, Y. Wei, L. Zvi, D. E. Clapham, and T. A. Krulwich. 2004. The voltage-gated Na^+ channel NaVBP has a role in motility, chemotaxis, and pH homeostasis of an alkaliphilic *Bacillus. Proc. Natl. Acad. Sci. USA* **101:**10566–10571.

Iyer, R., T. M. Iverson, A. Accardi, and C. Miller. 2002. A biological role for prokaryotic ClC chloride channels. *Nature* **419:**715–718.

Javelle, A., E. Severi, J. Thornton, and M. Merrick. 2004. Ammonium sensing in *Escherichia coli.* Role of the ammonium transporter AmtB and AmtB-GlnK complex formation. *J. Biol. Chem.* **279:**8530–8538.

Jiang, Y. X., V. Ruta, J. Y. Chen, A. Lee, and R. MacKinnon. 2003. The principle of gating charge movement in a voltage-dependent K^+ channel. *Nature* **423:**42–48.

Kloda, A., and B. Martinac. 2001a. Mechanosensitive channels in Archaea. *Cell Biochem. Biophys.* **34:** 349–381.

Kloda, A., and B. Martinac. 2001b. Molecular identification of a mechanosensitive channel in archaea. *Biophys. J.* **80:**229–240.

Kloda, A., and B. Martinac. 2001c. Structural and functional differences between two homologous mechanosensitive channels of *Methanococcus jannaschii*. *EMBO J.* **20:**1888–1896.

Kloda, A., and B. Martinac. 2001d. Mechanosensitive channel of *Thermoplasma*, the cell wallless archaea: cloning and molecular characterization. *Cell Biochem. Biophys.* **34:**321–347.

Krulwich, T. A., M. Ito, R. Gilmour, and A. A. Guffanti. 1997. Mechanisms of cytoplasmic pH regulation in alkaliphilic strains of *Bacillus*. *Extremophiles* **1:**163–169.

Kuo, M. M. C., Y. Saimi, and C. Kung. 2003. Gain-of-function mutations indicate that *Escherichia coli* Kch forms a functional K^+ conduit in vivo. *EMBO J.* **22:**4049–4058.

Kustu, S., J. Hirschman, and J. C. Meeks. 1985. Adenylylation of bacterial glutamine synthetase: physiological significance. *Curr. Top. Cell. Regul.* **27:**201–213.

Levina, N., S. Totemeyer, N. R. Stokes, P. Louis, M. A. Jones, and I. R. Booth. 1999. Protection of *Escherichia coli* cells against extreme turgor by activation of MscS and MscL mechanosensitive channels: identification of genes required for MscS activity. *EMBO J.* **18:**1730–1737.

MacLean, M. J., L. S. Ness, G. P. Ferguson, and I. R. Booth. 1998. The role of glyoxalase I in the detoxification of methylglyoxal and in the activation of the KefB K^+ efflux system in *Escherichia coli*. *Mol. Microbiol.* **27:**563–571.

Maduke, M., D. J. Pheasant, and C. Miller. 1999. High-level expression, functional reconstitution, and quaternary structure of a prokaryotic ClC-type chloride channel. *J. Gen. Physiol.* **114:**713–722.

Martinac, B., M. Buehner, A. H. Delcour, J. Adler, and C. Kung. 1987. Pressure-sensitive ion channel in *Escherichia coli*. *Proc. Natl. Acad. Sci. USA* **84:**2297–2301.

Mayer, M. L., R. Olson, and E. Gouaux. 2001. Mechanisms for ligand binding to GluR0 ion channels: crystal structures of the glutamate and serine complexes and a closed Apo state. *J. Mol. Biol.* **311:** 815–836.

McLaggan, D., M. J. Selwyn, and A. P. Dawson. 1984. Dependence on Na^+ of control of cytoplasmic pH in a facultative alkalophile. *FEBS Lett.* **165:**254–258.

McLaggan, D., M. Keyhan, and A. Matin. 1990. Chloride transport pathways and their bioenergetic implications in the obligate acidophile *Bacillus coagulans*. *J. Bacteriol.* **172:**1485–1490.

McLaggan, D., J. Naprstek, E. T. Buurman, and W. Epstein. 1994. Interdependence of K^+ and glutamate accumulation during osmotic adaptation of *Escherichia coli*. *J. Biol. Chem.* **269:**1911–1917.

Meury, J., and A. Kepes. 1982. Glutathione and the gated potassium channels of *Escherichia coli*. *EMBO J.* **1:**339–343.

Meury, J., S. Lebail, and A. Kepes. 1980. Opening of potassium channels in *Escherichia coli* membranes by thiol reagents and recovery of potassium tightness. *Eur. J. Biochem.* **113:**33–38.

Miller, S., W. Bartlett, S. Chandrasekaran, S. Simpson, M. Edwards, and I. R. Booth. 2003. Domain organization of the MscS mechanosensitive channel of *Escherichia coli*. *EMBO J.* **22:**36–46.

Miller, S., R. M. Douglas, P. Carter, and I. R. Booth. 1997. Mutations in the glutathione-gated KefC K^+ efflux system of *Escherichia coli* that cause constitutive activation. *J. Biol. Chem.* **272:**24942–24947.

Miller, S., L. S. Ness, C. M. Wood, B. C. Fox, and I. R. Booth. 2000. Identification of an ancillary protein, YabF, required for activity of the KefC glutathione-gated potassium efflux system in *Escherichia coli*. *J. Bacteriol.* **182:**6536–6540.

Moe, P. C., P. Blount, and C. Kung. 1998. Functional and structural conservation in the mechanosensitive channel MscL implicates elements crucial for mechanosensation. *Mol. Microbiol.* **28:**583–592.

Moe, P. C., G. Levin, and P. Blount. 2000. Correlating a protein structure with function of a bacterial mechanosensitive channel. *J. Biol. Chem.* **275:**31121–31127.

Muller, V., and A. Oren. 2003. Metabolism of chloride in halophilic prokaryotes. *Extremophiles* **7:**261–266.

Munro, A. W., G. Y. Ritchie, A. J. Lamb, R. M. Douglas, and I. R. Booth. 1991. The cloning and DNA sequence of the gene for the glutathione-regulated potassium-efflux system KefC of *Escherichia coli*. *Mol. Microbiol.* **5:**607–616.

Nakamaru, Y., Y. Takahashi, T. Unemoto, and T. Nakamura. 1999. Mechanosensitive channel functions to alleviate the cell lysis of marine bacterium *Vibrio alginolyticus*, by osmotic downshock. *FEBS Lett.* **444:** 170–172.

Ness, L. S., and I. R. Booth. 1999. Different foci for the regulation of the activity of the KefB and KefC glutathione-gated K^+ efflux systems. *J. Biol. Chem.* **274:**9524–9530.

Ness, L. S., G. P. Ferguson, Y. Nikolaev, and I. R. Booth. 1997. Survival of *Escherichia coli* cells exposed to iodoacetate and chlorodinitrobenzene is independent of the glutathione-gated K^+ efflux systems KefB and KefC. *Appl. Environ. Microbiol.* **63:**4083–4086.

Nicholls, D. G. 2003. Bioenergetics and transmitter release in the isolated nerve terminal. *Neurochem. Res.* **28:**1433–1441.

Nikaido, H. 1996. Outer membrane, p. 29–47. *In* F. C. Neidhardt, R. Curtiss III, J. L. Ingraham, E. C. C. Lin, K. B. Low, B. Magasanik, W. S. Reznikoff, M. Riley, M. Schaechter, and H. E. Umbarger (ed.), Escherichia coli *and* Salmonella typhimurium: *Cellular and Molecular Biology*, vol. 1. ASM Press, Washington D.C.

Nikaido, H., and M. Vaara. 1985. Molecular basis of outer membrane permeability. *Microbiol. Rev.* **19:**1–32.

Nimigean, C. M., J. S. Chappie, and C. Miller. 2003. Electrostatic tuning of ion conductance in potassium channels. *Biochemistry* **42:**9263–9268.

Nottebrock, D., U. Meyer, R. Kramer, and S. Morbach. 2003. Molecular and biochemical characterization of mechanosensitive channels in *Corynebacterium glutamicum. FEMS Microbiol. Lett.* **218:**305–309.

Padan, E., and S. Schuldiner. 1994. Molecular physiology of Na^+/H^+ antiporters, key transporters in circulation of Na^+ and H^+ in cells. *Biochim. Biophys. Acta* **1185:**129–151.

Padan, E., D. Zilberstein, and H. Rottenberg. 1976. The proton electrochemical gradient in *Escherichia coli* cells. *Eur. J. Biochem.* **63:**533–541.

Park, Y. K., B. Bearson, S. H. Bang, I. S. Bang, and J. W. Foster. 1996. Internal pH crisis, lysine decarboxylase and the acid tolerance response of *Salmonella typhimurium. Mol. Microbiol.* **20:**605–611.

Pisabarro, A. G., M. A. de Pedro, and D. Vazquez. 1985. Structural modifications in the peptidoglycan of *Escherichia coli* associated with changes in the growth state of the culture. *J. Bacteriol.* **161:**238–242.

Popham, D. L., and K. D. Young. 2003. Role of pencillin-binding proteins in bacterial cell morphogenesis. *Curr. Opin. Microbiol.* **6:**594–599.

Preston, R. R., Y. Saimi, B. Martinac, and C. Kung. 1992. Genetic analysis of ion channels of prokaryotes and lower eukaryotes. *Curr. Opin. Genet. Dev.* **2:**780–784.

Quintela, J. C., M. A. de Pedro, P. Zollner, G. Allmaier, and F. Garcia-del Portillo. 1997. Peptidoglycan structure of *Salmonella typhimurium* growing within cultured mammalian cells. *Mol. Microbiol.* **23:**693–704.

Ren, D. J., B. Navarro, H. X. Xu, L. X. Yue, Q. Shi, and D. E. Clapham. 2001. A prokaryotic voltage-gated sodium channel. *Science* **294:**2372–2375.

Roe, A. J., D. McLaggan, I. Davidson, C. O'Byrne, and I. R. Booth. 1998. Perturbation of anion balance during inhibition of growth of *Escherichia coli* by weak acids. *J. Bacteriol.* **180:**767–772.

Ruffert, S., C. Berrier, R. Kramer, and A. Ghazi. 1999. Identification of mechanosensitive ion channels in the cytoplasmic membrane of *Corynebacterium glutamicum. J. Bacteriol.* **181:**1673–1676.

Ruta, V., Y. X. Jiang, A. Lee, J. Y. Chen, and R. MacKinnon. 2003. Functional analysis of an archaebacterial voltage-dependent K^+ channel. *Nature* **422:**180–185.

Saimi, Y., S. H. Loukin, X. L. Zhou, B. Martinac, and C. Kung. 1999. Ion channels in microbes. *Methods Enzymol.* **294:**507–524.

Schleyer, M., R. Schmid, and E. P. Bakker. 1993. Transient, specific and extremely rapid release of osmolytes from growing cells of *Escherichia coli* K-12 exposed to hypoosmotic shock. *Arch. Microbiol.* **160:**424–431.

Schrempf, H., O. Schmidt, R. Kummerlen, S. Hinnah, D. Muller, M. Betzler, T. Steinkamp, and R. Wagner. 1995. A prokaryotic potassium-ion channel with 2 predicted transmembrane segments from *Streptomyces lividans. EMBO J.* **14:**5170–5178.

Schultz, A. G., N. L. Wilson, and W. Epstein. 1962. Cation transport in *Escherichia coli.* Intracellular chloride concentration. *J. Gen. Physiol.* **46:**159–166.

Sesti, F., S. Rajan, R. Gonzalez-Colaso, N. Nikolaeva, and S. A. N. Goldstein. 2003. Hyperpolarization moves S4 sensors inward to open MVP, a methanococcal voltage-gated potassium channel. *Nat. Neurosci.* **6:** 353–361.

Skulachev, V. P. 1992. The laws of cell energetics. *Eur. J. Biochem.* **208:**203–209.

Soupene, E., L. H. He, D. L. Yan, and S. Kustu. 1998. Ammonia acquisition in enteric bacteria: physiological role of the ammonium/methylammonium transport B (AmtB) protein. *Proc. Natl. Acad. Sci. USA* **95:** 7030–7034.

Soupene, E., H. Lee, and S. Kustu. 2002. Ammonium/methylammonium transport (Amt) proteins facilitate diffusion of NH_3 bidirectionally. *Proc. Natl. Acad. Sci. USA* **99:**3926–3931.

Stokes, N. R., H. D. Murray, C. Subramaniam, R. L. Gourse, P. Louis, W. Bartlett, S. Miller, and I. R. Booth. 2003. A role for mechanosensitive channels in survival of stationary phase: regulation of channel expression by RpoS. *Proc. Natl. Acad. Sci. USA* **100:**15959–15964.

Stroud, R. M., L. J. W. Miercke, J. O'Connell, S. Khademi, J. K. Lee, J. Remis, W. Harries, Y. Robles, and D. Akhavan. 2003. Glycerol facilitator GlpF and the associated aquaporin family of channels. *Curr. Opin. Struc. Biol.* **13:**424–431.

Sukharev, S. I., P. Blount, B. Martinac, F. R. Blattner, and C. Kung. 1994. A large-conductance mechanosensitive channel in *E. coli* encoded by MscL alone. *Nature* **368:**265–268.

Szabo, I., V. Petronilli, and M. Zoratti. 1992. A patch-clamp study of *Bacillus subtilis. Biochim. Biophys. Acta* **1112:**29–38.

Vollmer, W., and J. V. Holtje. 2001. Morphogenesis of *Escherichia coli. Curr. Opin. Microbiol.* **4:**625–633.

Vuilleumier, S. 1997. Bacterial glutathione S-transferases: what are they good for? *J. Bacteriol.* **179:**1431–1441.

Vuilleumier, S., and M. Pagni. 2002. The elusive roles of bacterial glutathione S-transferases: new lessons from genomes. *Appl. Microbiol. Biotechnol.* **58:**138–146.

Wei, Y., T. W. Southworth, H. Kloster, M. Ito, A. A. Guffanti, A. Moir, and T. A. Krulwich. 2003. Mutational loss of a K^+ and NH_4^+ transporter affects the growth and endospore formation of alkaliphilic *Bacillus pseudofirmus* OF4. *J. Bacteriol.* **185:**5133–5147.

Yue, L. X., B. Navarro, D. J. Ren, A. Ramos, and D. E. Clapham. 2002. The cation selectivity filter of the bacterial sodium channel, NaChBac. *J. Gen. Physiol.* **120:**845–853.

Zablotowicz, R. M., R. E. Hoagland, M. A. Locke, and W. J. Hickey. 1995. Glutathione-S-transferase activity and metabolism of glutathione conjugates by rhizosphere bacteria. *Appl. Environ. Microbiol.* **61:**1054–1060.

Zakharian, E., and R. N. Reusch. 2004. *Streptomyces lividans* potassium channel KcsA is regulated by the potassium electrochemical gradient. *Biochem. Biophys. Res. Commun.* **316:**429–436.

Zimanyi, L., and J. K. Lanyi. 1989. Halorhodopsin: a light-driven active chloride transport system. *Ann. N. Y. Acad. Sci.* **574:**11–19.

INDEX